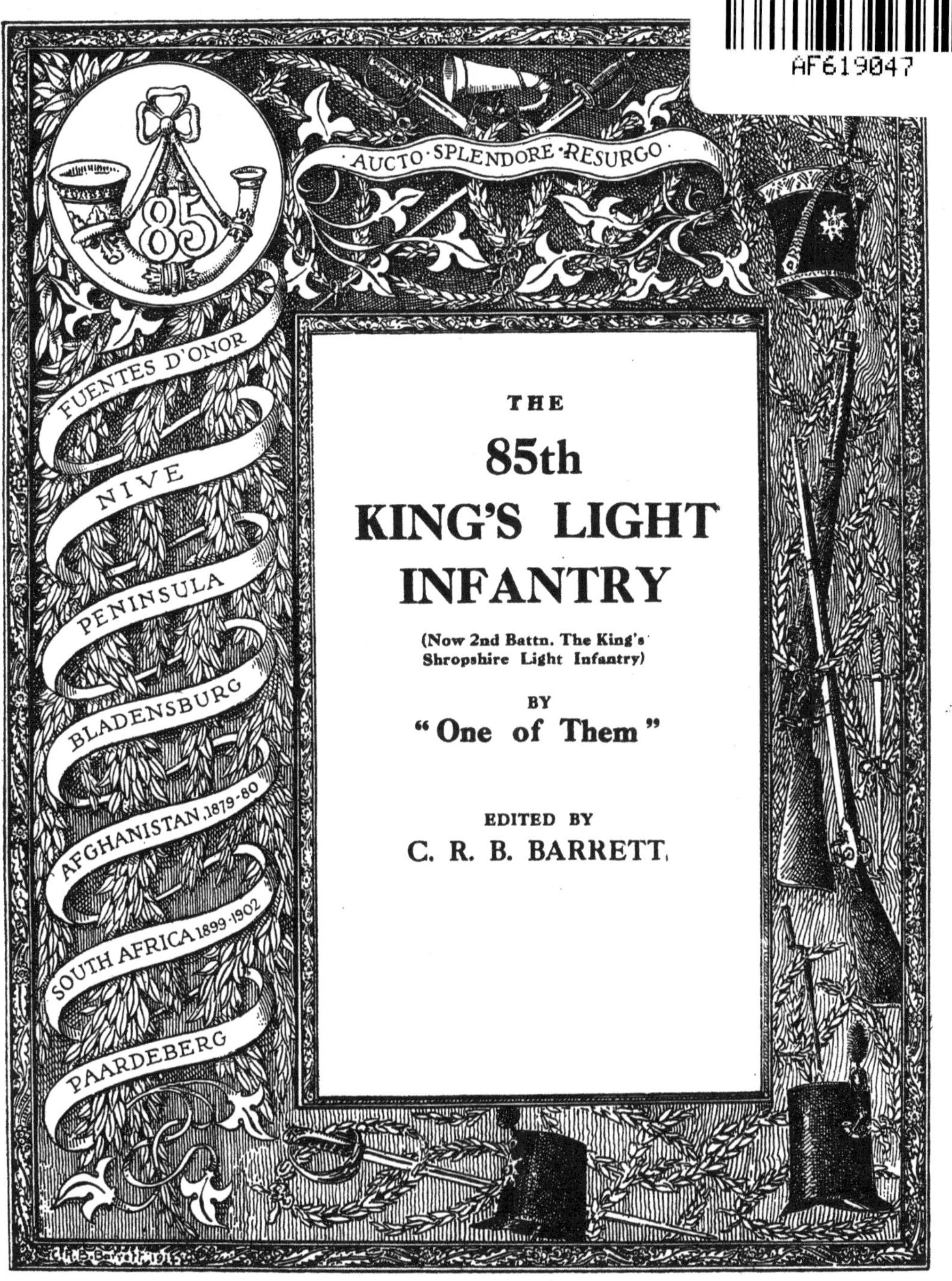

THE

85th KING'S LIGHT INFANTRY

(Now 2nd Battn. The King's Shropshire Light Infantry)

BY

"One of Them"

EDITED BY

C. R. B. BARRETT

FOREWORD

In presenting these pages, which it has been my task to edit, there are but few words to add. Therein will be found a narrative of the services at home and abroad, whether in times of peace or in the more strenuous periods of war, of the 85th King's Light Infantry, now designated the 2nd Battalion The King's (Shropshire Light Infantry).

As there have existed three Regiments in the British service which have borne the number '85,' of which the main subject of the present volume is the survivor, it has been decided, and correctly, to include histories of the two previous Regiments now long ago disbanded.

The present volume will be found to be very fully illustrated both in colour, half-tone, and line.

The selection of maps which have been inserted will be found of interest and of use to the reader. Some of these maps have not hitherto been published.

In one respect the History of this distinguished Regiment is most fortunate, in that the personal note derived from documents, private letters, and war diaries is very marked—more marked than is usual in books of this kind.

As regards the chief illustrations, 'Nivelle' is the work of Mr. R. Simkin, and the spirited monochrome of Paardeberg is from the brush of Mr. Harry Payne. The two views of San Sebastian and that of Los Passages have never previously been reproduced.

Five Appendices contain the names, dates of commissions, and the services in brief of the officers of the three Regiments.

The Editor's personal thanks are due to Lieut.-Col. A. Leetham, of the Royal United Service Institution, for the valuable assistance and advice given to him and the kindness shown to him during the course of his work.

Also to the librarian, Major WYLLY (late South Staffordshire Regiment), and to Mr. HARPER (late 5th Dragoon Guards), the clerk, who were of the greatest assistance in the selection of the maps. To A. D. CARY, Esq., Parliamentary librarian of the War Office, he is also much indebted for information upon points which were referred to him.

Lastly he has to express his thanks to Colonel W. B. CAPPER, to the Rev. T. L. SLACK, and to the MARQUESS OF EXETER for great assistance in endeavouring to trace the circumstances attending the Hoggins-Hylton duel.

That there may be omissions and shortcomings in this book goes without saying; but an honest endeavour has been made to avoid such blemishes as far as possible, and the greatest care and labour has been expended in verifying as far as may be all statements of a historical nature.

C. R. B. BARRETT.

STREATHAM HILL, S.W.
February, 1913.

PREFACE

THE author desires to express his acknowledgments and thanks to the late Mr. S. M. MILNE, of Calverley House, Leeds, for invaluable advice and the loan of pictures. He had promised to write the chapter on Uniform, but his untimely death prevented his intention from being carried out.

The task was, however, kindly undertaken by Mr. P. W. REYNOLDS, whose pictures of uniform help so much to adorn these pages, and whose knowledge of military dress and equipment is second only to that of the late Mr. S. M. MILNE.

Diaries have been kindly lent by Mrs. WARRENNE BLAKE, which were kept by her father, Lieut.-General THOMAS EDMUND KNOX, C.B.; by Mr. T. WILSON, that of his father Ensign Wilson, several of whose excellent caricatures have been reproduced; and by Colonel A. E. GLEIG, son of the author of 'The Subaltern,' and himself now in his ninety-third year.

The author has endeavoured as far as possible to bring a human tone into the history; and these diaries have been of great assistance in this way, by bringing the personal element and human nature into play, and thus toning down the hard incidents of war, the dull routine of garrison life, and the official note of the Regimental Digest of Services.

He is indebted to *The Times* 'History of the War in South Africa' for the general description of that campaign, and the earlier chapters and history of the previous 85ths have been extracted from the 'United Service Institution Journal.'

Finally, thanks are due to several past and present officers of the regiment who have rendered great assistance in those campaigns in which they had personal experience.

THE AUTHOR.

February, 1913.

CONTENTS

CHAPTER I

THE ROYAL VOLONTIERS, 1759–1763

CHAPTER II

THE 85TH (WESTMINSTER VOLONTIERS), 1779–1783

CHAPTER III

THE 85TH (BUCKS VOLUNTEERS), 1793–1808

CHAPTER VII

MARCH THROUGH FRANCE. NORTH AMERICAN CAMPAIGN. BLADENSBURG AND WASHINGTON, 1814

CHAPTER VIII

NEW ORLEANS, 1814

CHAPTER IX

PEACE, 1815–1834

HOME SERVICE, 1815—MALTA, 1821—GIBRALTAR, 1827—MALTA, 1828—ENGLAND, 1831—IRELAND, 1834

CHAPTER X

FOREIGN SERVICE, 1836–1846

NOVA SCOTIA, 1836—NEW BRUNSWICK, 1837—CANADA, 1838—WEST INDIES, 1843–1846

CHAPTER XI

HOME SERVICE, 1846–1853

CHAPTER XII

MAURITIUS, 1853—CAPE OF GOOD HOPE, 1856–1863

CHAPTER XV

HOME SERVICE, 1881-1899

CHAPTER XVI

THE SOUTH AFRICAN WAR TO THE CAPTURE OF PRETORIA

CHAPTER XVIII

A SHORT DIARY OF THE 85TH COMPANY OF THE 4TH MOUNTED INFANTRY IN SOUTH AFRICA, JANUARY 1900—JUNE 1902

CHAPTER XIX

INDIA, 1903

CHAPTER XX

APPENDICES

NIVELLE.
November 13th, 1813.

OFFICER AND PRIVATE,
1793.

OFFICER AND PRIVATE,
1809.

OFFICER,
1818-20.

OFFICER AND SERGEANT,
1839.

PRIVATE,
1851.

From a Sketch by R. Ebsworth.

BUGLE BAND.
1851.

From a Sketch by Ensign Wilson.

LANDING AT PORT ELIZABETH, ALGOA BAY.
JUNE 7th, 1856.

OFFICER.
1827.

BANDSMEN,
1851.

PLATES

The coloured plates in this Naval & Military Press reprint are placed between pages xviii and xix.

MAPS

THE 85TH KING'S LIGHT INFANTRY

CHAPTER I

THE ROYAL VOLONTIERS, 1759–1763

THERE have been three regiments in the British army that have been numbered as the 85th.

The first was raised in 1759 and disbanded in 1763. Its official designation was 'The 85th Light Infantry Regiment or Royal Volontiers.' It was also known as 'Crauford's Regiment of Foot,' from the name of its first and only Colonel and in accordance with the custom of those days.

The second regiment was raised in 1779 and disbanded in 1783. Its official designation was 'The 85th (Westminster Regiment).'

The third regiment was raised in 1793 and still exists. Its titles have, however, been varied. When raised it was designated as 'The 85th (Bucks Volunteers)'; later it became 'The 85th Light Infantry Regiment'; subsequently, 'The 85th (or Duke of York's Own) Regiment of Light Infantry'; afterwards, 'The 85th or the King's Light Infantry Regiment.'

It is now styled officially the '2nd Battalion King's Shropshire Light Infantry,' the 1st Battalion being what was once the 53rd Foot.

In this chapter we shall deal with the history as far as it is known of the 85th (Royal Volontiers), *i.e.* from 1759 to 1763.

THE 85TH LIGHT INFANTRY REGIMENT OR ROYAL VOLONTIERS, 1759–63

The formation of this Regiment is announced in the *London Gazette* for August, 1759, as follows:

'WHITEHALL, *4th August*, 1759.'

'The King has been pleased to constitute and appoint John Crauford Esq. to be Colonel of a regiment of Light Infantry, or Royal Volontiers, to be forthwith raised for His Majesty's service. William, Lord Viscount Pulteney to be Lieutenant-Colonel of the said regiment. Sir Hugh Williams Bart., to be Major, N. Heywood, W. Skinner, C. Cooper and J. Langham, Esquires, C. Cornwallis Esq., commonly called Lord Brome, Hugh Percy Esq., commonly called Lord Warkworth, Peter Bathurst, J. Moore, Edmund Nugent and Lockhart Gordon, Esquires, to be Captains.

'*First Lieutenants*.—G. Maxwell, J. Gwynne, G. Sandys, R. Shipley, Pryse Donaldson, Holland Williams, George Salusbury Townshend, C. Nesham, H. Dawson, J. Steward, J. Lloyd, W. Cawthorne, J. Dawson, G. Sinclair, A. Bruce, Woodford Rice, R. Barry, J. Rooke, ——J. Adams, H. Williams, ——Acton, Miles Allen, ——Cooper, C. Jones, ——Nugent, A. Wood, W. Green, ——Glynn, Gentlemen.

'*Second Lieutenants*.—Owen Meyrick, —— Humphrey, Roger Price, J. Johnes, ——Vaughan, R. Boycott, R. Lucas, C. Ware, ——Bathurst, ——Lambert, N. Merson, F. Thackeray, Gentlemen.

'*Adjutants*.—R. Gough, G. Maxwell.

'*Quartermaster*.—J. Roberts.'

In a subsequent *Gazette* of August 7, 1759, it is announced:

'The King has been pleased to add four companies to the Regiment of Light Infantry, or Royal Volontiers, commanded by Colonel Crauford, and to appoint the following gentlemen to command the same: St. John Jeffreys, Temple West, C. Egerton and W. Forester, Esquires.'

The Audit Office Enrolment Book for 1759 contains the following entry:

'GEORGE R.

'Whereas we have thought fit to order a Regiment of Light Infantry, or Royal Volontiers, to be forthwith raised under the command of Colonel John Crauford, which is to consist of ten companies of four Sergeants, four Corporals, two Drummers and one hundred private men in each, besides commissioned officers, with two Fifers to each of the two Companies of Grenadiers, and the said Colonel John Crauford having represented to us that to supply the said non-commissioned officers and private men with swords, leather accoutrements, and other species of clothing, and to defray the charge thereof, it will require twenty months off-reckonings of the said non-commissioned officers and private men etc. etc. : ' wherefore, ' a Warrant has been duly signed directing the Paymaster-General of the Land Forces to pay to the said Colonel John Crauford, his heirs, executors and assigns, the sum of £117 0*s.* 7*d.* for the said services.'

A similar notice, dated the same day, records the issue of a like warrant for the payment of £880 0*s.* 9*d.* for twenty months off-reckonings for the four additional companies.

The rendezvous of the Regiment was Shrewsbury ; but recruits were not only sought for from round the Wrekin, for it is stated in an old Birmingham paper in October 1759 that

' Two hundred and fifty recruits raised in the neighbourhood of London for the regiment of Royal Volontiers, now forming at Shrewsbury, in charge of three officers made a halt in passing through our town.'

The army gentlemen, it is added, exerted themselves with great zeal during their stay and carried off a large addition to their numbers.

The ceremony of presenting the colours was carried out with unusual éclat a few days before Christmas, 1759 :

' A Regiment of Foot was raised and rendezvou'd here—they were called the Royal Volontiers. Colonel Crauford commanded them.

' On Friday 21st December, St. Thomas' Day, the Colours were received by the Regiment with great pomp, being carried in procession to St. Chad's Church where a sermon was preached by the Rev. Rowland Chambre, M.A., from Ephes. VI. 10 (Finally, my brethren, be strong in the Lord and in the power of His might).'

The church here referred to was old St. Chad's, which was afterwards destroyed by the fall of its tower in 1788.

The new church, of the same name, was rebuilt in a different locality; but a fragment of the old building (stated to have been the chancel) still exists. The preacher—who, by the way, was a relative on the mother's side of the late General Lord Hill—enjoyed some reputation as a pulpit orator in his day. A specimen of his eloquence on the above occasion may not be out of place here. After descanting upon the subject of the death of Wolfe, then fresh in the public mind, he continues:

'But while I endeavour to signalise the good endeavours of men exerted in acts of a public nature, how should I congratulate the Commanders of this assembly, on the merit of raising these battalions with an alacrity incredible to their countrymen, formidable to their enemies—how congratulate my countrymen, in Commanders acquainted with the genius of its People, and equal to such unexpected Success—how congratulate you whose easier service it is to obey, in having men of politeness and candour in command? May the same emulative zeal which has united these forces conduct them in the Day of Battle and be the Presage of Future Honours. In that Day, (as the Sun once rested upon Gibeon, and the Moon was stayed in the valley of Ajalon), may the God of armies assist you with favourable incidents, till ye, like Joshua, be avenged on your adversary—and may these Laurels obtained in War, be the lasting emblems of your bearing in Peace proclaiming the actions that merited them to succeeding ages.'

A copy of this sermon, apparently the sole literary production of the reverend author, is in the British Museum Library. A thin quarto, profusely annotated, after the fashion of the period, with marginal quotations from the Latin poets, and dedicated to the officers of the Royal Volontiers, 'at whose request it was preach'd and publish'd.'

Another account states:

'The 85th was raised in 1759. This light infantry regiment was known as Craufurd's Royal Volunteers. Craufurd commanded as

CHANCEL OF OLD ST. CHAD'S CHURCH, SHREWSBURY.
Where Colours were presented to the first 85th Regiment, 1759.

Lieutenant-Colonel, and his regimental Staff numbered fifty-eight officers. Much local interest was excited by its embodiment, and the ceremony observed at the consecration of the colours testifies to the patriotic feeling that prevailed in the country at the time. A service was held at Shrewsbury, the rendezvous of the Corps, on January 2nd, 1760, at which they were consecrated. During the prayers and sermon the colours were held over the heads of the Lieutenant-Colonel Commandant and his Acting Lieutenant-Colonel, Lord Pulteney. The regiment was then marched to the Quarry where the colours, held by these two officers, were saluted and kissed, after which Craufurd entertained his staff at dinner, from which they repaired to a ball given by Lord Pulteney.'

In a ' State of the English Land Forces ' for the year 1760 given in Lloyd's Lists the uniform of the Regiment is given as ' red faced with white,' and its strength as ' 67 officers, 88 N.C.O.'s, 1486 R. and F., in a single Battalion of fourteen Companies.'

The regiment would appear to have moved to Newcastle-on-Tyne shortly after, as in the *Newcastle Magazine* for 1760 we find, under date January 21, that ' the regiment of Royal Volontiers were reviewed on the Town Moor by General Whitmore.' As the notes made by the Inspecting Officer at this Review are luckily still in existence, we are able to give with considerable detail an account of the uniform, arms and equipment of the officers and men of the ' Royal Volontiers.' The paper is dated March 10, 1760, and is as follows :

' The officers and men had swords, the officers armed with fuzees, and have cross buff belts. They wear their sashes round the waist. Salute differently from the rest of the army—Uniform a red coat without lapels, with blue cuffs and capes, silver loops lined white—double breasted short waistcoats of white cloth and breeches of the same—hats cocked in manner of King Henry VIII with a plume of white feathers. The arms much lighter than those of the infantry. Officers and men have hangers—the men short but young—accoutrements new, pouch belt much narrower than what is used by the infantry, the waist belt worn accross the shoulder. The men have red coats without lapels, blue cuffs and capes with white loops lined white—doubled breasted short waistcoats of white cloath, breeches of the same.—Hats cocked in the manner of King Henry VIII with a narrow white lace, and plume of white feathers, no white or black gaiters, but a black leather gaiter which comes half way up the leg.'

Evidently, from the reference to 'the infantry,' there were peculiarities in the uniform of the 'Royal Volontiers' which differentiated it from that of other corps.

Again, according to another authority we find that on the '10th March, the regiment of Royal Volontiers, stationed at Newcastle, consisting of fourteen hundred men, was reviewed by General Whitmore in a field near the North Shields road about three miles from Newcastle, and although a very young regiment gave great satisfaction to the General, their own officers and a large concourse of spectators.'

Shortly after this date, we learn from Brande's 'Newcastle,' a sum of £315 was voted by the Town Council to furnish an additional bounty to each 'braw pit-laddie' willing to engage for the 36th or 85th Regiments of Foot.

The Editor of the *Newcastle Magazine* of that day is responsible for a paragraph which tells us that on '31st March 1760, the regiment of Royal Volontiers marched to Killingworth Moor where in the presence of their Colonel, their officers, and a great number of gentlemen and spectators, they pitched their tents and formed a camp; upon which the Colonel ordered parties to go through the various forms of maroding, foraging etc., and when they returned, the whole corps struck tents etc. in the most expeditious manner, and returned to their quarters here in the evening, having given great satisfaction to all who saw them.' The use of the word 'maroding' is, we should be inclined to think, an error on the part of the person who wrote the account here quoted from the *Newcastle Magazine.*

'Maroding' (otherwise *marauding*) was practically brigandage on the part of troops. It was practised in foreign armies where discipline was slack, but in the British Service—even as far back as the year 1694—was a military crime *punishable by death.* Let us quote an instance.

On August 6, 1694, when the British Brigades commanded by Eppinger were on the march from Mont St. Andrée to Rousselaer, to prevent disorders on the march and better still to minimise the chances of small bodies of the troops from being cut off by the enemy, an order was issued which was commanded to be read at the head of every regiment.

In this order the following offences were forbidden, any breach being *punishable by death* :

1. Maroding.
2. Foraging without orders.
3. Molestation of victuallers or any persons who came to camp with provisions.

In the issue of May 15, 1760, is the paragraph :

'The Regiment of Royal Volontiers, which has lain here during the winter, marched for the South in four divisions.'

Their destination was Devonport, or (as it was then styled) Plymouth Dock, where the regiment arrived six weeks later and occupied the newly built barracks, which covered a part of the site of the present Raglan Barracks.

A paragraph in the Plymouth intelligence in the *Public Advertiser* for July, 1760, records that a soldier of Crauford's Light Infantry 'underwent the punishment of the gantlope, a few days ago, and received a very severe twigging.'

The punishment of the 'gantlope' (otherwise *running the gauntlet*) was very severe, as indeed were all military punishments in those days.

The men of a regiment, nay even sometimes of a brigade, were drawn up in two lines facing one another and armed with rods, sometimes with belts. The prisoner was compelled to run between the lines, being smitten if possible by each man as he passed. The victim was nearly nude—certainly nude to the waist. Until quite recently this punishment still existed in the Russian army, where green rods were used.

In the October following we hear of the arrival, by sea, of the Regiment at Portsmouth, to take part in the expedition then fitting out against the coast of France, under Lord Ligonier and Admiral Keppel. It appears to have disembarked at Gosport, and marched to billets in Southampton, at that time a depôt of French prisoners-of-war.

In November the Regiment marched back from Southampton to Gosport, and embarked with the other troops on board the fleet, which dropped down to St. Helens. Here the men-of-war acting as a convoy and the transports were detained until the middle of December. The season being then too far advanced, the troops were brought back to Portsmouth, landed and marched into country quarters for the winter.

In those days, when barracks were scarce in England, it was customary to billet troops on the inhabitants of towns large and small, usually at the public-houses.

This custom had long been a public grievance and for more reasons than one.

It was alleged that when troops were billeted on a publican his ordinary customers abandoned his house. This may or may not have been the case; but it is asserted by the Bishop of Wells, in a petition praying that Wells may be relieved of the soldiers then quartered there, that several publicans had been compelled to shut up their houses. That the amount of billeting money was insufficient to recompense the inn-keeper was admitted. Strangely enough, the building of barracks was, though often proposed in Parliament, invariably opposed there and negatived. Walpole tried hard during his last administration (1721–1742) to introduce a barrack-building scheme. It was opposed both by Ministerialists and the Opposition, the view taken being that the building of barracks was 'unconstitutional and inimical to the rights of the people,' and that the mixing of soldiers and civilians in billets or quarters was 'the best guarantee of the security of the Constitution as against the danger of Standing Armies.' Why, one cannot see! So it was that for another forty years (till, by a certainly unconstitutional act, barrack-building was started) billets and quarters were the order of the day.

The rules and regulations for the conduct both of officers and men while in billets or quarters were laid down with much care and precision. Civility to landlords was enjoined very particularly. Payment of all bills by officers with punctuality was strictly ordered, and severe measures were taken with any officer who was lax in this matter.

Particularly before moving quarters, or going on active service abroad, every account had to be settled and vouchers given.

Usually men occupied the same billet for a month and were then shifted, but occasionally the period was considerably longer. The system, though, was no doubt a bad one; and the Governments, one after the other, knew it to be bad. But they equally knew that it was vain to attempt to alter it legally. So all that could be done to minimise the defects of the system and to obviate oppression was (as far as regulations go) effected by law. Soldiers had privileges as regards arrest and could ordinarily only be taken into custody by military force. Officers' servants, strangely enough, were not held to be privileged (as 'not being soldiers'), yet they were in the ranks!

Early in the spring, preparations were made for resuming the enterprise, the command of which was given to General Studholme Hodgson; with Colonel Crauford, 85th Light Infantry, as second in command.

Belleisle, the destination of the enterprise—a destination hitherto kept secret—has been thus described:

'This once noted stronghold, a barren rock off the iron-bound coast of Brittany, is about twelve leagues in circumference, and then, as now, contained a population of some five thousand people, mostly fisher-folk. It had a certain adventitious importance as a rendezvous for the homeward bound French East-Indiamen, but its reduction in a military point of view was of very small importance. Nevertheless (Lord Mahon tells us in his "History of England," 1759–60) "the conquest, as Pitt had foreseen, was a signal humiliation to France, and exerted a material influence upon the conduct of the subsequent negotiations for peace," which contributed so greatly to the aggrandisement of our Colonial Empire.'

Early in January, 1761, the same regiments had again been warned for active service abroad. Next, orders were received for them to proceed to Portsmouth, and having marched thither they embarked at Spithead on March 29th.

A start was made, but contrary winds delayed the ships, and it was not until April 6th that the voyage really began. Belleisle was reached on the next day.

The accounts of the composition of the force are somewhat conflicting, but we believe at least the balance of evidence points to it having been made up as follows.

Either as whole regiments, or detachments of regiments, we find the 9th, 19th, 30th, 34th, 36th, 67th, 69th, Morgan's (then the 94th),

Stuart's (then the 97th), Grey's (then the 98th), and three companies of Artillery.

Either in whole or in part the 3rd Buffs, Boscawen's (then the 75th), and Crauford's (the 85th) sailed a few days later, as also did two troops of the 16th Light Dragoons (the entire regiment, we now know, did not go).

This second force arrived at Belleisle during April—one account says during May, and another during June. On April 8 the fleet of men-of-war and the transports sailed round the island in an endeavour to fix upon a suitable place for disembarkation, but it was impossible to discover any undefended, spot. Eventually Port La Maria was chosen, a spot on the south-eastern side of the island.

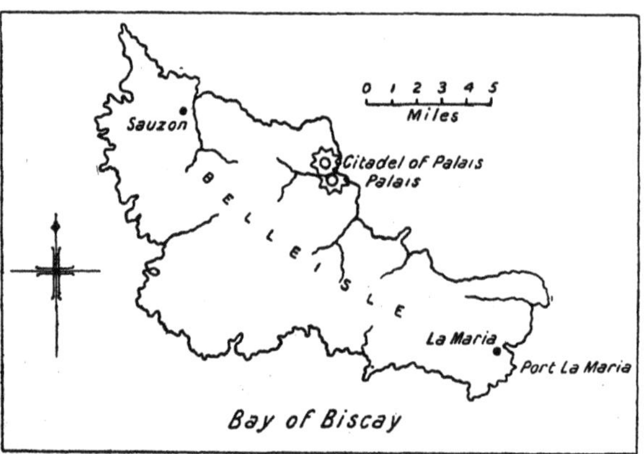

Flat-bottomed boats had been brought for landing the troops and these were speedily filled. The attack was threefold, two attacks being feints and one real. Part of the fleet (with Stuart's, Grey's, and Marines) in transports were sent against St. Andro, a little port; and three ships with bomb-vessels were told off to silence a fort and some of the enemy's works in a sandy bay, near Point Locmaria.

The attack then began. The guns of the *Achilles* first silenced a four-gun battery which commanded the landing-place of the real attack, and the flat-bottomed boats reached the shore. Here, however, they found the enemy very strongly entrenched on each side of a hill which was excessively steep and the foot of it scarped so that it was impossible to scale it, and the entrenchments therefore could not be stormed.

About 260 men landed under Major Purcell and Captain Osborn. A tremendous fire was opened on them by the enemy and many fell, including both these officers. About sixty men of Erskine's Grenadiers managed to scramble up and formed on the top of the hill when, after a gallant but fruitless endeavour to maintain their post, they were cut off all but twenty, who managed to regain the boats.

Under cover of fire from the British vessels the remains of the attacking force succeeded in re-embarking after losing five hundred men either killed, wounded, or prisoners. Keppel, the Admiral in command, appears to have been very doubtful whether it would be possible for any force landing to maintain itself. Still, another attempt was made.

Again two feint attacks were carried on simultaneously with the real one. This time La Maria and Sauzon were selected for the false attacks, while the real one was directed against some rocks between them which (being so very steep) had not been fortified.

Meanwhile reinforcements had been sent from Plymouth in the shape of Lord Robert Manners' regiment and the remainder of Crauford's (85th). The attack, when it developed, was quite successful. A force under Brigadier Lambert, during the confusion caused by the false attacks, seized a rock near Point Locmaria, Beauclerk's Grenadiers actually performing this desperate service, headed by a Captain Patterson. Immediately they were attacked by some 300 men, but succeeded in forming behind a stone wall and then kept the enemy at bay till the rest of Lambert's Grenadiers had scrambled up. The French then bolted, leaving three brass field-pieces as the spoils of the victors. But many had fallen, indeed thirty were killed, and amongst the numerous wounded were Colonel Mackenzie and Captain Murray of the Marines. It is amusing to read that the Marines 'climbed the precipice with astonishing intrepidity, and were no wise behind the regulars in valour and activity'!

The British Army then landed. Monsieur de St. Croix, the French commandant, drew in his force to a camp just beneath the walls of Palais, the capital of Belleisle. This camp he fortified, and he further augmented the force by calling out the militia of the island to the number of 4000.[1] He meant fighting and he fought well. Again, however, in the attack there was delay from stress of weather.

Keppel found it impossible to land the artillery, and for some days General Hodgson had the mortification of being compelled to look on idly while St. Croix, with admirable skill, erected no less than six redoubts. These protected all the avenues to the town.

Meanwhile the two troops of the 16th Light Dragoons had arrived and had managed to land. These were detached to take post at Sauzon and a body of infantry was sent to a village called Bordilla, where they proceeded to entrench themselves. Down on them, however, came the grenadiers of the enemy and dislodged them. However, the greater part of the British force managed to entrench itself while awaiting the advent of the guns. Immediately these arrived and with them some mortars as well, St. Croix withdrew into the citadel, a castle which had been formerly the property of the Marquis de Belleisle, but later had, on falling into royal hands, been fortified by Vauban.

By May 2 the British were able to break ground for a regular siege.

Next night the enemy made a sortie and attacked the trenches with much vigour. The picquets on the left were thrown into confusion, the works were destroyed, many men were killed and wounded, and Crauford and his two aides-de-camp were taken prisoners, Crauford being badly wounded. The attack was not pressed by the enemy on the right and they returned to the citadel with their prisoners.

Having repaired the damage, on the morrow a redoubt was built on the right to prevent any similar surprise.

From this time the siege was carried on with vigour. The enemy made frequent sallies and many men fell on both sides. On June 13 General Hodgson ordered the French redoubts to be attacked. The attack began at daybreak with four guns and thirty cohorns[2] which, concentrating their fire on the right redoubt, opened a way for a

---

[1] This is the number given, but in the face of the statement that the total population of the island does not exceed 5000 there would appear to be a mistake somewhere. Probably 400 is the correct number.

[2] Cohorns were mortars.

detachment of Marines, sustained by a part of Loudoun's regiment, to reach the parapet. Once there, they drove in the enemy at the point of the bayonet and took possession of the post. Similar scenes took place at the other five redoubts which were carried by the same detachment, though reinforced by Colville's regiment under the command of Colonel Teesdale and Major Nesbit. The enemy then fled into the citadel after having been badly cut up. Their flight was somewhat precipitous and the British pursuing smartly entered the town pell-mell with the French, took possession of it, made many prisoners and, what is more, released the captive British.

The citadel alone remained to be taken. It was strong both in situation and in its fortification and its commander was a man of resolution. Hodgson was of a similar build, but left nothing to chance. He set to work quite in the orthodox manner; parallels, barricades and batteries were constructed and an incessant fire from his mortars and artillery, not omitting the three captured brass guns, was kept up for thirteen days and nights.

St. Croix, on his part, replied with vigour until May 25, and then his fire began to slacken.

Meanwhile the British were rather in straits for lack of provisions. On the island there was nothing to be had; St. Croix had very wisely seen to that. All provisions therefore had to come from England and these could not always be landed. But Hodgson kept on pounding away at the fortifications and by the end of May had made a breach in the citadel, though not sufficient to be practicable. Feverishly by day and night the defenders endeavoured to repair the damages, but in vain. By June 7 a practicable breach was made, and St. Croix, after his gallant defence, surrendered. Of relief from the mainland he had no hopes, for the British held command of the sea. One likes to think that the garrison were permitted to march out 'through the breach' with all the honours of war.

So ended the famous siege of Belleisle, a gallant and most toilsome enterprise. The losses of the British force were heavy—thirteen officers and three hundred men killed, twenty-one officers and four hundred and eighty non-commissioned officers and men wounded.

Colonel Crauford, now General, was appointed Governor of the new garrison.

In the spring of 1762 the 85th were again ordered on active service, this time to Portugal as forming part of a British contingent sent to that country under the command of Lord Tyrawley. Lord Tyrawley was, however, not long afterwards replaced by the Earl of Loudoun.

This force consisted of the two troops of the 16th Light Dragoons (Burgoyne's), the 3rd Buffs, the 67th (Lambert's), Boscawen's, the 85th (Crauford's), two battalions, while the 83rd (Armstrong's) and the 91st (Lord Blayney's) were sent from Ireland. A number of British officers were also sent to take up commands in the Portuguese Army.

Arrangements had been concluded between France and Spain by which the former undertook to prosecute the war in Westphalia, whilst the latter overran Portugal. Spanish armies had already possessed themselves of the provinces of Tras-os-Montes and Beira, and a third was forming on the frontier of Estramadura, with the intention of penetrating into the Alentejo. The force from Belleisle arrived in the Tagus early in May, but it was not till June that the four regiments from Ireland put in an appearance. With them came the Count de Lippe Bückeburg, the destined commander of the Allies. Lord Pulteney, and subsequently Lord Loudoun, commanded the British. The position was this: as soon as the Spaniards captured Almeida (which they did early in August), de Lippe Bückeburg was compelled to act on the defensive, and so covered Lisbon at the line of the Tagus.

The Count de Lippe Bückeburg determined that the Spaniards should be attacked in camp before they entered Portugal, and Brigadier-General Burgoyne was ordered on this duty. Although at a distance of five days' march, and in spite of many obstacles, this gallant officer rapidly struck across the mountains of Castel da Vida, and succeeded in effecting a complete surprise of the enemy on September 27th at Valencia d'Alcantara, taking prisoners the Spanish Major-General and a large number of officers and men.

The infantry, on this occasion, composed of detachments of Grenadiers from various regiments, mounted on dragoon horses, were under the command of Lieut.-Colonel Lord Pulteney of the 85th Light Infantry.

The exploit drew forth the following General Order:

Extract from General Order, August 24, 1762.

'The Field-Marshal thinks it his duty to acquaint the army with the glorious conduct of Brigadier-General Burgoyne, who, after having marched fifteen leagues without halting, has taken Valencia d'Alcantara sword-in-hand, destroyed the Spanish regiment of Seville; captured three standards, a colonel, many officers of distinction, and a great number of soldiers.

'The Field-Marshal makes no doubt but the whole army will rejoice at this event, and that every one will, in proportion to his rank, strive to imitate so glorious an example.'

A part of the 85th was, subsequently, with Burgoyne in the dashing affair at Villa Velha:

'The Spanish troops, at Castel Branco, had made themselves masters of several important passes, and were preparing to cross the Tagus at the village of Villa Velha, but Lippe, moving with great rapidity, got to Abrantes before them, and sent out strong detachments under Burgoyne and the Count de St. Jago to obstruct the passage of the river. The latter was drawn from the pass of Alvito by the enemy, when (the Spanish Commander) D'Arando attacked the old Moorish Castle of Villa Velha and took possession of it in a few days, in spite of Burgoyne, who was posted with the intention of disputing it. Observing, however, that the troops kept no vigilant look out and were unprotected in rear and flank, the Brigadier-General formed the design of sending Colonel Lee with fifty Light Dragoons, one hundred Grenadiers and two Companies of the 85th Light Infantry, across the river to effect a surprise. This officer fell upon the enemy on the night of 6th October, while Burgoyne attacked at the same time, and made a considerable slaughter. He dispersed the whole party, destroying the magazine with scarcely any loss to himself' (see Cust's 'Annals').

The loss of the 85th on this occasion was six rank and file wounded (see 'Extract from Despatches' in *London Chronicle*, 1762).

A fuller account of Burgoyne's exploit is as follows:

'Burgoyne passed the Tagus at midnight on August 23rd with 400 men. At an appointed rendezvous he was joined by a detachment of all the British grenadiers belonging to the various regiments, eleven companies of Portuguese grenadiers, two small guns and two small howitzers. After a fatiguing march the force reached Castel da Vida, where another small accession of strength came to him in the shape of 100 regular Portuguese Foot, 58 irregular cavalry and about 40 armed peasants. Burgoyne then made his final arrangements. He marched during the night of the 25th and expected to surprise Valencia d'Alcantara before daybreak, but his guides played him false by deceiving him as to the distance. Day broke an hour before he could reach the town; so, halting the rest of his force, he pushed on with his dragoons only, and with the grenadiers from the various regiments mounted on dragoon horses—these under the command of Lord Pulteney of the 85th.

'Fortune favoured Burgoyne. The Advance Guard being unmolested and finding the entrance to the town clear, rushed the place sword in hand, cutting down or making prisoners of the guard on the square before they could offer resistance. A few parties rallied and attempted opposition, but were speedily driven off or slain or captured. From some of the houses desultory firing now began, but a threat to burn and sack the town put an end to this. It is said that a sergeant and six dragoons meeting with a party of 25 Spanish dragoons under an officer, while on patrol duty, killed six, captured the rest and also all their horses. Burgoyne did not find in Valencia d'Alcantara the magazines and stores he had expected, but he raised a contribution from the town and destroyed many arms and much ammunition there which he could not remove. What he could he did and with other captures in the way of a Major-General, his aide-de-camp, one Colonel, his Adjutant, two Captains, seventeen subalterns, fifty-nine privates and three Colours he returned to headquarters. The British loss was Lieutenant Bank, one sergeant, and three privates killed; two sergeants, one drummer, and eighteen privates wounded.'

Burgoyne's victory had but little effect on the campaign, as the Spaniards then pushed forward towards Lisbon. De Lippe Bückeburg did not dare to risk a battle in which the Crown of Portugal would have been at stake. Added to this, too, the French were sending considerable reinforcements to the Spanish. The unfortunate commander had to do what he could, and quitting his strong camp at Ponte de Murcella in the Beira returned to Estremadura and encamped at Abrantes. Lord Loudoun encamped at Sardoal, and Lippe also undertook the task of guarding as best he might with his army every road and pass to Lisbon. To Burgoyne was charged the duty of defending the pass over the Tagus at Villa Velha, and he took post on the south side of the river, facing the town just between Nissa and the Tagus. Here he had part of his own regiment, the Royal Volontiers (Crauford's 85th) and the English Grenadiers. Another detachment of four battalions, six companies of grenadiers and a regiment of cavalry (all Portuguese), under the command of the Count de St. Jago, occupied the pass of Alvito. A considerable detachment was posted at Perdrigal to prevent the advance of the enemy through the mountains in his front, and on his right at Villa Velha were 150 Portuguese under the command of a captain. The pass of Alvito was supposed to be impregnable, nevertheless the Count de Maceda with 6000 Spaniards did not think so.

On October 1 he tackled the Count de St. Jago and his force, attacked the old Moorish Castle of Villa Velha on his right and a small post commanded by a major at St. Simon on his left.

The castle was covered by Burgoyne's fire and held out for a time, but St. Simon's party were easily routed, and this placed the Spaniards in a position to attack St. Jago and his men both in front and rear. St. Jago, therefore, drew off, his retreat being covered by Lord Loudoun who was sent hastily thither by de Lippe for that purpose with four British battalions and four field guns. Loudoun advanced with great rapidity. As soon as St. Jago's outposts were withdrawn Loudoun had the defences in the pass levelled, as they might have been used against him. And here, crowning the heights of Astalliardes, Loudoun's force remained till the retreating Portuguese battalions had filed off along the road of Sobrina Formosa. Loudoun's force had meanwhile been augmented by six companies of Portuguese Grenadiers, fifty of Burgoyne's Dragoons and as many Portuguese cavalry.

St. Jago's retreating force was attacked in the rear by the enemy, but Loudoun's men by their presence saved the situation, and the four guns which were actively and cleverly fought by Major M'Bean enabled them to retire towards Cardigas without further loss. Meanwhile the

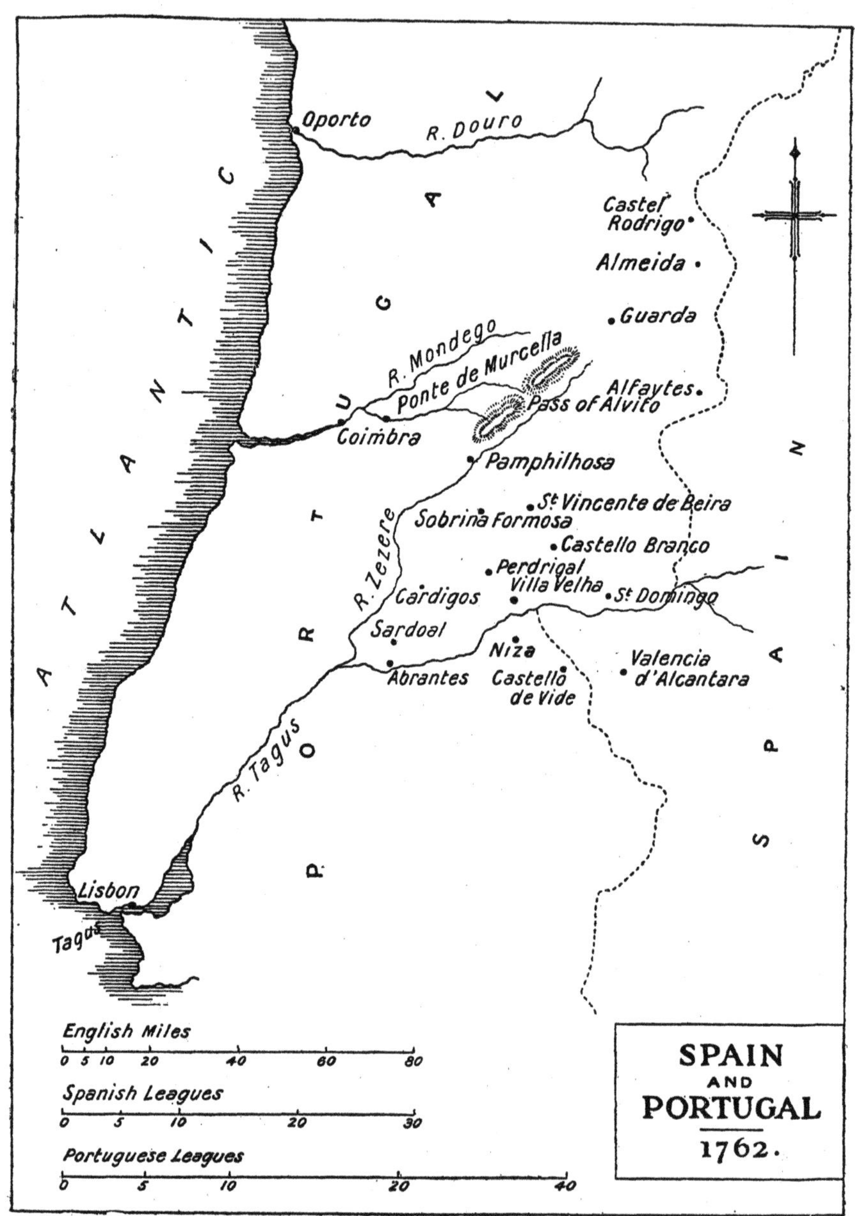
ATLANTIC
PORTUGAL
SPAIN
Oporto
R. Douro
Castel Rodrigo
Almeida
Guarda
R. Mondego
Ponte de Murcella
Alfaytes
Pass of Alvito
Coimbra
Pamphilhosa
St Vincente de Beira
Sobrina Formosa
Castello Branco
R. Zezere
Perdrigal
Villa Velha
St Domingo
Cardigos
Sardoal
Niza
Valencia d'Alcantara
Abrantes
Castello de Vide
R. Tagus
Lisbon
Tagus
English Miles
0 5 10 20 40 60 80
Spanish Leagues
0 5 10 20 30
Portuguese Leagues
0 5 10 20 40
SPAIN AND PORTUGAL 1762.

troops that had thus been employed by the enemy had weakened the force near Villa Velha and Burgoyne discerned the situation.

It offered him a most favourable opportunity for entering into contact with the troops and artillery which remained there.

Consequently, ordering a detachment of 100 British Grenadiers, 200 of Crauford's (85th) and 50 Light Dragoons under Lieut.-Colonel Lee to ford the Tagus on the night of the 5th, he determined to surprise the Spanish cavalry camp at Villa Velha.

Colonel Lee's force had the fortune to get into the encampment without being perceived.

Some firing then took place and a good many of the enemy were killed in their tents. The British Infantry used the bayonet without firing a shot. The only part of the Spanish force that made a stand was a body of horse. These were charged by Lieutenant Maitland at the head of the Dragoons and were routed with considerable slaughter. Most of the Spanish officers (who, to their credit be it observed, did their best bravely to rally their men) were killed and among them a Brigadier-General. Four guns were spiked, two had been removed. The magazines were burned, sixty artillery mules were brought off, a few horses, a captain and two subalterns of Horse, a subaltern of Artillery, a sergeant and fourteen privates and a considerable quantity of valuable baggage and loot. The British loss was one corporal killed, two men wounded, four horses killed and six wounded. Many of the men had had their helmets cut through, but without damage to themselves.

After this dashing little affair there was little more fighting owing to the advanced state of the season.

The 85th passed the winter in quarters among the Portuguese peasantry.

Peace was declared in the spring of 1763.

The regiments were all sent home, among them of course the 85th.

On arrival in England the regiment was very shortly afterwards disbanded, the majority of the officers being relegated to half-pay.

Such was the brief but by no means inglorious career of the first 85th Regiment in the British Army.

The following document gives information as to the establishment and pay of the regiment in 1762:

ESTABLISHMENT OF THE 85TH REGIMENT OF FOOT (OR ROYAL VOLONTIERS)

Field and Staff Officers.

1 Colonel @ 12*s*. } 14*s*. 2*s*. in lieu of servant }	2 Adjutants, each 4*s*. 0*d*. } 4*s*. 8*d*. 8*d*. }
Lieut.-Colonel 7*s*.	Surgeon, 4*s*.
Major 5*s*.	2 Surgeons' Mates, each 3*s*. 6*d*.
Chaplain 6*s*. 8*d*.	

One Company.

Captain 8*s*. } 10*s*. 2*s*. }	2 Dragoons, each 8*d*. Private Men ,, 8*d*.
2 First Lieutenants 8*s*. 0*d*. } 9*s*. 4*d*. 1*s*. 4*d*. }	Allce. Widows 1*s*. 4*d*. *Colonel* 1*s*. 2*d*.
Second Lieutenants 3*s*. 0*d*. } 3*s*. 8*d*. 8*d*. }	*Captain* 1*s*.
4 Sergeants, each 1*s*. 6*d*.	
4 Corporals ,, 1*s*.	Agent 6*d*.

11 *Companies more of same*, £5. 5*s*. 8*d*.

1 *Company of Grenadiers.*

Captain	2 Fifers, each 1*s*.
3 First Lieutenants	

A few words of explanation are needed here with regard to 'Allce. Widows.'

In those days on the roll of every company in an Infantry Regiment were borne what were therein entered as 'Widows' Men.'

The same custom obtained in every troop of a Cavalry Regiment.

These 'Widows' Men' existed only on paper. Their pay was, however, regularly drawn as if they were realities.

The cash which thus accrued formed a pension fund for the widows of officers. It was a curious method of supplying a need and of course has long since been done away with.

The total establishment of the regiment in 1762 was as follows :

85th.

1 Colonel without Company	13 Captains more
1 Lieut.-Colonel	30 First Lieutenants
1 First Major	12 Second Lieutenants
1 Second Major and Captain	1 Chaplain

2 Adjutants
14 Companies in all
1611 Men (Officers included)
1 Quartermaster
1 Surgeon
3 Mates
56 Sergeants
56 Corporals
4 Fifers
1400 Private Men.

N.B.
The Second Major has only pay as Captain on establishment.

At the time the regiment was disbanded in 1763 its strength was as follows, according to an inset in the Army List for that year.

It is entered thus and shows the Regiment to have been at full strength :

Rank	Uniform	Cos	Commd Officers	Non-Commd	Private Men
85	Red White Facings	14	67	88	1456

Total	Rise (*i.e.* Date of Raising)
1611	21 July 1759.

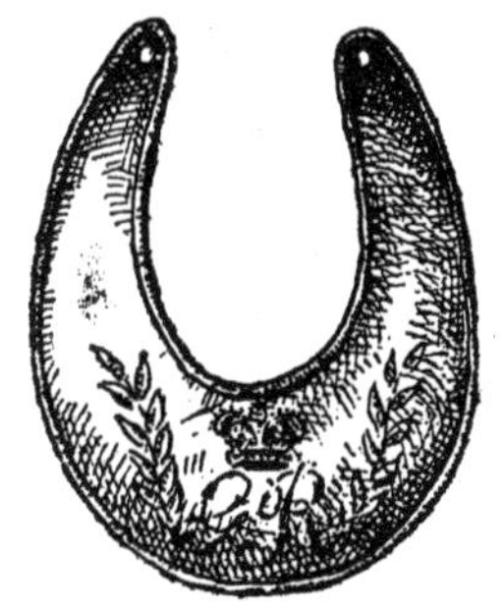

CHAPTER II

THE 85TH (WESTMINSTER VOLONTIERS), 1779–1783

IN the summer of the year 1779 another 85th Regiment of Foot made its appearance upon the rolls.

Sixteen years of European peace and comparative prosperity had culminated in another era of strife; and England was not only engaged in a bitter struggle with her rebellious Colonies, but also involved in a warm contest with France and Spain. Increased armaments and additional levies were again the order of the day.

Many new corps sprang into brief existence at this period; amongst which may be named the 72nd or Manchester Volontiers, who fought at the famous defence of Gibraltar; the 79th or Liverpool Volontiers, more commonly known as the 'Liverpool Blues'; the 80th or Edinburgh Volontiers; the 83rd or Glasgow Volontiers, etc.; but if we may credit the public prints of the day, few, if any, of the new corps—'the loyalty corps' as they were designated—surpassed in appearance a battalion of foot raised in Westminster through the united efforts of the Earls of Harrington and Chesterfield and popularly known as the 'Westminster Volontiers.'

The regiment was formed in the summer of 1779, and was soon afterwards numbered as the eighty-fifth of the Line.

It does not appear to have borne the title of 'Westminster Volontiers' officially. The Colonelcy of the corps was bestowed upon the Earl of Harrington, a young and popular officer of the Guards, who, as Lord Petersham, had already acquired some military experience in America.

The appointments to the regiment appear in the *London Gazette* of September 14, 1779, as follows:

'85th Regiment of Foot: *To be Lieut.-Colonel Commandant,* Colonel the Earl of Harrington, from 3rd Foot Guards.—Majors: Hon. H.

Phipps from 1st Foot Guards, Richard Crewe.—*Captains*: Lieutenants Hely from the 50th, William Brown from 14th, Francis Grose from 52nd, S. Pole from 56th; T. Mawhead from 19th; Lord Henry Fitzgerald from 66th, Philip Earl of Chesterfield; Captain-Lieut. John Row from 9th.—*Lieutenants*: Ensign Henry Webb on half-pay late 47th Foot, Ensign Michael Dobblyn 33rd Foot; Ensign Richard Brook 34th Foot; Ensign Alexander, Baron de Salans 9th Foot; 2nd Lieut. W. Wilkinson 23rd Foot; Ensign Thomas Steele 63rd Foot; Ensign J. Sackville Higgins 47th Foot; Ensign Richard Dickson 48th Foot; Ensign the Hon. Vere Poulett 62nd Foot; Wm. Parsons.—*Ensigns*: J. F. Hill, The Hon. F. St. John; Edward Lascelles, Robert Molesworth, John Duke; John Finney; Moses Kinkead and John Gould.—The Rev. John Tickle, *Chaplain*; John Forster Hill, *Adjutant*; Alexander M'Dowall, *Quartermaster*; and John Rutherford, *Surgeon*.'

The total strength being: one Lieutenant-Colonel Commandant, two Majors, seven Captains, one Captain-Lieutenant, ten Lieutenants, eight Ensigns, one Adjutant, one Chaplain, one Quartermaster, one Surgeon, one Assistant-Surgeon.

The uniform of the corps was red, faced and turned up with bright yellow. A Royal Warrant authorising the payment of a sum of £756 5*s*. in lieu of twenty months off-reckonings 'to provide swords, leather accoutrements, and other species of clothing' for the non-commissioned officers, drummers and privates of the Regiment shows the strength as 'ten companies, with four sergeants, four corporals, two drummers, and seventy private men per Company; besides commissioned officers, and two fifers for the Grenadier Company.'

The Regimental Agents were Messrs. Cox and Muir of Craig's Court.

On its formation, the Regiment was stationed at Chatham. Here it appears to have embarked for Jamaica early in the year 1780—'one of the finest corps that ever left the shores of England for the Antilles.'

Its strength on embarkation is given in the *Abstract of State Papers* in the 'Annual Register' of the following year as follows:

'Three field-officers, seven captains, fifteen subalterns, four staff, twenty-nine sergeants, thirty-nine corporals, twenty-one drummers, 644 privates. Total: twenty-nine officers and 733 non-commissioned officers and rank-and-file.'

The arrival of the 85th at its destination is duly chronicled in the *Jamaica Royal Gazette* of the day amongst the sparse items of local intelligence, which intermingle in its columns with scraps of 'Europe' news, advertisements of runaway slaves, and grandiloquent semi-official announcements touching a certain mysterious 'grand secret expedition now in preparation against the possessions of His Catholick Majesty' which eventually resolved itself into the paltry attempt—paltry in respect only of the inadequacy of the force—against the Spanish Honduras, under Nelson, then the youthful captain of the *Hichenbroke*.

'Arrived on Tuesday 1st August (1780) at Kingston from Torbay via St. Lucia, under convoy of a squadron commanded by Commodore the Hon. E. Walsinghame the outward bound fleet including with transports, 148 vessels.

'The following reinforcements for the Island are on board: 85th (yellow facings), Earl of Harrington's; 92nd (buff facings) Stewart's; 93rd (yellow facings), McCormick's; 94th (white facings) late Dundas'. Colonel Dundas died on the voyage.

'The Right Honourable the Earl and Countess of Harrington, the Honourable Colonel Stewart, Brigadier-General Garth, Colonel McCormick, and a number of distinguished officers are on board.'

Governor Dalling was at this juncture replaced by Major-General (afterwards Sir A.) Campbell, K.B., and very active measures were put in force for the effective defence of the island, but the details are of mere local interest, whilst those specially relating to the individual corps are scanty in the extreme.

The garrison of the island consisted of the 1st Battalion 60th (Royal Americans), 79th (Royal Liverpool Volunteers), 85th, 87th, 88th, 92nd, 93rd, 94th and the Loyal Irish Regiments of Foot; to which were shortly after added the new 99th, or Jamaica Regiment, from England.

Lord Harrington was appointed to command a flying brigade of Grenadiers, formed of the Grenadier Companies of the various corps.

The 'hat-men' of the 85th remained under Colonel the Honourable H. Phipps, afterwards Lord Mulgrave, at Up Park Camp.

('Hat-men' was a term long retained in some regiments for the battalion companies who wore hats as distinguished from the Grenadiers and Light Company, of whom the former wore caps and the latter leather helmets.)

Like most other regiments on the station, the 85th appears to have suffered severely from sickness. A set of company muster rolls, dated at Up Park Camp, Jamaica, December 24, 1780, embracing a period of 300 days—from February 29 to December 24, both days inclusive—shows that on the day of muster there were sick in hospital and quarters four officers and 165 non-commissioned officers and men, or 30 per cent. of the strength; and that in the three-hundred-day period aforesaid, which must have included the voyage out, the regiment had lost two officers and 130 non-commissioned officers and rank and file. Eighteen months later the regiment received orders for home. Accordingly, in July, 1782, it embarked in the fleet and convoy collected at Bluefields under the command of Admiral Graves, and thus it came to participate in one of the most terrible disasters which have ever befallen the British navy.

Admiral Graves' fleet consisted of the *Ramillies*, 74, bearing the flag of that officer; the *Ville de Paris*, once the pride of the French navy, mounting 110 guns; also the *Glorieux*, *Le Caton* and *Hector*, 64—all, the *Ramillies* excepted, prizes of Rodney's action; the *Canada*, 64; *Ardent*, 64; *Jason*, 64; *Pallas*, frigate, 36, and about 100 sail of merchantmen and transports. The *Ville de Paris*, it may be remembered, was the flagship of the Count de Grasse and was captured by Admiral Rodney, April 12, 1782, in the great naval victory in the West Indies. In this battle the British lost 1000 men, and their adversaries 2000 killed and 4000 wounded.

From the want of uniformity in the Muster Rolls of the Regiment it is difficult to learn the exact distribution of the troops; but it would appear that the following officers of the 85th were actually embarked on board the *Ville de Paris* together with a proportion of non-commissioned officers and privates of whom the number is not specified:—Major Poole, Captain Baron A. Salans, Lieutenants Brooke, Duke, Dobblyn, Finney, Gould, and Parsons; Quartermaster Booth, and

Surgeon Phillips. The names of two other officers, Lieutenants Maxwell and Molesworth, have been given as drowned; but their names appear in the Reduced Half-Pay List for 1785.

At the last moment the *Ardent* and *Jason* were detained at Bluefields as unfit for sea. The number of men-of-war was thus reduced to seven.

On August 26 the fleet and convoy weighed anchor and stood out to sea. As part of the convoy was destined for New York, the Admiral was obliged to shape his course in a more northerly direction than otherwise would have been necessary.

Some days later *Le Hector*, 64, a prize, half-manned, and with her rigging in bad condition, dropped astern and parted company. The cruel sufferings of her people a few weeks later, when encountering two French frigates, who hauled off after a sharp encounter, and left her disabled and waterlogged in mid-Atlantic, are no part of our story.

Later again there was heavy weather on September 8; and the *Pallas*, frigate, and *Le Caton*, 64, showing signs of distress, were signalled by the Admiral to proceed to Halifax, Nova Scotia. On September 16, being then off the Bank of Newfoundland, in lat. 42° 15′ N. and longitude 48° 55′ W., with the wind at E.N.E., the fleet encountered a tremendous gale, which lasted through the night, making sad havoc, more especially amongst the larger vessels.

About 3 a.m. on the 17th there was a temporary lull, which was succeeded by a still more furious gale from the opposite quarter. The fleet had encountered the vortex of one of those vast rotary storms so little understood at the time, and long afterwards believed to be peculiar to the regions of the tropics. Scarce a vessel in the convoy but suffered severely, while on the line-of-battle ships, which were short-handed, ill-found, and old, the effects were exceptionally disastrous.

After hopeless efforts to repair the damages sustained by his own vessel, Admiral Graves, on the third day after the gale, signalled the few merchantmen which had been able to keep company—about nineteen sail—to receive the crew of the *Ramillies*. This arrangement having been carried out, she was set on fire and abandoned with fifteen feet of water in her hold. Shortly afterwards she blew up.

Three days later the *Centaur* was abandoned in a sinking state.

The *Ville de Paris* was seen by the *Centaur* on the morning after the gale with all her masts standing and having, apparently, full control of her helm. She, however, took no notice of the latter vessel's signals of distress. Both the *Ville de Paris* and *Le Glorieux* were fallen in with on subsequent days by other ships of the convoy; but when the remains of the convoy reached home both vessels were missing. Their fate was for some time considered doubtful.

Months afterwards an English seaman was forwarded to the Admiralty by the French authorities, in a neutral (a Russian) vessel. He had been picked up insensible upon a piece of wreck in the middle of the Atlantic by a Danish barque returning from the West Indies. For some days he had remained unconscious, and when he had sufficiently recovered his faculties to be able to give some account of himself, it was found that his limbs were paralysed from the effects of long immersion. He had, therefore, been put on shore at Havre de Grace, where the French people showed him much kindness. On his recovery, his case had been brought to the notice of the French King, the kind-hearted Louis XVI., by whose orders he was provided with a free passage to England.

His tale ran that he was a seaman of the *Ville de Paris*; that she went to pieces the day after the gale; he, at the last, had clung to a piece of wreck, but had been so overwhelmed with terror that he could remember nothing more. He had a recollection, however, that *Le Glorieux* went down in sight some days before the *Ville de Paris*. And so it was learned that these too must be added to the number of the lost.

Of their living freight of many hundred gallant red- and blue-jackets no tidings were ever heard; their names and their numbers even are now unknown.

Later in the year 1782 the scattered remains of the convoy which had outlived the storm reached various ports on the English and Irish coasts; one transport, Plymouth; another, Ilfracombe; a third, the Thames, and so on. The detachments of troops were landed at the first ports in each case, and the remains of the 85th were collected at Dover under the command of Lieutenant-Colonel Lord Henry Fitzgerald.

The subsequent history of the corps has little of note. We hear of recruiting parties having been sent by the regiment to Nottingham and Northampton—of a change of quarters between the 85th in Dover town and the Cinque Ports Regiment in the Castle; also of a draft being given by the 85th to the 45th Regiment, then stationed at Tiptree camp in Essex. The 45th Regiment had just returned from a protracted tour of forty years' continuous service in the West Indies. In those days long periods of foreign service were by no means rare, but the forty years' service in the West Indies of the 45th is almost a record.

At Dover the 85th continued during the latter end of 1782, and the beginning of the following year; and at Dover Castle, on May 14, 1783, it was disbanded.

Lord Harrington was transferred to the Colonelcy of another corps, Lord H. Fitzgerald and the rest of the officers were placed upon half-pay, and His Majesty's 85th Regiment of Foot again ceased to exist.

A set of Muster Rolls taken on the day of disbandment and attested before 'J. Stringer, Mayor of Dover,' shows the strength on that day as 31 Officers, 30 Sergeants, 30 Corporals, 21 Drummers and 240 Privates.[1]

It is not stated what the sickness was that so seriously impaired the strength of the regiment while in the West Indies, but it was probably yellow fever.

In those days, and indeed later, the methods employed to combat this scourge were very ineffectual. In 1796–7 the 13th Light Dragoons suffered terribly. The mortality among the officers was enormous and, after supplying a few drafts of men fit for service to other regiments on the station, but fifty-two non-commissioned officers and men returned to England, and these were speedily invalided out of the service. So much for the ravages of the dreaded 'Yellow Jack.'

That the state of health of many of the non-commissioned officers and men of the 85th on their arrival at Dover and after the disbandment of the regiment was far from satisfactory is proved from the following document addressed to the Lieutenant-Colonel, Lord Henry

---

[1] The only documents among the War Office papers in the Public Record Office referring to this regiment (besides a few entries in the enrolment books and marching orders) are three sets of Muster Rolls—one taken at Up Park Camp, Jamaica, for 1780, and two in Dover, both for 1783.

Fitzgerald—and it may be remarked that, owing to deaths among the officers, either by disease or by drowning, Lord Henry, from the rank of a junior captain, had now arrived at the command of the regiment. He appears to have appealed to the authorities to permit the sick to remain in hospital after the disbandment and to have begged that their 'subsistence' might be continued. The granting of this request, a request as kindly made as justly due, was conveyed to him in the document here quoted:

'W.O., *10th May*, 1783.

'MY LORD,

'I have received the honor of y^{r} L^{d}ship's letter of the 8th inst. sending a Return of the sick men of the 85th Regt. of Foot under your L^{d}ship's Command, who are unable to quit the Regtl. Hospl. on being disbanded, and requesting further instn. relative to the sd. men. In answer thereto I have the honor to acq. yr. L^{d}ship that humanity requires that the Men in this situation shd. still be taken care of. I think the absence of their subsistence beyond the day of the Regt. being disbanded may be charged in the Contingent Bill of the said Regiment. I have the honor &c.

'(Signed) R. FITZPATRICK.

'Lt.-Col. Commdt.
'Ld. Hen. FitzGerald,
'85th Foot,
'Dover Castle.'

This is the story of the ill fate of the second 85th Regiment of Foot during its brief but singularly unfortunate existence from 1779 to 1783. And again we find the officers relegated to half-pay.

The position of the officer placed on half-pay (on reduction or disbandment) was by no means enviable. His income was a mere pittance, and unless he had private means and also powerful friends it was almost hopeless for him to expect ever again to be actively employed, at any rate until more 'new' regiments had to be raised. A glance at the old Army Lists tells the tale.

CHAPTER III

THE 85TH (BUCKS VOLUNTEERS), 1793–1808

TEN years after the period of the army reductions consequent upon the peace of 1783, which led to the disembodiment of the 85th (Westminster Volontiers), and numerous other corps, war was declared against the newly established French Republic; and a large increase in the number of line regiments speedily ensued.

Amongst other corps raised at this period was a battalion of 'Bucks Volunteers' formed by the late Field-Marshal Sir George Nugent, at the time a Lieut.-Colonel in the Guards and with the army in Flanders, and through the interest of the first Marquis of Buckingham, his cousin. This regiment shortly afterwards became the 85th (Bucks Volunteers) Regiment of Foot.

A few cursory passages in the 'Grenville Correspondence' lead to the inference that the scheme for the formation of the regiment was not at first viewed with much favour by the authorities.

Writing to the Marquis of Buckingham under date Walmer Castle, October 11, 1793, Lord Grenville remarks:

'George Nugent has written to me twice on the subject of his proposal, and I sent him Lord Amherst's (Commander-in-Chief) answer, which is negative, at least for the present. He seems to have an invincible dislike to new corps, I fancy, from the badgering he got upon that subject last war. He now makes the plea of scarcity only, that the number to be raised is filled up with older Lieutenant-Colonels.'

Colonel Nugent appears, however, to have succeeded in getting permission to carry out his project, as six weeks later (November 21) we find Lord Grenville writing:

'I have spoken to Pitt about G. Nugent being appointed Aide-de-Camp (to the King), if the promotion mentioned should take

GEORGE TEMPLE, MARQUIS OF BUCKINGHAM.

place . . . I have reason to think, that at present, no idea exists of that promotion. On the whole, it would, perhaps, be better that he should go on with his corps.'

An entry in the War Office Marching Order Books dated November 22, 1793, and addressed to Colonel Nugent, directs that 'the corps now raising under your command be quartered at Buckingham and Aylesbury, until further orders.'

With the local influence he possessed, Colonel Nugent's task was no doubt a fairly easy one. We read, in the journals of the day, that six hundred men were enrolled within less than three months from the date of the 'beating order.' Enlistment by 'beat of drum' is here meant. It was one of the forms of raising recruits when required. An officer accompanied by an N.C.O. and sundry men and a drummer would issue a flamboyant proclamation. Plenty of rub-a-dub-dub collected the young men in the district. Once collected, the eloquence of the Recruiting Officer, added to the hardly veracious tales of the N.C.O., and lastly, an unstinted distribution of beer and other fluids, procured the harmless and necessary recruits. There were other methods. Debtors in prison could purchase their freedom by enlistment. Felons also, but they had to enlist for life, and were sent to unhealthy stations and could not choose their own regiments. Both of these last classes of recruit proved unsatisfactory, as may well be imagined. Besides those named, you had the voluntary recruit because he wished to be a soldier, and the recruit who enlisted because he had got into some trouble, or because his sweetheart had thrown him over.

The above statement is confirmed by an entry in Marching Orders under date December 4, 1793, directing the quarters of the corps now raising at Buckingham and Aylesbury 'to be enlarged to Winslow, and other adjacent places as may prove necessary.'

It may be mentioned that all these men engaged for life service.

A subsequent entry dated February 22, 1794, directs 'Colonel Nugent's corps to move in two divisions from Buckingham, *via* Birmingham and Wolverhampton, to Newport (Salop) and Whitechurch, there to await orders from the officer commanding the troops at Chester, respecting embarkation for Ireland.'

In March the same year the regiment was numbered the 85th (or Bucks Volunteers) Regiment of Foot. The appointments of the officers are announced in the *London Gazette* of March 1, 1794, as follows:

'85th Regiment of Foot: *To be Lieut.-Colonel Commandant,* Lieut.-Colonel G. Nugent from Coldstream Guards; *Majors*: Honourable E. Bligh, 1st Foot Guards, C. Dawkins, 1st Dragoon Guards; *Captains*: Captain F. Chichester Garston from late 89th Foot, M. Jenour half-pay 11th Foot, J. Ross half-pay, a late Independent Company, Lieutenant T. Talbot 24th Foot, Fred Sitwell 11th Dragoons, R. Benson Independent Company.—*Captain-Lieutenant*: Lieut. R. W. Ottley from 53rd Foot.—*Lieutenants*: Ensign Pigott from 68th Foot; Hon. E. Forbes from 14th Foot; F. Buckstone, Shipley, and P. Dayrell from Independent Companies; R. P. D'Arcy from 82nd Foot.—*Ensigns*: Lieutenant Young, from the Bucks Militia, T. Bowen, C. Whyte, J. Ross, W. Ross, P. Fortescue.—*Chaplain*: Rev. R. Marshall.—*Adjutant*: Captain-Lieutenant Ottley.—*Quartermaster*: Sergeant-Major Tomlin from Coldstream Guards.'

The uniform of the corps was red, faced with lemon-yellow. The coats and hats of the officers were laced with silver, and the coats of the non-commissioned officers and privates with coloured worsted.

The regimental agent was Mr. Donaldson of Caddick's Row, Whitehall. Some years later the agency was transferred to Messrs. Cox & Co., by whom it has since been retained almost uninterruptedly.

About the middle of March, 1794, the 85th moved into Chester, and there embarked for Ireland. The earliest 'Adjutant's Roll' to be found amongst the War Office records of the corps is dated July 18, 1794, and shows it to have been then at Cork with a strength of 3 Field Officers, 7 Captains, 1 Captain-Lieutenant, 13 Subalterns, 5 Staff, 32 Sergeants, 30 Corporals, 22 Drummers, and 742 Privates, having just received a draft from some Scotch regiment—probably one of the corps of Highland Fencibles—the name of which is not given. The stay of the regiment in Ireland was apparently of short duration as, under date July 26, we have an order to 'the officer commanding the 85th Regiment off Bristol' directing him forthwith to disembark the troops under his command, and march them into billets at Chippenham

Painswick and the adjacent villages. Subsequent orders of August 1st, 3rd, and 8th, direct the removal of the regiment from the neigh-

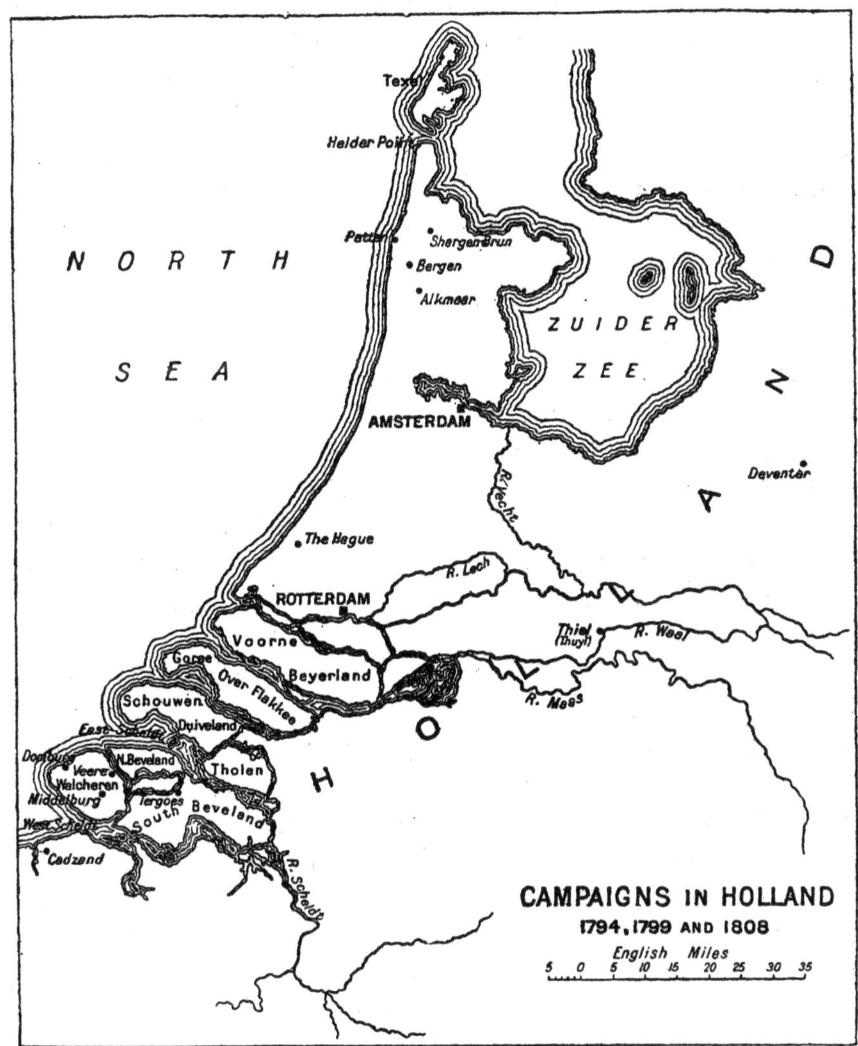

CAMPAIGNS IN HOLLAND
1794, 1799 AND 1808

bourhood of Bristol to Southampton, there to await instructions from the officer commanding at Netley Camp.

At the end of 1792, the Austrian Netherlands were in the undisputed possession of France. It was evident that neither England nor Holland could permit this to continue, and that, if France meant to keep possession of these provinces, war with England, Holland, and Austria as well, could hardly be avoided.

It was, therefore, wiser for the French to declare war with Holland, at any rate before that country was prepared to resist. Accordingly France declared war against England and Holland in February, 1793. There was, however, a party in Holland violently opposed to the Government, and of this the French made considerable use. They persuaded the people of the Seven United Provinces that they were not at war with them but with their Government. Dumourier was the first French general employed. He failed, and being summoned to Paris—a summons meaning execution—fled to the Austrian headquarters.

General Pichegru succeeded him after Dampierre, another French general, had met with a fatal wound. Against the invaders were British, Dutch, Hanoverian and Austrian troops.

At Southampton the regiment appears soon afterwards to have embarked for the Continent with other reinforcements. It landed at Walcheren late in the month of August 1794; and in the October following it joined the Duke of York's army, then in position on the Waal.

It was first engaged under fire in some of the minor affairs which occurred about this period, under command of Major the Hon. E. Bligh. This officer was subsequently promoted to the 5th Fusiliers, which regiment he commanded with great distinction in Holland in 1799. He died a General in 1840. He was described by Dr. Ferguson, Inspector-General of Hospitals, as 'a man of gigantic stature and of the most romantic courage.' In brigade with the 27th, 28th and 53rd Regiments, the 85th bore a share in the actions at St. Andrea, Thuyl, and Guldermazen. No details of importance appear to be forthcoming with regard to the affairs at St. Andrea or St. André and Thuyl. But of the defence of Guldermazen or, as it should be more properly written, Guelder Malsen, a brief account may here be given.

The British post of Guelder Malsen was held by General Dundas with the regiments or parts of the regiments named, the remainder being held in reserve. On January 5, 1795, the British force was attacked by a large body of cavalry supported by *tirailleurs*. The attack succeeded, and Dundas was driven back with the loss of two guns. Reinforcements came up from his reserve. The British then became the attacking force. Both guns were recovered, the French were repulsed and the post preserved. The Duke of York then returned home, leaving Walmoden in command. The outcome of this campaign had shown the impossibility of preserving the United Provinces from becoming the prey of France. Pichegru had 70,000 men under arms; and the Allies a force which, compared with that, was insignificant, and suffered severely in the terrible winter retreat to Bremen, where it arrived early in 1795. Colonel Nugent had meanwhile been appointed Brigadier-General.

A brief account of this disastrous retreat is here inserted.

'The allies could defend the passage of the Waal before the frost had congealed its waters; but in consequence of the severity of the winter, the French marched over the river, and transported their artillery across the ice with ease.

'At the beginning of the winter the Duke of York despairing of the preservation of Holland, as the natives were unwilling to resist the invaders, returned to England; and his departure encouraged the party adverse to the Stadtholder to act openly in behalf of the French.

'The English and Subsidiaries hastened towards the Leck, Pichegru ordered an attack; but after four assaults, his troops gave way. In the next battle he triumphed by the force of numbers. A retreat of the most disastrous kind ensued.

'For some time the British troops had been ill-fed, ill-clothed, harassed by sickness, and shamefully neglected both by the commissaries and the medical attendants of the army. Many had died in the hospitals, and not a few perished on their way to these receptacles of misery. But the retreat to Deventer was more strikingly unfortunate. The sick and wounded soldiers being removed in open waggons, amidst an intense frost, a considerable number died. Fatigue, hunger and cold destroyed many others on their march; and the survivors were involved in severe distress. They were insulted by the Dutch and treated as aliens, rather than as friends or fellow-warriors. Being pursued they did not remain long at Deventer, where they destroyed their artillery and military stores. They proceeded to the Vecht, sometimes (from a sudden thaw) through mud and water, at other times amidst ice and snow. From that river they continued their march to the Ems, still harassed by the enemy; and at length they reached the Duchy of Bremen, when they were conveyed to England.

'[The expulsion of the British troops from Holland, and the consequent dispersion of the Dutch army, opened the gates of Amsterdam to the French General; and seven flourishing provinces submitted to the sway of the Parisian Convention.]'

When the breaking up of the ice opened the Elbe ports once more, the regiment embarked at Bremen Lee on April 14, 1795, with Major-General Coates' brigade, landed at Harwich, and marched to Wells in Somersetshire, where it appears to have been quartered some months. It may here be mentioned that of all cities in the west of England, Wells, perhaps, was one of the most uncomfortable for troops. The inhabitants hated men being billeted on them, and there are many petitions praying for the removal of these unwelcome guests which spread over a long period of years.

In August 1795 the 85th moved to Andover, and immediately after to Southampton, whence it proceeded to Gibraltar under the command of Lieut.-Colonel Dawkins. A small depot detachment, left at Southampton, was moved first to Camberwell, and afterwards to Chatham. At Gibraltar the regiment remained until the summer of 1797, receiving in the interim drafts of men from the 64th and 108th Regiments, the latter of which was broken up in that garrison.

In July, 1797, the 85th, under Lieutenant-Colonel John Ross, returned to England in a skeleton state (the whole of the privates having been drafted into other corps), landed at Chatham, June, 1797, and marched to Derby to recruit from the Militia Regiments in the northern districts (viz. the Derby, Nottingham, Leicester and Staffordshire Supplementary Militia).

In February, 1798, it proceeded to Bridgnorth, Salop, sending out officers' recruiting parties to Buckingham, Bristol, Newport (S. Wales) and Manchester. During the summer of this year, two companies were completed to 100 rank and file each, and despatched to Portsmouth to join an expedition fitting out against Ostend. But the force had sailed prior to their arrival, and they were accordingly brought back again to Bridgnorth without having the opportunity to participate in that rather inglorious enterprise.

In July, 1798, the regiment left Bridgnorth, marched to Southampton, embarked for Jersey, and took over quarters at St. Oban's. In July, 1799, it re-embarked at St. Heliers, and proceeded to Lymington, Hants, remained there a fortnight and then moved into barracks at Parkhurst, Isle of Wight, at that time the chief rendezvous of the French emigrant and other foreign corps in British pay.

The army had, by this time, partially rallied from the effects of the terrible disasters of a few years previous, and the Government of the day found itself once more in a position to indulge its mischievous proclivities for ill-matured military enterprises. Upon this occasion the move was a 'secret' expedition to Holland, where it was assumed—on utterly insufficient grounds, as it subsequently proved—that the Dutch population would rise *en masse* against their Republican masters.

The whole of the troops who had been recovering health and discipline in breezy encamping grounds about Southampton and the New Forest, under the vigilant eye of Sir Ralph Abercrombie, were moved by forced marches, via Fareham, to the south-east coast and concentrated in the neighbourhood of the Downs. The 85th received orders to join this force, and, accompanied by their women and children—who then, as a matter of course, were allowed to encumber every movement, in or out of the field—embarked on board a little fleet of revenue cutters at Cowes and sailed for the Downs, disembarking at Deal, and marching to Ramsgate, where they were posted to Sir Eyre Coote's brigade, of which the headquarters lay at Margate.

The situation politically was this. The Dutch had been ready enough to submit to the French, and their complaisance had been rewarded by a series of insults and humiliations. A revolutionary army had been quartered in their country and such alterations of Government as the rulers of France chose to dictate were effected without difficulty. The British Ministry therefore concluded that the Dutch could not but hate their oppressors, their apparent acquiescence notwithstanding. The British Ministry also, and upon most insufficient grounds, became impressed with the notion that the Dutch would in consequence be prepared to rise, if a British force landed in Holland. Surely some of that spirit which had rescued the States from the tyranny of Philip II must remain, or at any rate might be expected to be capable of revival. How erroneous the views of the British Ministry actually were, the subsequent events proved.

Nor was British assistance alone available. Russia, too, joined in the fray, and to Holland despatched a body of troops to co-operate with the British and to deliver the Dutchmen who at the time desired to be left undelivered; nay more, who were prepared to turn round and rend their self-elected deliverers.

On July 11, 1799, the regiment embarked on board transports and sailed out of harbour. On the 13th the squadron under the command of Vice-Admiral Mitchell appeared in sight. The transports immediately joined them and the whole put to sea. The fleet did not reach its destination until the afternoon of July 21, but the weather being unfavourable it was again obliged to put to sea, and did not again make land until the afternoon of the 26th, when it anchored near the Helder Point. The troops composing the expedition to Holland, under General Sir Ralph Abercrombie, received orders to hold themselves in readiness to disembark at daylight the following morning. One point about the composition of this force needs notice. It possessed neither field artillery nor cavalry. The lack of the former was, however, somewhat counterbalanced by the fleet and its guns, which could co-operate along the shore.

The regiment having been formed into brigade with the 2nd, 27th and 29th regiments, under the orders of Major-General Sir Eyre Coote, one hour before daybreak commenced their disembarkation into flat-bottomed boats, brought with the fleet for that purpose.

On a signal being given that the whole of the troops were on board the boats, a line was formed, and they proceeded towards the shore, covered by some small vessels and gunboats belonging to the fleet, and soon after effected a landing, although keenly opposed by the enemy, who had strongly posted themselves on the shore, with sand hills in well-covered situations, and a six-gun battery directly opposite the place of landing.

Lieutenant-Colonel Ross, having formed the regiment, although much impeded by the enemy's shot and shell, immediately advanced to meet them, drove them from the heights with the bayonet, and during the whole day the regiment behaved in a most gallant and soldierlike manner.

The success of this day's action was principally to be attributed to the conduct of Major-General Coote's brigade, who, instead of waiting the attacks of the enemy, invariably advanced to meet them, in situations where the smallest piece of ground could not be lost without danger.

The regiment continued in action from daylight until about four o'clock in the afternoon without the smallest respite, and received the thanks of the Commander of the Forces as also those of Major-General Sir Eyre Coote in the orders issued on the occasion.

The regiment suffered severely, and on the enemy retreating took the outpost duty for the night.

Officers wounded: Major Robert Ottley, Lieutenants Barry and Travers severely; Captain M'Intosh, slightly.

Under arms on July 28, the day after the action, when Major-General Sir Eyre Coote personally returned thanks to the corps for their steady conduct the preceding night while on outpost duty, the regiment kept its ground on the Sand Hills, the inclemency of the weather being such that the army could not advance, nor could any communication be had with the fleet.

On September 1 the regiment, with the brigade, advanced about five or six miles without opposition, and occupied some farm houses on the road leading to Evert Sluis. On the morning of the 6th it moved on the road leading to Schargen Brun, and occupied a church in that village.

On the 10th, the enemy having made a vigorous attack on the British lines, the brigade were under arms and proceeded to ground pointed out by the Quartermaster-General, but the regiment not having been brought into action returned to its former quarters.

The regiment, with the brigade, marched on September 19 upon the village of Patten with a view of attacking the enemy, but (through the failure of a Russian column under the command of Lieut.-General De Hemnems) the brigade returned to their former quarters, not having been brought into action.

On the afternoon of October 1 it again marched on the village of Patten, arriving that night and laying by its arms.

On October 2 the brigade moved along the seashore and just at daybreak came in contact with the enemy's outposts, attacked them and drove them back to their main body, who had strongly posted themselves on an ascent leading to the Sand Hills.

The 85th, being on the right and considerably advanced, became warmly engaged with the enemy, who showed a disposition to come upon the right of the brigade.

Lord Chatham's brigade (with the exception of the 31st Regiment) was now ordered from the plain to the Sand Hills, to reinforce and form on the regiment's right, and the 31st to move close under the hills, parallel with Major-General Coote's brigade. These movements were well executed, and Lord Chatham's brigade took its station some distance behind the 85th, outflanking it by about two battalions. The line being thus formed the whole received orders to advance at a brisk pace to gain the heights. The men had to proceed for about three-quarters of a mile through a scrubby wood, and then by an ascent to the Sand Hills, the enemy disputing every inch of the ground.

The regiment at this time, coming into contact with a considerable body of the enemy, made a most desperate charge with the bayonet, headed by Lieut.-Colonel Ross. The enemy were forced from their position with great slaughter. They continued advancing until they obtained a favourable situation below these heights, which both blocked up and commanded the avenue and great road leading through Bergen.

At this time a considerable body of the enemy advanced along the avenue, and made a most spirited attack on the posts held by the regiment, with an intent to regain the heights, but were met with the point of the bayonet and repulsed with a severe loss.

The regiment gallantly maintained its position during the rest of the day and throughout the night against several other attacks of the enemy; and occupied some houses near Bergen on the following morning.

The 85th suffered severely during the above action and received the particular thanks of H.R.H. the Commander-in-Chief for its gallant services during the actions of the day.

Lieutenant Nestor killed; Lieut.-Colonel John Ross wounded

when leading the regiment to the charge; Captains Bowen and M'Intosh and Lieutenant Reilly wounded.

The regiment moved along with the brigade, on the morning of the 6th, upon Alkmaar; on passing through which His Royal Highness reviewed the brigade, after which the whole proceeded forward to gain the front of the army, which was at this time warmly engaged with the enemy at the village of Anher Sluis; but, not arriving until near dark, the regiment was not engaged. It took up, however, its quarters for the night in some empty houses in that village. The aspects of the enterprise had now become sadly changed and the end was nigh at hand. The strategical skill which had marked the preliminary operations under the guidance of Sir Ralph Abercrombie had since become conspicuous by its absence; a contingent of Russian troops had joined, but want of harmony existed between the allied commanders. The anticipated co-operation of the Dutch had proved a myth. Moreover, the reinforcements sent out from England were of the worst description—disorderly levies of young and raw militia men, whose limited ideas of military discipline had been readily overset by the extravagant bounties paid to them for volunteering, and their hurried transfer to regiments they had never seen before.

The following is a specimen of the class of advertisement that was used to attract these recruits. It was displayed in Yorkshire at the end of the eighteenth century, the officer referred to having been invalided home from the Low Countries and placed in charge of the recruiting service in his district—he died in 1801, aged twenty-four. We here give the extract from the *Yorkshire Post* of the same year:

'Now or Never

'G. [Royal Arms] R.

'Wanted a number of bold aspiring Yorkshire Lads to serve as Gentlemen Soldiers, in His Majesty's 85th or Young Bucks, Regiment of infantry; whose hearts beat high at the sound of the Drum, and who have an inclination above servile employment. Let them repair with the spirit of heroes to their countryman, Captain Kirkby, or

[Blank for insertion of Recruiting Sergeant's name]

'where they will enter into present Pay and Good Quarters.

'Now is the Time, my Lads; step forth, the War will soon be over!

Consider your advantages, to be then free in any town in His Majesty's Dominions, together with your wives and children, enjoy the pleasures of Military Life, only perhaps, for a few Months. Consider, my Bucks, what a liberal Bounty you'll receive merely to go on a Party of pleasure ! ! !

'God save the King.

'HUZZA ! HUZZA ! HUZZA !

'☞ Friendly advice Gratis.

'[Pierson, Printer, Sheffield.]'

The result was a precipitate retreat coastwards, in the wettest of seasons, during which the 85th, in common with other regiments, had to leave its women, baggage, and all who could not keep up, in the hands of the enemy. Then the Convention of Alkmaar was concluded, by which it was agreed to evacuate Holland immediately. In accordance therewith, the allied troops were brought back as hastily as they had been thrown over—the British being landed *en masse* on the Norfolk coast, and the Russians sent round to winter in the Channel Islands, until the breaking up of the ice in the Baltic should admit of their return home.

The 85th arrived from Texel in the *Trusty* ship of war and went into temporary barracks then existing at Great Yarmouth on November 29.

Matchett's *Norwich Remembrancer*, 1822, states:
(October 28, 1799.)

'The Guards and other regiments to the number of 25,000 men were landed at Yarmouth. The Grenadier Brigade of Guards entered Norwich by torchlight, and were followed by 20,000 troops on the succeeding days. . . . Owing to the indefatigable exertions of J. Herring Esq., Mayor, and the innkeepers and inhabitants generally, these brave men received every accommodation their circumstances required. Mr. Herring was subsequently presented to the King and received an offer of knighthood, which he declined.'

The route was soon after received for Silver Hill Barracks near Robertsbridge, Sussex, and on December 7 the regiment began its march. But at the first halt—Norwich—forty men were stricken down with fever—and at Thetford, the day after, fifty more were added to the list. Matters now looked ominous, and Colonel Lee, the Commanding Officer, resolved upon halting to await instructions from headquarters. The Commander-in-Chief immediately issued orders that the regiment should be distributed in detachments amongst the adjacent villages. Headquarters at Swaffham, detachments at East Dereham and Walton, with temporary hospitals at Norwich and Thetford. Bedding and other requisites, and additional medical officers were despatched from London, and although, with few exceptions, every man in the regiment had the disorder, the deaths were comparatively few and the health of the corps was soon completely restored.

Early in the ensuing year, Major-General Nugent, who was then Adjutant-General in Ireland, addressed a circular letter to the commanding officers of the several corps of Irish Militia, requesting them to use their influence to procure volunteers for his late regiment, the 85th.

In this way about 1800 fresh men were soon obtained. It may here be observed that this connection with the Irish Militia was maintained, to a certain extent, until the peace. By far the larger moiety of the men who joined the 85th during the latter years of the French war were volunteers from the Irish Militia serving in England.

The regiment was brought together again in February, 1800, at Norwich Barracks, and some days later the Irish volunteers joined.

In March the same year it marched in four divisions to Colchester, where orders were received to form a 2nd Battalion. Additional arms and new clothing were forthwith issued to both battalions.

The appointments to the 2nd Battalion appear in the *London Gazettes* of May 13 and 24, 1800, as under :

'85th Regiment of Foot (2nd Battalion) : *To be Colonel Commandant*, Major-General Sir Charles Ross, Bart, from the late 116th (Perthshire) Highlanders.—*To be Lieut.-Colonel* : Brevet Lieut.-Colonel Wm. Douglas from 74th Highlanders.—*Majors* : Brevet Lieut.-Colonel R. Lee from half-pay of the late 124th Foot ; George Jackson, half-pay late 96th Foot.—*Captains* : Lieutenants R. D'Arcy ; Cookson, half-pay of a late Independent Company ; Wilkins, half-pay, of Macdonald's late Corps; Brock from 5th West India Regiment; Mulhouse, 1st West India Regiment; Price, 13th Foot.—*Captain-Lieutenant*: Lieutenant J. Emerson from half-pay of the late Dublin Regiment.'

In June 1800 the two battalions marched from Colchester to join the Camp of Instruction formed near Windsor on Bagshot Heath, under the command of H.R.H. the Commander-in-Chief and General Sir David Dundas, for the practice of the much-abused 'eighteen manœuvres' and field movements generally. These 'eighteen manœuvres' were the work of the celebrated General Dundas, who had compiled them in accordance with the orders of the Horse Guards.

The force amounted to over thirty-five thousand regular troops. Both battalions of the 85th were brigaded with the 2nd Battalion Royals, the 64th Regiment and the 5th Dragoon Guards, under command of Major-General Lord Charles Somerset, and were encamped close to Swinley, where the Duke of York had a small house.

The troops were repeatedly exercised before the King, and an extract from the *Times* of July 18, 1800, gives presumably a fair idea of a grand divisional field day of that period.

'Yesterday (17th) His Majesty reviewed on Winkfield Plain the whole of the cavalry and infantry encamped at Swinley and Bagshot Heath, amounting to 32,000 men.

'The King, attended by the Prince of Wales, the Duke of York, his aides-de-camp, and a numerous suite of officers, arrived on the Plain at half-past ten, followed by the Queen and the Princesses Augusta, Elizabeth and Mary. The Princess of Orange and the Countess of Harrington went in sociables and the Stadtholder in another open carriage.

'On the King's arrival a salute of twenty-one guns was fired to the right of the ground where His Majesty entered.

'The troops were stationed in two lines on the whole length of the plain, from Mr. Watson's house to the Forest, a distance of a mile and a half in length, the cavalry being stationed at each extremity of the plain, with the artillery and the infantry forming the centre; the Foot Guards being placed opposite Cranbourne Lodge.

'His Majesty having passed and repassed in front of the line, proceeded along the centre between the front and rear, the bands of the different regiments playing as he passed.

'The King then took his station in the middle of the plain; a Royal salute being given on the left of the ground, the whole line of cavalry and infantry at once fired in quick succession from one extremity to the other, beginning with the cavalry on the left in front, and ending with the cavalry on the right in rear; the battalion guns of each regiment firing as signals, and the music playing after each salute. This part of the review, which was three times repeated, had a striking military effect.

'A signal was then fired for the cavalry and infantry to form single companies, in order to march off the ground before His Majesty; the cavalry on the left preceding the infantry, the cavalry on the right bringing up the rear; the close of the mass finishing with the regiment of York Hussars and a small squadron of French *émigrés*.

'H.R.H. the Duke of York as Commander-in-Chief marched on the ground at the head of the Guards; he wore in his hat a large branch of laurel.

'After the review, the Royal Family went to dinner to Cumberland Lodge, and in the evening a grand ball was given to the General Officers and their ladies. The magnificent tents presented some time since by Mr. Warren Hastings to the King were prepared for their reception.

'The review was over by half-past one o'clock; but the number of persons who had come from town was so great, that accommodation could not be procured for them at the neighbouring inns, and many of them slept in their carriages.'

The camp was broken up in October, and the troops marched to country quarters.

The 85th proceeded via Winchester and Southampton to Jersey, to replace the 88th Connaught Rangers. The 1st Battalion went into quarters at Granville, and the 2nd Battalion at St. Heliers.

The regiment (the 1st Battalion being under the command of Brevet Lieut.-Colonel Lee) was removed to Cowes, Isle of Wight, in

February, 1801, and on June 25, 1801, the 1st Battalion received orders to prepare for foreign service. It was accordingly completed to 1100 rank and file by drafts from the 2nd Battalion, to which all sick and weakly men were transferred in return.

On June 27 the 1st Battalion 85th, under the command of Lieut.-Colonel Willoughby Gordon, and a company of the Foreign Corps of Artillery started with a naval squadron under sealed orders. The destination of the expedition was popularly reported to be Brazil; but, in reality, it was intended for the Island of Madeira.

The English Government had decided on the occupation of that Island (and also of the Portuguese possessions in India) in consequence of France and Spain having stipulated, in a treaty recently concluded with Portugal at Badajoz, for the exclusion of English vessels from Portuguese ports.

The surrender of the island was successfully negotiated by the officer commanding the expedition—Colonel, afterwards General Sir William Clinton—and the troops landed and entered into peaceful possession of the forts commanding the harbour.

The 1st Battalion 85th remained at Madeira, under command of Colonel Willoughby Gordon, until the end of the year, and on January 18, 1802, proceeded to Port Royal, Jamaica, arriving there on February 13, and occupying Stony Hill Barracks. The 2nd Battalion meanwhile had moved from the Isle of Wight to Lymington, Hants, thence to Fort Cumberland, and afterwards to Portsmouth.

At Portsmouth it embarked under command of Lieut.-Colonel Lord Aylmer for Jamaica, and arrived at Port Royal in February, 1802.

In the reductions consequent on the peace of Amiens, the 2nd Battalion was broken up. Sir Charles Ross and the officers were placed upon half-pay, the effective N.C.O.'s and men were retransferred to the 1st Battalion and the 85th became again a single-battalion corps.

The six remaining years during which the 85th remained in the island of Jamaica present few if any events worth recording in connection with the history of the corps.

The garrison of the island consisted of the 1st, 2nd, 4th and 6th Battalions, 60th Foot; the 55th, 85th, and 2nd West India regiments, some Artillery, Invalid Companies, etc. Like other white corps in the command, the 85th had a small detachment for recruiting purposes in England, with the General Army Depôt at Parkhurst, Isle of Wight. Like them too it had its attached quota of black slave pioneers—by whom all fatigue work was performed—whose names, ranging from 'Apollo' to 'Duke of York,' figure at the end of each monthly pay-list.

In 1803 Colonel Willoughby Gordon was appointed Deputy Barrack-master General in Great Britain, and was on July 16 replaced in the command of the regiment by Lieut.-Colonel David Mellifont. On June 25, 1806, Lieut.-Colonel David Mellifont returned to England on leave. The ship in which he embarked foundered in the Gulf of St. Lawrence in August and all perished. He was succeeded by Lieut.-Colonel Henry Cuyler, from the Buffs. For the first four or five years the regiment was at Kingston and Spanish Town, and the health of the corps would seem to have been average good for the station—the deaths amounting to five or six a month, although, in certain exceptional cases, they appear to have risen to forty per month. Desertions, too, were nearly as plentiful. During the latter portion of its stay, the regiment was at Fort Augusta, and at Up Park Camp, with the sick on board a hospital ship. During this period the sanitary conditions, so far as any criterion is afforded by the regimental accounts, would seem to have improved considerably.

Several foreign officers (amongst them a Lieutenant Ali Dey) served in the 85th during its sojourn in Jamaica.

On April 20, 1808, the 85th embarked at Port Royal, under command of Lieut.-Colonel Cuyler, and sailed for England, arriving at Portsmouth on June 28, and occupying barracks at Hilsea. The strength of the regiment on landing in England was as follows: one Major, four Captains, four Lieutenants, Ensigns (none), one Staff, sixteen Sergeants, thirteen Corporals, twelve Drummers, thirty-one Privates. Total, eighty-two!

From Hilsea it marched to Buckingham on July 14, 1808. The Depôt under Colonel Henry Cuyler had previously arrived from Lichfield. Here it was joined by a draft of 200 recruits from Lichfield, and the accommodation in the town of Buckingham proving insufficient, three companies were detached to Stony Stratford and Newport Pagnell.

Early in September these companies were brought back

again, and the regiment marched in two divisions for Shorncliffe Camp.

While *en route*, a Horse Guards order was received directing the transformation of the regiment into a Light Infantry Corps,[1] Major-General Baron de Rottenburg, a German officer, who had served in the Jäger Battalion of the 60th Rifles and in the Swiss regiment of Roll, being deputed to superintend its instruction in light manœuvres.

The strength at this period was as follows: one Colonel, two Lieut.-Colonels, two Majors, ten Captains, twenty-two Lieutenants, eight Ensigns, one Paymaster, one Quartermaster, one Adjutant, one Surgeon, one Assistant Surgeon, one Sergeant-Major, one Quartermaster-Sergeant, one Paymaster-Sergeant, one Armourer-Sergeant, forty Sergeants, forty Corporals, twenty-two Buglers, 570 Privates.

[1] The following is the order of seniority of the present Light Corps: 90th Foot made Light Infantry when raised in 1794; 52nd, January, 1803; 43rd, June, 1803; 68th and 85th, 1808; 51st and 71st, 1809; 13th, December, 1822; 32nd, 1859 (after the defence of Lucknow); 105th, brought into the British Line as a Light Infantry Corps in 1860, having been raised as Light Infantry in 1839. (See 'Records and Badges of the British Army'). The first 85th Light Infantry or Royal Volontiers, raised in 1759 and afterwards disbanded, was apparently the first Light Infantry Corps raised.

LIEUT.-GENERAL SIR GEORGE NUGENT, BART.
Who raised the Regiment, 1793.

CHAPTER IV

85TH (BUCKS VOLUNTEERS) LIGHT INFANTRY, 1808–11

OF the system of light manœuvres introduced into the regiment by Baron de Rottenburg, no detailed accounts have been preserved, but it is believed to have been nearly identical with that sketched in the old editions of the revised 'Field Exercises' of 1833, which method, with a few unimportant differences of detail, was practised by the corps for many subsequent years as 'Colonel Thornton's Light Drill.'

After a sojourn of some months at Shorncliffe, the 85th Light Infantry moved to Brabourne Lees, a locality situate about six miles from Westonhanger, upon the Canterbury side of the present line of the South Eastern Railway. It had been used as a site for encampments during more than one of the perennial invasion panics of the eighteenth century; and, at the period in question, barracks for a couple of battalions of infantry and a squadron of cavalry erected at an enormous cost, but which have long since been pulled down, existed there. These had been some of the temporary barracks erected there between 1792 and 1804.

At Brabourne Lees the 85th was brigaded with the 68th, another newly organised light corps. There it stayed until June, 1809, when it moved to Gosport and encamped, and on July 16 embarked at Blockhouse Point on board the *Resolution* and *Plover* men-of-war for the Scheldt; the strength of the regiment being as follows: three Field Officers, five Captains, twenty-seven Subalterns, three Staff, forty Sergeants, nineteen Buglers, 570 Rank and File.

The political situation has been thus described:

'In the hope of assisting the Austrian Emperor by an enterprise which might alarm the French King, and also with a view of destroying a French fleet in the Scheldt, which might otherwise be used for invasive purposes, the King (George III), listening to the suggestions of Lord Castlereagh rather than to the dictates of his own judgment, ordered a great armament to be prepared.'

Narrating the events of the campaign, the account proceeds thus :

'The military and naval commanders selected for the expedition were the Earl of Chatham and Sir Richard Strachan. Above 38,000 men were employed under the former; while the Admiral directed the operations of thirty-seven ships of the line and a considerable number of smaller vessels.

'The original intention was to proceed up the West Scheldt ; and, with that view, two preparatory operations were planned. The Marquis of Huntly and Commodore Owen were directed to destroy the batteries on the Isle of Cadsand ; and Sir John Hope and Rear-Admiral Keats were ordered to make a descent on South Beveland, and storm or seize those batteries which might otherwise obstruct the progress of the fleet ; but the violence of the wind drove the whole armament towards the East Scheldt.

'It was then proposed that a disembarkation should be attempted near Domburg, on the Isle of Walcheren ; and when this was prevented by a heavy swell it became expedient to take shelter in the road of Veer, where the constant succession of gales for many days rendered it impossible to recur to the former scheme. The fleet being skilfully conducted through a narrow and difficult passage, the troops effected a landing, near a fort which was quickly abandoned by the garrison. As the chief town of the island was readily given up to the invaders, the reduction of Flushing was the only object which retarded the advance of the fleet.

'It was deemed hazardous by the Commander-in-Chief to leave such a post uncaptured, and it was therefore attacked ; but the siege had been carried on for some days before the wind would allow a naval blockade to be formed ; and the Earl seemed then disposed to leave the conduct of the attack to one of his officers and proceed with a great part of the army to Batz, where Sir John Hope was stationed with his division anxiously waiting for assistance. After some delay, a flotilla proceeded by the Sloe passage into the West Scheldt, and other divisions of the fleet followed.

'Flushing was taken after a siege of eleven days ; and when the

Commandant proposed a capitulation the town was burning in many parts from the effect of a vigorous bombardment.'

On July 30, the 85th under Lieut.-Colonel Cuyler as a part of the light brigade of its old commander Sir Eyre Coote's division of the army, landed near the heights of Briscard on the Isle of Walcheren; and on the following day advanced to the village of Grypskirk. On August 1, when the brigade formed the advanced guard, the regiment performed its first service under fire since its formation as a light corps, when the French outposts were driven in upon Flushing. In the execution of this duty it suffered some loss, viz. one Sergeant and two Rank and File killed; Lieutenant Brock, two Sergeants, one Bugler, twenty-two Rank and File wounded; four Rank and File missing.

It remained in camp near the village of West Souberg, in the neighbourhood of Flushing, taking a share in the siege operations during their continuance. The enemy surrendered August 16, 1809. In the middle of August, the regiment crossed the Scheldt, and with the brigade (68th and 71st, 85th Light Infantry, and two companies, 95th Rifle Corps, under Major-General de Rottenburg) took up quarters at Tergoes in South Beveland. Upon the evacuation of South Beveland the regiment returned to Grypskirk in the Island of Walcheren, the headquarters of the brigade being at Tervere, eleven miles distant.

On November 4 the regiment marched to Scrooskirk, and relieved the 1st Light Battalion, King's German Legion, and a fortnight later joined the brigade at Tervere.

The capture of Flushing, though a victory, proved to be a mischievous conquest; for the fever arising from marshy exhalations soon began to spread among the troops—a disease which is said to be more malignant in the Isle of Walcheren than in any other spot except Batavia.

'The Ministers received early intelligence of this calamity; but in the vain hope of supporting the interest of the Emperor of Austria, who had not then concluded the treaty of Vienna, they ordered the island to be retained, long after the preparations of the French for the security of their towns, forts, and shipping, had confounded the schemes of the invaders, whose views were baffled by the original delay of the equipment and by the dilatory proceedings of the armament.

'Even a whole month was suffered to elapse, after the intelligence of the conclusion of peace between France and Austria had reached England, before the pestilent spot was abandoned.'

Captain Cooke of the 43rd Light Infantry, in his 'Personal Narrative,' thus describes the country and the fever:

'The whole face of the country was perfectly flat and was intersected with ditches, covered with a thick ooze or vegetable matter and high dykes rising on each side of the way. We were much struck with the cleanliness of the cottages and at the contented air of the well-dressed peasantry. The females were decorated with silver or gold ornaments about their persons and many of them wore a plate of the same metal across their foreheads. The little boys of five or six years old held pipes in their mouths, smoking with all the gravity of men, and wore their hair long behind, broad-brimmed hats, brown jackets, short breeches, shoes and silver buckles, precisely similar to the elders.

'The humble dwellings of the peasantry bore an air of comfort, and the abundantly supplied dairies, paved with well washed tiles, presented a freshness seldom exhibited among the poorer classes of other countries.

'We were surrounded with abundance—after the surrender of Flushing—our days were occupied in the sports of the field, and pay was issued almost to the day it was due. Provisions of all descriptions were offered for sale at a very low rate; tea, sugar and coffee were not half the price of the same in England; wines, brandy, hollands and liqueurs, might be purchased for a mere trifle; and fat fowls or ducks for 10*d*. the pair.

'In this land of plenty we were lulled into a fatal security; for, about the 20th, the soldiers fell ill, staggered and dropped in the ranks, seized by the dreadful fever (the sailors on board ship did not suffer much from the malady), and with such rapidity did this malady extend that in fourteen days, 1286 soldiers were in hospital on board ship or sent to England; the deaths were numerous and sometimes sudden; convalescence hardly ever secure; the disorders ultimately destroying the constitution and causing eventually the destruction of thousands in far distant climes.

'The natives now became ill, and informed us that one-third of them were confined to their beds every autumn until the frosty weather set in, which checked the exhalations from the earth, and gave new tone to their debilitated frames, and thereby stopped the progress of the complaint. Independently of the records of the unhealthiness of these islands, where every object depicts it in the most forcible manner, the bottom of every canal that has communication with the sea is thickly covered with an ooze, which, when the tide is out, emits a most offensive effluvium; and every ditch that is filled with water is loaded with animal and vegetable substances. If persons living in these islands from their infancy, who practise a cleanliness that cannot be excelled and live in good houses, cannot prevent the effects of the climate, it may readily be supposed how much more a foreign army must suffer. The inhabitants informed us that, in the preceding autumn, two hundred French troops were quartered in the village, out of whom one hundred and sixty had the fever, and seventy of them died.

'The town of Flushing, after the siege, presented a deplorable appearance, with many houses burnt down, and most of them unroofed, and scarcely supplying sufficient covering for the sick soldiers, who continued to increase so fast, that ten inhabitants to each regiment were requested to assist as attendants in the hospitals. The medical officers were extremely harassed; numbers of them became incapable of attending on their patients, being themselves seized by the same fatal malady, so that, as the fever gained ground, the doctors diminished in numbers. At one period 498 soldiers died in a fortnight in Walcheren, which place the Austrians were very solicitous our troops should continue to occupy as long as any chance remained for them against Napoleon, who was at this time in the very heart of their Empire.'

'At length the docks and the arsenal of Flushing were destroyed; and after a severe loss, the enfeebled survivors returned.'

On December 18, 1809, a small remnant, consisting of one Field Officer, two Captains, six Subalterns, two Staff, eleven Sergeants, one Bugler, ninety-nine Rank and File—all the Frisian swamps had spared of the fine battalion which had quitted England six months before—marched to Flushing and embarked on board the transports *Nile* and *Friendship*.

REFERENCES TO PLAN

(Flushing, March 1st, 1810.)

A. Batteries, *en crémaillère* for a direct fire on the point attacked.
B. The Seamen's Battery, for enfilading the Left Face of the Ravelin No. 16; and of the Bastion No. 2, also for forming on the part attacked.
C. Battery to enfilade the Left Face of the Demibastion No. 3 and the Front attacked.
D, E. Batteries for enfilading the Left Face of Bastion No. 4, and the Front of Attack in General.
F. Mortar Battery begun 3rd of August.
G. Batteries for the false attack, to draw attention of the Besieged to their Right.
H. Battery begun the 14th August for Six 68 Pr. Carronades to breach the Face of Bastion No 1.
I, K. Rocket Stations.
L. Battery begun the 14th August for two 10-inch Howitzers to co-operate with Battery H.
M, N. Traverses for covering the advanced Sentries, when the Besieging Army first took up its ground.
O. Communications.
P. Battery begun for the false attack; the 14th August.
Q. French Advanced Work, carried by assault on the Night of the 14th August, afterwards abandoned.
R. Bridge of Communications across the Canal, made by the Quartermaster General's Department.
S, T. Batteries begun the 2nd of August against Fort Rammekens.
V. French Advanced Works and Traverses.
a. Advanced Post of the British Army, on the 13th August.
b. Entrenchment of the French Picket, stormed and carried by the English, afterwards enlarged and filled by them with *Chevaux de Frize*.

NOTE.—At A and G are two Forts, begun by order of the French Emperor. That at G somewhat in an advanced state.

Dover was reached on the 28th of the same month, and the 85th landed and marched to its old quarters at Brabourne Lees barracks the same day.

With regard to the military and naval chiefs of this unfortunate expedition, the failure of John, second Earl of Chatham, at Walcheren, as a military commander, was absolute and complete. On his return to England a storm of recrimination arose. Chatham unwisely, as it was absolutely contrary to all military etiquette, presented to the King, at a private audience, a partisan report. This communication should of course have been forwarded through Castlereagh, if forwarded at all—certainly it should have been presented to that nobleman, who was at the time Secretary of State.

An inquiry was held which greatly damaged the military reputation of Chatham. That officer laid the blame on the naval

Facsimile of a Map in the Royal United Service Institution. [*By kind permission.*]

FLUSHING.

Position of the 85th marked in centre of the three Regiments near the top of the map.

commander, Admiral Sir Richard John Strachan, Bart., K.B., whom he accused of delay and inertia in the earlier movements. Hence an epigram, for the friends and supporters of Strachan retorted with a *tu quoque*. The epigram is as follows:

'Great Chatham, with his sabre drawn,
Stood waiting for Sir Richard Strachan;
Sir Richard, longing to be at 'em,
Stood waiting for the Earl of Chatham!'

To Chatham by the friends of Strachan was also applied the nick-name of 'the late' Earl of Chatham, owing to his alleged want of punctuality.

Chatham was nevertheless promoted to General in the Army, January 1, 1812, and in 1820 was made Governor of Gibraltar. He died September 24, 1835. Strachan, who had seen much service afloat and had had a distinguished professional career, the details of which do not concern this History, was never again employed. His argument in reply to Chatham's accusations was that his ships had done all that they had been asked to do, all that from the nature of things they could do. He was, however, promoted Vice-Admiral July 31, 1810, and Admiral July 19, 1821. Sir Richard Strachan died February 3, 1828, when his baronetcy (to which he had succeeded in 1777) became extinct. For his services at Trafalgar he had received a pension of £1,000 per annum and a 'K.B.'

In the month of January, 1810, a very sad tragedy took place at Brabourne Lees. We give the only account of it obtainable. This is derived from the *Kentish Gazette* of Tuesday, January 9, 1810.

'Fatal Duel:—

'On Friday morning last (Janry. 5), a meeting took place in the vicinity of Brabourne Lees, between Capt. H——g——s [Hoggins] and Lieut. H——n [Hylton], both of the 85th Regt.; when the former was shot through the body, and expired in a few hours after being conveyed to the barracks. His antagonist immediately left the ground and has since absconded. The cause of this unfortunate event arose out of some reflections used between the parties with respect to each other's conduct while in Walcheren.'

Captain Thomas Hoggins was the younger of two brothers, William and Thomas, whose sister was Sarah, Countess of Exeter, the story of whose romantic marriage to the Earl (while Mr. Henry Cecil) was celebrated by Tennyson in his well-known poem.

Presumably Mr. Cecil had his two brothers-in-law educated for the army. That he did so in the case of William, who entered the Cameronians (26th Foot), is proved by a letter in the War Office which states that 'he has prepared himself some time for the Profession.'

William Hoggins eventually became a Captain in the Regiment, and was drowned with all hands in the *Aurora* transport on the Goodwin Sands in 1805 while on the voyage to the Texel. The present Marquess of Exeter possesses a miniature of this unfortunate officer.

The younger brother, Thomas, fell, as has been mentioned above, in a duel.

The photograph here reproduced is that of a portion of his hair-trunk with an engraved name-plate. This came into the possession of Colonel Capper, formerly of the regiment. It was purchased among the effects of an aged woman who died about 1880 at Brabourne Lees, and was presented to Colonel Capper, then serving in the 85th, as a regimental relic, by a Miss Perry Ayscough, the daughter of the then Rector of Brabourne. The local tradition was that there was something unfair in the conduct of the duel, and a painted board with an inscription which formerly marked the grave of Captain Hoggins is stated to have borne the words 'cruelly done to death.' This board has, however, apparently vanished. The grave was near the north porch of the church. A very aged woman, now many years dead, was wont to relate that the duel took place near some 'pollarded trees,' which she could then point out, and that she 'saw the gentleman' killed and his adversary escape by mounting a coach which at once drove off.

Careful search has been made in the Kentish and other papers to ascertain, if possible, what (if any) proceedings were instituted against the then principal in the duel, the seconds, etc.; but in vain, for none seem to be recorded, at any rate in print.

This we know, that Lieutenant John Hylton was promoted Captain

THE NAME PLATE OF CAPT. HOGGINS' HAIR TRUNK
Now in possession of the Regiment.

in the 85th on June 27, 1811, and this seems to contradict the idea that there was any foul play. He was cashiered in 1813, being one of the officers who were court-martialled at that time and under the circumstances mentioned hereafter. It does not, however, appear that the duel with Captain Hoggins was referred to on that occasion, and the serious charges then made against Captain Hylton were in no way connected therewith.

Through the kindness of the Rev. T. Slack, the Vicar of Brabourne, the following documents have been obtained. After an exhaustive search no other details are obtainable, and the question of the fairness or unfairness of the duel must remain unsolved.

BRABOURNE REGISTER—1800

'1810

'Thomas Hoggins Esquire of the 85 Reg^t^: 11 Jan^y^: *

'* Brother of Sarah, wife of Henry—1st Marquess of Exeter, shot in a duel with John Hilton Gent. against whom a verdict of wilful murder was returned on the Coroner's Inquest.

'J.B.'[1]

'The above is a true copy of the Entry under "*Burials*" in the "*Brabourne Register* 1800," extracted this tenth day of December in the Year of our Lord One Thousand Nine Hundred and Twelve

'By me, THOS. LINDSAY SLACK, *Vicar.*'

BRABOURNE REGISTER—1800

'1810

'KENT, TO WIT,

'WHEREAS I with my Inquest the Day and Year hereunder written have taken a view of the Body of Thomas Hoggins Esquire who was wilfully murdered by John Hilton Gentleman and now lies dead in your Parish and have proceeded therein according to Law THESE are therefore to certify that you may lawfully permit the Body of the said Thomas Hoggins to be buried and for your so doing this is your

[1] Believed to be the Rev. J. Bradshaw.

Warrant: GIVEN under my Hand and Seal the Tenth Day of January 1810.

'ROBT. HINDE, *Coroner.*'

'To the Minister and Churchwardens
of the Parish of Braborne
in the said County of Kent.

[*Entered thus in pencil*

'Capt Hoggins was buried 11 Jany
Thomas Hoggins Esquire of the 85th Regiment 11th Jany.]

'The above is a true Copy of the Coroner's Order for the Burial of Thomas Hoggins Esquire, and is now affixed to the Brabourne Register dated 1800 opposite the Entry.

'Given this Tenth Day of December, 1912,

'By me, THOS. LINDSAY SLACK, *Vicar.*'

OLD BRIDGE OVER THE NIVELLE

CHAPTER V

THE PENINSULA CAMPAIGN, 1811, FUENTES OÑORO

WHILE at Brabourne Lees the establishment of the regiment was raised.

It now consisted of ten companies of the following strength:

1 Colonel, 1 Lieut.-Colonel, 2 Majors, 10 Captains, 22 Lieutenants, 8 Ensigns, 6 Staff, 54 Sergeants, 50 Corporals, 22 Buglers, and 760 Privates. Total: 936 officers, non-commissioned officers and men. On December 3 and 4, 1810, the regiment marched in two divisions from Brabourne Lees to Hailsham in Sussex. Here it remained in quarters until January 23, 1811, when, having received orders to proceed on foreign service, the 85th marched to Portsmouth and embarked on board His Majesty's ships *Ganges* and *Orion* for Portugal on January 27, 1811.

The strength of the portion which thus sailed from England amounted to 27 officers and 459 non-commissioned officers and men.

This practically represented five companies, of which Lieut.-Colonel Cuyler was in command. The remaining five companies formed the Regimental Depôt at Hailsham, and were composed of 'Officers and men absent from the Corps, those on the Recruiting service, sick, and those left in charge of the Heavy Baggage, consisting on the whole of the numbers herein specified, and under the command of Major Mein.' That is to say, 5 companies, 3 Field officers, 15 Captains, 16 Subalterns, 17 Sergeants, 8 Buglers, 135 Rank and File.

On April 10, 1811, as from December 25, 1810, it is also stated that the regiment was then reduced to the following strength, ten companies, 1 Colonel, 1 Lieut.-Colonel, 2 Majors, 10 Captains, 22 Lieutenants, 8 Ensigns, 6 Staff, 44 Sergeants, 22 Buglers and 610 Rank and File. Why this reduction took place does not appear. In the consideration later of certain events which took place, it is rather important that we should know actually how many commissioned officers were serving with the regiment in the Peninsula during 1811 and 1812. On arrival at Lisbon on March 5 the 85th remained there for five days, and then marched to join the army of Lord Wellington at the front. The regiment was ordered to join the 7th Division then lying at Ponte de Murcella, a place near Coimbra. Here it arrived on March 19, and was formed in brigade with the 51st Foot and the regiment now disbanded but then known as the 'Chasseurs Brittanniques.' The brigade was under the command of Lieut.-Colonel Cuyler of the 85th. This, too, is a point which must be remembered later. The 51st Regiment had quite recently arrived in Portugal, 650 strong, under the command of Lieut.-Colonel Mainwaring. The Regimental Record of the 51st makes no mention at this time either of the 85th or of Ponte de Murcella; but states that it left Lisbon 'early in March to join Wellington's army then in pursuit of Massena.' It also mentions that the 51st passed through Leiria 'while the flames that consumed it still raged and presented a frightful picture of ruin and desolation.' The line of march was marked by the putrefying corpses of dead French soldiers, stretched beside the wreck accumulated by their wanton and shameful outrages. The 51st joined Wellington at the village of Carrapinha. Leiria is about seventy-five miles from Lisbon and thirty-five from Ponte de Murcella. Now on March 21 we find the 85th encamped with the brigade at Carrapinha, a small village, the camp being on the heights beyond it. Carrapinha was known as 'Starvation Camp,' owing to the sufferings there of the troops from want of food. For nearly a week they were without either bread or spirits, having but the 'lean and tough ration of beef killed and served out *instanter* to the troops, popped half alive into the pot; and happy was the individual who could add thereto an onion, or

INFANTRY: CHASSEURS BRITTANNIQUES

the slightest vegetable to put into the water in which this carrion was boiled and miscalled soup.'

But such hardships were common enough during the Peninsula Campaigns. Still the British troops endured them, nay, endured worse, and yet fought again and again ; hungry, thirsty, weary, war-worn and sick.

A very brief account of the 'Chasseurs Brittanniques,' as the regiment no longer exists, may not be out of place here.

This was a regiment of foreigners and was mainly officered by foreigners, though at the time the Lieut.-Colonel was British and by name William Eustace (afterwards Sir William). The Chasseurs Brittanniques were originally a part of the Prince of Condé's army of emigrants. The regiment was raised in 1801 and first saw service in Sicily. It was sent thence to the Peninsula. Five companies arrived there in November, 1810, after being detained for some time at Cadiz.

Prisoners of war were enlisted in its ranks and also many deserters from the French army. But in the ranks of the Chasseurs Brittanniques, as also in the Corps of the Duke of Brunswick-Oels, desertion was very rife. It is said that as many as twenty-six men deserted in one night at Cadiz. In consequence an order came out that deserters were not to be employed 'near the enemy.'[1] In the Peninsula, the regiment fought at Fuentes Oñoro, San Cristoval (Badajos), Salamanca, the capture of Madrid, at Burgos and in the Pyrenees.[2] At Fuentes Oñoro it was on the extreme right and was distinguished for great steadiness during the fighting. After the war the regiment served on board the *Ramillies* in America. It was disbanded in 1815, its then strength being eleven companies. The Chasseurs Brittanniques wore uniforms similar to those of the British Infantry, as did all foreign regiments except the King's German Legion, the Duke of Brunswick-Oels' Corps and the Duke of York's Greek Light Infantry. This last regiment, it may be stated, looked more like comedy opera brigands than British soldiers.

[1] The 'deserters' here alluded to were French deserters who had been enlisted into our army.

[2] In the second storming of St. Cristoval a young officer of the Chasseurs Brittanniques, a lieutenant, succeeded in mounting almost to the top of the breach when he was bayoneted. He was a Frenchman, by name Dufief. He fell wounded, calling to his men, 'Je monte, suivez moi.'

To resume the story of the campaign. The 7th Division was under the command of Major-General Houstoun.

From the bivouac near Carrapinha the division moved into cantonments in the village of Villa Mayor, where it arrived on April 9. Meanwhile Massena had fallen back from the lines of Torres Vedras, and having been beaten at Pombal, Redinha, Casal Nova and Foz d'Arouce had concentrated his force in the Sierra de Moita, Lord Wellington being in active pursuit. At Moita the bulk of the British and Allied Armies assembled on March 19, awaiting provisions from Lisbon, the pursuit of Massena being carried on by the 3rd and 6th Divisions. On the 28th the troops at Moita moved on to Celerico. The 7th Division under Houstoun here joined for the first time.

DUKE OF YORK'S LIGHT INFANTRY

The next move was on May 2, when the regiment with the brigade and division advanced from Villa Mayor and took up a position on the great Caroll Road leading to Almeida. Up to this time there had been no fighting, a projected attack on Guarda (a nearly impregnable mountain town) falling through. Here a great struggle had been expected in which the 1st and 7th Divisions were to have formed the centre column in the attack. The enemy, however, retired without firing a shot. At Sabugal, on the Coa, Wellington attacked the French also on April 3; but the 7th Division was on this occasion in reserve. It was a fierce fight and terminated in the utter defeat of the French, who two days later crossed the frontier and cleared out of Portugal.

Then it was that the 7th Division went into the cantonments at Villa Velha, and there they remained till Massena attempted to relieve Almeida in the early days of May.

On May 3 the battle of Fuentes Oñoro began. The 7th Division was stationed as part of the left (towards the centre) of the line.

The French attacked the right of the line. Next day the 85th was ordered out on advance picquet duty to the village of Villa Poza near Fuentes Oñoro. The regiment was under the command of Major M'Intosh of the 85th. The division was here occupying the hill of Nava d'Aver and supported Julian Sanchez, the partisan chief.

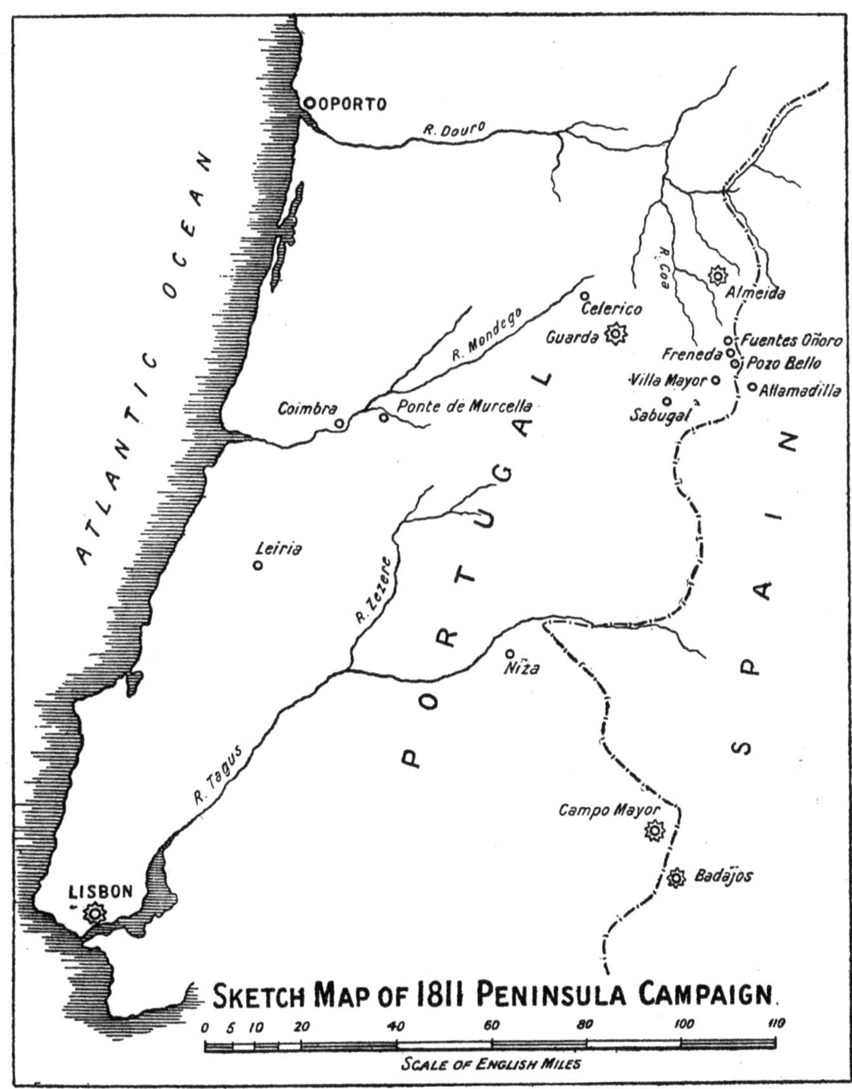

SKETCH MAP OF 1811 PENINSULA CAMPAIGN.

The main body of the enemy was encamped quite near, and it was a very vulnerable part of the British position as it was particularly open to attack from Rodrigo. The attack duly took place, and at first the left wing of the division was driven back out of Poço Velha, otherwise Villa Poza, with some loss. Reinforcements were, however, sent up and the fight was successfully renewed. Meanwhile General Montbrun had disposed of the partisans under Sanchez and had then turned the right of the 7th Division. Against him were sent some British cavalry, but the disparity in numbers between the French and English was too great and the British cavalry could not prevail. The French cavalry then made a dash at the 7th Division. Though all or almost all were young and unseasoned soldiers, both the 85th and the 51st stood firm and by their steady volley-firing repulsed the attack of the enemy. The 85th at this moment were forming a 'chain'[1] in a wood close to the village. Having repulsed the French, the division retired from the wood from the right by companies and reformed in line on some rising ground, where they were immediately subjected to a severe artillery fire. In this affair Lieutenant Samuel Holmes and twelve rank and file were killed, Lieutenants Brock and Hogg, three sergeants and twenty-one rank and file wounded; two sergeants, one bugler and forty-six rank and file taken prisoners, and Major M'Intosh recommended for reward.

The following account of the battle was written home by a young officer of the 85th Regiment.

BRITISH CAMP, VILLA FORMOSÀ,
May 6.

'We left Villa Mayor about five days since, and took up a position in the neighbourhood, in consequence of Massena shewing himself in large force, and seemingly with an intention to attack us. On the 3rd he commenced by attempting to drive in our picquets from a village in front of our line, where we had a number of men killed and wounded, but he did not effect his purpose. Our Regiment were merely spectators of the business, and there was certainly some very pretty skirmishing, particularly of cavalry. It was our turn to take the advanced posts yesterday; and about dusk, the evening before, we went to our station, a village in a wood, about a mile and a half from the main army; a plain of half a mile across separated us from the enemy. We were well aware of what we had to expect in the morning, as we observed the French making movements which

[1] Presumably extended order, at least the letter quoted below leads to that conclusion.

indicated their intention of driving us from our station. At daylight in the morning, the enemy's cavalry (consisting of about 5000) moved to our right, out of a wood opposite to the one we occupied, and formed in columns. Some squadrons of them charged a few cavalry videttes of ours, and drove them in. Captain Nixon's Company, in which I am, covered their retreat into the wood; they (the enemy) attempted to charge us, but as soon as they were within 20 or 30 yards, we gave them such a reception as made them retreat as fast as they advanced. As soon as we fired, they wheeled about and went off, horses and riders tumbling in every direction—they were Bonaparte's Imperial Guards, very finely dressed; we had about 40 men opposed to them; they consisted of at least 400; we had the advantage of the trees. At the same moment the cavalry retired, three columns of infantry advanced from the wood opposite us, in double quick time, shouting and drums rolling—one column took the right of our regiment, another the centre, and the third the left. I must inform you that we were in extended order from right to left of our wood. We waited coolly till they came within 50 paces, then fired and retreated for several hundred yards, took up a new position behind an old wall, and kept them in check for about an hour afterwards, till we were relieved by the Rifle Corps; we unfortunately lost a number of men—Lieut. Holmes was killed; Lieutenants Hogg and Brock and Captain Nixon (my Captain) wounded. Most of the wounded were unavoidably taken, which are accounted for as missing in the return; we had but 220 in the field, and of them 101 are killed, wounded, and missing, including Officers. Almost all had balls through their cloaths somewhere or other—for a short time there was the hottest fire ever experienced by so small a body of men: we had the fire of about 2000 men upon us—they must have expected to have found a much larger force than we had—we had the assistance of about 300 caçadores after we got out of the wood, which was of some service to us. Lord Wellington, and several other Generals, observed us from a hill at a distance and gave us the highest praise for our conduct. General Houston told us that no men could have behaved so well unless they were induced to it by the energy and gallantry of their Officers. We afterwards fell back upon the army; the French cavalry and artillery advanced; the cavalry charged part of our division, but were driven off with loss. Our regiment was then ordered into some enclosures on the right wing of the army, where we remained all day spectators of the operations of the two armies. There was no cessation of firing from the time it began with us, at five in the morning, till dusk in the evening ; there was no general attack made, but a great deal of manœuvring ; the cavalry and light infantry were principally engaged. I have not been able to ascertain the number killed on either side ; but am certain the French have lost twice the number we did. The despatches will give you all particulars which I cannot. Our division is now placed upon a hill, on the right wing of the army. To-morrow we expect a general attack, as Massena seems determined to relieve Almeida ; he will have tough work. We have not been under cover since we left Villa Mayor. I have no convenience for writing ; so shall make no further excuse for my paper, pens, ink, etc. I am in good health, high spirits, and anticipating a glorious victory ; shall write in a day or two ; so adieu for the present.'

[NOTE.—It is to be regretted that the name of the writer of this most characteristic letter should be unknown.]

On May 8, the brigade with the division occupied the village of Frenada.

This occurred in consequence of the British right being thrown back. The movement was accomplished by crossing the river Turones and moving down the left bank till the village was reached. Lord Wellington says that this movement was 'well conducted, although under very critical circumstances.'

At Frenada, the brigade held the village and defended the ford below the bridge. In the village itself much use was made of numerous stone walls thereabouts which were lined with troops.

Four days later, Lord Wellington, having determined on the siege of Badajos, despatched the brigade and division thither, and Badajos was reached on May 23. On this duty the 3rd Division was also sent and the 2nd Hussars (King's German Legion) as well. The investment of Fort Christoval began on the evening of May 24. Fire was opened on the morning of June 2 from the four batteries which had been erected on the right bank of the Guadiana.

St. Christoval stood on the north bank of the Guadiana, opposite to the main fortress and town of Badajos.

It was heavily armed, well fortified, and occupied by a resolute enemy. General Phillipon, who was the Commandant at Badajos, was a brave and skilful officer, perhaps as brave and skilful as any in the armies of Napoleon. From its position, it was absolutely imperative that Fort St. Christoval should fall before the capture of Badajos could possibly be brought about.

Hence it was that the attention of the Allies was first devoted to the

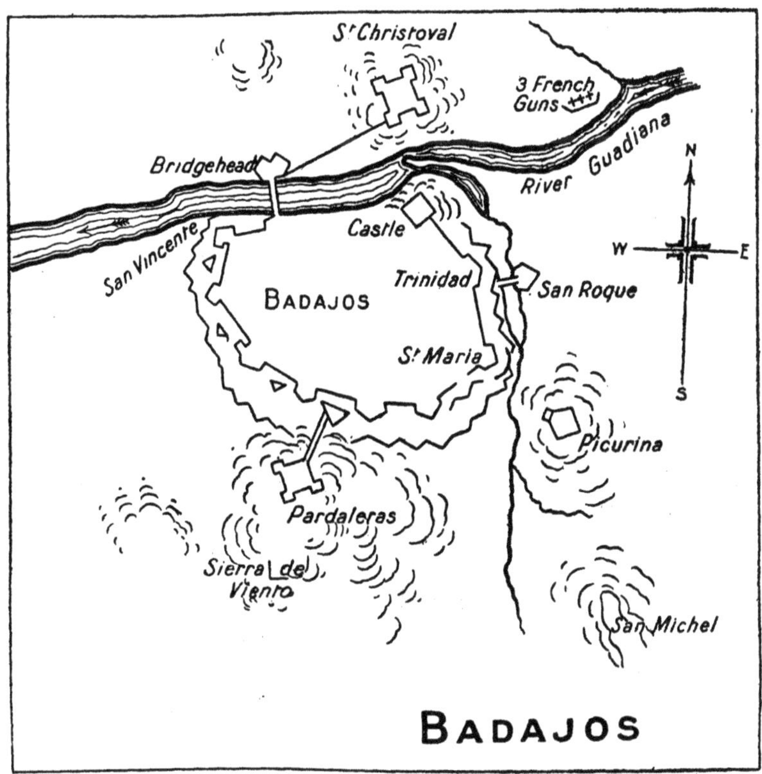

reduction of this powerful adjunct to the defences of the besieged stronghold of Badajos.

For the first assault the regiments from which volunteers were picked were the 51st (1st Battalion), the 85th (2nd Battalion), and the 17th Portuguese. A Lieutenant of the Engineers, by name Forster, who on the previous night had boldly explored the ditch, guided the party. The forlorn hope of twenty-five men was led by Lieutenant Dyas of the 51st. The main storming party consisted of 155 grenadiers, led by Major M'Intosh of the 85th, was divided into two companies.

Ten ladders, fifteen feet long, were carried by the leading company.

Detachments from the trenches were sent out to prevent reinforcements from reaching the fort from the bridge-head. At midnight the storming party left the trenches, crossing the 400 yards between them and the breach at a double and uphill. It was pitch dark and the little band suffered scarcely any loss, despite the heavy fire opened upon them. The counterscarp was crossed and the ditch reached, but there the party was brought to a standstill; for not only was there a seven-foot perpendicular bank or wall to surmount, but the gap in the breach had been stopped up with carts, tubs, *chevaux de frise* and sandbags. The forlorn hope was therefore withdrawn. Now, however, the main body of the stormers came up and entered the ditch at the spot but four feet deep. Attempts were made with the ladders in several directions, but all in vain. Meanwhile the defenders were keeping up a continuous musketry fire and also rolling down live shells from above. The assault was a failure, though attempts to scale were made in several places; and after losing 12 killed and 80 wounded, out of the 180 men engaged during the hour the combat lasted, the remains of the attacking force withdrew.

The list of killed, wounded and missing is as follows: 51st Regiment—3 killed, 1 officer and 35 men wounded, 3 missing; 85th Regiment—2 officers and 6 men wounded; 17th Portuguese—9 killed, 2 officers and 26 men wounded; Chasseurs Brittanniques and Brunswick-Oels—7 wounded; Engineers—1 officer mortally wounded.

The enemy, it is reported, lost but 1 killed, and 5 wounded.

For three days and nights after this gallant but unsuccessful assault the bombardment was continued, and nightly the French repaired as best they might the damage done; and repaired it, too, despite the constant fire which throughout that period was directed against the breach, or rather the breaches—for a second one had by this time been made. With daylight each morning the shot-torn parapet was discerned to have been renewed as far as possible by sandbags, gabions, *chevaux de frise,* and packs of wool. By nightfall all these obstructions had been swept away again by the besiegers' fire. General Phillipon had, however, from the bridge-head sent out a reinforcement of another company to the beleaguered fort, thus doubling its garrison. Consequently there were a larger number of defenders; and what is more, each man was furnished with three muskets in addition to the grenades, fire-balls, and live shells of which a large store was ready at hand for use in case of a second assault.

On the 9th a second assault was decided on, and at 9 P.M. the storming party left the trenches.

For the second assault of St. Christoval 400 men were detailed, and they were supported by 100 picked shots, who lined the outer edge of the ditch and kept up a fire on the enemy in the breach. The storming party was guided by Lieutenant Hunt, an engineer officer; the whole being under the command of Major McGeechy of the 17th Portuguese. Again Lieutenant Dyas, of the 51st, led the forlorn hope and again he returned unscathed.

The storming party was composed of volunteers from Sontag's Brigade; that is to say, from the 51st, 85th, Chasseurs Brittanniques and Brunswick-Oels Corps, with a party also from the 17th Portuguese.

The stormers were divided into two parties, one to attack one breach (the smaller), and the other the larger, in the salient angle and the curtain respectively. With the first party were six and with the other ten ladders.

Immediately on leaving the trenches both Hunt and McGeechy were killed, and many others also fell beneath the terrific fire brought to bear on them by the garrison. The forlorn hope reached the ditch and entered it, making at once for the breaches. It was soon followed by the remainder of the party.

INFANTRY: BRUNSWICK-OELS CORPS

In vain were the ladders brought forward and reared. Nearly every man who attempted to mount them was shot down, and the few that succeeded in getting a footing on the breach were bayoneted. Those in the ditch itself were subjected to missiles of all kinds both explosive and otherwise. For nearly an hour the contest continued till 5 officers and 49 men had been killed outright, and 8 officers and 77 men wounded. Again, despite the gallantry displayed, the attack had failed, and again the stormers had to withdraw.

The losses were thus apportioned : 51st Regiment—1 officer and 23 men killed, 2 officers and 31 men wounded ; 85th Regiment—1 officer and 7 men killed, 1 officer and 10 men wounded, 1 officer missing ; Chasseurs Brittanniques—8 men killed, 1 officer and 13 men wounded, 2 men missing ; Brunswick-Oels—1 man killed, 1 officer and 5 men missing ; 17th Portuguese, 2 officers and 10 men killed, 1 officer and 16 men wounded and 1 officer missing. To which list must be added Lieutenant Hunt, Royal Engineers.

On the morrow, the siege of the Castle of Badajos, which had all this time been proceeding, was continued with vigour, but with Christoval a six hours' truce was asked for and granted that the many wounded who lay between the trenches and the fort and near the breach itself might be brought in. They were numerous as the list shows, and their very numbers had prevented the British guns from covering the retirement of the storming party on the preceding night. The French had no wounded to collect, and wisely employed the time in repairing the breach unmolested. The plan of the fortress here given shows how absolutely necessary it was to take Christoval before any serious attack on the Castle of Badajos could be expected to succeed.

The siege of Badajos was now converted into blockade. It was found after the failure of a second attempt to storm Christoval that the place appeared simply impregnable. The fact was that the British artillery train was not of sufficient weight for the operations in hand. The entrenching tools were very faulty and the Corps of Engineers and Sappers very much lacking in strength.

On June 16 the regiment, with the division, withdrew to Campo Mayor where it remained until July 19. It then proceeded to Niza, arriving there on July 23.

Leaving Niza on August 2, Villa Mayor was reached on August 10. Here the regiment was detached from the brigade and posted at Arafau, a small village about two miles distant.

At Arafau they remained stationary until September 17, when the regiment again rejoined the brigade, which had meanwhile proceeded to Escrobalhadas.

On September 23, having rejoined the division, the Spanish frontier was crossed, and for three days a halt was made at the village of Allamadilla.

A retreat to Sabugal now began, and from thence was continued to Penamacor. This place was reached on September 30.

Orders were now received for the regiment to return to England to recruit, and it marched from Penamacor for Lisbon on October 5.

On arrival at Lisbon, the regiment (now reduced to 20 officers and 246 men) embarked on H.M. ship *Agincourt*, and the *Hartford* transport for Portsmouth. At Portsmouth the vessels arrived on December 2.

It should be noted that it was the practice of Lord Wellington, when a regiment became so reduced in strength as to lose its fighting efficiency as a regiment, to order it home to recruit its strength. This was almost invariably done. Lord Wellington adopted this method in preference to getting drafts of militia out from England. His view was that it was better to send the numerically weakened regiment home to recruit from the militia at home and not to fill up its ranks abroad on active service by a fortuitously collected draft. By this means, by the time a regiment returned to the Peninsula the new element in its ranks had settled down, regimental spirit had been acquired, and as an arrangement it was certainly wise. Immediately on its disembarkation the regiment marched for its old quarters at Brabourne Lees, arriving there on December 13, 1811. On the march the regiment halted at Lewes in Sussex, as will be seen from the following extract.

The mention of tea and coffee is interesting, especially when we consider that tea in particular was most costly in those days and far beyond the reach of the soldier's wife.

'Lewes, Dec. 16th, 1811. The reception which the 85th Regiment of Light Infantry, under the command of Lieut-Colonel M'Intosh met with, and the kindness they have experienced in quarters, since their disembarkation from the Peninsula, must be truly gratifying to them, but more particularly the humanity displayed in this town, towards the poor widows and wives of the soldiers, arising from a very liberal subscription set on foot for that purpose. Five companies of the regiment, it will be remembered, embarked last January, and although they returned not more than 90 effective men, yet it is to be hoped, their conduct has so far confirmed their long established name, that with the assistance of the next volunteers and the recruits at the depot, they will again be able to pay their respects to the French Voltigeurs. The men on being drawn up to proceed on their route, on Tuesday morning, were plentifully regaled with strong beer, etc., for which they expressed their thanks and gratitude, by alternate huzzas and tunes from their bugles, as they marched through the town. The women, before they mounted the baggage waggons, had tea and coffee served to them for their breakfasts.'

Here it remained until May 5, 1812, when it was moved to Hythe.

The 85th was stationed at Hythe until July 21, 1813.

On May 27 in that year the establishment of the regiment was again augmented to the following strength: 1 Colonel, 1 Lieut.-Colonel, 2 Majors, 10 Captains, 22 Lieutenants, 8 Ensigns, 6 Staff, 55 Sergeants, 22 Buglers, and 810 Rank and File.

While stationed at Brabourne Lees, near Ashford, Kent, considerable dissension amongst the officers came to light and led to a series of courts-martial.

Accusations, for the most part false, were made, challenges to duels—horse-whipping and a pugilistic encounter between an officer and a sergeant—even the accusation was made against an unfortunate subaltern of embezzling the money stopped from the pay of the men for the purchase of coffins when they were dead. Although, as stated, most of these accusations turned out to be false, yet the spirit amongst the officers was such that the Commander-in-Chief decided that it would be best for all parties if all the officers were moved to other regiments and their vacancies filled up by others selected from different regiments in the service.

This drastic remedy was accordingly carried into effect.

To the old officers of the 85th who were thus moved to other and new regiments a full explanation was given (by the order of the Prince Regent) that it was adopted as 'a general measure of expediency, and not intended as an imputation against any individual.'

Brevet Lieut.-Colonel M'Intosh, formerly the senior major, and Major Mein, the junior major, were to be specially informed that 'the measure does not lessen the feeling of approbation under which their services are appreciated.' The Colonel (General Thomas S. Stanwix) and the Lieutenant-Colonel (Lieut.-General the Hon. H. A. Bennett) were not displaced.

One serving officer alone, who had been gazetted to the regiment but had never joined it up to date, was allowed to remain in it. This was Captain Brown (afterwards Sir George Brown), who in one swoop went from junior to senior captain.

[By kind permission of 'The Regiment.'

THE ELEGANT EXTRACTS

All the new officers were selected from other corps, and from this affair arose the nickname of the regiment—a nickname which has endured till the present day.

For the new officers were at once known as the 'Elegant Extracts.'

On February 3, 1813, the undermentioned officers were appointed to the regiment in succession to those removed to other corps.

Lieut.-Colonel Thornton, Duke of York's Greek Light Infantry	*Vice* Cuyler
Major Deshon, 43rd Regiment	,, Mein

Major Ferguson, 79th Foot	*Vice*	M'Intosh
Captain Gubbins, 24th Foot	,,	Soden
,, Ball, 37th Regiment	,,	Nixon
,, MacDougall, 53rd Regiment	,,	Watson
,, Schaw, 60th Regiment	,,	Stannard
,, De Bathe, 94th Regiment	,,	Campbell
,, Knox, 40th Regiment	,,	White
,, Hamilton, 52nd Regiment	,,	Meredith
,, Grey, 52nd Regiment	,,	Hylton
,, Cottingham, 28th Regiment	,,	Glen
Lieutenant Fairfax, Rifles	,,	Gammell
,, Wilkinson, 43rd Regiment	,,	Imlach
,, Williams, African Corps	,,	Stevenson
,, Wellings, 57th Regiment	,,	Powell
,, Hamilton, 52nd Regiment	,,	Mitchell
,, Fisher, 60th Regiment	,,	Perham
,, Burrell, 6th Foot	,,	Brock
,, Bennett, 6th Foot	,,	Gell
,, Charleton (first appointment)		Orr
,, Watts, 21st Fusiliers	,,	Spooner
,, Forster, 12th Foot	,,	Cash
Ensign Urquhart, 6th Foot	,,	Swiney
,, Belstead, 6th Foot	,,	Kelson
,, Green, 90th Regiment	,,	Rothwell
,, Gleig, 3rd Garrison Battalion		Dutton
,, Hickson, 46th Regiment	,,	Maxwell
,, Blake (first appointment)	,,	Busteed

It has already been mentioned that the regiment had the title of the 'Bucks Volunteers.' They were, however, more familiarly known as 'The Young Bucks.'

This was to distinguish the regiment from the 14th Foot, or 'Old Bucks.'

Yours ever G. R. Gleig

AUTHOR OF "THE SUBALTERN".

CHAPTER VI

THE SECOND PENINSULA CAMPAIGN, 1813–14

WE left the regiment at Hythe, where it arrived in May, 1812, from Brabourne Lees. It will be remembered that the strength of the regiment was much reduced on its return home from its first Peninsula Campaign.

Recruiting was, however, immediately and energetically carried on.

The old establishment of the regiment was quickly reached, so much so that an augmentation was ordered and completed by May 27, 1813. The establishment now consisted of 1 Colonel, 1 Lieut.-Colonel, 2 Majors, 10 Captains, 22 Lieutenants, 8 Ensigns, 6 Staff, 55 Sergeants, 22 Buglers, and 810 Rank and File.

As considerable extracts are about to be given from the works of the late Rev. George Robert Gleig, Chaplain-General of the Forces and for some time an officer in the regiment, a notice of his career from the *Army and Navy Gazette*, July 14, 1888, is here inserted.

'The Rev. G. R. Gleig, who for nearly thirty years was Chaplain-General of the Forces, died on Monday at Stratfield Turgis, near Winchfield.

'Born in 1796, the son of a Scotch bishop, George Robert Gleig was educated for a few years at Glasgow, and at the early age of fifteen he was sent to Balliol College, Oxford. In 1812 he entered the army, and joined the forces of the Duke of Wellington in Spain in 1813. Many years afterwards he published his experiences of the campaign under the title "The Subaltern." Mr. Gleig served in the campaigns in the Peninsula of 1813 and 1814 in the 85th Light Infantry, including the siege of San Sebastian, the passage of the Bidassoa, the battle of the

Nivelle (where he was wounded twice), the battle of the Nive (where he was again wounded), and the investment of Bayonne. For his services in the war he received the medal with three clasps. He afterwards served in the American war at Bladensburg, Baltimore, New Orleans, the capture of Washington, and Fort Bowyer, and was thrice wounded in the course of his service. He afterwards returned to keep his terms at Oxford, where he took his degree in 1819. In the following year he was ordained, and shortly afterwards was presented to the perpetual curacy of Ash, and then to the rectory of Ivy Church, in Kent. In 1844 Mr. Gleig was made Chaplain of Chelsea, an appointment which he held for only two years, when he was promoted to the position of Chaplain-General of the Forces, which he held till 1875.

'Soon after his becoming Chaplain-General he was appointed Inspector-General of Military Schools, this office being given him in consequence of a scheme devised by him for the education of soldiers. He was also nominated in 1848 to the Prebendal Stall of Willesden, in the Cathedral of St. Paul's. Among his literary works may be mentioned the following: "Campaigns at Washington and New Orleans," "The Lives of Military Commanders," "The History of India," "The Story of the Battle of Waterloo," the "Lives" of Lord Clive, Warren Hastings, and the Duke of Wellington, "Memoirs of Sir Thomas Munro," "Traditions of Chelsea Hospital," "A Military History of Great Britain," and "The Soldier's Help to Divine Truth."'

A hitherto unpublished experience of Mr. Gleig is worth insertion at this point. While in residence as an undergraduate at Oxford after he had left the service his old regiment chanced to march through the city. Gleig doubtless knew it was about to do so and naturally was one of those eager to see his old comrades-in-arms. As the regiment marched through the streets he was standing among the crowd. He was recognised by the men of his old company and heartily cheered, an experience for an undergraduate of Oxford which was probably unique.

The 85th were stationed at Hythe when orders were received to proceed to the Peninsula for the second time. On July 21, 36 officers and 360 men, under command of Lieut.-Colonel Thornton, marched from Hythe at 4 A.M., and by midday were all embarked at

From an Oil Painting by G. Carpenter, circa 1835.

LOS PASSAGES.

Dover on board the two transports *Isabella* and *Bellona* with the gunboat *Dymas* as escort. The following is an example of the kit taken by an officer, who also took some £17 in cash, which he considered 'no bad reserve for a subaltern officer in a marching regiment, though it happened to be a crack one':

'1 Regimental jacket with appendages of wings lace etc., 2 pairs gray trousers, sundry waistcoats, white, coloured and flannel, a few changes of flannel drawers, 6 pairs of both worsted and cotton stockings, 6 shirts, 2 or 3 cravats, a dressing case competently filled, 1 undress pelisse, 3 pairs boots, 2 pairs shoes, pocket handkerchiefs etc., in proportion, the whole being about 1 mule load.'

The transports did not put to sea until midday on July 22, and had just sighted the point of Dungeness, when the hitherto favourable breeze veered round to a gale in the opposite direction, with the result that they remained in this locality for about a week. Another week was spent at Plymouth, when they put in for provisions and water, Spain not being sighted until August 13, and the Biscayan port of Passages not reached until the 19th. The last six days of the voyage were not without excitement; for on the 14th the escorting gunboat chased and captured an American privateer schooner; and on the 18th, while lying becalmed close in to the besieged town of San Sebastian, the guns of that fortress opened fire on the transports, from which fire only an opportune breeze enabled them to escape without damage, and so proceed on their voyage to Passages.

Apparently the transports and their convoy must have sailed past their port of debarkation, and it is not easy to see why the vessels should first have approached San Sebastian. Passages, or Los Passages, is a seaport situate on the land-locked estuary at the mouth of a small river; and is distant about three miles due east of the fortified town of San Sebastian.

But to resume. At 9 A.M. on August 19, the transports entered the port of Passages, but the regiment did not disembark until evening, which delay was accounted for by Gleig as follows:

'Soldiers are, as everybody knows, mere machines. They cannot think for themselves or act for themselves on any point of duty and as no orders had been left here respecting us, no movement could be made, until intelligence of our arrival had been sent to the general commanding the nearest division.'

With the help of all the boats in the harbour, the disembarkation was completed after dark, and the night was spent in bivouac at Rentaria, about two miles from Passages, when the regiment was brigaded with the 62nd, 76th and 84th regiments under command of Major-General Lord Aylmer, who had previously commanded the 2nd Battalion of the 85th during its brief existence.

The regiment remained in this camp until August 27, spending the intervening days in erecting huts, no tents having been issued, and purchasing horses and mules, which were brought in great numbers to the camp by the country people, and for which considerably more than their worth was given. All idea of a general mess for the officers seems to have been abandoned henceforth, 'the officers dividing themselves into small coteries of two, three, or four, according to friendship rather than convenience.'

By August 26, 1813, all arrangements were completed for opening the batteries in the second siege of San Sebastian. The besieging force contained the finest train of heavy ordnance which had ever at that time been placed at the disposal of an English General. There were no fewer than sixty pieces of artillery, some thirty-two and none less than eighteen-pounders; in addition to 20 mortars, all of which, as Gleig says:

'Were prepared to scatter death among its defenders, and bade fair to reduce the place itself to ashes.'

On the night of August 26, two companies, Captain Brown's and Captain Gubbins's, under Major Deshon, were embarked on board the cruiser H.M.S. *Diadem* to capture a redoubt on one of the islands in the harbour, which service was performed without loss, only a few shots being fired before the small garrison of one officer and thirty men were made prisoners.

The capture of this redoubt was considered a necessary precaution, in that, to a certain extent, from it an enfilade fire could be delivered against the besiegers' trenches. This redoubt was captured at night, and it speaks well for those engaged in the venture that historians report the night of the 26th 'to have passed by in quiet.'

From an Oil Painting by G. Carpenter, circa 1835.

SAN SEBASTIAN.

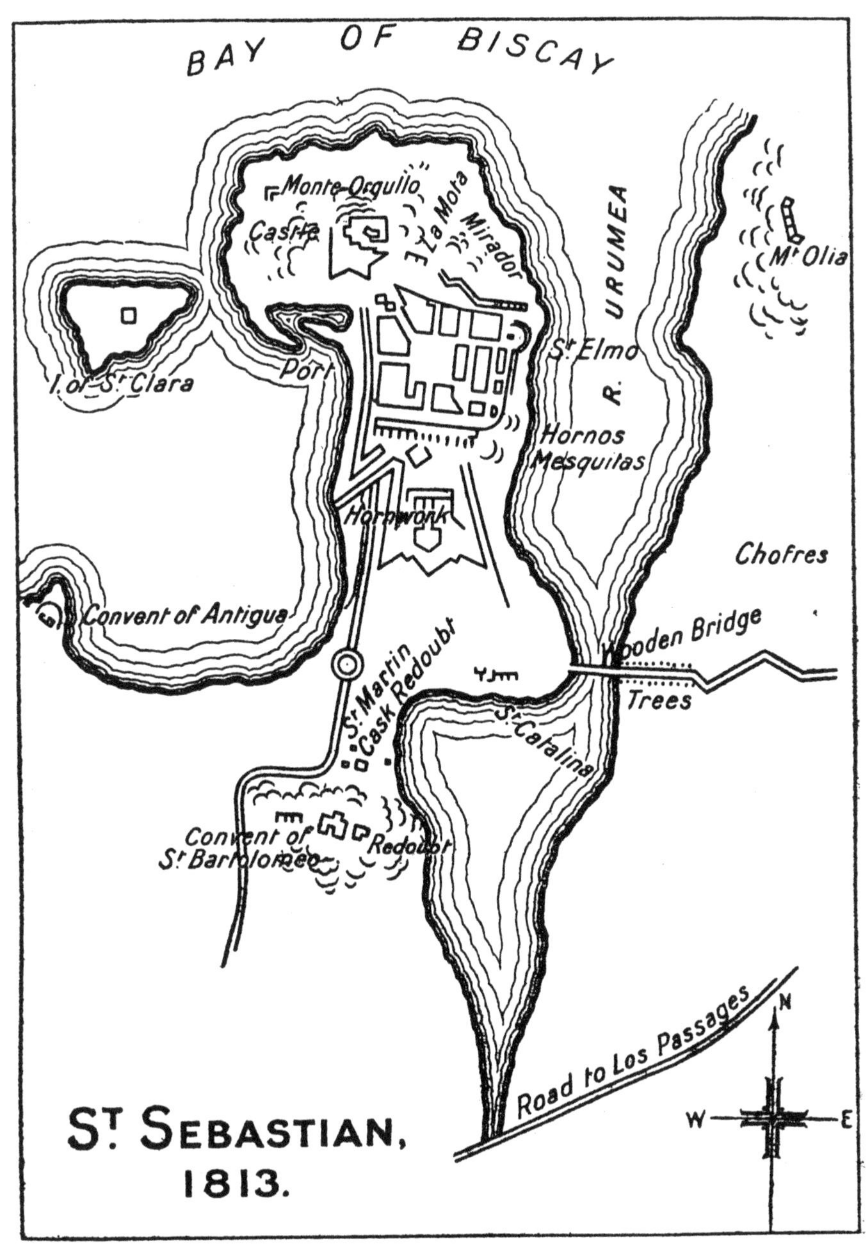
BAY OF BISCAY
Monte Orgullo
Castle
La Mora
Mirador
URUMEA
R.
Mt Olia
I. of St Clara
Port
St Elmo
Hornos
Mesquitas
Hornwork
Chofres
Convent of Antigua
Wooden Bridge
St Martin
Cask Redoubt
Trees
St Catalina
Convent of
St Bartolomeo
Redoubt
Road to Los Passages
N
W
E
St SEBASTIAN,
1813.

The fortified town of San Sebastian is peculiarly situated. It stands on a conical peninsula with an inlet of the sea on the southern side and the river Urumea on the northern. The only land approach is by means of a low sandy isthmus. Here a line of works had been erected to defend the crossing, having a large hornwork in the front. The remainder of the peninsula, being washed either by the sea or by the mouth of the Urumea, was merely protected by a high wall broken here and there by a few small towers. On the right bank of the Urumea are a range of sand hills known as the Chofre which flank the defences of the town. Facing the hornwork where the isthmus begins, and distant from it about 900 yards, is some high ground on which stood the fortified Convent of St. Bartolomeo. Behind the town, on the top of the peninsula, is the Castle ; the way up to which was very steep naturally and had been rendered steeper as to its sides by artificial scarpings. The harbour is on the south side.

At low water the Urumea is fordable.

The wooden bridge was destroyed by the defenders.

The rocky island crowned by a tower is known as the Isle of St. Clara.

The little ruined tower in the pencil sketch is not marked in any map.

The remaining five companies, commanded by Captains Bell, McDougall, Schaw, de Bathe and Knox, had been moved on August 25 to Oyazum, but returned to Rentaria after two days, and on August 27 marched to Irun, and there joined the Division occupying the Pass. Here they remained in quiet until August 30, encamped in a barren valley surrounded completely by steep and rugged mountains, and occupying huts previously erected for them.

At 3 A.M. on August 30, orders were received for the 85th to immediately join the army besieging San Sebastian, and by 7 A.M. these orders had been carried out.

The 85th was then selected to embark on the boats of the fleet, with a view to assaulting the Castle at the moment that the main body moved from the trenches. Such was the design of Sir Thomas Graham ; but as it appeared that it could not be carried out, without disaster to the party so utilised, it was abandoned.

On the 31st, the regiment retired to Irun, and spent the succeeding

Pencil Sketch by G. Carpenter, circa 1835.

SAN SEBASTIAN

Taken for the Panorama or Diorama of the Peninsula War, subsequently executed by the elder Telbin.

day under arms, awaiting some movement on the part of Soult, who had been prevented in his effort to relieve the defenders of San Sebastian.

What occurred at this juncture to Soult is briefly as follows:

On the last day of August, San Sebastian fell; and Soult, unaware of the fact, made another effort for its relief—an effort which, as his former efforts had done, failed. He attacked the Spaniards under General Freyre who occupied the left bank of the Bidassoa and covered from the strong heights of St. Marcial the high road to Bayonne.

The Spanish right extended in front of the Haya mountain, and they were thus enabled to keep observation on the various fords of the river. Their left was supported by the 1st Division, and Lord Aylmer's brigade was in rear of Irun, while General Longa's Spanish Division was posted in rear of their right. The Spanish routed the French when the attack did come off. Soult was more fortunate in an attempt he made against the Portuguese stationed on the right of the Haya mountain. Here, in spite of the support given to them by General Inglis, the Portuguese were compelled to abandon the heights between Lezaca and the Bidassoa. Inglis withdrew to the ridge in front of the Convent of St. Antonio, where the remainder of the 7th Division shortly after came up to his support. The position thus taken up was too formidable for Soult to attack, and having failed at St. Marcial he recrossed the Bidassoa, not without considerable loss from General Sherret's brigade at the Bridge and Pass of Vera.

On September 4, the regiment was advanced towards the base of one of the mountains overhanging the Irun valley. The road was so precipitous and winding as to cause all sort of order to disappear, and the battalion occupied over three-quarters of a mile from front to rear, although under 600 strong. The march was exceptionally trying, on account of great heat, bad roads, and heavy loads. After five hours' marching, on reaching the summit of an isolated green hill, the regiment was met by Lord Wellington, with three of his staff. He was immediately recognised by many of the men, who had served under him in the earlier Peninsula Campaign, and who greeted him with cries of 'Douro,' the familiar title given by soldiers to the Duke. To this Lord Wellington replied by taking off his hat, and bowing, subsequently conversing with the Colonel and commending the appearance of the regiment. On Lord Wellington's advice a halt was made on this ridge; and, on the arrival of the baggage, tents were pitched and breakfast cooked.

After halting here for two days, camp was struck on the morning of September 6, and in the afternoon of the same day the 85th marched to an eminence immediately over the Bidassoa, which was within sight of the enemy's camp. This march and the subsequent wait for the baggage were rendered most unpleasant by torrents of rain. The 85th was now the advanced line of the army, having only a few Spanish picquets between it and the enemy. Accordingly the regiment stood to arms an hour before daybreak in close column, and until sunrise remained still, when they piled arms, and to get warm ran round them, not daring to go far from them. Here the regiment remained for five days, standing to arms before daybreak each morning and being dismissed soon afterwards. On September 11, the regiment was moved back again to the foot of the mountains previously occupied and about two miles from Irun, which was one of the most agreeable posts as yet assigned to them, and in which they remained until the advance of the army into France. The time in this camp was spent in working at redoubts, until about 37 were completed, which commanded the most assailable points in the French line, between Fuenterrabia and the Foundry.

The following letter, written home by Captain R. Gubbins, 85th Light Infantry, is here given in part, and is full of interest:

'CAMP NEAR IRUN,
'*September* 15, 1813.

'We have had a great deal of marching about from place to place all within a circle of 10 miles. Our present camp is a most beautiful spot exactly facing the entrance of the river (the Bidassoa) which divides France from Spain; the Spanish towns of Irun and Fonterabia on one side; on the other St. Jean de Luz and the French camp, all within 3 miles of us.

'I went yesterday with Colonel Thornton and Mardanale to the remains of the bridge burnt by the enemy, the only connection between the two countries near here. We went within 60 yards of the French picquet. This is the place where the enemy crossed a few days ago and were beaten back by the Spaniards with a loss of about two thousand.

'Lord Wellington's dispatch will give you every particular of the fall of St. Sebastian.

'I went with part of the Reg[t] with an intention to storm the place on the sea side, but we received orders to make a diversion only.

'The shot and shells flew about us pretty thick, but, thank God, the only injury we sustained was a sprinkling from the water thrown up by the shells that fell near the boats.

'We are now employed in fortifying the heights all along the position, but what the Lord's intentions are it would be presumptive of me even to imagine.

'The town of St. Sebastian was by far the best in this part of Spain, the houses were large and furnished in a superior style, many of them from England, the inhabitants were rich and lived well. One house connected with England in trade lost above half a million in specie—plundered by the troops that stormed the place.

'The loss of killed and wounded on both sides was immense, and not *one* house in the town has escaped, the whole being a mass of flames for several days; and nothing now remains but a convent and the bare walls of some of the strongest houses.'

A perfect understanding seems to have existed between the two opposing forces. Sentries and picquets were never molested or attacked unless a general action was to follow. Similarly, from this camp officers would wade across the Bidassoa when fishing, into the French picquet line, who would watch their sport without interference.

Soult was apparently expecting the Allies to advance, and thought to deter them from entering Gascony by circulating printed proclamations through the camp by means of market people. The purport of these proclamations was a threat of instant death to all who entered Gascony, and a declaration that the Gascons had risen *en masse* to prevent such invasion.

On October 6, orders were received to ford the Bidassoa, thus entering France, and to attack the heights above it on the 7th. The long period of comparative inactivity in camp, combined with the prospects of invading France, rendered this order most welcome.

The night of October 6 was oppressively hot until the atmosphere was cooled by a violent thunderstorm, which had, however, cleared off before 4 A.M. (on 7th) when the regiment fell in. It was pitch dark when the 85th, having been joined by two other battalions, total strength 1,500, marched towards the Bidassoa. Tents were left standing and the mules and transport remained in protection of a guard. This was done with a view to deceiving the enemy, which was the more necessary as the river would not have been fordable until about 7 A.M., by which time it would have been broad daylight. The object of the early march was to gain unobserved a hollow on the left bank of the Bidassoa under cover of darkness. This was accomplished in perfect silence, and without creating the smallest alarm. Halting in this hollow, the brigade was joined by other divisions and three 18-pounder guns; which latter, advancing beyond the brigade, climbed the rising ground between it and the river, and took up their position in battery. From this spot the guns commanded the ford, and although within musket shot of the enemy's picquets, the position had been occupied without arousing them.

Of the total force destined to attempt the passage of the Bidassoa, about 20,000 could be clearly seen from the position occupied by the 85th, which Gleig describes as 'a very beautiful and animating sight.'

It will here be necessary to briefly describe the position of Soult's army. His extreme right flank rested on the sea, the central brigades occupying a chain of heights sufficiently steep to check the progress of an advancing force, and well adapted to cover the defenders from the fire of the assailants. Along the face of these heights was the straggling village of Hendaye, and immediately under them ran the Bidassoa, which, so near its mouth, was only fordable at two places, one opposite Fuenterrabia, and the other near the main road. Of these two fords, the former was commanded from the French side by a fortified house or *tête-de-pont*, completely filled with infantry, and the latter, which was only approachable by the main road, was protected by overhanging precipices, from which 100 resolute defenders would have been more than a match for 1,000 adversaries. Yet these two fords were the most assailable points, all the ground towards the right flank of Soult's position being little else than perpendicular cliffs.

The tide of the Bidassoa was fast going down, and soon the advance began, backed up by the fire of the 18-pounders; whose opening fire was the signal for the general advance. With comparatively small loss (not more than 500), the enemy's picquets were driven in, and our troops established on the other side of the river. Although the alarm was given, and columns were moved up from the rear of the French position to beat back the assailants, a panic seemed to have seized the French army, 'who fired their pieces and fled, without pausing to reload, nor was anything like a stand attempted till all their works and much of their artillery had fallen into our hands.' The 85th took a conspicuous part in the capture of the village of Hendaye, and the day was practically gone when the order was given to halt. Troops bivouacked in brigades and divisions along the heights they had taken from the French.

The baggage from the camp at Irun did not arrive until the next morning, and then was just in time to allow of the tents being pitched before a heavy rain set in which lasted for two entire days. The position was very exposed and bare, saving for furze, which served the double purpose of firewood and fodder.

The use of furze for fodder was frequently employed in the Peninsula campaigns when forage was, as it so often was, exceedingly scarce. We are not sure whether the custom is a Spanish one or whether it was imported by our men. If the latter, it was probably so imported by Irish troops; for in Ireland, in Ulster, it was a few years ago quite common to see a rude half-circle of stones at one end of a cabin in which the spiky foliage of the gorse or furze was pounded for fodder, the implement employed to pound it being a heavy stone through which a hole had been laboriously bored to hold a handle.

The heavy pounding makes a sort of paste of the furze, but as an article of diet for horses it is by no means a success. Several cavalry regiments that were perforce compelled to fall back on pounded furze as a diet for lack of other forage found that farcy supervened. This was, however, possibly owing to the low condition of the animals from previous starvation and fatigue.

Here the 85th remained from October 8 to November 9, a period devoid of any interesting circumstance. The weather was during this time most inclement, and it was feared that winter quarters would be taken up at Irun or Fuenterrabia. It was well known that the protracted defence of Pampeluna hindered our further advance, news of the surrender of this important town not reaching us till November 3rd.

From another letter from Captain Gubbins we gather the following information:

'CAMP ABOVE UROGNE,
'SOUTH WEST OF FRANCE.
'*October 29th*, 1813.

'. . . . Since I last wrote to you the Corps (Sir John Hope's) of the army to which we belong has entered France.

'On the 8th of this month we received orders to be under arms at 2 o'clock the next morning, and soon after daylight we forced the passage of the Bidassoa and established ourselves in a fine bold position between Adaye and Urogne in the S.W. of France. Our advance was so rapid and daring that the enemy gave way almost everywhere before our troops could get near him and the few places that were disputed were soon carried by the bayonet. Our loss was about four or five hundred which, considering the great object obtained, was very small.

'We have changed country much for the better in appearance, but provisions of any sort are scarce and of course much dearer—not a single villager remained behind, and indeed they are quite right not to trust themselves to the honour of either the Spaniards or Portuguese, for they most likely would be murdered by them.

'The French army is close to us, we see them plainly with the naked eye at drill. Our sentries at the outposts are only twenty yards from the enemy's. I frequently ride between them. They never molest any officer—but once I was desired to go farther away.

'Pampluna is by this time in our possession as commissioners on both sides were to meet the day before yesterday to agree upon terms—and give in they must as the garrison is in a dreadful state of starvation.

'It is said we are immediately to advance with the whole of the army—our Left being still in Spain—this in all probability will be a tough battle and most likely the last of this campaign.'

During the period spent in this camp a vast number of desertions seem to have occurred.

'Owing to the nature of the ground covered by our picquets in the Pyrenees it was usually impossible to place sentries in pairs, and it is thought that these numerous desertions were due to the fact that sentries were frequently posted near the mangled remains of some soldier who had fallen in the recent fighting, and that, knowing the serious punishment which awaited them if they should go back to their picquets, they elected to go over to the enemy rather than to remain alone to endure the pangs of diseased imaginations. Certainly it was remarked, that, on the army descending from the hills, and the sentries being doubled, desertion ceased.'

The Battle of the Nivelle

Soult's position after his failure to prevent the passage across the Pyrenees by the Allies was this.

He had formed a strong line of defence over a front of nearly twelve miles, which covered the town of St. Jean de Luz and extended from the sea across the River Nivelle and as far as the heights behind Ainhoe. His whole front was fortified, particularly on the right, where there were not only several formidable redoubts but also an inner line of entrenchment which was very strong.

The centre extended along the left bank of the Nivelle, which there winds round a mountain called La Rhune, and proceeded along a range of heights which were covered on the left by the Sarre. The line then crossed the Nivelle and passed along the strong ridge behind Ainhoe where it was covered by several more redoubts.

It was on November 8 that orders were at last received to advance. The period in camp had been very trying, and the welcome news contained in these orders, namely to attack on the next day, had been eagerly anticipated. During the night of the 8th, these orders were cancelled on account of the bad state of the roads; but they were reissued on the evening of the 9th. The task allotted to the 85th was to take and hold the village of Urogne, a village containing a church and some 100 houses, but strongly held; so much so that, as Gleig writes:

'The 85th was selected, in preference to many others, for this service perilous though honourable.'

Preparations for the advance having been completed, the regiment fell in before daylight on November 10. Gleig comments on the behaviour of the women, of whom only six per company were allowed to accompany each regiment to the Peninsula. He considered them 'sadly unsexed,' a condition which he attributed partly to their being inured to dangers, and partly to the fact that, if they became widows, it was not a state in which they long remained.

It was still dark when the regiment moved silently out of camp, and it had only proceeded about 1½ miles along the main road when it was considered necessary to halt, in order to make dispositions for the attack, and also to await sufficient daylight to enable surrounding objects to be distinguished. During this halt three companies were detached a little to each side of the road in order to surprise two of the enemy's picquets, the remaining companies being extended so as to cover the whole breadth of the road, and ready to rush the village at the double. The village of Urogne was strongly barricaded and filled with infantry, but it was hoped by the suddenness of the attack to avoid much damage from the enemy's fire.

These dispositions had been completed some half an hour before the signal to advance was given. The detached companies, though they could not surprise the French picquets, drove them in, in gallant style, so that the remaining companies could press forward. They were greeted with a heavy fire from the barricades and houses, which caused several casualties. During the rush, no firing was done by the regiment, though a covering fire was kept up by two 9-pounder guns with grape and cannister. In two minutes the regiment had reached the base of the barricades, and in another they were on top of them, when the enemy, appearing panic-stricken at such sudden tactics, abandoned their defences and fled. The chase continued down the street of Urogne, beyond the extremity of which the 85th had orders not to proceed.

The losses suffered by the regiment in this attack were 1 officer (Lieutenant Johnson) and 13 non-commissioned officers and men killed, and 7 wounded.

Here we give interesting extracts from two more letters home, written by Captain Gubbins.

GATARY, NR. ST. JEAN DE LUZ,
November 15, 1813.

'I have a few minutes allowed to tell you that we have just given the French a complete thrashing and they are making off as fast as they can.

'The 85th Reg[t] had a very honourable post and performed their duty in admirable style. We had the misfortune to lose a fine young officer killed, Lieutenant A. Johnson, and 13 men—I will write all particulars immediately—I hope this will reach you with Lord Wellington's dispatches.'

St. Jean de Luz, France,
November 24, 1813.

'In the late action the 85th Reg[t] had the honour to be particularly selected to attack the town of Urogne and in half an hour we got complete possession of it and kept it during the whole day under a fire of artillery and musketry. We lost a very fine young man (Lieutenant A. Johnson) who only joined the Reg[t] the day we left Hythe, he was shot through the heart.

'I was very lucky in not being touched as both the men on my right and left were wounded and what is rather remarkable they were both my servants.

'The day after the action we advanced to St. Jean de Luz, the enemy having retired to the neighbourhood of Bayonne.

'The country we passed through was very fine and the innumerable redoubts, batteries etc. we saw in every direction proves the admirable style in which Lord Wellington directs the movements of his army, to force an enemy to give up such a formidable position.

'The bridges over the River Nivelle which runs through St. Jean de Luz having been burnt by the French we were obliged to wait till low water when we forded the river and went forward a few miles and encamped for the night.

'The inhabitants of the town remained in their houses, relying upon being well treated by the English—they even appeared rejoiced to see us, they said they had been plundered by their own army and that they had lived in a most miserable state for many years, they hated their government and prayed for the English to come and give them peace.

'The people immediately on each side of the Pyrennees are called Basques and have a peculiar language of their own which is very difficult to understand.

'The country and the houses on this side are far better than on the other. I am now in a house with my company and two others of the regiment, altogether ten officers and 150 men—my men have a chapel that is attached to the house and I and my two subs have the vestry, not at all resembling English ones.

'I have just given you a sketch that you may see how well it is proportioned—there being no chimney I built one at A, B is the door, C window, D a large dresser quite rotten from damp, O door into the chapel. The window when first we took possession was not of material use as the roof admitted plenty of light. Where you see the X is exactly the spot where I place my *arm*chair and where I am now writing close to a comfortable fire made with part of the old dresser.

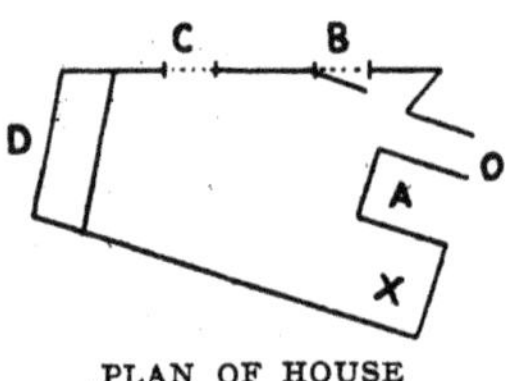

PLAN OF HOUSE

'I have so beautified and repaired this said vestry that if ever I should visit this place again I make no doubt that my name in gold letters on a blue board just like those in Epsom Church will be stuck up in some conspicuous place.

'I rode to the front yesterday to witness our Light Division drive the enemy from some heights we wished to occupy. At one time I was not more than three miles from Bayonne and without a glass distinctly saw thousands of workmen repairing the fortifications. I am afraid a mail has been lost, I found several English letters in a house near Urogne that had been the H[d] Quarters of the French army. Two of the letters were for officers of the 85th Reg[t].

'My Quarters are about a mile out of St. Jean de Luz which is Lord Wellington's headquarters. We get supplies of all kinds, but extravagantly dear.'

To return to the narrative of the campaign.

Of the French position, the right of which had just been won, Lord Wellington remarked that he 'had never beheld anything more formidable.' It extended over about three miles of country naturally suited to defence, and had been much strengthened by Soult's army during the last four weeks. In fact no serious attack was attempted against the enemy's right, other than that against Urogne, and this village the 85th were ordered to keep at all hazards, thus deterring Soult from detaching any of his army from his right to his left flank. Lord

Wellington's object was to turn the French left, and this he succeeded in doing after twelve hours' severe fighting.

As soon, therefore, as the 85th had driven the French out of Urogne, the barricades were removed to the other end of the town, and the regiment entrenched itself behind them. This was done without loss of time as an attack from a mass of French infantry, collected on a hill opposite the village, appeared about to be developed. Meanwhile the village was swept by fire from the enemy's guns, which fortunately did small damage, beyond destroying the church and houses. From the entrenchments, at about 11 A.M., three companies were sent out on the left flank, to watch the movements of the enemy more closely. Having gained a hollow road, with only one harmless salutation from the enemy's artillery, these companies joined the line of skirmishers sent out from Colonel Halkett's corps of Light Germans. Before long, it became evident that a large force of *tirailleurs* were eager to catch this skirmishing line in the hollow road, and accordingly the skirmishers clambered up the acclivity, and rushed forward to meet the *tirailleurs*. These were driven back over a few fields, and the skirmishing line returned to the hollow road, perceiving that to pursue further would bring them under the enemy's artillery fire.

Soult, hitherto prevented from sending reinforcements from his right to his left, made a determined effort to do so at about 3 P.M. A heavy column marched from the French lines, covered by a large force of skirmishers. These skirmishers were driven back by the 85th, who now took up a more advanced position, from which frequent attempts were in vain made by the enemy to turn them out. By holding this position the reinforcements from Soult's right to his left were kept back until dusk, when all advanced troops were recalled.

The regiment spent the night in the church, which was the only edifice left which could afford any protection from the occasional fire from the enemy's guns. The drum and band instruments were placed in the pulpit,[1] the arms in the side and the men in the centre aisles; the officers lying round the altar, the whole church being dimly lighted by small rosin tapers.

After about four hours' sleep the regiment was fallen in, inside the church, news having been received that the enemy was moving. By

[1] This is interesting as being the first mention of the existence of a band.

dawn the whole of the enemy's outposts and right wing had been withdrawn, and shortly afterwards their entire force retired in good order. Soult had perceived that his reinforcements from his right to his left had arrived too late to be of use; which it will be remembered could not proceed thither until the 85th were withdrawn. Had these reinforcements been able to get through, the result of the battle of the Nivelle might have been very different, and it is largely due to the series of feints and sallies performed by the 85th from Urogne that these reinforcements could not reach Soult's left until they were too late and the position had been won.

Seeing that Lord Wellington's objective was to overwhelm the French left, and that, as stated above, this was only accomplished after twelve hours' fighting, it seems that the 85th, who took so important a part in first capturing Urogne, and then delaying Soult's reinforcements until Lord Wellington's object was achieved, should have deserved the addition of Nivelle to their battle honours. It is hoped, however, that this honour may yet be granted.

On the morning of November 11, having buried the dead in the churchyard of Urogne and eaten a hasty breakfast, the regiment started in pursuit. At about 9 A.M., on arriving at St. Jean de Luz, which had been the headquarters of the French army during the previous day's fighting, it was found that the bridge over the Nivelle had been destroyed. To prevent this the 12th and 16th Light Dragoons had pushed forward, but had arrived too late. Consequently the pursuing army of some 25,000 men were collected in St. Jean de Luz, pending the rebridging of the river. By about noon, the tide serving, and only infantry crossing by the repaired bridge, the whole of the left column had crossed the river. The pursuit was continued, despite a heavy fall of cold rain. Towards evening, the advance guard of the pursuing army came up with the rear of the retreating French. These were posted round the village of Bidart. Here the force camped for the night, without cover, and in great discomfort owing to wet ground and rain. Tents had not arrived, the baggage being still some fifteen miles in the rear. As soon as the rain began, the Spanish and Portuguese troops left the ranks and scattered over the countryside, perpetrating many horrible crimes, for which about twenty were hanged.

During this pursuit, the 85th was the leading infantry regiment, having a brigade of cavalry in front, and a Portuguese brigade n rear.

During the morning of November 12 the tents arrived, and finding that Soult had retired from his position at Bidart towards Bayonne, the army encamped. The ground allotted to the 85th was comparatively good, and in this camp the regiment remained until November 17.

Some forty-five officers and servants took up their quarters in a farm-house near the camp, which, despite the incessant rain, seems to have been less comfortable than a tent would have been owing to the numerous insects.

The incessant rain, and inadequate protection from it, caused much sickness in the army; so the orders issued on November 17 were received with universal rejoicing. Their purport was that the troops would go into winter quarters; and accordingly, on November 18, the 85th marched into their new camp, at a place some three miles from St. Jean de Luz, and six miles from Bidart. The ground allotted to the regiment was a species of common, dotted about with farm-houses and comparatively dry.

The time spent in these winter quarters was employed in occasional skirmishes with the enemy, improving the camp generally and, by the officers, in hunting and shooting. Lord Wellington's foxhounds always hunted twice a week, and Gleig remarks that the Duke

'took the field regularly as though he had been a denizen of Leicestershire, ceasing to be the Commander-in-Chief of three nations, and becoming the gay merry country gentleman, who rode at everything.'

The boundaries of this position in which the army was quartered extended from Garrett's House on the right, through Guthory or Guétary and Arcangues, to Bidart on the left, a distance of six miles. It was found necessary to take the field again on December 9, up till which date the army remained in the above position.

The occupation of winter quarters was found necessary before Lord Wellington had originally intended, and they were consequently in a not altogether suitable locality. The Nivelle interfered with the communications between the right and left flanks. The entire left wing was consequently advanced on December 9, with a view to securing the passage of the river, which was not completed till nightfall. The 85th returned to camp at 9 P.M., having been under arms since 4 A.M. without opportunity for eating; 'nor had we,' says Gleig, 'been permitted to unbend either our minds or bodies in any effectual degree.'

On the following day, December 10, heavy firing was heard in the direction of Bidart, towards which place the 85th was hurried, passing *en route* several wounded, who gave terrifying accounts of the enormous numbers of the enemy. After passing through Bidart, the view

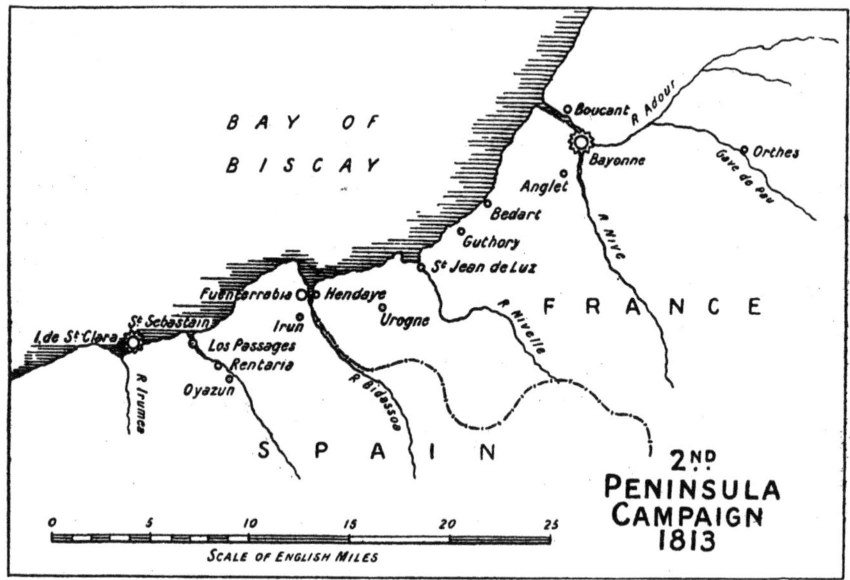

obtained made it apparent that a mere handful of British troops was opposed to 'a mass of men so dense and extended as to cover the whole of the main road as far as the eye could reach.'

Previous to the arrival of the column mentioned below, the Fifth Division was fast losing ground, near Bidart, before vastly superior numbers of the enemy. That they were fighting well was proved by parties of French prisoners passing to the rear, with escort about one-tenth their number.

The approach of the column including the 85th was greeted by a well delivered fire from ten or twelve guns, which caused some loss. It appears that both sides were struggling for the possession of a thick wood, which, at the moment when the column arrived on the scene, was held by two British and one Portuguese Battalion, against a perfect swarm of *tirailleurs*. The 85th were ordered to assist in the defence of this wood, and were engaged until about 3 P.M., frequently hand to hand. At this time the French began to draw back from their attack, leaving unguarded a chateau belonging to the Mayor of Biarritz. To this chateau, General Sir John Hope, with only a few A.D.C.'s and orderlies, repaired to observe the French movements; seeing which, the chateau was rushed by a large party of French infantry, Sir John only escaping with the greatest difficulty, his horse being killed, and three musket balls passing through his hat. This was the signal for the renewal of the fight, the French making repeated and determined efforts to rid the wood of its defenders. As dark came on the combatants withdrew, and it became apparent that neither side had gained any ground.

The scene was one of great disorder, the allied troops being completely mixed together. French, English, German, Dutch, Spaniards, Portuguese, were all calling in their various languages for their scattered comrades. 'So complete, indeed, was the confusion that neither one party nor the other availed themselves of it for military purposes.' Finally the French, having gradually collected their scattered battalions, retired; and the British force took up its positions for the night,the 85th rejoining their brigade with much diminished numbers.

Before daylight on December 11, the whole force, which had spent the bitterly cold night in small parties, each with its own fire, stood to arms without any premonitory bugle or trumpet-call, and in absolute silence. Apparently the French army did likewise; but it was not until some time after dawn that any movement was made by either side, each force expecting the initiative to be taken by the other. Finally Lord Wellington ordered the advance of three Portuguese battalions with the sole object of bringing matters to a crisis. These regiments were gallantly met, and, after a good deal of fighting, repulsed. This was followed by a determined attack upon the wood, the attackers coming on to within 100 yards distance, but slowly and in silence; then, raising a loud discordant yell, they rushed forward with their usual spirit and dash. 'Their attack was met with such coolness that the whole affair might have been a piece of acting,' says Gleig. No man quitted his ground or discharged his musket without orders to do so. All efforts on the part of Soult to take either the mayor's house or the wood, or to drive back the British force at any point, were in vain, so that his whole column, 'which covered the high road as far as the eye could reach, was perforce obliged to halt and remain idle.' About noon the enemy were seen to be retiring, a movement which was correctly diagnosed by Lord Wellington as a feint, so that he was not tempted to pursue—whereby he would have been compelled to fight on unknown ground, and on ground, moreover, which would have been advantageous to the French. Lord Wellington's orders were, on the contrary, to prepare dinners. Before these were ready, 'two ponderous masses of infantry,' covered by twelve pieces of cannon, rushed forward towards the village of Arcangues, and for a time carried all before them. A Portuguese and a British battalion were driven back, and the French force were for the first time in possession of the wood. The confusion in the rear was now great—artillery galloping to the front, ambulance waggons galloping to the rear, plunderers flying in all directions, and the half-prepared rations being flung into the fires just as the desire to eat them was about to be satisfied. This was particularly hard on the Portuguese battalions, who had been for nearly two days without food. Two squadrons of cavalry were next ordered out, who were to check the retirement of the fugitives as well as the advance of the French. The infantry line then advanced with a tremendous and overwhelming rush. The enemy stood nobly, but were gradually beaten back into the wood. Here the fight soon became split up into one amongst small parties, all compact order being lost—the gradual retreat of the enemy encouraged the allies to pursue, until finally the disorder of the pursued was not greater than that of the pursuers. The 85th followed to a point considerably beyond the wood, and far in front of their brigade, A sudden cavalry charge, delivered without sufficient warning to enable them to form to meet it, compelled them to retire through the wood and so rejoin their brigade.

So for a time the attack was repulsed, and the different regiments had time to re-form, and take up their various positions. The position allotted to the 85th was somewhat advanced, together with a Portuguese and two British regiments, the Portuguese being the most advanced of all. Noting this somewhat weak defence, Soult lost no time in directing another vigorous assault. The Portuguese, being the first met, were not only much reduced numerically, but were weak with fighting and, as previously mentioned, had been unfed for nearly two days; they gave way almost immediately, and fled the field. The two other British regiments, who became engaged simultaneously with the 85th, were overpowered:

'And even we, whose ranks had hitherto been preserved began to waver; when Lord Wellington himself rode up. "You must keep your ground, my lads," he cried, "there is nothing behind you, charge! charge!" The effect was electrical. We poured in but one volley and then rushed in with the bayonet. The enemy would not stand it: their ranks were broken, and they fled in utter confusion!'

The chateau and wood were soon recaptured, and in the now gathering darkness the fight ended for the day. The ground occupied by both forces was almost identical with that of the previous day. The same confusion prevailed, and the same arrangements of lighting fires, and lying down to sleep by them in small parties. The 85th, however, moved to the right of the line and took over the outposts.

Soult's losses during these few days' fighting were not confined to the 4,000 or so casualties, but included about 2,000 German veterans, who deserted and were shipped direct to Germany.

On the morning of December 12, leaving two companies as skirmishers, the rest of the 85th retired a short distance behind the outpost line, where they were more prepared to resist attack. The day, however, passed quietly, and the 85th remained on outposts for the night of the 12th. So close were the two forces that the sentries are said to have been not more than thirty paces apart. During this night, Soult retired through his entrenched camp, and massed his force against the right of the Allies under General Rowland Hill. In the action on December 13, the 85th took no part. It will be remembered that the French were totally defeated, leaving 5,000 killed and wounded on the field.

The night of the 13th was spent in collecting units, the 85th camping round a cottage, which apparently contained more than half the regiment. On the 14th, they marched to camp on a ridge between

Bidart and Arcangues in pouring rain, remaining there until 2 P.M. on the 15th, when they marched back to the camp previously occupied by them on November 10. Here they remained until December 18, when their turn for outposts came. The outpost system was for regiments to be relieved after three days, so that each regiment only held the outposts for four periods during the winter—in all, twelve days.

The winter of 1813–14 was extremely cold. The camp occupied was on the scene of the fighting of 11th and 12th; and the men, when not on outposts, were employed in felling the wood round the mayor's house, throwing up breastworks, and 'constructing a square redoubt capable of holding an entire battalion.'

On December 21 the 85th went into cantonments at Guthory, where they remained until January 2, 1814, a period of great monotony, only relieved by Christmas Day, and auction sales, in which the clothing and effects of those who had fallen were disposed of, under the sergeant-major as auctioneer.

Soult's army was much reduced by the numbers of veterans he had been compelled to send to help Napoleon. He accordingly enlisted every available man or boy, and frequently the British camp was alarmed by the noise of the instruction in firing given to Soult's recruits.

At 6 A.M. on January 3, the whole of the left column, including the 85th, advanced towards the mayor's house, and occupied a position near the Church of Arcangues. The advance was continued on the 4th, the position finally adopted being so close to that of the French that not more than a quarter of a mile separated the sentries of the two forces. These movements had become necessary on account of Soult's threatening attitude. Considering the hardships entailed by such frequent moves in bitter weather, and frequently on short rations, it is surprising to note the almost amicable feeling which existed between the two forces. Gleig tells a story which well illustrates the intimacy between the picquets, and the conveyance, by this means, of money and letters to prisoners. It was not until January 6 that the British army returned to its winter quarters. From positions near Arcangues the whole French army, estimated at 120,000, could be seen; and it was in this position that the 85th spent the night of the 5th, previous to taking over the picquets on January 6. This appears to have been an occasion when the three-day principle of outposts duty was violated as it was not until January 11 that the 85th handed over to the Brigade of Guards and retired to winter quarters at Bidart.

While at Bidart, excursions to Biarritz seem to have been the chief amusement; for in addition to the charming ladies, whom Gleig describes as 'having about them all the gaieties and liveliness of Frenchwomen, with a good deal of the sentimentality of our own fair countrywomen,' there was also the risk of being caught by the French. In fact, on one occasion, only the opportune arrival of a cavalry patrol was able to prevent the capture of several 85th officers by a party of French hussars.

From the 23rd to the 25th of January, the 85th were again on outposts, returning on the 25th to their quarters, where they remained until February 21.

The allied army began to leave winter quarters on February 16. On the 20th, orders were received by the 85th to prepare to do so on the 21st. The 20th was accordingly spent in repairing tents, and making good deficiencies of ropes, pegs, etc., and by 3 A.M. on the 21st the regiment, having breakfasted, was accoutred and in line of march. Gleig denounces the setting off on 'any military excursion, without having in the first place layed in a foundation of stamina to work upon.' For, as he says, 'An empty stomach so far from being a provocation is a serious antidote to valour.'

The night of 21st was spent in a chateau about half a mile distant from the River Nive. At first orders had been to bivouac on high ground overlooking the river; but it was later decided that it was too early for bivouacking, and accordingly no picquets were employed, only a line of sentries being kept between the two armies.

On February 23, the 85th left camp at 3 A.M., marching towards Bayonne, thus giving justification to the rumours which had been rife overnight, of a proposed investment of that town. By 8 A.M. the regiment halted about three miles from the works of Bayonne. At the other end of the field in which the 85th halted was a French picquet, which immediately lined the ditch, apparently expecting to fight. There seems no account of the further actions of this picquet, which was presumably nonplussed by the conduct of the 85th, who 'were content to form into column, pile arms, and await orders'!

The fighting on 23rd, round Bayonne, seems to have been confined to the right and left of Wellington's army. At least the 85th, who were about the centre of the line, took small part; while the Spanish infantry division on the left and Portuguese brigade (one battalion *caçadores*, two battalions heavy infantry) resisted several spirited attacks.

During the early part of the night of the 23rd, the 85th, having been relieved by a Spanish battalion, then marched towards the left of the line, inclining always to the front, and so drawing nearer to the sand hills overlooking the River Adour, finally halting for the remainder of the night in the village of Anglete. At dawn the march was continued for some three miles to a small plain, protected from the outworks of Bayonne, only three-quarters of a mile distant, by a low ridge of sand hills. The baggage arrived some hours later and a hearty breakfast was then made off 'slices of beef hastily and imperfectly broiled, mouldy biscuits and indifferent tea,' which was the regiment's first meal for over forty hours.

In the fighting on 24th and following days, up to the victory at Orthes on 27th, the 85th were only spectators. During this period, the 1st and 5th British Divisions, two or three brigades of Portuguese, and 'a crowd of Spaniards' invested the citadel of Bayonne, under Sir John Hope.

The town of Bayonne stands upon a sandy plain, and the citadel on a rock closely overhanging the town. Between the two runs the Adour, a sluggish, narrow and shallow stream. Gleig's description of the town is as follows:

'Both town and fortress are regularly and strongly fortified, and on the present occasion a vast number of field-works, of open batteries, flèches and redoubts were added to the more permanent masonry which formed the ramparts. . . . Various sluices were cut from the river, by means of which, especially in our immediate front, the whole face of the country could be inundated at pleasure to the extent of several miles; whilst ditches, deep and wide, were here and there dug, with a view of retarding the advance of troops, and keeping them exposed to a heavy fire from the walls, as often as the occurrence of each might cause a temporary check. The outer defences began in all directions at the distance of a full mile from the glacis.

'The roads were everywhere broken up and covered with abattis and other encumbrances. Nothing, in short, was neglected, which

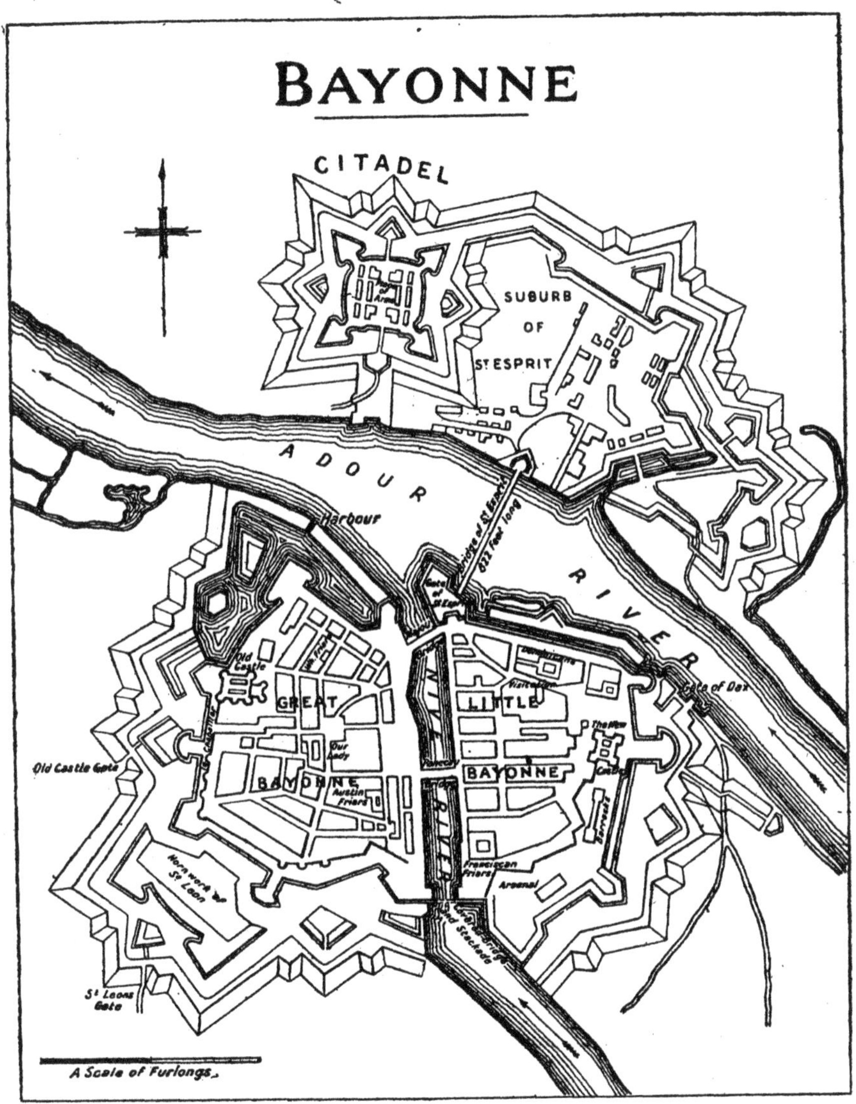

FEB. 10, 1814.

promised in any degree to contribute to the strength of a place which is justly regarded as the key to the southern frontier of France.'

The surrounding country was generally flat, and slightly wooded for about four miles in all directions. The citadel itself was 'a place of prodigious strength—the more assailable of its faces being that on which the shape of the ground has permitted the engineer to bestow the largest share of his attention.' Quite close to the walls of the citadel was the village of St. Esprit, commanded not only by the guns of the citadel, but also by a redoubt, which the French governor, General Thouvenot, had erected on the only conveniently level place; which was, incidentally, the only spot 'across which the besiegers could hope to push a sap with any success or facility.'

Up to February 24, the besiegers' line was in some places as much as five miles from the works, and practically no restraint had been put upon the besieged army and the inhabitants of the town. On this day, between 14,000 and 20,000 Spaniards took post on the faces overlooking Helletre and Joyeuse, the left of the line resting on the Adour, at a point where the stream was sufficiently narrow to enable pontoon bridges to be thrown across. By this means communication was soon established between the forces on either side of the river.

On the 25th, some battalions of Guards, the King's German Legion, as well as some cavalry and guns, joined the besieging army and effectively closed all remaining gaps in the investing line. A slight skirmish took place at the village of Boucaut, which the French evacuated and withdrew their sentries to a ridge about half a mile in rear of the village. All chance of communication between Soult and the besieged force was now gone, as well as the opportunity for renewing supplies.

The garrison of Bayonne under General Thouvenot was composed of the best troops in the French army, including about 15,000 picked men. The besieging force numbered some 30,000, under Sir John Hope; but only about 15,000 of these could be relied upon, and the line drawn round Bayonne, which this force was holding, measured something over four miles.

On the 26th, this line was contracted towards the city, with a view to taking the citadel. This did not entail an advance on the part of the 85th, who were already within musket-shot of the fort; and who, during the day, were 'treated to a cannonade, from the effects of which the nature of the soil secured us, the shells either burying themselves in the sand to the extinction of the fuse, or exploding when we were all snugly laid flat, and therefore safe from their fragments.' The fighting, as the day went on, became more severe, and resulted in the French troops, which had hitherto occupied positions outside, being driven within the citadel. By nightfall, the besieging sentries and advanced posts were established as near as the village of St. Esprit, which was only about 'half a pistol-shot from the nearest redoubt.' The 85th returned to their tents for the night, having lost 1 killed (Sergeant McDermot) and 3 wounded.

On the next day, February 27, the siege of Bayonne began in earnest, and continued till April 28 following, when the war was concluded.

During these months the 85th were employed chiefly in erecting batteries, or on outposts or such occupations, 'which have in them nothing of excitement, and a great deal of danger.' Such occupations were, of course, frequently accompanied by the fire of the enemy's cannon and mortars, which Gleig thus discusses :

> 'If it be simply a cannon shot, you either toil on without heeding it or, having covered yourself as well as you can till the ball strikes, you start up again and seize your tools. If it be shell, you lie quite still till it bursts.'

At first it might seem a matter of great difficulty to be able to decide which course to adopt after each explosion, but Gleig tells us that, if the work was not very pressing, one of the party was generally stationed to watch. On seeing the explosion, this man would call out 'shot,' or 'shell,' as the case might be. It seems that the higher trajectory, as well as the visibility of the fuse, were the betraying marks of a shell ; while the sharp report of a gun was always distinct from the explosion of a mortar. It would appear that the work of these men placed to watch was extremely good. Although such fatigue work lasted for several weeks, and generally under fire of some sort, only five men were killed. Such slight loss was also very largely due to the sandy nature of the soil. The position that the 85th occupied was of very great importance, as is shown by the almost continuous fire to which they were exposed. The French realised that could this position be forced they might easily gain the bridge over the Adour before fresh troops could be brought to oppose them : and, had they achieved this, they would thereby have reopened communication with Soult.

During this period of the siege, alarms were not infrequently caused by deserters. Gleig narrates one such, when after 'the Commanding Officer had read prayers to the battalion,' it being a Sunday night, a French officer had arrived in camp, with the information that a sortie was planned for midnight. Accordingly the night was spent in readiness, the men lying down with 'knapsacks buckled up, and pouches and bayonets slung on.' Nor does this particular alarm appear to have been false, as the subsequent firing of cannon goes to show, though it is thought that the attack was stopped on account of the French officer being missed; it being rightly conjectured that he had betrayed the plan to the enemy.

On the 1st of April all the 85th, excepting the picquets, were moved back to a pine wood, some two miles distant from the sand hills, which they had occupied for the last six weeks. The move was a very welcome one, the shade of the woods in the new camp affording much relief from the great heat of the sun experienced in the old one. But this enjoyment only lasted for three days, and on April 4 the regiment was moved up again into the line of trenches, pitching camp near the village of Boucaut. While here, the inhabitants seem to have flourished; and it is reported that:

'Not a single instance of violence to a native in person or property occurred; indeed both men and women scrupled not to assure us that they felt themselves far more secure under our protection than they had been while their own countrymen were among them.'

The regiment encamped at Boucaut was relieved every three days, the duties incumbent upon it being very arduous, and including not only strong daily fatigues at the batteries and redoubts under construction, which were mainly 'within half musket-shot of the enemy,' but also the finding of the picquets around St. Esprit.

The regiment was accordingly relieved on April 7, by a brigade of Portuguese, who took over the camp, while they returned to the camp they occupied from the 1st to the 4th of April.

An amusing incident resulted from an almost tragic origin on April 8. The day was very dark and foggy. Two brigades had been ordered to move to a certain spot, to witness the punishment of a sentry who, on the previous night, had been drunk on duty. Arrived at what was supposed to be the appointed spot the troops formed square, and the mist suddenly lifting, it was discovered that they were within point-blank range of the enemy's guns! Such sudden appearance of two brigades naturally caused the French to expect an assault, and the moments lost by the enemy in preparing, and their consequent hesitation to attack, were utilised by the British to get away before many shots had been fired into their close formation.

On the same day (April 8), the regiment marched to Anglête, where they were billeted, or at least accommodated in houses, for the next four days.

News of the abdication of Buonaparte, and of the entry of the Allies into Paris, was received by the army investing Bayonne on April 11. It may be imagined with what feeling of relief at the near approach of peace such information was received. It appears, too, that this feeling of relief was accompanied by slackness, which resulted in a complete surprisal of the outposts round St. Esprit on the 14th. The news above referred to had been immediately communicated to General Thouvenot, who, it will be remembered, was in command of the besieged. The French General absolutely refused to credit such news, and consequently no suggestion of a cessation of hostilities could be made by either side. Accordingly the regiment at Boucaut was relieved again on April 12 by the 85th, who assumed the fatigues and sentries round St. Esprit, with the exception that the building of batteries, which in all likelihood would never contain guns, was discontinued. It was confidently expected also that no active hostilities would occur. At 3 A.M., however, on April 14, General Thouvenot made a desperate sortie. The whole picquet line was warmly engaged. The enemy had come out in two columns, one against St. Esprit, and the other against Boucaut. The severity of the fighting may be realised from the fact that 1000 French and 900 Allies were killed, and many more wounded, including General Hope, who was succeeded by General Colville. A truce was concluded on April 15, the day being spent by both armies in burying the dead. This was the actual end of the siege so far as fighting was concerned, though sentries were posted until April 20, on which day the war was formally declared to be at an end.

On April 28, the Allied and French armies were assembled to witness the hoisting of the *Drapeau Blanc* over Bayonne. The standards of Britain, Spain, Portugal and the Bourbons were already flying over the Allied Camp, and hitherto a Tricolor had flown over Bayonne. The ceremony of hauling down the Tricolor and hoisting the *Drapeau Blanc,* though regarded as a signal of triumph by the Allies, was regarded with great hatred by the French. Even the country people, who had shared the discomforts of the siege, showed clearly their attachment to their former master, rather than any enthusiasm at the conclusion of the war and its attendant perils and hardships. Amongst the French army, and especially the officers, this day was the signal for a distinctly less friendly attitude to be adopted towards the Allies than had been the case even during the war. The French officers, prevented from continuing the war collectively, took every opportunity to bring about duels, and so do so individually. Although balls and parties were arranged in all the regimental camps, which were attended by the French, both civil and military, no sort of friendship was established between the French and English officers, who 'embraced every opportunity of bringing on personal quarrels, and as I have already hinted, though reluctant to accede to the matter at the outset, we came by degrees to see the necessity of indulging them: and I am inclined to think that they sometimes got enough of it.'

In this way things continued until May 8, 1814, on which day the regiment struck tents and marched towards the rear. Here they remained for a few days until orders were received which took them to Bordeaux.

Forty-six years after the publication of the 'Subaltern' and nearly sixty after the Peninsular war, Gleig revisited the scene of the campaign.

In the preface of one of the editions he gives his impressions of the changes to be observed in the various spots.

He found Bayonne much as it had been in 1813; but 'the intrenched camp, of which Marshal Soult made it the key' had disappeared. Large permanent additions had, however, been made to the fortress. He continues, 'Of the works thrown up by the English during the last siege, not a vestige remains. The Blue House, as we used to call a château standing in the suburb of St. Pierre, and in the garden of which we established our most formidable mortar battery, retains no traces of the fire to which it was then exposed. . . . The graves of the British officers who fell in the sortie are well kept up.'

Outside Biarritz he found an old post-house with its stables and outbuildings, though fast falling into decay. It was from along this road that Gleig and his two friends 'rode for their lives when chased by French cavalry out of Biarritz.'

The hollow road in which the French formed for their last rush at the 'Mayor's House' on December 11, 1813, appeared unaltered, and 'between it and the house lies the triangular field, just outside the wood, where the carnage was fiercest.' Here the slain were buried in great numbers.

'Beyond it, over the hill, and sloping towards Bayonne, extends the scraggy copse or belt of wood through which the fighting was close and desperate, and near is the lake into which many of the French horse and men were driven. They were driven not without great promptitude of action on our part; for there the French cavalry fell upon us with such rapidity and determination, that being scattered in loose order, we had just time, favoured by the swampy nature of the ground to throw ourselves into circles and receive them.'

After returning to the great road he reached the 'common or broken plain where, on December 10, 1813, the other corps of the left column of the army came up in support of the hard-pressed and hard-fighting fifth division. It was a wild bleak spot in winter, sixty years ago; it is little changed now. The redoubt which we threw up, and named after one of the mayor's daughters, has indeed disappeared; but the belt of wood round which the squadrons of the 12th and 16th Light Dragoons made their charge keeps its place exactly as it did then; and the hollow in which the brigade to which the "Subaltern" was attached bivouacked that night, lacks only the embers of their watch-fires to make it precisely what it was when they left it.'

In the neighbouring villages of Bedart and Ganthony, some two or three miles nearer St. Jean de Luz, 'the clusters of houses which afforded winter quarters to the 85th Light Infantry, appear as if they had been evacuated but yesterday.'

At St. Jean de Luz 'the bridge over the Nivelle is in exactly the state in which we left it; the piers of the arches which the French blew up stand where they did, as well as the beams and planks with which we replaced the broken roadway. It is pointed out to all who are curious in such matters as the *Wellington Bridge.*' Urogne 'lies as it lay sixty years ago, in the low ground, overlooked by the heights which Soult had fortified and which, he flattered himself, would stop our further progress. There, too, before descending the hill-side on the right of the old carriage-road stands the château, in the library of which we found, on the 11th of November 1813, a captured English mail. It seems to be, as far as outward appearances can be trusted, precisely what it was on that day.' Gleig carried off from it two pieces of plunder—a Spanish grammar which he had kept all those years, and a small and prettily enamelled pair of bellows which he had lost; though its use on the campaign had been considerable.

At Urogne, too, the three-gun battery 'the fire from which swept the main street of the town, and struck against the old church-wall without passing through it' had vanished. It was in this church that some hundreds of men (Gleig included) took shelter during the night of the 10th and 11th; and where, in the churchyard outside, they had buried their fallen comrades.

As Gleig neared the frontier the evidences of the long-ago war were greater. Hendaye was still not rebuilt. Fuenterrabia 'Houses, wherever you encounter them—the château equally with the cottage and farmhouse—seem to be in a state of dilapidation.'

Here and there relics of the great Duke's army are to be discerned 'in the mounds which mark the spots where, before advancing into France, he took the precaution to block the gorges of passes with a succession of redoubts.'

Passages, Gleig remarks 'seems to be not only unchanged but unchangeable.'

San Sebastian, on the other hand, was different:

'There all things are new. The fortifications have disappeared, and where bastion and curtain once stood, long boulevards are drawn out. The town has been rebuilt with great regularity; and along the banks of the Urumea, where our storming-parties crossed, the process of construction is still going on.'

CHAPTER VII

March through France. North American Campaign. Bladensburg and Washington, 1814

On May 8, 1814, the regiment, which formed a part of the brigade commanded by Major-General Lord Aylmer, marched from Boucaut and encamped near St. Jean de Luz. The brigade had been augmented by the addition of the 37th Foot on April 17.

The strength of the regiment was about 600 men; that of the 4th and 44th Foot nearly 800 each.

The cause of the trouble with America, according to a declaration issued by the Prince Regent, was as follows:

'The origin of the present contest would be found in that spirit which had so long unhappily actuated the councils of the United States; in that marked partiality which prompted them to palliate and even assist the aggressive tyranny of France; and in their systematic endeavours to inflame their people against the defensive measures of Great Britain.' He then took notice of the arbitrary conduct of France toward the Americans, and of their ready and abject submission to all the acts of violence and injustice which were perpetrated by their pretended friends. From their community of origin and interest with Great Britain, and from their professed principles of freedom and independence, the United States were, he said, the last power in which he could have expected to find 'a willing instrument and abetter of French tyranny.'

The actual cause of the war was the right of search on American vessels claimed by British seamen.

Remaining in camp near St. Jean de Luz until the 15th, the brigade then resumed its march, and passing along the trunk road through

Bayonne, Castets, and Belin, reached Bordeaux, a distance of nearly 120 miles, by May 23.

In his 'Campaigns of the British Army at Washington and New Orleans' Lieutenant G. R. Gleig thus describes the march. The book in question has been extensively drawn on for the use of this history; but the reader is recommended to obtain a copy for himself (a cheap edition is now published) as it is in itself a full history of the regiment during the period that is to be related and is a continuation of 'The Subaltern,' which describes the doings of the regiment during the Peninsular war, and which is by the same author.

'It was on the evening of the 14th May, 1814, that the route was received, and on the following morning, at daybreak, we commenced our march. Behind us rose the Pyrenees in all their grandeur, forming, on that side, a noble boundary to the prospect; and on our left was the sea; a boundary different, it is true, in kind, though certainly not less magnificent. The houses being all in a ruinous and dilapidated condition, reminded us forcibly of the scenes of violence and outrage which had been lately acted among them. As far as the powers of vision extended, we beheld cottages unroofed and in ruins, châteaux stripped of their doors and windows; gardens laid waste, the walls demolished, and the fruit-trees cut down; whole plantations levelled, and vineyards trodden under foot. Here and there, likewise, a redoubt or breastwork presented itself; whilst caps, broken firelocks, pieces of clothing, and accoutrements scattered about in profusion, marked the spots where the strife had been most determined, and where many a fine fellow had met his fate. Our journey lay over a field of battle, through the entire extent of which the houses were not only thoroughly gutted, but for the most part were *riddled* with cannon shot. Round some of the largest, indeed, there was not a wall nor a tree which did not present evident proofs of its having been converted into a temporary place of defence, whilst the deep ruts in what had once been lawns and flower-gardens, showed that all their beauty had not protected them from being destroyed by the rude passage of heavy artillery.

'Immediately beyond the village of Bedart such spectacles were particularly frequent. It was here, it may be remembered, that in the preceding month of December there had been fighting for four successive days; and the number of little hillocks now within our view, from under most of which legs and arms were beginning to show themselves, as well as the other objects which I have attempted to describe, sufficiently attested the obstinacy with which that fighting had been maintained.

'We halted, about two hours after noon, at the village of Anglet.

'We found this village in the condition in which it was to be expected that a place of so much importance during the progress of the late siege would be found—in other words, completely metamorphosed into a chain of petty posts. Being distant from the outworks of Bayonne not more than a mile and a half, and standing upon the great road by which all the supplies for the left of the British army were brought up, no means, as may be supposed, had been neglected, which art or nature could supply, towards rendering it as secure against a sudden excursion of the garrison as might be. About one hundred yards in front of it felled trees were laid across the road, with their branches turned towards the town, forming what soldiers, in the language of their profession, term an abattis. Forty or fifty yards in rear of this a ditch was dug, and a breastwork thrown up, from behind which a party might do great execution upon any body of men struggling to force their way over that impediment. On each side of the highway again, where the ground rises into little eminences, redoubts and batteries were erected, so as to command the whole with a heavy flanking fire; while every house and hovel lying at all within the line of expected operations was loop-holed, and otherwise put in a posture of defence. But upon the fortification of the church a more than ordinary degree of care seemed to have been bestowed. As it stood upon a little eminence in the middle of the hamlet, it was no hard matter to convert it into a tolerably regular fortress, which might serve the double purpose of a magazine for warlike stores and a post of defence against the enemy. With this view the churchyard was surrounded by a row of stout palings, called in military phraseology stockades, from certain openings in which the muzzles of half a dozen pieces of light artillery protruded. The walls of the edifice itself were, moreover, strengthened by an embankment of earth to the height of perhaps four or five feet from the ground, above which narrow openings were made in order to give to its garrison an opportunity of levelling their muskets; while on the top of the tower a small howitzer was mounted, from which either shot or shell could be thrown with effect into any of the lanes or passes near. It is probably needless to add that the interior arrangements had undergone a change as striking as that which affected its exterior. Barrels of gunpowder, with piles of balls of all sizes and dimensions, now occupied the spaces where worshippers had often crowded; and the very altar was heaped up with sponges, wadding, and other implements necessary in case of an attack.

'I have been thus minute in my description of Anglet, because what has been said of it will apply more or less exactly to every village, hamlet, or cluster of cottages, within the compass of what were called the lines. It is true that neither here nor elsewhere, excepting at one particular point, and that on the opposite side of the river, were any serious intentions entertained of breaching or storming the place; and that the sole object of these preparations was to keep the enemy within his work, and to cut him off from all communication with the surrounding country. But to effect even this end, the utmost vigilance and precaution were necessary, not only because the number of troops employed on the service was hardly adequate to discharge it, but because the garrison hemmed in was well known to be at once numerous and enterprising. The reader may accordingly judge what appearance a country presented which, to the extent of fifteen or twenty miles round, was thus treated; where every house was fortified, every road blocked up, every eminence crowned with fieldworks, and every place swarming with armed men. Nor was its aspect less striking by night than by day. Gaze where he might, the eye of the spectator then rested upon some portion of one huge circle of fires, by the glare of which the white tents or rudely constructed huts of the besiegers were from time to time made visible.

'The march of the regiment was continued next day; and as it would have been considerably out of our way to go round by the floating bridge, permission was applied for, and granted, to pass directly through Bayonne. With bayonets fixed, band playing, and colours flying, we accordingly marched along the streets of that city; a large proportion of the garrison being drawn up to receive us, and the windows crowded with spectators, male and female, eager to behold the troops from whom not long ago they had probably expected a visit of a very different nature. The scene was certainly remarkable enough, and the transition from animosity to good-will as singular as it was sudden: nor do I imagine that it would be easy to define the sensations of either party on being thus strangely brought in contact with the other. The females waved their handkerchiefs, whilst we bowed and kissed our hands; but I thought I could discover something like a suppressed scowl upon the countenance of the military. Certain it is, that in whatever light the new state of affairs might be regarded by the great bulk of the nation, with the army it was by no means popular; and at this period they appeared to consider the passage of British troops through their lines as the triumphal entrance of a victorious enemy.

'As soon as we had cleared the entrenchments of Bayonne, and got beyond the limits of the allied camps, we found ourselves in a country more peaceful and more picturesque than any we had yet traversed. There were here no signs of war or marks of violence. It is impossible to describe the feeling of absolute refreshment which such a sight stirred up in men who, for so long a time, had looked upon nothing but ruin and devastation.

'The road along which we proceeded had been made by Napoleon, and was remarkably good. It was sheltered, on each side, from the rays of the sun, by groves of cork-trees mingled with fir. Our march was, therefore, exceedingly agreeable, and we came in, about noon, very little fatigued, to the village of Ondres, where the tents were pitched, and we remained till the morrow.'

From this date we propose to quote extensively from the original diary kept from day to day by Lieutenant Gleig, and which has been kindly lent by his son, Colonel A. C. Gleig, R.A. ret^{d}, himself now in his ninetieth year. Gleig's habit of keeping a diary he learnt from his friend and constant companion, Captain Grey. The latter was always known in the regiment as 'Old Grey.' He was but thirty at this time; but he had lost his hair from the effects of the Walcheren fever, which gave him a more venerable appearance than his age warranted. The diary is in very good preservation, and we think the reader will agree that, for a boy of eighteen, it is a most creditable and interesting production.

May, 1814.

Friday, 20th.—'Marched at 4 o'clock A.M. Our route to-day lay through the same immense tracts of pinewood with here and there large heathy commons, some miles in extent. We saw fewer seats and villages this day than we have yet seen, and the country is more barren. Were very much amused by the sight of countrymen walking upon stilts, at least five feet from the ground, and they appeared to walk, too, as easily as they could do on foot. It is not amusement but necessity which compels them to do so, for the country is so very flat that had they not this contrivance they never could watch their sheep. Came in at 11 o'clock to a neat village called La Bohane. Burrell caught a hare on the road which we had for dinner, and Col. Wood dined with us. Bathed in nice mill dam.'

Saturday, 21st.—'Marched at 3 over much the same kind of country as that we passed through yesterday. I never saw such extensive pine forests in the whole course of my life, nor a finer country for Cavalry to act in. Came in at 10 to a village called Muret. It is more in the English style than any I have yet seen, and conveys the idea of rural simplicity and happiness to the mind most completely. Grey and I took a very long walk together in the evening.'

Sunday, 22nd.—'Moved from our camp ground at half past two. Our march to-day was longer than usual. It was upwards of 6 leagues. The country if anything shewed rather more signs of cultivation, and we passed through several very pretty, well-inhabited villages. We halted at one where we bought a couple of fowls for 4 shillings and eggs for a shilling per dozen. We halted at 12 o'clock at a very small village called Le Barp. It is a very insignificant one, but there is a church in it. I was very anxious to go to church, but unfortunately for me there was no service in the evening. I went into the churchyard to read the epitaphs, but the only thing they mark the graves of their friends with is a little wooden cross at the head of each. The people were all delighted with the new order of things.'

Monday, 23rd.—'Marched at 3 o'clock from Le Barp. The whole of the country we passed through was barren and uncultivated until we came to a place called Belle Vue where it suddenly changed from the wildest to the finest country I ever beheld. Belle Vue is a league and a half from Bordeaux, and all the country on each side of the road from the former to the latter place is nothing but one continued vineyard interspersed with cornfields and meadows of the richest pasturage. The seats and villages are as lovely as it is possible to conceive, peeping from amid woods or rather groves of vast poplar and other kinds of trees. At 12 o'clock we began to enter the suburbs of this famous city, where we halted to refresh the men, after which we again moved on at attention. We passed through several very fine streets, and near the triumphal arch which was built for Buonaparte, but now instead of having *Vive l'Empereur*, it has *Vive le Duc d'Angoulême* on it. It is the finest thing of the kind I ever saw, and all hung with garlands of different flowers. We passed it and at last halted inside the square of the Military Hospital. It is a very fine modern building, and well fitted up for the reception of some thousands of sick. The 37th and our Regiment got quarters in it, but there were none for the Officers, so they procured billets upon the inhabitants. In the houses on which we got billets there was no room, so Grey, Burrell and I went to a Hotel and ordered dinner. While it was getting ready we walked about and saw the place. We visited the Cathedral, which is one of the finest Gothic Buildings I ever saw, not excepting those in Oxford. The town is an uncommonly fine one, and cannot be described. We dined in the Hotel, but returned to sleep in the Hospital in a large room with Col. Wood and several other Officers.'

Tuesday, 24th.—'We were in hopes we should have had one day's halt in Bordeaux to see the place, but were disappointed. At 8 o'clock we marched from our barracks, with bayonets fixed, band and bugles playing, and were not a little stared at by the inhabitants as we passed through. We went about three leagues out of the town, and encamped in a large plain near the village of Moco where the whole American force is assembling. We pitched our tent near a cottage, the people of which were very kind to us. We were mustered to-day.'

Wednesday, 25th.—'Grey and I got up at 6 o'clock and went out shooting but met with nothing. Passed a great number of Companies, in which there cannot be less than 5,000 men all on this plain. Another Brigade joined the Army to-day. On our return from shooting we observed an Ass lying dead about 100 yards from our tent, and to our no small surprise learned that it was killed last night by the wolves, which the country people say swarm about this place. We go to-night to watch for them by the dead donkey, G. and I with firelocks. . . . Watched till 12 o'clock without seeing anything, when Grey fired, and on running up to see what he had shot we found it was only a dog. We then went home.'

Thursday, 26th.—'While we were on parade a Staff Officer rode up, and told the Col. that we should embark immediately (in two days) and that the Officers were to sell their horses as soon as possible. Sent our animals off immediately but could get nothing for them.'

Friday, 27th.—'Got two months' pay, and along with Grey rode into Bordeaux. Purchased a pair of overalls and a blue great-coat for 30 dollars, and sold my horse for 13.'

Saturday, 28th.—'Got under arms at 5 o'clock and marched at 6. The country was one continued vineyard through which we passed, and we halted at 11 in the village of La Moe. It is a very pretty one and not far from the Château Margaux, famous for its wine. I got a billet on a delightful family, the Master of which had been two years a prisoner in England. He had a wife, a sister, two beautiful daughters, one 16 and the other 18 years of age, and two little boys which composed the family. They were most attentive to us (Burrell and myself), and we got the band to amuse them in the evening. Bathed in the Garonne.'

Sunday, 29th.—'To my great sorrow we marched this morning at 5 o'clock. I felt more sorrow at leaving La Moe than I have felt since I left home. We came in at 9 to the Town of Paulliac, and immediately embarked on board two transports, but the wind would not allow us to sail. Came ashore again to breakfast and employed myself all day in selling Grey's and Burrell's horses. Paulliac is a small town situated on the banks of the Garonne, which is here three leagues in breadth. There is nothing worthy of notice in it except the money of the people who gave us 8 dollars for horses which cost us 120. Slept on board to-night.'

Monday, 30th.—'All Officers came on board this morning, but we did not sail. We were very much crowded, so much so that half were obliged to sleep on deck. During the night we set sail.'

Tuesday, 31st.—'By the time breakfast was finished we found ourselves very near the Men-of-War in which we were to sail, and at

10 o'clock we came to, and boats came alongside to convey us from the transports. At one we found ourselves settled, baggage and all on board H.M. Ship *Diadem* of 64 guns. There are 40 of us in our cabin; so the crowd is immense. However, anything is better than a dirty transport. Dined and passed the night on board.'

[Captain Richard Gubbins was more lucky and writes as follows:—

'St. Michael's of the Western Isles,
'20 *June* 1814.

'We are just going to anchor at St. Michael's for 4 hours to take on water and fresh provisions.

'We sailed from Bordeaux immediately after I wrote to you, and notwithstanding the extremely crowded state of the ship, have had a tolerably pleasant passage. From this place we go to Bermuda where a Force is assembling for the purpose of acting somewhere against the Americans, but in what part or in what manner has not been made known to us yet.

'We are fortunate in having an agreeable and gentlemanly man as Captain and indeed all the officers of the ship are universally civil. The first Lieutenant has been kind enough to give me half his cabin—and I assure you that two in a cabin is somewhat preferable to 40 in a cabin.']

June 1814.

Wednesday, 1st.—'A dead calm so we cannot move. The Captain seems to be a very gentlemanly man so I daresay our time will pass pleasantly enough. Walked the deck all day.'

Thursday, 2nd.—' A delightful day but scarcely any wind. About 4 P.M. a breeze sprung up and we weighed anchor. The sight was a fine one, 9 ships of war sailing together. The breeze continued all night, and next morning we were out of sight of land.'

Friday, 3rd.—' Fell in early with a brig who made signal that she had something to communicate. The Admiral (Malcolm) bore away after her, but we continued our course. About 3 P.M. the Admiral returned but did not tell us any news, though he made us beat about without making a mile, and dispatched a brig back to Bordeaux for orders from Lord Keith. What this means I know not, but we do not proceed a mile on our course. A seaman punished.'

Saturday, 4th.—'We continue to beat about in the same way, the sea pretty rough, and the weather squally. Felt a little sick to-day, but did not eat. The Capt. dined with us to-day, but I was too ill to eat much. Went to bed.'

Sunday, 5th.—'All the Seamen and Marines paraded in their best attire, and Capt. Hanchett read prayers to them in a very solemn manner. Col. Thornton likewise read prayers to us. The day very fine. A great deal of larking in the evening.'

Monday, 6th.—'A rainy morning, cleared up by mid-day. Nothing particular except that we have scarcely any wind, and make very little way.'

Tuesday, 7th.—'A wet morning, soon cleared and a nice breeze sprung up. What sameness in a seafaring life, am thoroughly tired of it, and anxious to be ashore again anywhere.'

Wednesday, 8th.—'A beautiful sunny day, with what little wind there is in our favour. A strange sail hove in sight, which was supposed to be a frigate following us from Bordeaux. The *Pomona* was sent to reconnoitre and returned to tell that it was but a merchant-man. Col. Thornton dined with General Ross, when he found that the brig which was seen on Friday had brought intelligence of four French Frigates acting the pirate on the coast of Portugal.'

Thursday, 9th.—'A nice fresh breeze going six knots. Amused ourselves by fishing for mackerel but without success. Spent a very pleasant evening in the cockpit.'

Friday, 10th.—On getting up in the morning we saw a Sloop of war bearing down for the Admiral. Was in great hopes that she brought an order for England, but was disappointed. Playing draughts.'

Saturday, 11th.—'The wind blowing very fresh but not at all fair. About 6 o'clock P.M. the Admiral made a signal for the Fleet to close, and immediately after made another to alter our course and steer for Halifax. So the important secret is at length disclosed, and all my hopes blown up. The wind is as fair as it can blow and we go along famously.'

Sunday, 12th.—'Orderly Officer for the day. The wind has chopped round and blows rather fresh but we go three knots still. A Frigate very nearly ran foul of us in the night.'

Monday, 13th.—'On guard. The ship labours dreadfully and a good many of the soldier Officers are sick, but I am not. The Admiral made signal that we were in Long. B. 10. Nothing particular but a strange sail seen towards dark. Rained hard in the night.'

Tuesday, 14th.—'Came off guard. A fine day with little wind, nothing to be seen but sea and sky.'

Wednesday, 15th.—'A lovely day with a nice little fair breeze. Made a foolish bet of five dollars that I would go up to the mizzen yard on a single rope. Tried it and lost. Lost 2 dollars and 2 shillings more at cards.'

Thursday, 16th.—'A fine wind, but we have altered our course and are now steering for the Western Islands. Lost 3 dollars more. Spent a pleasant evening in the cockpit.'

Friday, 17th.—'A fine fair wind going five knots right before the wind.'

Saturday, 18th.—'A nice little breeze. We expect by to-morrow to be in sight of St. Miguel's. No strange sail nor anything particular during the day.'

Sunday, 19th.—'The breeze still continues, but rather less of it. Two Frigates were dispatched by the Admiral to purchase water for the fleet. No appearance of land to be discerned.'

Monday, 20th.—'The first thing that met our eyes this morning was the high land of St. Miguel's about 15 miles ahead of us. But the breeze had almost entirely died away; so we were afraid we should not be able to reach it to-day. About 11 the wind sprung up again, and we soon neared the land. It is very high and mountainous but appears to be in a very fine state of cultivation. We continued to sail along the shore near enough to view the beauties of the place till four o'clock when five others and myself agreed to go ashore in the boat, and go round by land to Ponta del Gada where we should meet the ship again. We accordingly got into the boat and after some difficulty (owing to the immense reefs of rock that ran along the coast) landed at a small town called Villa Franca. As the boat drew near the shore the beach was covered with the inhabitants who flocked down in great numbers to look at us. Immediately on our landing the Governor of the town came and asked us to go with him in English, which he spoke very fluently, though he had never been out of St. Miguel's. There was no such thing as an Inn in the place, so we very gladly went with him, and as we walked up the street, the people crowded after us as if we had been wild beasts, never having seen British Officers before. The Governor conducted us to the house of an old Gentleman who likewise spoke English, where we supped and passed the night. We tried to buy stock here but there was nothing to be had for money, so we resolved to wait till we got to Ponta del Gada, but how we were to travel was a matter of no small difficulty, as there was nothing in the shape of animal to be hired but Asses. We accordingly made up our minds to start next morning at daybreak on donkeys.'

Tuesday, 21st.—'At 5 o'clock the Asses were at the door and we set off, eleven people, eleven donkeys, and eleven drivers to go a distance of 15 miles. It was the most ludicrous scene ever I witnessed to behold us mounted on *Buros* with pack saddles and no bridles, sitting sideways with a fellow behind each of us spurring up our donkeys with a long pole and a nail at the end of it. The country through which we went was romantic and beautiful in the extreme, the road in some places winding up the highest hills cultivated to the top, sometimes down the steepest ravines, sometimes by the edge of perpendicular precipices which made one shudder to look down, and at other times out through the bosom of the mountains. Everything bears the strongest marks of the Island being entirely a volcanic production, for the strata of Lava, and even the ashes, are everywhere discernible. After one of the pleasantest rides I ever had in my life we alighted at half-past nine o'clock at the only Inn in Ponta del Gada, kept by an English woman of the name of Mrs. Currie, where we breakfasted, after which we set out to see the place. Ponta del Gada is a town which contains about twelve or fourteen thousand inhabitants. Its situation is very pretty on the seashore, and at the foot of some hills, covered with verdure to the summit. There are a great many Nunneries and Monasteries in it filled with Nuns and Monks, but none of them at all remarkable for fineness of architecture. There is no theatre nor any place of public amusement here, nor is there any considerable trade carried on. The only thing they do trade in is their oranges, which are the finest in the world, and in great abundance, 100 for a shilling. I walked about all day, visited every place worth seeing, chatted with the Nuns through the grate who sang to me and gave me flowers, went through a convent of Augustine Friars, and in short was not a single moment unemployed. There is a small fort here which I went to see, but the sentries would not admit us, and I returned on board ship at night completely fagged. The fleet is at anchor off the town.'

Wednesday, 22nd.—'Came ashore immediately after breakfast and went through the same round of adventures I had gone through the day before. Dined at 2 o'clock, and mounted a donkey to go and see a Volcano, but when we had got about a mile out of town we found that it was 27 miles distant, too far for us to go, so we merely rode a little way into the country and came back. Went to see the Nuns again, and stayed with them some time at the grate, laid in a stock of oranges and other luxuries for the voyage, and after performing various other necessary little actions, returned on board ship to sleep.'

[The nuns appear to have received a number of visitors, as Captain R. Gubbins writes as follows :

'BERMUDA,
26 *July* 1814.

'I was two days on shore at St. Michael's. This place is wretched in point of comfort and accommodation, but the country is remarkably fine and particularly well cultivated, the population is very great with a monstrous proportion of Religious of both sexes. The Friars and Monks covered the streets something like their pigs, grunting, fat, and quite as dirty.

'The poor Nuns looked through their gratings like so many wax dolls in a toyshop window, a few of the *oldest* ones were allowed to converse with us, one of them gave me a nosegay and some sweetmeats.

'I rode a good deal about the country on an ass and altogether was pleased with my trip. Our mess got a pretty good supply of fresh provisions, but, excepting oranges, fruit was not in bearing.

'From the crowded state of our ship we have suffered much from the heat, and at one time rather an alarming fever broke out on board. but with timely exertions it has been got the better of, with the loss of only one man.

'After a long tedious passage we reached the Bermudas the 24th inst.

'Our ship (*Diadem*, 64) is a miserable old tub, it has been several times ashore without being repaired and sails worse than a collier.

The officers I am sorry to tell you have not proved so good as I at first told you; they have, Captain and all, quarreled amongst themselves and some of the best leave the ship here.

'The approach to this place is extremely dangerous and in working in we ran smack on a rock and stuck fast for several hours, but the night being very fine we got off eventually with but little injury.

'The service we are going upon has not been made public, but everyone says the Chesapeake is our destination—from all I have been able to collect there are several detached forces of about 3 or 4000 men each to act in different places on the coast of America and in their principal Rivers one of which is the Chesapeake leading to Norfolk, Washington etc. We expect to reach the scene of action in about ten days or a fortnight.'

'*30th July.*

'Our sailing is put off for a day or two in consequence of a fleet with 4000 troops having just appeared in sight from Gibraltar—it is now said we must do all we have to do by the 20th August as after that time the season is so unhealthy that troops could not exist in the Chesapeake. We shall winter in all probability at this place or Halifax.']

Lieutenant Gleig's Diary Continued.

Thursday, 23rd.—'Came ashore once more for an hour or two. Wrote a hasty letter home, bought some other little articles, and returned on board at 12 o'clock again. In half an hour the Admiral hoisted the blue peter and all hands came on board, and in an hour more we were under weigh, and bade adieu to St. Michael's. We had a nice little fair breeze, and before dark the land was sinking from our view. There are two Frigates, the *Pomona* and *Menelaus*, left behind to bring up any dispatches which may follow us.'

Friday, 24th.—'On coming up on deck this morning not a vestige of land was discernible. Once more all is sea and sky. We have a fine steady breeze and go about four knots an hour. The regiment was mustered.'

Saturday, 25th.—'A very warm day with very little wind, but what is is fair. We go about two and three knots, with no strange sail nor anything else to interrupt the sameness that prevails.'

Sunday, 26th.—'The Ship's crew were assembled and prayers read to them by Capt. Hanchett. After which we likewise paraded and had Divine Service read to us by the Colonel. In the afternoon the *Pomona* and *Menelaus* appeared in view. We were all anxiety to know what news they brought, but no signal is made; so we take it for granted that they have nothing to communicate. A delightful refreshing shower in the evening.'

Monday, 27th.—'A fair wind going five knots. Saw a shark which kept following the ship. Set about preparing hooks and lines to catch him, but by the time everything was ready he was gone.'

Tuesday, 28th.—'A fair wind with weather intensely warm, the Ship sailed so badly that the rest of the Fleet ran away from us and by the evening we lost sight of them altogether.'

Wednesday, 29th.—'There was not a sign of the Fleet till about midday when the *Pomona* was seen moving down towards us, and making signal that the Admiral was astonished we did not make more sail. While we were at dinner a whirlwind took us and carried away two of our studded sail booms in a moment. The Ship was immediately put about and thus escaped further mischief. A squall then came on which sent us ten knots and the *Pomona* having now taken the transport in tow before dark we saw the Fleet lying-to for us.'

Thursday, 30th.—'A fever has broke out in the Ship which proves to be the Typhus. Six soldiers and young Eden have got it, but luckily not very dangerously. We made up with the Fleet this morning when Col. Thornton and Capt Hanchett went on board the Admiral to consult what was best to be done with the sick, when it was agreed to put them on board the Rocket Ship. The wind continues fair but we do not make the most of it.'

July, 1814.

Friday, 1st.—'The Fleet this morning lay to and the Captain sent a boat on board the Horse Ship. Upon which the Admiral ran alongside and ordered him to recall it putting him in mind pretty smartly that it was the Rocket Ship, No. 446, he had mentioned. The boat was instantly recalled and on its way back picked up a man who had fallen overboard from the *Gram*. The sick were then sent on board the 446, six in number. The Captain has opened his cabin doors for us to read in every morning. Read Smith's "Wealth of Nations."'

Saturday, 2nd.—' The fever increases, four more men have been taken ill and sent on board the Hospital ship [the Rocket Ship mentioned above]. The wind continues fair and we go about 6 knots. Amused myself with " Count Fathom." '

Sunday, 3rd.—' No divine service to-day, but why I do not know. The wind begins to waver and I fear will blow foul. 6 more men ill to-day. It is determined that the men's place shall be whitewashed to prevent if possible further contagion, and it is to be done to-morrow.'

Monday, 4th.—' The wind is dead foul. They have whitewashed the place below but that will not do. 7 more men and a woman are taken with the fever. They have now resolved to knock down the standing berths and make the men sleep in hammocks which work they have to-day commenced. We now lose instead of make way.'

Tuesday, 5th.—' A dead calm, weather oppressively sultry. Only one soldier and two sailors taken ill to-day; so the knocking down of the berths seems to have been of use. Fishing for Dolphins to-day and one was caught. The change of colour when it is dying is uncommonly beautiful though not so fine as expected. Read " Humphrey Clinker." '

Wednesday, 6th.—' Not a breath of air to alleviate the intense heat which annoys us. The awning is of very little use against the powerful beams of the sun. Nothing particular to-day except 4 more patients.'

Thursday, 7th.—' Weather the same as yesterday. One Sergeant and a man taken ill. The Launch is taken away and is now in tow of another ship, and in the place where it was they have erected a hospital with spars and sail, where the sick are swung in hammocks. The poor dogs are sent off to the Horse Ship. Playing cards in the gun-room in the evening.'

Friday, 8th.—' Still calm and very sultry. A great number of Dolphins playing about the Ship, but they were too wary to take any of the numerous baits that were thrown out to them. Played again and as usual lost. I shall play no more.'

Saturday, 9th.—' A dead calm. Fishing all the morning but without success though the Dolphins were playing about in crowds. A boat was lowered at three P.M. and I with several others bathed overboard, the water was delightful. When we were going to bed an unusual bustle arose upon deck the cause of which was three men falling overboard. Both life-buoys were cut away, but the men having laid hold of ropes were fortunately pulled up without injury.'

Sunday, 10th.—'Divine service as usual. A hot calm day. Nothing whatever to-day worthy of notice.'

Monday, 11th.—'A meeting of the Mess was proposed to-day but was broken off by the Capt. appearing on parade and informing the Colonel that a sailor had been robbed of a £5 note, several shirts and some pairs of trousers. A search was immediately set on foot and after a great deal of trouble the thief was discovered, a man of Cottingham's Company. Col. Wood, Knox, De Bathe and Sir Peter Porter of the *Menelaus* and Capt. Alexander of the *Devastation* came on board and dined with the Capt. The band played upon the poop and the *Menelaus* coming alongside its band played likewise and continued to do so till 10 o'clock most beautifully.'

Tuesday, 12th.—'Immediately after parade all the Officers adjourned to the ward-room where a Mess meeting was held, which occasioned a great deal of disputation and no small amount of amusement. This occupied us till dinner time. Light unsteady breeze but we make a little way.'

Wednesday, 13th.—'Some Dolphins making their appearance at the head of the Ship the Capt. took his grains (harpoon with five prongs) and struck at them from the Bowsprit but without touching any. Read "Don Quixote" in his cabin. No wind till about nine o'clock when a five knot breeze sprung up and continued all night.'

Thursday, 14th.—'We were this morning well up with the Fleet, but gradually fell astern again. Two strange sail appeared ahead which were stopped and boarded by the Admiral, but all that we could learn respecting them was that one was a Spanish Ship. We were too far astern ourselves to speak any of them. We go about three knots.'

Friday, 15th.—'Orderly Officer for the day. Busy all day in getting the men's deck washed and well scrubbed with the holy-stone. Towards night the breeze freshened again and put us on about five.'

Saturday, 16th.—'On guard. The most troublesome piece of duty I ever performed. Walked the deck all day and night without closing my eyes. The Col. and Capt. dined with the Admiral to-day and on their return told us that the Ship we could not make out was an English

Merchant-man from Leghorn to the West Indies, which brought intelligence of five sail of the line, some Frigates, and a large fleet of transport coming from the Mediterranean to join us.'

Sunday, 17th.—'A number of water-spouts this morning but none near us. At 10 o'clock a heavy squall came on and took the Admiral all aback. The Ships ahead felt it all so that they were obliged to alter their course and we to avoid the danger did the same. The squall soon blew over but the wind blows right against us. Divine service as usual.'

Monday, 18th.—'A little breeze and rather fair. The whole day taken up with a Mess meeting. A mild serene evening. We knocked up a ball in the ward-room and danced with each other till a late hour, Col. Thornton joining us.'

Tuesday, 19th.—'A nice fair wind going 5 knots. In the evening the Admiral lay to and a number of Officers from all the Ships went on board the Royal boat to see a play. It was "The Apprentice" and went off remarkably well. The *Dramatis Personæ* were Officers of the Navy and Artillery. The quarter deck was fitted up as the stage and illuminated with lamps. The fête concluded with a ball, which was kept up till midnight, when a blue light was hoisted at the masthead, when the boats reported to the Admiral and took us all back to our respective ships.'

Wednesday, 20th.—'The wind freshened very much, going at the rate of 7, 8 & 9 knots. The heat is most oppressive and the sky lowering and cloudy, and a new moon. Everything seems to portend squally and unpleasant weather. No strange sail nor anything particular to-day.'

Thursday, 21st.—'Several very violent squalls and the sky so threatening that the people could scarcely leave the deck all day, but no injury done. A number of the Officers sick and among the rest myself. Dined with the Capt. In the middle of night I was awoke by the ship giving a violent heave, succeeded immediately by a tremendous flash of lightning and clap of thunder. The storm continued a very short time, but while it did last it was most violent. One flash fell so near us that I was sure we were struck. It blew over however without our suffering the smallest injury.'

Friday, 22nd.—'The wind has fallen again and we have another calm, so much so that we scarcely go a knot. The *Menelaus* has got a great way ahead, and the *Regulus* is sent on to Bermuda to prepare things for our arrival.'

Saturday, 23rd.—' No less than 7 strange sails in sight this morning, which prove to be the Mediterranean Fleet. We were informed by a stranger from Bermuda that Sir T. Cochrane waits there for us. There is very little wind and we lay to all night.'

Sunday 24th.—' Land in sight. We were mustered as usual. While we were yet a good way at sea we observed a transport with the rockets and our sick on board running right upon a reef of rocks. A signal was made for her to tack, but she took no notice. Again we fired, then two together without effect, when a whole broadside was fired at her upon which she put about. We were now drawing fast towards the land, the appearance of which is beautiful. It is very low and covered with cedars, which look uncommonly well at a distance. The navigation is without exception the most intricate I ever met with, so much so that though we had a good pilot on board, a clear day and nice little breeze, we first struck and then went aground. Most fortunately for us the wind did not increase. We fired guns of distress, and soon had a number of boats about us, but in spite of all the exertions that were made, it was four in the morning before we got off. The rocket ship went aground also, and did not get off as soon as we.'

Monday, 25th.—' Anchored this morning opposite the Tanks and about four miles from the town of St. George's. Grey and Burrell went on shore, and I remained on board all day. The prospect of the land to eyes which have so long seen nothing but sea and sky is delightful, adorned as it is with numberless white villas peeping from amid groves of cedar, but what above all things took my fancy most was the number of Blacks and all speaking English.'

Tuesday, 26th.—' Went ashore at 6 A.M. with Col Thornton. When the boat once enters what is called the ' Ferry,' the sail all the rest of the way to St. George's is truly picturesque. The little white houses, the groves of cedar mixed with bare rocks, which present themselves on each side of you are beyond description. St. George's is a small but very pretty town, every house in which is as white as the driven snow, with nothing in it whatever worthy of description. Breakfasted at the Hotel along with a number of Officers, after which I went and brought the dogs ashore. The heat is so oppressive there is no stirring out till after sunset with any degree of comfort. Went in the evening up to a hill on which the signal station is and got a view of all the Islands of which there is no smaller number than 365. Dined and slept at the Inn.'

Wednesday, 27th.—' Got up in the morning and bathed. Went to see the Tanks which are worth seeing. What an inhuman thing is slavery. Almost every Black person here is a slave, and as the people in Bermuda are too indolent to do anything to procure their own livelihood those miserable wretches are obliged to support them with the produce of their labour at the oar, or any other work which they do. Bermuda is the worst cultivated most barren place I was ever in. It produces nothing but cedar trees, and even those very small. All the fruit and everything they have the people get from America or the West Indies. The place does not even produce grass enough to support the few horses that are on it. The Garrison is composed of a detachment of 6 or 700 of the 4 Government Battalions, 30 Artillerymen and a few Artificers. I stayed to lunch and then called for my bill, which came to £6, a complete imposition. Paid it and went on board.'

Thursday, 28th.—' I did not intend going ashore again, when I was ordered on a watering party. Went ashore to the wells, filled the boats and came off again, bringing with me a little black boy who had run away from his Master. Wrote home, and passed the rest of the day quietly on board. Bathed in the evening.'

Friday, 29th.—' The Mediterranean Fleet off the Island. It consists of six Frigates and 30 sail of Transports having the 21st, 29th and 62nd Regiments on board. A man came on board looking for a runaway slave. The Ship was searched and my boy found, but he proved not to be the one, though the man knew his mistress, so am sure to be found out, and if so, fined £1000. Remained on board all day.'

Saturday, 30th.—' The Fleet this morning began to enter the roads, and at 4 o'clock were all anchored round us. About this time our Company was ordered to get ready to go on board another Ship. This was soon done and by 7 in the evening we found ourselves on board the *Golden Fleece* Transport, No. 247. There are two Officers of Engineers, Capt Blanchette and Lieut. West, with a Company of Sappers on board. Went ashore and walked about for an hour by moonlight, came back again and slept on board.'

Sunday, 31*st.*—' Another Company came here this morning with the Officers, making our Mess 8 in number. Bathed before breakfast. I was obliged to send my boy ashore to save myself from being fined. Passed the day on board and bathed again in the evening.'

August, 1814.

Monday 1*st.*—' Spent the morning on board and took an early dinner with the intention of going to see the caverns.'

Wednesday 3*rd.*—' When we got up this morning they were warping the Ship out as there was no wind whatever, but about 12 o'clock when we had cleared the harbour or creek we cast anchor again. We did not remain long quiet; for the signal was made for the whole fleet to weigh, which was soon done, and we stood to sea. Farewell Bermuda, I quit you without one single pang of regret. Your soil is uncultivated, your shores are barren and your inhabitants a set of lazy indolent rogues who live by preying upon every stranger that visits there. May I never see you again! The breeze, if so it may be called, was so very light, that though the land is very low we still saw it just before dusk. Grey harpooned a small shark about 3 feet long, which was given to the men. In the cool of the evening we amused ourselves by rowing about for an hour.'

Thursday, 4*th.*—' Nothing to be seen this morning of the vile shores of Bermuda. The *Dictator* and two Frigates are missing, the rest of the Fleet is well together. The sky this morning looked very treacherous, and it was not long before we experienced one of the Mucrean squalls as they are called. We saw it coming, so were prepared and suffered nothing. This sort of weather continued all day, and the ship rocked about so much that I had great difficulty in keeping from getting sick.

Friday, 5*th.*—' A pretty stiff breeze at first against us, but about 9 o'clock it became fair and sent us at the rate of 5 knots. The day is very rainy, so there is very little comfort in the ship. The only one we have is the pleasing idea that our distance shortens very considerably every hour.'

Saturday, 6*th.*—' A fine fair steady five-knot breeze. Nothing particular occurred all day except that the *Dictator* and the two missing Frigates made their appearance again and soon rejoined the

Fleet. While we were at dinner a message came for the Doctor, as one of the Sailors had fallen from the fore-yard, and had hurt himself very severely. We all ran up, and found the poor fellow perfectly insensible and apparently in the agonies of death. The Doctor bled him copiously, which soon brought him to himself again. He had no external signs of violence, so he was put to bed and is doing well.'

Sunday, 7th.—' Thank God the fair wind still continues, and we are not much more than 300 miles from the mouth of the Chesapeake. Divine service was read by Capt Blanchett. The Master and Agent for Transports dined with us.'

Monday, 8th.—' The fair wind still continues, and everything goes on in the same manner day after day.'

Tuesday, 9th.—' The wind has fallen and a perfect calm prevails, but the appearance of the sky is truly alarming. It is not that of a passing squall, but of a determined hurricane. The sea is troubled and black and the waves are breaking in white foam although there is not a breath of wind. This appearance continued till the evening when to our agreeable surprise it went gradually off without our experiencing anything of what we expected, and a nice little fair breeze sprung up. A shark of an immense size came close to the stern of the ship, but we did not get the bait out in time enough, so he got away.'

Wednesday, 10th.—' Again it is a calm, and oppressively hot beyond anything I ever felt. Nothing particular all day. A little wind in the evening.'

Thursday, 11th.—' The breeze continued all day, by which means we made a considerable way. The whole of our life now consists in eating drinking and sleeping. The wind died away again at night.'

Friday, 12th.—' A dead calm, heat excessive. This is the Prince of Wales' birthday and first day of grouse shooting. We kept it up accordingly and most of us drank rather too freely. Nothing worth mentioning during the day.'

Saturday, *13th.*—' A great many fish about the head of the Ship. We tried to harpoon them but without effect. Grey at last struck his grains into a sunfish but it was too strong, broke the grains and got off.'

Sunday, 14th.—' Very calm in the morning but towards noon a breeze sprung up. We have no less than 7 cases of fever on board. While we were at dinner there was a cry of land, and on running up we saw the low coast of America plain enough. Continued our course till dark when we brought to close to the shore.'

Monday, 15th.—' A fine steady breeze which carries us proudly up the Chesapeake. The sight is glorious, an English Fleet standing up an enemy's Bay with all sail set, filled with troops panting to meet that enemy. We could see very little of the shore, the Bay is so wide; however we could discern a village or two and some windmills. The coast is very woody and sandy. At night all the Fleet once more anchored.'

Tuesday, 16th.—' Weighed again at daylight, and by the time we got up we were again at anchor beside Admiral Cockburn. We had now a very fine view of the mouth of the Potomac and of the coast beside it, the beauty of which was heightened by a Frigate coming down with all sail set. A signal was very soon made to weigh again which was done, and we set sail for the Patuxent with a fine fair wind. The sky looked lowering but the storm blew over and we went on till dark when we again anchored off the Patuxent.'

Wednesday, 17th.—' The wind this morning was very foul, so we could only beat off and on without attempting to go in. This is most provoking, but there is no help for it.'

Thursday, 18th.—' The wind this morning rather better, so we stood in to the river. The sail up surpasses even that up the Thames, the woods are so fine, the cottages so beautiful, and the cultivation so rich. We now got everything ready for landing, our men accoutred, and our baggage packed, but by the time the ships had gone as far up as they could, the day was too far spent to do anything. Hamilton and Gascoigne came on board and passed the night with us.'

Friday, 19th.—' At daybreak we were roused up as the boats were alongside to carry us ashore. We got up, took each his haversack containing three days' provisions, a spare shirt, pair of stockings, towel &c., myself with a blanket over the other shoulder prepared for disembarkation. Unluckily there were not boats enough, so we were left on board till 7 o'clock when they at last procured for us a punt, and with a little boat to tow us we set out to pull a distance of 15 miles. The tide was running strong against us and what with that and the weight of the punt, we never should have got to our journey's end had we not met a launch of the *Sovereign's* which took us on board. As it was, it was one o'clock before we landed. We landed at a small deserted village called St. Benedict's where we remained till the rest of the Company with Burrell came ashore. We then set out and marched about a mile to some heights where the light Brigade were barracking. Procured three ducks on which together with our salt meat we made a very hearty dinner. The afternoon was devoted to a proper distribution of the force; which was divided into three brigades, in the following order:—

'The first, or light brigade, consisted of the 85th, the light infantry companies of the 4th, 21st, and 44th regiments, with the party of disciplined negroes, and a company of marines, amounting in all to about eleven hundred men; to the command of which Colonel Thornton of the 85th regiment, was appointed. The second brigade, composed of the 4th and 44th regiments, which mustered together fourteen hundred and sixty bayonets, was intrusted to the care of Colonel Brooke, of the 44th; and the third, made up of the 21st, and the battalion of marines, and equalling in number the second brigade, was commanded by Colonel Patterson, of the 21st. The whole of the infantry may, therefore, be estimated at four thousand and twenty men. Besides these, there were landed about a hundred artillery-men, and an equal number of drivers; but for want of horses to drag them, no more than one six-pounder and two small three-pounder guns were brought on shore. Except those belonging to the General and staff-officers, there was not a single horse in the whole army. To have taken on shore a large park of artillery would have been, under such circumstances, absolute folly; indeed, the pieces which were actually landed, proved in the end of very little service, and were drawn by seamen sent from the different ships for the purpose. The sailors, thus employed, may be rated at a hundred, and those occupied in carrying stores, ammunition, and other necessaries, at a hundred more; and thus, by adding these, together with fifty sappers and miners, to the above amount, the whole number of men landed at St. Benedict's may be computed at four thousand five hundred.

'This little army was posted upon a height which rises at the distance of two miles from the river. In front was a valley, cultivated for some way, and intersected with orchards; at the farther extremity of which the advanced piquets took their ground, pushing forward a chain of sentinels to the very skirts of the forest. The right of the position was protected by a farm-house with its enclosure and outbuildings, and the left rested upon the edge of the hill, or rather mound, which there abruptly ended. On the brow of the hill, and about the centre of the line, were placed the cannon, ready loaded, and having lighted fusees beside them; whilst the infantry bivouacked immediately under the ridge, or rather upon the slope of the hill which looked towards the shipping, in order to prevent their disposition from being seen by the enemy, should they come down to attack. But as we were now in a country where we could not calculate upon being safe in rear, any more than in front, the chain of piquets was carried round both flanks, and so arranged, that no attempt could be made to get between the army and the fleet, without due notice, and time given to oppose and prevent it. Everything, in short, was arranged with the utmost skill, and every chance of surprise provided against; but the night passed in quiet, nor was an opportunity afforded of evincing the utility of the very soldier-like dispositions which had been made.

'Such is the little army with which we invade America. Burrell, Codd and I slept together under a tree in B.'s cloak and my blanket.'

Saturday, 20th.—'Got up an hour before daylight and fell in to march, but something happening that rendered a delay necessary we were dismissed again. After breakfast Codd and I set out in search of something to eat and walked two miles beyond our pickets. Went into several houses which were all deserted, except one, where we found one unfortunate woman completely plundered. We took one fowl for which we gave her a quarter dollar and came back to the tree where we found a pig, a goose and a couple of fowls. Just as we had sat down to a nice dinner the bugle sounded the assembly and we were obliged to leave our meal and fall in. The Brigade formed and moved to the road, when General Ross passed us, and we cheered him as he passed. We then moved on, and marched a distance of 4 miles.'

From the History.

'The march was conducted with the caution and good order that had marked the choice of ground for encamping and the disposition of the troops in position. The advanced guard, consisting of three companies of infantry, led the way. These, however, were preceded by a section of twenty men, moving before them at the distance of a hundred yards ; and even these twenty were but the followers of two files, sent forward to prevent surprise, and to give warning of the approach of the enemy. Parallel with the head of the three companies marched the flank patrols ; parties of forty or fifty men, which, extending in files from each side of the road, swept the woods and fields to the distance of nearly half a mile. After the advanced guard, leaving an interval of a hundred or a hundred and fifty yards, came the light brigade ; which, as well as the advance, sent out flankers to secure itself against ambuscades. Next to it, again, marched the second brigade, moving steadily on, and leaving the skirmishing and reconnoitring to those in front ; then came the artillery, consisting, as I have already stated, of one six and two three-pounder guns, drawn by seamen ; and last of all came the third brigade, leaving a detachment at the same distance from the rear of the column, as the advanced guard was from its front.

' In moving through an enemy's country, the journeys of an army will, except under particular circumstances, be regulated by the nature of the ground over which it passes : thus, though eight, ten, or even twelve miles may be considered as a short day's march, yet if at the end of that space an advantageous position occur (that is, a piece of ground well defended by natural or accidental barriers, and at the same time calculated for the operations of that species of force of which the army may be composed), it would be the height of imprudence to push forward, merely because a greater extent of country might be traversed without fatiguing the troops. On the other hand, should an army have proceeded eighteen, twenty, or even twenty-five miles, without the occurrence of any such position, nothing except the prospect of losing a large proportion of his men from weariness ought to induce a General to stop until he has reached some spot at least more tenable than the rest. Our march to-day was extremely short, the troops halting when they had arrived at a rising ground distant not more than six miles from the point whence they set out ; and having stationed the piquets, planted the sentinels, and made such other arrangements as the case required, fires were lighted, and the men were suffered to lie down.

' It may seem strange, but it is nevertheless true, that during this short march of six miles a greater number of soldiers dropped out of the ranks, and fell behind from fatigue, than I recollect to have seen in any march in the Penisula of thrice its duration. The fact is that the men, from having been so long cooped up in ships, and unused to carry their baggage and arms, were become relaxed and enervated to a degree altogether unnatural; and this, added to the extreme sultriness of the day, which exceeded anything we had yet experienced, quite overpowered them. The load which they carried, likewise, was far from trifling, since, independent of their arms and sixty rounds of ball-cartridge, each man bore upon his back a knapsack, containing shirts, shoes, stockings, &c., a blanket, a haversack, with provisions for three days, and a canteen or wooden keg filled with water. Under these circumstances, the occurrence of the position was extremely fortunate, since not only would the speedy failure of light have compelled a halt, whether the ground chanced to be favourable or the reverse, but even before darkness had come on scarcely two-thirds of the soldiers would have been found in their places.

'The ground upon which we bivouacked, though not remarkable for its strength, was precisely such as might tempt a General to halt, who found his men weary, and in danger of being benighted. It was a gentle eminence, fronted by an open and cultivated country, and crowned with two or three houses, having barns and walled gardens attached to them. Neither flank could be said to rest upon any point peculiarly well defended, but they were not exposed; because, by extending or condensing the line, almost any one of these houses might be converted into a protecting redoubt. The outposts, again, were so far arranged differently from those of yesterday, that, instead of covering only the front and the two extremities, they extended completely round the encampment, enclosing the entire army within a connected chain of sentinels; and precluding the possibility of even a single individual making his way within the lines unperceived.'

Diary Continued.

'But the weather was so immensely hot, and I was so loaded, that I was more fagged with that short march than ever I was before. I went on picket when we came in. A tremendous thunderstorm with heavy rain.'

Sunday, 21st.—'Marched at daylight a distance of 10 miles over a very sandy road. Our Company happened to form the Advanced-guard, but were afterwards changed into the flank-guard. We had a great deal of scampering through the woods, and took two militia men.'

This incident is more fully described in the book as follows :

'In the course of this day's march a little adventure occurred to myself, which, in the illiberality of my heart, I could not but regard as strikingly characteristic of the character of the people to whom we were now opposed, and which, as at the time it had something in it truly comical, I cannot resist the inclination of repeating, though aware that its title to drollery must in a great measure be lost in the relation. Having been informed that in a certain part of the forest a company of riflemen had passed the night, I took with me a party of soldiers, and proceeded in the direction pointed out, with the hope of surprising them. On reaching the place, I found that they had retired, but I thought I could perceive something like the glitter of arms a little farther towards the middle of the wood. Sending several files of soldiers in different directions, I contrived to surround the spot, and then moving forward, I beheld two men dressed in black coats, and armed with bright firelocks and bayonets, sitting under a tree ; as soon as they observed me, they started up and took to their heels, but being hemmed in on all sides, they quickly perceived that to escape was impossible, and accordingly stood still. I hastened towards them and having arrived within a few paces of where they stood, I heard the one say to the other with a look of the most perfect simplicity, "Stop, John, till the gentlemen pass." There was something so ludicrous in this speech, and in the cast of countenance which accompanied it, that I could not help laughing aloud ; nor was my mirth diminished by their attempts to persuade me that they were quiet country people come out for no other purpose than to shoot squirrels. When I desired to know whether they carried bayonets to charge the squirrels, as well as muskets to shoot them, they were rather at a loss for a reply ; but they grumbled exceedingly when they found themselves prisoners, and conducted as such to the column.'

Diary Continued.

'Halted on a pretty little green hill, well wooded, where we set a-cooking. Mustasek caught a little leveret which we boiled for dinner, but as usual had just sat down to it, when we were obliged again to fall in and march. A few shots were fired at the advanced guard and we were all in hopes that the Yankeys would make a stand but they disappointed us. We came in at dusk to a good-sized village called Nottingham, where the enemy's flotilla had been, but on our approach they moved higher up the river. The boys caught a parcel of turkeys and geese on which we made a capital supper and slept under the shade of an old barn full of Tobacco.'

Monday, 22nd.—' Fell in an hour before daylight but did not march till past 7 o'clock. The road was well wooded, so the rays of the sun did not oppress us so much as usual. Our advanced people had some little affair with the enemy's Cavalry, in which no one was killed. Came in at one o'clock to a nice village called Marlboro, where we bivouacked in a large green field. Got plenty of fowls and for once eat a hearty dinner undisturbed. Chose a snug situation with some trees, where we brought some hay, and passed the night very comfortably.'

Tuesday, 23rd.—' Under arms as usual before day. Walked up into the village where I procured some tea and sugar, and got a bottle of milk from a gentleman of the name of Dr. Bean. He remains quietly in his house, has his property respected and does not seem inimical to us. About twelve o'clock the assembly sounded and we advanced, our Company again forming the advanced guard. We went on without interruption till we had gone about four miles, when some Americans whom we could not see fired a volley at the General, the balls of which came about us. We pushed on and saw a party drawn up on some heights, from which they threw two round shot very correctly at us. They did not wait for us, but made off, the moment we advanced. We remained for some time on the heights they had occupied, about 200 strong, all the rest of the army having filed to left, and then filed to the left also, and overtook them encamped at a place called Wood-yard. Our company went on picket, and Grey and I got into a suspect house where all the inhabitants remained. They provided a capital supper for us with tea, and feather beds to sleep in. But the latter we never enjoyed. For scarcely had we lain down, when Burrell came and told us the enemy were surrounding us. Our situation was most ticklish. We were fully two miles from the camp and with only 16 men We had received orders to remain where we were till daylight, but if we did so there seemed to be every chance of our being taken. We were under arms the whole of the night, and the anxiety I endured is past description, alarmed as we were first by the arrival of Col. Wood who had been in the rear, and afterwards with that of some sailors with rum. At last to our no small relief the hour of departure arrived.'

Wednesday, 24th.—'At 3 in the morning we fell back to Burrell's picket, and at daybreak joined the regiment. Tired as we were without having had any sleep all night we set out upon a march, and I was never so tired at the end of a march as I was at the beginning of this. Moved off the ground about 5 o'clock and went on without halting till 9 when we halted for an hour and then went on again. When it was drawing towards 12 o'clock we could perceive heavy clouds of dust rising, and in half an hour more we saw the Yankeys drawn up on the heights above the village of Bladensburg. We had marched at least a distance of 14 miles exposed to the rays of a scorching sun, and we were so tired that literally it was with difficulty we could drag one leg after the other. However, notwithstanding this, the General resolved to attack immediately. In front of the enemy's line ran a branch of the Potomac. The only way by which we could cross was a narrow wooden bridge exposed to the fire of a two-gun battery. To this we advanced, Thornton leading us on in the most gallant style. The moment our front Company shewed itself, they opened upon it and knocked down seven men of the 85th, but the rest of us moved as fast as weariness would admit, and got across without losing a man. Our Company then filed to the left, and after throwing away our packs advanced in as good a line as the ground would admit. We moved on through some wood without receiving a shot, till of a sudden we came to an open sloping field at the top of which at least one Battalion was drawn up in line. The moment we shewed ourselves they fired a tremendous volley at us, which they continued while we advanced but slowly, firing only now and then. Here we remained full half an hour without being able to advance and scorning to retire with about 20 men opposed to at least 700. Here fell poor Codd, gallantly leading the way, and here poor Hamilton was killed, in short every one of our Officers were hit in this field. There was one time when I thought we should not succeed and I literally shed tears of vexation to see our poor fellows exposed to such a fire, and totally unsupported. But at this critical moment the General arrived with more men, and the enemy began to give way as their left was turned. At that we set up a shout and rushed on the Yankeys now flying in all directions before us. In this field I got the left arm of my jacket torn with a ball, and received a slight wound in the thigh. I followed the enemy for upwards of a mile when I dropped down with fatigue and thirst. Luckily there was a pool of water near, and after resting for an hour or so I collected what men I could, and marched back to join the Regiment. On this day fell the flower of the 85th Officers and men. We lost the two best Sergeants, McDermott and Higgins, and 18 of our best men out of the Company. By the time the scattered remains of the Company was collected, and everything prepared it was dark; but this did not prevent our moving, for we marched directly for Washington. On coming near it the spectacle was very grand. The Naval stores, a large Frigate and the Capitol were all on fire. We passed the night on a large green a little way out of the town exposed to a tremendous thunderstorm which came on towards morning.'

A general description of the battle is here given, extracted from the History:

'The position occupied by the Americans was one of great strength and commanding altitude. They were drawn up in three lines on the brow of a hill, having their front and left flank covered by a branch of the Potomac, and their right resting upon a thick wood and a deep ravine. This river flowed between the heights occupied by the American forces and the little town of Bladensburg. Across it was thrown a narrow bridge, the road from which passed through the very centre of their position; and its right bank, above which the enemy was drawn up, was covered with a narrow strip of willows and larch trees, whilst the left was altogether bare, low, and exposed.

'Amongst the willows and larch trees the Americans had stationed strong bodies of riflemen, who, in skirmishing order, covered the whole front of their army. Behind this plantation, again, the fields were open and clear, intersected, at certain distances, by rows of high and strong palings. About the middle of the ascent, and in rear of one of these rows, stood the first line, composed entirely of infantry; at a proper interval from this, and in a similar situation, stood the second line; while the third, or reserve, was posted within the skirts of a wood, which crowned the heights.

'The artillery, again, of which they had twenty pieces in the field, was thus arranged: on the high road, and commanding the bridge, stood two heavy guns; and four more, two on each side of the road, swept partly in the same direction, and partly down the whole of the slope into the streets of Bladensburg. The rest were scattered with no great judgment, along the second line of infantry; while the cavalry showed itself in one mass, within a stubble field, near the extreme left of the position. Such was the formidable position in which they awaited our approach; amounting, by their own account, to nine thousand men (8000 infantry, 700 cavalry, and 22 guns), a number exactly doubling that of the force which was to attack them.

'As soon as we arrived in the streets of Bladensburg and within range of the American artillery several of their guns opened upon us, no time was allowed for the stragglers to regain their places, or for a ford across the river to be discovered, but the order to attack was given, and the Light Brigade (consisting of the 85th Light Infantry under Lieut.-Colonel Wood and the Light Companies of the 4th, 21st and 44th; the whole force being under Colonel Thornton) pushed on at double-quick time towards the head of the bridge, the advance guard being under the command of Major Brown of the 85th.

'While we were moving along the street, a continued fire was kept up with some execution from those guns which stood to the left of the road; but it was not till the bridge was covered with our people that the two-gun battery upon the road itself began to play. Then, indeed, it also opened, and with tremendous effect; for at the first discharge almost an entire company was swept down; but whether it was that the guns had been previously laid with measured exactness, or that the nerves of the gunners became afterwards unsteady, the succeeding discharges were much less fatal. The riflemen likewise began to gall us from the wooded bank with a running fire of musketry; and it was not without trampling upon many of their dead and dying comrades that the Light Brigade established itself on the opposite side of the stream. When once there, however, everything else appeared easy.

From the Original Map in the Possession of the Royal United Service Institution.

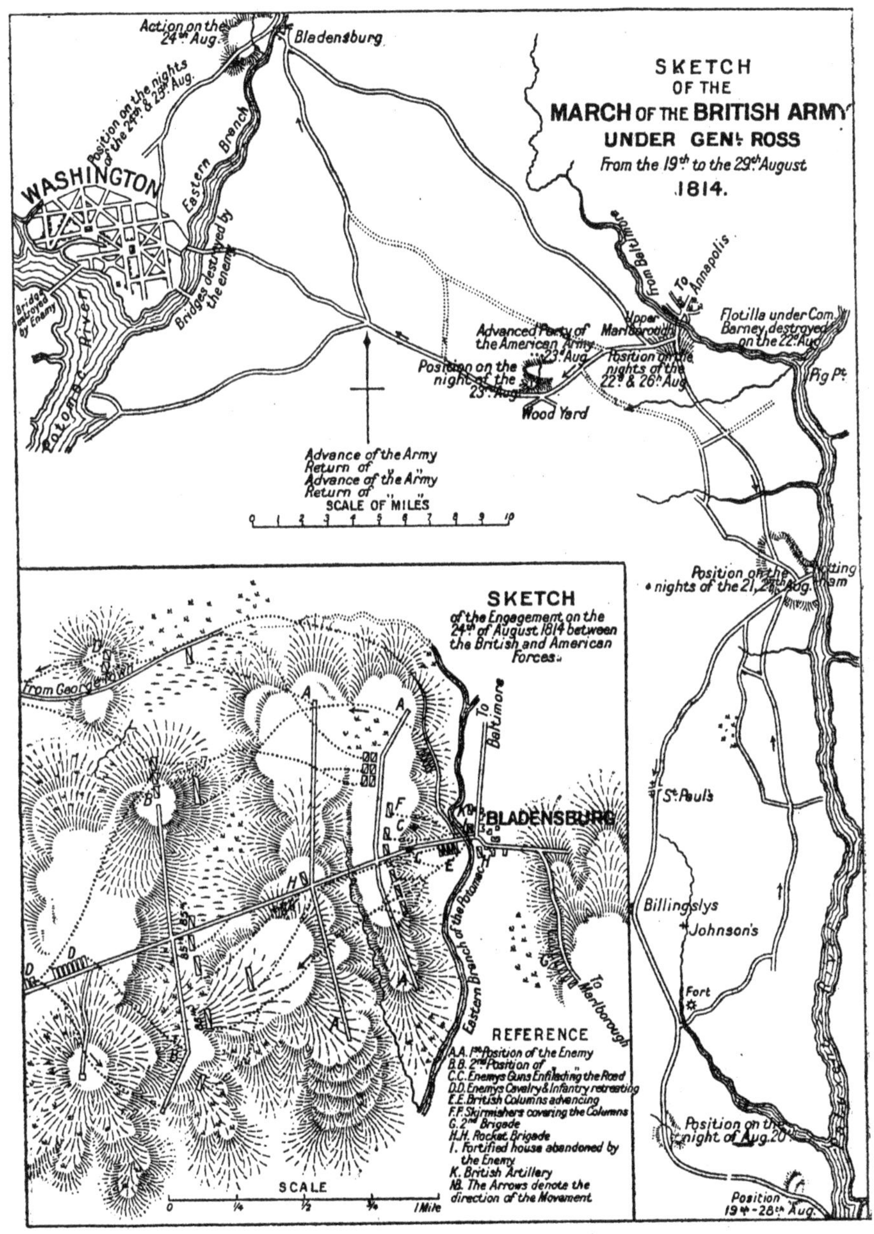

Wheeling off to the right and left of the road, they dashed into the thicket, and quickly cleared it of the American skirmishers; who, falling back with precipitation upon the first line, threw it into disorder before it had fired a shot. The consequence was that our troops had scarcely shown themselves when the whole of the line gave way and fled in the greatest confusion, leaving the two guns upon the road in possession of the victors.

'But here it must be confessed that the Light Brigade was guilty of imprudence. Instead of pausing till the rest of the army came up, the soldiers lightened themselves by throwing away their knapsacks and haversacks; and extending their ranks, so as to show an equal front with the enemy, pushed on to the attack of the second line. The Americans, however, saw their weakness, and stood firm, and having the whole of their artillery, with the exception of the pieces captured on the road, and the greater part of their infantry in this line, they first checked the ardour of the assailants by a heavy fire, and then, in their turn, advanced to recover the ground they had lost. Against this charge the extended order of the British troops would not permit them to offer an effectual resistance, and they were accordingly borne back to the very thicket upon the river's bank, where they maintained themselves with determined obstinacy, repelling all attempts to drive them through it, and frequently following, to within a short distance of the cannon's mouth, such parts of the enemy's line as gave way.

'In this state the action continued till the second brigade had likewise crossed and formed upon the right bank of the river; when the 44th Regiment moving to the right, and driving in the skirmishers, debouched upon the left flank of the Americans, and completely turned it. In that quarter, therefore, the battle was won; because the raw militia-men who were stationed there, as being the least assailable point, when once broken could not be rallied. But on their right the enemy still kept their ground with much resolution; nor was it until the arrival of the 4th Regiment, and the advance of the British forces in firm array to the charge, that they began to waver. Then indeed, seeing their left in full flight, and the 44th getting in their rear, they lost all order, and dispersed, leaving clouds of riflemen to cover their retreat; and hastened to conceal themselves in the woods, where it would have been madness to follow them. The rout was now general throughout the line. The reserve, which ought to have supported the main body, fled as soon as those in its front began to give way ; and the cavalry, instead of charging the British troops, now scattered in pursuit, turned their horses' heads and galloped off, leaving them in undisputed possession of the field and of ten (twelve ?) out of the twenty pieces of artillery.

'This battle, by which the fate of the American capital was decided, began about one o'clock in the afternoon and lasted till four. The loss on the part of the English was severe, since, out of two-thirds of the army which were engaged, upwards of five hundred men were killed and wounded. Amongst the former were Captain Hamilton, Lieutenant Codd, 1 Sergeant and 12 Rank and File of the 85th, and amongst the latter Colonel Thornton, Lieut.-Colonel Wood, Major Brown, Lieutenants Williams, Burrell, Maunsell, O'Connor, Gascoigne, Hickson, Gleig, Crouchley, 2 Sergeants, and 51 Rank and File, and 6 Rank and File missing.'

'Our troops being now worn down from fatigue and of course ignorant of the country, while the Americans were the reverse, the pursuit could not be continued to any distance. The defeat, however, was absolute and Washington was now at our mercy. The third brigade which had been in reserve pushed on rapidly towards Washington. Halting outside the city a flag of truce was sent in to dictate terms. The party bearing the flag, accompanied by the General himself, had scarcely entered the street when they were fired upon from the windows of one of the houses, the General's horse being killed. Such unjustifiable conduct roused the indignation of every individual. The troops forthwith advanced into the town, having first put to the sword all who were found in the house from which the shots were fired. The house was reduced to ashes and they then proceeded to burn and destroy everything in the most distant degree connected with the Government, including the Senate House, the President's Palace, an extensive dockyard and arsenal, barracks for two or three thousand men, several large storehouses filled with naval and military stores, some hundreds of cannon of different descriptions, and nearly twenty thousand stand of small-arms. There were also two or three public ropewalks which shared the same fate, a fine frigate pierced for sixty guns, and just ready to be launched, several gun brigs and armed schooners, with a variety of gun-boats and small craft. The powder-magazines were set on fire, and exploded with a tremendous crash, throwing down many houses in their vicinity, partly by pieces of the walls striking them, and partly by the concussion of the air; whilst quantities of shot, shell, and hand-grenades, which could not otherwise be rendered useless, were cast into the river. In destroying the cannon a method was adopted which I had never before witnessed, and which, as it was both effectual and expeditious, I cannot avoid relating. One gun of rather a small calibre was pitched upon as the executioner of the rest, and being loaded with ball and turned to the muzzles of the others, it was fired, and thus beat out their breechings. Many, however, not being mounted, could not be thus dealt with; these were spiked, and having their trunnions knocked off were afterwards cast into the bed of the river.

'All this was as it should be, and had the arm of vengeance been extended no further, there would not have been room given for so much as a whisper of disapprobation. But unfortunately it did not stop here; a noble library, several printing-offices, and all the national archives were likewise committed to the flames, which, though no doubt the property of Government, might better have been spared. It is not, however, my intention to join the outcry which was raised at the time against what the Americans and their admirers were pleased to term a line of conduct at once barbarous and unprofitable. On the contrary, I conceive that too much praise cannot be given to the forbearance and humanity of the British troops, who, irritated as they had every right to be, spared, as far as possible, all private property, neither plundering nor destroying a single house in the place, except that from which the General's horse had been killed.

'Whilst the third brigade was thus employed, the rest of the army, having recalled its stragglers, and removed the wounded into Bladensburg, began its march towards Washington. Though the battle came to a close by four o'clock, the sun had set before the different regiments were in a condition to move, consequently this short journey was performed in the dark. The work of destruction had also begun in the city before they quitted their ground; and the blazing of houses, ships, and stores, the report of exploding magazines, and the crash of falling roofs, informed them, as they proceeded, of what was going forward. It would be difficult to conceive a finer spectacle than that which presented itself as they approached the town. The sky was brilliantly illumined by the different conflagrations; and a dark red light was thrown upon the road, sufficient to permit each man to view distinctly his comrade's face. Except the burning of St. Sebastian's, I do not recollect to have witnessed at any period of my life a scene more striking or more sublime.

'Having advanced as far as the plain, where the reserve had previously paused, the first and second brigades halted; and forming into close column, passed the night in bivouac. At first this was agreeable enough, because the air was mild, and weariness made up for what was wanting in comfort. But towards morning a violent storm of rain accompanied with thunder and lightning came on, which disturbed the rest of all who were exposed to it. Yet in spite of the inconvenience arising from the shower, I cannot say that I felt disposed to grumble at the interruption, for it appeared that what I had before considered as superlatively sublime, still wanted this to render it complete. The flashes of lightning vied in brilliancy with the flames which burst from the roofs of burning houses, whilst the thunder drowned for a time the noise of crumbling walls, and was only interrupted by the occasional roar of cannon, and of large depots of gunpowder, as they one by one exploded.

'I need scarcely observe, that the consternation of the inhabitants was complete, and that to them this was a night of terror. So confident had they been of the success of their troops, that few of them had dreamt of quitting their houses or abandoning the city; nor was it till the fugitives from the battle began to rush in, filling every place as they came with dismay, that the President himself thought of providing for his safety. That gentleman, as I was credibly informed, had gone forth in the morning with the army, and had continued among his troops till the British forces began to make their appearance. Whether the sight of his enemies cooled his courage or not I cannot say, but according to my informant, no sooner was the glittering of our arms discernible, than he began to discover that his presence was more wanted in the senate than in the field; and having ridden through the ranks, and exhorted every man to do his duty, he hurried back to his own house, that he might prepare a feast for the entertainment of his officers, when they should return victorious. For the truth of these details I will not be answerable; but this much I know, that the feast was actually prepared, though, instead of being devoured by American officers, it went to satisfy the less delicate appetites of a party of English soldiers. When the detachment sent out to destroy Mr. Maddison's house, entered his dining parlour, they found a dinner-table spread, and covers laid for forty guests. Several kinds of wine in handsome cut-glass decanters were cooling on the sideboards; plate-holders stood by the fire-place, filled with dishes and plates; knives, forks, and spoons, were arranged for immediate use; everything in short was ready for the entertainment of a ceremonious party. Such were the arrangements in the dining-room, whilst in the kitchen were others answerable to them in every respect. Spits loaded with joints of various sorts turned before the fire; pots, saucepans, and other culinary utensils stood upon the grate; and all the other requisites for an elegant and substantial repast were in the exact state which indicated that they had been lately and precipitately abandoned.

'The reader will easily believe that these preparations were beheld, by a party of hungry soldiers, with no indifferent eye. An elegant dinner, even though considerably over-dressed, was a luxury to which few of them, at least for some time back, had been accustomed; and which, after the dangers and fatigues of the day, appeared peculiarly inviting. They sat down to it, therefore, not indeed in the most orderly manner, but with countenances which would not have disgraced a party of aldermen at a civic feast; and having satisfied their appetites with fewer complaints than would have probably escaped their rival *gourmands*, and partaken pretty freely of the wines, they finished by setting fire to the house which had so liberally entertained them.

'I have said that to the inhabitants of Washington this was a night of terror and dismay. From whatever cause the confidence arose, certain it is that they expected anything rather than the arrival among them of a British army; and their consternation was proportionate to their previous feeling of security, when an event, so little anticipated, actually came to pass. The first impulse naturally prompted them to fly, and the streets were speedily crowded with soldiers and senators, men, and women, and children, horses, carriages, and carts loaded with household furniture, all hastening towards a wooden bridge which crosses the Potomac. The confusion thus occasioned was terrible, and the crowd upon the bridge was such as to endanger its giving way. But Mr. Maddison, as is affirmed, having escaped among the first, was no sooner safe on the opposite bank of the river, than he gave orders that the bridge should be broken down ; which being obeyed, the rest were obliged to return, and to trust to the clemency of the victors.

'In this manner was the night passed by both parties ; and at daybreak next morning the light brigade moved into the city, whilst the reserve fell back to a height about half a mile in the rear. Little, however, now remained to be done, because everything marked out for destruction was already consumed. Of the Senate House, the President's Palace, the barracks, the dock-yard, &c., nothing could be seen, except heaps of smoking ruins ; and even the bridge, a noble structure upwards of a mile in length, was almost entirely demolished. There was, therefore, no further occasion to scatter the troops, and they were accordingly kept together as much as possible on the Capitol Hill.

'I have stated above that our troops were this day kept as much together as possible upon the Capitol Hill. But it was not alone on account of the completion of their destructive labours that this was done. A powerful army of Americans already began to show themselves upon some heights, at the distance of two or three miles from the city ; and as they sent out detachments of horse even to the very suburbs, for the purpose of watching our motions, it would have been unsafe to permit more straggling than was absolutely necessary. The army which we had overthrown the day before, though defeated, was far from annihilated ; it had by this time recovered its panic, began to concentrate itself in our front, and presented quite as formidable an appearance as ever. We learnt, also, that it was joined by a considerable force from the back settlements, which had arrived too late to take part in the action, and the report was that both combined amounted to nearly twelve thousand men.

'Whether or not it was their intention to attack, I cannot pretend to say, because it was noon before they showed themselves ; and soon after, when something like a movement could be discerned in their ranks, the sky grew suddenly dark, and the most tremendous hurricane ever remembered by the oldest inhabitant in the place came on. Of the prodigious force of the wind it is impossible for one who was not an eye-witness to its effects to form a conception. Roofs of houses were torn off by it, and whirled into the air like sheets of paper; whilst the rain which accompanied it resembled the rushing of a mighty cataract rather than the dropping of a shower. The darkness was as great as if the sun had long set, and the last remains of twilight had come on, occasionally relieved by flashes of vivid lightning streaming through it; which, together with the noise of the wind and the thunder, the crash of falling buildings, and the tearing of roofs as they were stript from the walls, produced the most appalling effect I ever have, and probably ever shall, witness. The storm lasted for nearly two hours without intermission, during which time many of the houses spared by us were blown down, and thirty of our men, besides several of the inhabitants, buried beneath their ruins. Our column was as completely dispersed as if it had received a total defeat; some of the men flying for shelter behind walls and buildings, and others falling flat upon the ground, to prevent themselves from being carried away by the tempest; nay, such was the violence of the wind, that two pieces of light cannon, which stood upon the eminence, were fairly lifted from the ground, and borne several yards to the rear.

'When the hurricane had blown over, the camp of the Americans appeared to be in as great a state of confusion as our own; nor could either party recover themselves sufficiently during the rest of the day to try the fortunes of a battle. Of this General Ross did not fail to take advantage. He had already attained all that he could hope, and perhaps more than he originally expected to attain; consequently, to risk another action would only be to spill blood for no purpose. Whatever might be the issue of the contest, he could derive from it no advantage. If he was victorious, it would not do away with the necessity which existed of evacuating Washington; if defeated, his ruin was certain. To avoid fighting was therefore his object, and perhaps he owed its accomplishment to the fortunate occurrence of the storm. Be that, however, as it may, a retreat was resolved upon; and we now only waited for night, to put the resolution into practice,

'There was, however, one difficulty to be surmounted in this proceeding. Of the wounded, many were so ill as to preclude all possibility of their removal, and to leave them in the hands of an enemy whom he had beaten was rather a mortifying anticipation. But for this there was no help ; and it now only remained to make the best arrangements for their comfort, and to secure them, as far as could be done, civil treatment from the Americans.

'It chanced that, among other prisoners taken at Bladensburg, was Commodore Barney, an American officer of much gallantry and high sense of honour. Being himself wounded, he was the more likely to feel for those who were in a similar condition, and having received the kindest treatment from our medical attendants, as long as he continued under their hands, he became, without solicitation, the friend of his fellow-sufferers. To him, as well as to the other prisoners, was given his parole, and to his care were our wounded, in a peculiar manner intrusted,—a trust which he received with the utmost willingness, and discharged with the most praiseworthy exactness. Among other stipulations, it was agreed that such of our people as were left behind should be considered as prisoners of war, and should be restored to us as soon as they were able to travel ; and that, as soon as they reached the ships, the Commodore and his countrymen would, in exchange, be released from their engagements.

'It being a matter of great importance to deceive the enemy and to prevent pursuit, the rear of the column did not quit its ground upon the Capitol till a late hour. During the day an order had been issued that none of the inhabitants should be seen in the streets after eight o'clock ; and as fear renders most men obedient, the order was punctually attended to. All the horses belonging to different officers were removed to drag the guns, no one being allowed to ride, lest a neigh, or even the trampling of hoofs, should excite suspicion. The fires were trimmed, and made to blaze brightly ; fuel enough was left to keep them so for some hours ; and finally, about half past nine o'clock the troops formed in marching order, and moved off in the most profound silence. Not a word was spoken, nor a single individual permitted to step one inch out of his place, by which means they passed along the streets perfectly unnoticed, and cleared the town without any alarm being given. Our pace, it will be imagined, was none of the most tardy, consequently it was not long before we reached the ground which had been occupied by the other brigades. Here we found a second line of fires blazing in the same manner as those deserted by ourselves ;

and the same precautions in every respect adopted, to induce a belief that our army was still quiet. Beyond these, again, we found two or three solitary fires, placed in such order as to resemble those of a chain of piquets. In a word, the deception was so well managed, that even we ourselves were at first doubtful whether the rest of the troops had fallen back.

'When we reached the ground where yesterday's battle had been fought, the moon rose, and exhibited a spectacle by no means enlivening. The dead were still unburied, and lay about in every direction completely naked. They had been stripped even of their shirts, and having been exposed in this state to the violent rain in the morning, they appeared to be bleached to a most unnatural degree of whiteness. The heat and rain together had likewise affected them in a different manner; and the smell which rose upon the night air was horrible.

'In Bladensburg the brigade halted for an hour, while those men who had thrown away their knapsacks endeavoured to recover them. During this interval I strolled up to a house which had been converted into an hospital, and paid a hasty visit to the wounded. I found them in great pain, and some of them deeply affected at the thought of being abandoned by their comrades, and left to the mercy of their enemies. Yet, in their apprehension of evil treatment from the Americans, the event proved that they had done injustice to that people; who were found to possess at least one generous trait in their character, namely, that of behaving kindly and attentively to their prisoners.

'As soon as the stragglers had returned to their ranks, we again moved on, continuing to march without once stopping to rest during the whole of the night. Of the fatigue of a night march none but those who have experienced it can form the smallest conception. Oppressed with the most intolerable drowsiness, we were absolutely dozing upon our legs; and if any check at the head of the column caused a momentary delay, the road was instantly covered with men fast asleep. It is generally acknowledged that no inclination is so difficult to resist as the inclination to sleep; but when you are compelled not only to bear up against that, but to struggle also with weariness, and to walk at the same time, it is scarcely possible to hold out long. By seven o'clock in the morning, it was found absolutely necessary to pause, because numbers had already fallen behind, and numbers more were ready to follow their example ; when, throwing ourselves upon the ground, almost in the same order in which we had marched, in less than five minutes there was not a single unclosed eye throughout the whole brigade. Piquets were of course stationed, and sentinels placed, to whom no rest was granted, but, except these, the entire army resembled a heap of dead bodies on a field of battle, rather than living men.

'In this situation we remained till noon, when we were again roused to continue the retreat. Though the sun was oppressively powerful, we moved on without resting till dark, when having arrived at our old position near Marlborough, we halted for the night. During this day's march we were joined by numbers of negro slaves, who implored us to take them along with us, offering to serve either as soldiers or sailors, if we would but give them their liberty ; but as General Ross persisted in protecting private property of every description, few of them were fortunate enough to obtain their wishes.

'We had now proceeded a distance of thirty-five miles, and began to consider ourselves beyond the danger of pursuit. The remainder of the retreat was accordingly conducted with more leisure ; our next march carrying us no farther than to Nottingham, where we remained during an entire day, for the purpose of resting the troops. It cannot, however, be said that this resting-time was spent in idleness. A gun-brig, with a number of ships' launches and long-boats, had made their way up the stream, and were at anchor opposite to the town. On board the former were carried such of the wounded as had been able to travel, whilst the latter were loaded with flour and tobacco, the only spoil which we found it practicable to bring off.

'Whilst the infantry were thus employed, the cavalry was sent back as far as Marlborough, to discover whether there were any American forces in pursuit ; and it was well for the few stragglers who had been left behind that this recognisance was made. Though there appeared to be no disposition on the part of the American General to follow our steps and to harass the retreat, the inhabitants of that village, at the instigation of a medical practitioner called Bean, had risen in arms as soon as we departed ; and falling upon such individuals as strayed from the column, put some of them to death, and made others prisoners. A soldier whom they had taken, and who had escaped, gave information of these proceedings to the troopers, just as they were about to return to head-quarters; upon which they immediately wheeled about, and galloping into the village, pulled the doctor out of his bed (for it was early in the morning), compelled him, by a threat of instant death, to liberate his prisoners; and mounting him before one of the party, brought him in triumph to the camp.'

Two private letters descriptive of the battle of Bladensburg are here inserted:

Letter from Captain R. Gubbins, 85th L.I.

'On Board the *Madagascar* in the
'River Patuxent, N. America,
'31 *August* 1814.

'I have just returned (and, Thank God, safe and sound, although greatly fatigued) from the destruction of the famous City of Washington.

'We left our ships on the 18th and on the 19th we landed at a place high up this river called Benedict. The following evening we commenced our march with about 4000 men and on the 24th we came to Bladensburg, about 8 miles from Washington where we perceived the American Army—from 8 to 10,000—drawn up for battle.

'Previous to landing I had been appointed Major of Brigade, in the Light Brigade commanded by Col. Thornton, and was with him at the commencement of the attack.

'About 100 of our Regt led by Col. Thornton, although fatigued with a 15 mi. march dashed on the bridge in front of the whole army and were closely followed by the 85th Regt and the remainder of the Light Brigade—altogether 800—with this small force and the 4th Regt, about 400, we banged the Yankees most completely, and by 6 o'clock the same evening reached Washington and set fire to their grand House of Congress—or Capitol. Genl Ross's despatches will of course give you full particulars of all our performances. It is with extreme sorrow I tell you, that our gallant Colonel Thornton is severely wounded and together with Col. Wood and Major Brown also severely wounded left at Bladensburg. We had two fine young men and most promising officers killed[1] and 12 officers wounded,

[1] Captain Hamilton and Lieutenant Codd.

mostly severely, and the 85th has suffered the greatest loss of any Reg[t].

'In consequence of the loss of Cols. Thornton and Wood and Major Brown, I succeeded to the command of the Reg[t] and which I am busily employed in putting in order for another expedition in this neighbourhood.'

Extract from a letter written by Captain Hon. John James Knox, 85th Light Infantry, dated Kingston, Jamaica, November 23, 1814, after Bladensburg.

'I think we shall have some promotion before our return to England —for if bullets do not give it the climate will.

'We have received the English papers with General Ross's dispatch. I am glad the news was so well received, for the 85th had the brunt of the day.

'I was never under such a heavy fire of cannon and small arms since I listed. When I saw three field officers down and 8 or 9 others of the 85th sprawling on the ground, before we had been a quarter of an hour under fire, thinks I to myself, thinks I, by the time the action is over the devil is in it if I am not either a walking Major or a dead Captain. I wish I could get a brevet majority before the war is at an end.

'Ask Edward if my father has paid any of my bills, for I would much rather remain on service than face my creditors.'

Continuation of Lieutenant Gleig's Diary.

Friday, 26th.—'At 7 o'clock we at last halted and I threw myself under the shade of a tree, and fell asleep instantly. At 12 they woke me to breakfast which we had just finished when we were ordered again to fall in and march. We continued to go on without any incident occurring till dark, when we once more came into Marlboro'. When everything was arranged for the Company, Grey and I strolled up to a house where our wounded Officers were conveyed. We found them all in great pain, but the three Field-Officers had been left in Bladensburg too ill to be moved. In this house we supped and slept, better than we had done since we landed.'

Saturday, 27th.—'At 6 o'clock we marched from Marlboro' and took the road for Nottingham. Our march to-day was very uninteresting nothing occurring worthy of mention all the way. We came in pretty early to Nottingham, the distance being only 10 miles, and took possession of our old shed again. Upon the very straw on which I this night slept poor Codd had rested. The idea was too melancholy to be borne, and I indulged myself in a soothing cry. We made a hearty dinner on some capital soup and slept sounder than your sons of luxury ever do on their beds of down.'

Sunday, 28th.—'We expected to have marched to-day but we did not. The day was passed in embarking the wounded on board a gun-brig which came up for their reception, and in carrying off stores of Tobacco to a very considerable amount. I walked through the town trying to get some bread but did not succeed. Almost every house was deserted and nothing whatever was to be got. At 3 o'clock the Company went on picket, and a most pleasant one we had. While we were sitting looking about us, to our no small surprise we saw our friend Dr. Bean brought in as a prisoner. On enquiring into the cause we learned that as soon as our troops had left the village he had armed his slaves, and sallied forth cutting off all our stragglers. As soon as the General heard of it, he sent back our Cossacks who took him out of his bed, and brought him off a prisoner.'

Monday, 29th.—'The wounded, the artillery, and plunder, being all embarked on the 28th, at daybreak on the 29th we took the direction of St. Benedict's, where we arrived, without any adventure, at a late hour in the evening. Here we again occupied the ground of which we had taken possession on first landing, passing the night in perfect quiet. The distance is twenty miles, which was very severe upon people knocked up as we were, however we contrived to do it without many men falling out. The fatigue was beyond description, but the soundness of our sleep after we came in made up for all. We took up exactly our old position, made a supper on fowl soup and slept under a tree.'

Tuesday, 30th.—'Got under arms as usual before light but did not move. All the troops and everything else are embarked but our brigade, so as we were the first on Yankey ground, we shall be the last to leave it. At twelve o'clock the bugle sounded and we fell in, and marched to the boats. When we came into the village a number of sailors collected together, and cheered us, which we returned, feeling highly gratified at the marked compliment as ours was the only Regiment cheered. The regiments, one by one, marched down to the beach. We found the shore covered with sailors from the different ships of war, who welcomed our arrival with loud cheers; and having contrived to bring up a larger flotilla than had been employed in the disembarkation, they removed us within a few hours, and without the occurrence of any accident, to our respective vessels. We very soon got into the boats, and in a short time found ourselves once more on board the *Golden Fleece*. The first thing I did was to wash myself thoroughly and put on clean things, which I never stood so much in want of, and the luxury I enjoyed in turning into clean sheets is beyond description, but I was too much fatigued to sleep sound.'

Wednesday, 31*st*.—'Here we are once more on board ship after having marched with a handful of men 50 miles up an enemy's country and destroyed their capital. Remained quietly on board all day, not knowing well what to do, I felt so completely at rest. Bathed in the evening. The ship dropped down to-day close to the mouth of the river.'

Commodore Barney and the 85th.

From an old newspaper the following extracts have been gathered, and they are of sufficient interest to warrant their inclusion here, concerning as they do both the Regiment and Commodore Barney:

'Barney, after he was wounded, was taken by a soldier of the 85th, whom the Commodore requested to stay by him and take care of him, expressing a wish at the same time to remain where his captured guns were. The man replied that his comrades appeared to be warmly engaged, and that he must go to assist them; and Barney, to induce him to comply with his wishes, pulled out his watch, and offered it to him, which the soldier immediately refused, saying, that if he would not remain with him without a reward, he would not take a bribe, and then left him. After the action was over, our officers visited the Commodore, who related the circumstance to them and requested that the soldier might be called out of the ranks to confirm what he had said. The soldier appeared, and Barney observed to him, "That as he would not receive his watch to induce him not to do his duty, would he do him the favour to receive it for having done it." To which the soldier replied, "No, Sir, I cannot; you are a prisoner."'

'If there had been any cavalry on the spot Maddison (the President) might have been taken prisoner, for the Officers of the 85th distinctly saw him mount his horse when the militia took to their heels; he was accompanied by two others.'

'Mr. Urquhart (Lieut. Beau Colclough Urquhart of the 85th) has got Maddison's fine dress sword, which he took out of his house.'

An American Account of Bladensburg

Extracted from *A Sketch of the Events which preceded the Capture of Washington by the British on August* 24, 1814. Philadelphia: Published by Carey & Hart, and also Chas. Marshall, 148 Chestnut St. 1849.

'It is proposed to review the method by which a portion of the war of 1812 was conducted, which led to the deep disgrace of the nation abroad, and its deeper mortification at home. The capture of Washington by a handful of men, after more than twelve months' notice to the proper authority of actual, impending peril, ought never to be forgotten, for the lesson it holds out to confident security and ill-judged procrastination.

'After 34 years, it is difficult to realise that 4500 infantry, without artillery, and under the effects of a climate deadly to European constitutions, should have marched 50 miles into a country peculiarly adapted for defence, whose inhabitants had heretofore been celebrated for bravery and their skill in irregular warfare, destroyed with every degree of wanton barbarism the capitol of the country, and been permitted to retire unmolested to their shipping to prosecute a new enterprise undertaken upon the impunity which attended such extraordinary success.

'War was declared against Great Britain on the 18th June 1812; and so early as December of that year notice was received in the United States from Bermuda that a British squadron had arrived at that place, having on board a considerable body of troops, with the requisite munitions, including Congreve rockets, destined for the attack of the southern cities of the United States. On the 4th February following, two ships of the line, three frigates, and some smaller vessels of this squadron, entered the mouth of the Chesapeake and came to anchor in Hampton Roads. The destruction of private property, the capture of negroes, and the burning of Frederick, Georgetown, Havre de Grace and Frenchtown and the exercise of almost incredible barbarities on the defenceless inhabitants, followed the arrival of this force, which was afterwards increased to seven ships of the line and 13 frigates, having on board 4000 infantry. An attempt was made by this armament to take Norfolk on 22 June 1813; but the timely organisation of a small militia force, and their efficient resistance, with the aid of a handful of seamen and marines, proved sufficient to defeat the attack and repulse the enemy with signal loss and they proceeded to North Carolina, to repeat at Ocracoke and Portsmouth the revolting acts they had previously perpetrated in Maryland. The attitude of the defenders of these two places induced Admiral Cockburn to return to the Chesapeake to resume his system of plunder.

'So early as March 1st, 1814, Admiral Cockburn, with one "74," two frigates, a brig and a schooner, arrived in Lynnhaven Bay and began his usual system of capture and plunder. Their presence was continually taken notice of and published in the city of Washington where the probable result of the campaign in Europe which soon left at the disposal of the British Government a large body of troops then serving in France, was publicly known, and seems to have excited neither attention, remark, nor preparation. Soon after, the fact was announced that 4000 troops, said to be destined for the United States, had, on January 20, arrived at Bermuda, where the preparations for their successful action were going on with the knowledge of every member of the Cabinet. The shadows of coming events grew stronger as the events themselves drew nearer; and the intelligence from France, little heeded, it would seem, soon became distinct and positive. The official gazette contained a paragraph taken from a London paper of April 20 announcing that "a number of the largest class of transports were fitting out with all possible speed at Portsmouth, as well as all the troopships at that port, for the purpose, it was supposed, of going to Bordeaux, to take the most effective regiments in Lord Wellington's army to America. On June 28th intelligence was received that a fleet of transports with a large force on board was sailing from Bermuda in one or two days for some port in the United States—probably for the Potomac. Such intelligence was not entirely without effect upon the long continued apathy of the administration, but the War Department treated the matter very lightly. A military district was formed under the command of Brigadier-General Winder, but it was only composed of two detachments of the 36th and 38th Infantry and a small detachment of Artillery, amounting in the whole, to 330 men although a requisition for 93,500 men was made upon the several States to be held in readiness to be called out for service; but General Winder was given no staff, so that the proper organisation of this force was impossible and his time was occupied with an oppressive mass of detail.

'He applied to the Secretary of War for 4000 militia to be called out without delay and pointed out that the English could be in Washington, Baltimore or Annapolis in four days after their arrival off the coast and that then it would be too late to disseminate through the intricate and winding channels the various orders to the militia, for them to assemble, have their officers designated, their arms, accoutrements and ammunition delivered, the necessary supplies provided, or for the commanding officer to learn the different corps and detachments.

'He pointed out that, if the enemy's force should be strong, sufficient numbers of the militia could not be warned and run together, even as a disorderly crowd, without arms, ammunition or organisation, before the enemy would already have given his blow—and that the utmost regular force at his disposal would not exceed 1000 men.

'To his application no answer was returned by the Secretary of War, General Armstrong, who appears to have thought an attack on Washington was unlikely.

'On Friday, August 18, it was known at Washington that the enemy was coming up the Bay in force; and the next day there was information of the determination of the Admiral "to dine in Washington on Sunday after destroying the flotilla."

'The English landed on the 18th August at Benedict. The force by which the hazardous expedition was undertaken amounted to but 4500 men, including sailors and armed negroes, entirely destitute of cavalry, and dragging with them by hand one six-pounder and two three-pounder "grasshoppers." The distance to march was upwards of 40 miles, through a country intersected with streams, and covered with woods, during excessively sultry weather, the effect of which upon men who had been relaxed by being long cooped up in ships and unused to carry their arms was so extremely severe that the second day the army marched but six miles, during which, says one of the number, "a greater number of soldiers dropped out of the ranks and fell behind from fatigue than I recollect to have seen in any march in the Peninsula of thrice its duration."

'It required very little military knowledge to dispose of an invading force of such a strength under such circumstances. To obstruct the roads by felling trees across them, and breaking down the bridges—to hover round the flanks and rearguard of the advancing corps, continually harassing them by a fire of musketry and rifles—a succession of attacks on the advance whenever a stream was to be crossed, or fallen trees to be removed or avoided—and occasionally a shell from a howitzer—would have obliged them to surrender at discretion.

'The time occupied in reaching Washington was five days, the retreat occupied more than four days—but no one was at hand even to harass them.

'Two thousand (or even one thousand) regular troops encamped, and ten thousand organised militia and volunteers properly equipped and brought into the field in time to receive sufficient instruction to act in concert against the enemy and held in readiness so to do, would have been amply sufficient for the purpose. But at this juncture the regular force was found to consist of 330 men of the 36th and 38th Regiments of Infantry and two troops Cavalry—125 men—under Lieut.-Colonel Laval—and one company of the 12th Infantry—80 men—who shortly after arrived at Washington.

'The militia consisted of a quota of the Brigade of General Stansbury who marched on August 20th from Baltimore with 1353 men; this force halted at the Stag tavern on the evening of the 21st, and on the 22nd advanced towards Bladensburg, near which place they encamped; and on the 23rd commenced moving towards Marlborough, the orders of General Winder being to take a position on the road not far from that place; on the 23rd he was joined by Lieut.-Colonel Sterret's command, consisting of the 5th Baltimore Regiment of Volunteers, Major Pinkney's rifle battalion, and two companies (Myers' and Magruder's) of Artillery. The fatigued state of Sterret's command induced General Stansbury to halt and remain during the night on the hill near Bladensburg. A false alarm, by which the command was roused and kept under arms until after two o'clock in the morning of the 24th, added to the exhaustion and distress of the troops.

'After making a movement towards Washington, General Stansbury's force was ordered by General Winder to return to Bladensburg, where they took post in the orchard near the mill; the artillery, consisting of six guns, being behind a small breastwork of earth, commanding the pass into the town, and the bridge south-west of it, which was also commanded by Pinkney's riflemen.

'In the meantime it was ascertained by General Winder that the enemy was proceeding towards Washington by the Bladensburg road instead of moving on Annapolis or Fort Washington, as his course for a time threatened, and he passed into the City of Washington that night over the bridge at the Eastern Branch. He had with him Commodore Barney, who joined him at the Old-Fields, with the flotilla men and marines, amounting to about 500 men, Laval's Dragoons, 125 in number, some volunteer cavalry, 260 strong, the 36th and 38th Regiments of United States Infantry, 330 men, the militia under Colonel Hood (600 or 700) and the brigade of militia of the district, 1070 men commanded by General W. Smith. These various bodies amounted in numbers to about 5100 "a mass suddenly assembled without organisation or discipline" and, with the exception of Commodore Barney and Major Peters, without any officers with the least knowledge of service, and wearied and exhausted by sudden exertion, and toilsome marches in very hot weather.

'Colonel Minor's force of 600 is not included in this estimate, as they were not on the battle-ground, having been detained from taking part in the action by the negligent, frivolous, and dilatory course pursued by the Secretary of War and the persons of his department whose duty it was to furnish them with arms and ammunition. One very young man, who had charge of the armoury, counted over again the flints after they had been counted by Colonel Minor's officers, and so cautiously dealt out the stores that the regiment was unable to get to the field in time.

'With 5000 men assembled in time to organise them, and allow their officers to become acquainted with each other and with his plan of operations, a general acquainted, as General Winder was, with the country through which his enemy had to penetrate, would undoubtedly have given a good account of him ; but to oppose in pitched battle the undisciplined valour and exertions of the same men to that of an equal number of regular, veteran troops, could only be justified by absolute necessity—a necessity which existed on the 24th of August 1814 ; the result of an improvident disregard, on the part of the Secretary of War, of continual warning against coming danger, which he had the means of resisting and subduing, with honour to his country and credit to himself, but did not call them forth in time.

'The formation of the American lines was scarcely completed when the enemy appeared. Colonel Beale, who was on the right, had just arrived with his regiment of Maryland Militia, fatigued and exhausted after a march of 16 miles, and had taken post on the high ground, near the battery of Commodore Barney, which latter had been posted so as to command the bridge and the road by which the enemy approached. To Barney's right extended the flotilla men and marines under Captain Miller, and on his left was Colonel Magruder's regiment (the 1st) of District Militia ; Lieut.-Colonel Scott with the United States Infantry, composed of portions of the 36th, 38th and 12th Regiments was in front of Magruder about 100 yards ; but his position was afterwards changed because in the way of the guns of Major Peters's battery, and the men fell back and formed in line with Magruder's regiment. Peters's battery (six 6 pounders) Davidson's light infantry and Stull's rifle corps were thus in advance, and Colonel Cramer was posted still further in advance, in the woods on the right of the road, with his battalion of Maryland Militia. The troops under Stansbury were to the west of Bladensburg, in an orchard, and on the left of the road to Washington, and formed, together with Beale's command, the first line ; their artillery was behind a small breastwork in front, and the infantry in the rear and to the left, to protect the position ; the other corps which we have mentioned formed the second line.

'The enemy first approached Stansbury's line, about half-past 12 o'clock, and their light troops were dispersed by the fire of the Baltimore Artillery, and taking shelter behind the houses of the village and trees, began to concentrate towards the bridge, and press across it and the river. Pinkney's riflemen now opened a very brisk fire upon them, which, added to the artillery, occasioned them a severe loss of men. They passed the bridge, however, and having deployed into line, advanced on the artillery and riflemen, and compelled them to retreat and join the troops of the first line. Their advance was annoyed by Captain Burch's company of Volunteer Artillery, belonging to the city, and a small detachment near it, who opened a sharp cross fire upon them. General Winder, who was on the left of the 5th Regiment, ordered it to advance and sustain the artillery, which it did with great promptness, and opened a steady well-directed fire on the enemy, in which it was followed by Ragan's and Schutz's regiments, forming the right and centre of the line. Some rockets thrown by the enemy, which passed very close over the heads of Ragan's and Schutz's regiments, created a panic in these raw troops, in action for the first time, and they broke. Their officers exerted themselves to rally them, in which they were aided by General Winder, who displayed great zeal, activity and personal bravery; but their efforts were ineffectual, and both regiments were broken and dispersed, leaving the 5th Regiment with its flanks exposed. This regiment, however, kept its place in line firmly, covered the retreat of Ragan's and Schutz's by a smart fire, and did not retreat until ordered by General Winder to do so, after the enemy had gained both its flanks.

'The first line having been dispersed, the left of the enemy's force advanced on the second line. Passing along the road in heavy column, they were encountered by the corps of militia under Colonel Cramer, whom they drove back after a short and sharp conflict, and who formed upon Beale's command, and their column deployed in the field on the right of the road, and became exposed to the fire of Peters's battery, which galled, but did not check their progress. The onward movement of the enemy brought them in front of Barney's position, where, for a moment, they made a halt; and then pushed forward upon him, but received such a destructive charge from an 18-pounder that the road was completely cleared of them; and a second and third attempt to advance was repulsed in the same effectual manner.

'To avoid the battery, the enemy turned to the left into a field, with a view to turn the right flank of the position, but the movement was promptly met by the marines under Captains Miller and Sevier, and the flotilla men acting as infantry, who charged them with such vigour that they broke the 85th and the 4th or "King's Own" and pursued them until they got into a ravine, leaving their officers, Major Brown, Lieut.-Colonel Wood and Colonel Thornton, all severely wounded, in possession of the Americans.

'The dispersion of Stansbury's troops left the ground on the left of the flotilla force undefended; and the enemy having pushed a body of 200 or 300 men against the militia under Beale, who was posted in a strong position on the right of Barney, dispersed them, and the British light troops gained both his flanks, and the Commodore himself was wounded severely, and also some of his best officers. The drivers, too, of his ammunition wagons, had gone off with them, in the confusion of the retreat of the militia; and deprived of their ammunition the power to resist any longer ceased, and the flotilla men and marines effected their retreat in good order; but the Commodore's wound rendered him unable to move, and he was made prisoner.

'The behaviour of the flotilla men and marines excited the highest admiration on the part of the enemy.

'The loss of the Americans did not exceed 10 or 12 killed and 40 wounded.

'The great body of American troops retreated after the battle towards Montgomery Court House; and there were, of course, no obstacles in the way of the enemy, who proceeded to march the portion of his army who had not taken an active part in the engagement into the City of Washington.

'It is a matter of history and of lasting reproach to the British nation, that in violation of all the rules of civilised warfare, General Ross proceeded to destroy and lay waste the public buildings, monuments, and property, including a valuable library, and some of the archives, in the most wanton manner, involving in their destruction many private dwellings and a great amount of private property.'

With regard to the destruction of Washington an extract from the despatch of Major-General Ross to Earl Bathurst and dated August 30, 1814, is here given.

From its tenor it would appear that the destruction of the public buildings in the city was designed. It must be remembered that this destruction has been stated to have been an act of retaliation for the treacherous firing on General Ross by Americans while a flag of truce was displayed. General Ross does not mention any act of the kind but writes :

' Having halted the army for a short time, I determined to march upon Washington, and reached that city at eight o'clock that night. Judging it of consequence to complete the destruction of the public buildings with the least possible delay, so that the army might retire without loss of time, the following buildings were set fire to and consumed :—the Capitol, including the Senate house, and House of Representation(Representatives), the Arsenal, the Dock-yard, Treasury, War-Office, President's Palace, Rope-walk, and the great bridge across the Potomac : in the dock-yard a frigate nearly ready to be launched and a sloop of war were consumed.'

There has been much controversy on the subject of the destruction of Washington, and the curious may be referred to the ' Refutation of Aspersions on " Stuart's Three Years in North America," by James Stuart, Esq. London, 1834.'

UROGNE CHURCH
(*see pages* 89 *to* 91)

CHAPTER VIII

NEW ORLEANS, 1814

Lieutenant Gleig's Diary Continued :
September, 1814.

Thursday, 1*st*.—' Remained in the same place all day. We have got two wounded Officers on board, Burrell and Crouchley, to whom every attention is paid. Wells has behaved uncommonly well to us, and taken us to mess with him which makes us much more comfortable than we formerly were.'

Friday, 2*nd*.—' Grey left us this morning to go on board the *Diadem* for the better convenience of medical aid. I went on shore with my gun and dogs, and shot two wild Turkeys and a Virginia Nightingale. Went four miles up the country with nobody with me but my black servant and Francis. Entered a house where there were two men who were much more afraid of us than we were of them. Got some bread and milk and returned on board.'

Lieutenant Gleig has also given us a fuller account of the same incident :

' I one day continued my walk to a greater distance from the fleet than I had yet ventured to do. My servant was with me, but had no arms, and I was armed only with a double-barrelled fowling-piece. Having wearied myself with looking for game, and penetrated beyond my former landmarks, I came suddenly upon a small hamlet, occupying a piece of cleared ground in the very heart of a thick wood. With this, to confess the truth, I was by no means delighted, more especially as I perceived two stout-looking men sitting at the door of one of the cottages. To retire unobserved was, however, impossible, because the rustling which I had made among the trees attracted their attention, and they saw me, probably, before I had seen them. Perceiving that their eyes were fixed upon me, I determined to put a bold face upon the matter; and calling aloud, as if to a party to halt, I advanced, with my servant, towards them. They were dressed in sailors' jackets and trowsers, and rose on my approach, taking off their hats with much civility. On joining them, I demanded to be informed whether they were not Englishmen, and deserters from the fleet, stating that I was in search of two persons very much answering their description. They assured me that they were Americans, and no deserters, begging that I would not take them away; a request to which, after some time, I assented. They then conducted me into the house, where I found an old man and three women, who entertained me with bread, cheese, and new milk. While I was sitting here, a third youth, in the dress of a labourer, entered, and whispered to one of the sailors, who immediately rose to go out; but I commanded him to sit still, declaring that I was not satisfied, and should certainly arrest him if he attempted to escape. The man sat down sulkily; and the young labourer coming forward, begged permission to examine my gun. This was a request which I did not much relish, and with which I, of course, refused to comply; telling the fellow that it was loaded, and that I was unwilling to trust it out of my own hand, on account of a weakness in one of the locks.

'I had now kept up appearances as long as they could be kept up, and therefore rose to withdraw; a measure to which I was additionally induced by the appearance of two other countrymen at the opposite end of the hamlet. I therefore told the sailors that, if they would pledge themselves to remain quietly at home, without joining the American army, I would not molest them; warning them, at the same time, not to venture beyond the village, lest they should fall into the hands of other parties, who were also in search of deserters. The promise they gave, but not with much alacrity, when I rose, and keeping my eye fixed upon them, and my gun ready cocked in my hand, walked out, followed by my servant. They conducted us to the door, and stood staring after us till we got to the edge of the wood; when I observed them moving towards their countrymen, who also gazed upon us, without either advancing or flying. The reader will readily believe that as soon as we found ourselves concealed by the trees, we lost no time in endeavouring to discover the direct way towards the shipping; but plunging into the thickets, ran with all speed, without thinking of aught except an immediate escape from pursuit. Whether the Americans did attempt to follow, or not, I cannot tell. If they did, they took a wrong direction; for in something more than an hour I found myself at the edge of the river, a little way above the shipping, and returned safely on board, fully resolved not again to expose myself to such risks, without necessity.'

Diary Continued.

Saturday, 3rd.—'Attended the sale of the kit of poor Hamilton and Codd. Three of our wounded Officers, and those of the men who were very bad were put on board the *Iphigenia* which goes to England with the dispatches. Burrell went to the *Diadem.*'

Sunday, 4th.—'The dispatches sailed this morning with a fair wind. Grey came on board to dinner and informed me that Burrell and Gascoigne were going to Halifax that evening. I went to see them after dinner and found them in the boat coming back as they could not be taken.'

Monday, 5th.—'Burrell and Gascoigne went on board the *Tonnant* which sailed with a fresh breeze followed by the *Majestic* (Rozu) and several Frigates. We remain quiet but expect to sail every hour.'

Tuesday, 6th.—'At daylight got under weigh and sailed towards the Chesapeake with a fine fair wind, but instead of running up as we expected, we saw Admiral Cochrane beating towards us, who made signals for us to anchor which was done accordingly. An order went through the Fleet by telegraph for every ship to give in a return of the number of seamen she could land with small arms besides Marines. This will be no inconsiderable reinforcement to our little army.'

Wednesday, 7th.—'Set sail to-day for the Potomac but for what reason we know not. Saw the *Menelaus* with her pennant half-mast high, a sign indicative of the death of Sir Peter Parker. The body is put into a Frigate to be sent it is supposed to Bermuda.'

Thursday, 8th.—'Stood up the Potomac in high hopes of having something dashing to do but anchored at dusk. The lightning was uncommonly vivid and beautiful.'

Friday, 9th.—' Weighed anchor again at an early hour and stood up the river. Admiral Cochrane left us very soon and went on by himself, and to our great mortification we kept beating about although all the time the wind blew as fair as it could. In the evening we saw the Admiral coming down again and a large fleet of vessels with him bearing the Yankee flag, and among them a large Frigate. This put us in great spirits, but they afterwards turned out to be only an English one leading the prizes. We now put about again all in a body and took the direction of Baltimore which is confidently stated to be the next point of attack.'

Saturday, 10th.—' Standing up the bay with a fine fair wind. I went on board the *Diadem* and saw a letter from poor Thornton in which he says that all the wounded are doing well and that the Americans behaved very kindly to them. Brought number of papers off with me in one of which is Wilkinson's promotion, but the ship was going so fast that it was with the utmost difficulty we could reach her.'

Sunday, 11th.—' The wind still favours us and by twelve o'clock the Line of Battleships came to an anchor. I went on board the *Diadem* again where I stayed dinner. The little *Fleece* in the meantime had got close in shore, and we had a pull of upwards of ten miles before we reached her. The enemy fired alarm guns all along the coast and showed every symptom of terror. We expect to disembark to-night, so every man sleeps with his clothes on.'

Monday, 12th.—' At 3 o'clock this morning we were roused out of bed by the boats coming alongside to carry us ashore. All that I take with me this time is a blanket and a haversack with three days provisions on my back. It was broad daylight before we landed, and when we did get ashore we found all the Brigade there before us. We remained here for about an hour when we moved forward, and halted again by a farmhouse three miles further on waiting for the rest of the army. Here three of the enemy's Light Horse were taken, who had been watching us, and sent to the ships. When the rest of the army arrived we pushed on, but had not gone above a mile when 2 mounted Officers came back to hurry us on, as the advanced guard was engaged. At this we stept out, but by and by McDougall came galloping back and called on Williams to follow him. We all dreaded something was wrong, and our fears were too soon realized, for in a short time we saw the General lying by the side of the road. There were five English soldiers wounded and three Yankees dead near him. We passed on without seeming to notice it, though there was not a man who felt not his loss, and soon came in sight of the army drawn up in a wood, with a high railing in front of him, and each flank protected by water. This position was strong, but an immediate attack was determined upon. The troops were drawn up as follows. Our Brigade in extended order all along the front of his line, the 21st in column by the road, the 4th away to turn his left, and the seamen, 44th and Marines in line in the rear as a support. We remained for about half an hour lying under the shade of some trees, and fanned by a nice little breeze which completely freshened us, when the order was given to advance and obeyed instantly. We sprung over some palings which were in front of us, and advanced in a cool orderly manner, notwithstanding the showers of grape with which the enemy plyed us, till we got within 150 yards of them, when they gave us a volley of musketry from right to left of their line. To this we returned a hearty cheer, and giving them back a volley rushed on at double quick. Previous to this I received a blow from a grapeshot in the groin which would certainly have killed me had not my haversack which the motion of walking had turned round intervened to save me. The enemy now kept up a very smart fire which we returned as briskly, but seeing us approaching to the charge, they gave way in all directions. The rout was most complete. Cavalry, Infantry and guns were all jambed together making the best of their way off. I fired two shots but I fear missed both. We desisted from pursuing them and reassembled the scattered army on the position the enemy had occupied this morning, where we passed the night. Our loss this day was very trifling, that of the enemy immense. I had the knee of my overalls cut with a musket ball.'

[A Captain Duncan MacDougall of the 85th had been left behind in bed with a severe attack of ague, but before the regiment had been engaged many minutes a shout was heard and up came MacDougall mounted on a pony which he was belabouring unmercifully lest he should not be in time to take part in the fray.

This officer, who afterwards became Sir Duncan MacDougall, was Sir de Lacy Evans's second-in-command of the British Legion which was raised in 1835 to fight for the Queen of Spain against the

Carlists. He was a most gallant soldier. He, in 1833, commanded the 79th Highlanders. He had a fiery (Highland) temper, which was very demonstrative when roused; but he was the kindest and most courteous of men for all that. He set himself against corporal punishment when it was both cruel and much in use. During the time he commanded the 79th there was not one instance in which it was inflicted. He was, however, a very strict disciplinarian and had his own way of expressing it, and that very successfully.

On one occasion a very young soldier went to the Colonel's quarters and rapped at his bedroom door, intending to make some complaint or representation of some kind, either wilfully or ignorantly avoiding the proper way of first going to his Captain.

The Colonel said ' Come in,' and for a moment the two looked at each other with astonishment—for the Colonel was in his dressing-gown and shaving himself, an aspect of him which presumably the soldier had never conceived of before, and the Colonel had never thought of such a thing as a private coming to interview him in such an unceremonious manner.

Then Colonel MacDougall jumped up and rushed at the man, razor in hand, and the man on his part turned and fled in terror from an apparent madman, who pursued him half-way across the barrack square, vituperating at the man, and his dressing-gown skirt flying behind him.]

Tuesday, 13th.—' It rained very hard this morning when we first fell in, and continued drizzling all day. When the troops were all formed and everything ready, we moved off in the highest spirits. Our Regiment was in front and our Company formed the flank-guard. We took several prisoners among the woods, but got completely wet wading through the long grass and bushes loaded with rain. After a march of several hours we arrived within sight of the main army of the enemy, twenty thousand strong, posted upon the heights above Baltimore. Those heights they had strengthened with breastworks, batteries and redoubts, in which were mounted one hundred pieces of cannon. Our little army consisted of at most four thousand men, with eight pieces of cannon, and in spite of such tremendous odds an attack was determined upon. As it would have been madness to have thought of assaulting those works in the daytime it was resolved we should do it that night, and it would certainly have been done had not a message arrived from the Navy to say they could not co-operate with us. This immediately altered all our plans, and I was not a little annoyed when I expected an attack to find the column formed left in front on the main road at 4 o'clock in the morning. During the night we had been on picket, and it rained as heavily as ever I witnessed. We had no shelter whatever, not even the small protection of my blanket, for that I had left behind along with the men's when we advanced this morning, so we had to be exposed to it all, and thoroughly soaked to the skin. However we bore it all patiently in the hope of giving the Yankees a good drubbing before morning.'

Wednesday, 14th.—' But our hopes were blasted, we commenced our retreat in excellent order. We marched on without the occurrence of a single incident till towards midday, when an alarm was given, and we formed line in a moment. I never saw anything done in such style, but it was needless as the alarm proved to be a false one. We broke once more into the line of march till we came to a nice position about a mile in rear of the place where we halted on the 12th, which we took up, and where we passed the night. In marching to-day over the field of battle the dead Yankees were lying in dozens stripped naked and bleached as white as snow with the rain. In this position, for the first time since we landed, did Grey and I get a meal cooked. We got a pot of capital soup which we enjoyed beyond measure, and making a tent of a couple of blankets slept under it as comfortably as possible.

Thursday, 15th.—' At 7 o'clock we recommenced our march, but it was a very short one, for we had not gone above a mile and a half before we came in sight of the shipping. Our Regiment now halted to cover the embarkation of the rest of the army, and during this halt we employed ourselves in chasing and killing pigs, of which a great number were running about. We had a very nice fry made which we had just finished when the order was given to fall in, when we marched to the boats and embarked immediately. We soon got on board the *Golden Fleece*, where the first thing I enjoyed was a thorough wash and change, then a good dinner, and to sum up all a sound sleep in clean sheets.'

Narrative of the Retirement from Baltimore.

'At an early hour on the 13th the troops were roused from their lairs, and forming upon the ground, waited till daylight should appear. A heavy rain had come on about midnight, and now fell with so much violence, that some precautions were necessary, in order to prevent the firelocks from being rendered useless by wet. Such of the men as were fortunate enough to possess leathern cases, wrapped them round the locks of their muskets, whilst the rest held them in the best manner they could, under their elbows; no man thinking of himself, but only how he could best keep his arms in a serviceable condition.

'As soon as the first glimmering of dawn could be discerned, we moved to the road, and took up our wonted order of march; but before we pushed forward, the troops were desired to lighten themselves still further, by throwing off their blankets, which were to be left under a slender guard till their return. This was accordingly done; and being now unencumbered, except by a knapsack almost empty, every man felt his spirits heightened in proportion to the diminution of his load. The grief of soldiers is seldom of long duration, and though I will not exactly say that poor Ross was already forgotten, the success of yesterday had reconciled at least the privates to the guidance of their new leader; nor was any other issue anticipated than what would have attended the excursion had he still been its mainspring and director.

'The country through which we passed resembled, in every particular, that already described. Wood and cultivation succeeded each other at intervals, though the former surpassed the latter in tenfold extent; but instead of deserted villages and empty houses, which had met us on the way to Washington, we found most of the inhabitants remaining peaceably in their homes, and relying upon the assurance of protection given to them in our proclamations. Nor had they cause to repent of that confidence. In no instance were they insulted, plundered, or ill-treated; whereas every house which was abandoned fell a prey to the scouts and reconnoitring parties.

'But our march to-day was not so rapid as our motions generally were. The Americans had at last adopted an expedient which, if carried to its proper length, might have entirely stopped our progress. In most of the woods they had felled trees, and thrown them across the road; but as these abattis were without defenders, we experienced no other inconvenience than what arose from loss of time; being obliged to halt on all such occasions till the pioneers had removed the obstacle. So great, however, was even this hinderance, that we did not come in sight of the main army of the Americans till evening, though the distance travelled could not exceed ten miles.

'It now appeared that the corps which we had beaten yesterday was only a detachment, and not a large one, from the force collected for the defence of Baltimore; and that the account given by the volunteer troopers was in every respect correct. Upon a ridge of hills, which concealed the town itself from observation, stood the grand army, consisting of twenty thousand men. Not trusting to his superiority in numbers, their General had there entrenched them in a most formidable manner, having covered the whole face of the heights with breastworks, thrown back his left so as to rest it upon a strong fort erected for the protection of the river, and constructed a chain of field redoubts which covered his right and commanded the entire ascent. Along the side of the hill were likewise *flèches* and other projecting works from which a cross fire might be kept up; and there were mounted throughout this commanding position no less than one hundred pieces of cannon.

'It would be absurd to suppose that the sight of preparations so warlike did not in some degree damp the ardour of our leader; at least it would have been madness to storm such works without pausing to consider how it might best be attempted. The whole of the country within cannon-shot was cleared from wood, and laid out in grass and corn-fields; consequently there was no cover to shelter an attacking army from any part of the deadly fire which would be immediately poured upon it. The most prudent plan, therefore, was to wait till dark; and then, assisted by the frigates and bombs, which he hoped were by this time ready to co-operate, to try the fortune of a battle.

'Having resolved thus to act, Colonel Brook halted his army; and, secured against surprise by a well-connected line of piquets, the troops were permitted to light fires and to cook their provisions. But though the rain still fell in torrents, no shelter could be obtained;

and as even their blankets were no longer at hand, with which to form gipsy-tents, this was the reverse of an agreeable bivouac to the whole army.

'Darkness had now come on, and as yet no intelligence had arrived from the shipping. To assail such a position, however, without the aid of the fleet, was deemed impracticable; at least our chances of success would be greatly diminished without their co-operation. As the left of the American army extended to a fort built upon the very brink of the river, it was clear that could the ships be brought to bear upon that point, and the fort be silenced by their fire, that flank of the position would be turned. This once effected, there would be no difficulty in pushing a column within the works; and as soldiers entrenched always place more reliance upon the strength of their entrenchments than upon their own personal exertions, the very sight of our people on a level with them would in all probability decide the contest. At all events, as the column was to advance under cover of night, it might easily push forward and crown the hill above the enemy, before any effectual opposition could be offered; by which means they would be enclosed between two fires, and lose the advantage which their present elevated situation bestowed. All, however, depended upon the ability of the fleet to lend their assistance; for without silencing the fort, this flank could scarcely be assailed with any chance of success, and, therefore, the whole plan of operations must be changed.

'Having waited till it was considered imprudent to wait longer, without knowing whether he was to be supported, Colonel Brook determined, if possible, to open a communication with the fleet. That the river could not be far off we knew, but how to get to it without falling in with wandering parties of the enemy was the difficulty. The thing, however, must be done; and as secrecy, and not force, was the main object, it was resolved to dispatch for the purpose a single officer without an escort. On this service a particular friend of mine chanced to be employed. Mounting his horse, he proceeded to the right of the army, where, having delayed a few minutes till the moon rising gave light enough through the clouds to distinguish objects, he pushed forward at a venture, in as straight a line as he could guess. It was not long before his progress was stopped by a high hedge. Like knights-errant of old, he then gave himself up to the guidance of his horse, which taking him towards the rear, soon brought him into a narrow lane, that appeared to wind in the direction of the enemy's fort ; this lane he determined to follow, and holding a cocked pistol in his hand, pushed on, not perhaps entirely comfortable, but desirous at all hazards of executing his commission. He had not ridden far, when the sound of voices through the splashing of the rain arrested his attention. Pulling up, he listened in silence, and soon discovered that they came from two American soldiers, whether stragglers or sentinels it was impossible to divine ; but whoever they were, they seemed to be approaching. It now struck him that his safest course would be to commence the attack, and having therefore waited till he saw them stop short, as if they had perceived him, he rode forward, and called out to them to surrender. The fellows turned and fled, but galloping after them, he overtook one, at whose head he presented a pistol, and who instantly threw down his rifle, and yielded himself prisoner ; whilst the other, dashing into a thicket, escaped, probably to tell that he had been attacked by a whole regiment of British cavalry. Having thus taken a prisoner, my friend resolved to make him of some use ; with this view he commanded him to lay hold of his thigh, and to guide him directly to the river, threatening, if he attempted to mislead or betray him into the hands of the Americans, that he would instantly blow out his brains. Finding himself completely in my friend's power, the fellow could not refuse to obey ; and accordingly, the man resting his hand upon the left thigh of the officer, they proceeded along the lane for some time, till they came to a part where it branched off in two directions, My friend here stopped for a moment, and again repeated his threat, swearing that the instant his conduct became suspicious should be the last of his life. The soldier assured him that he would keep his word, and moreover informed him that some of our ships were almost within gun-shot of the fort ; a piece of information which was quickly confirmed by the sound of firing, and the appearance of shells in the air. They now struck to the right, and in half an hour gained the brink of the river : where my friend found a party just landed from the squadron, and preparing to seek their way towards the camp. By them he was conducted to the Admiral, from whom he learnt that no effectual support could be given to the land force ; for such was the shallowness of the river, that none except the very lightest craft could make their way within six miles of the town; and even these were stopped by vessels sunk in the channel, and other artificial bars, barely within a shell's longest range of the fort. With this unwelcome news he was accordingly forced to return; and taking his unwilling guide along with him, he made his way, without any adventure, to our advanced posts; where, having thanked the fellow for his fidelity, he rewarded it more effectually by setting him at liberty.

'Having brought his report to head-quarters, a council of war was instantly summoned to deliberate upon what was best to be done. Without the help of the fleet, it was evident that, adopt what plan of attack we could, our loss must be such as to counterbalance even success itself; whilst success, under existing circumstances, was, to say the least of it, doubtful. And even if we should succeed, what would be gained by it? We could not remove anything from Baltimore, for want of proper conveyances. Had the ships been able to reach the town, then, indeed, the quantity of booty might have repaid the survivors for their toil, and consoled them for the loss of comrades; but as the case now stood, we should only fight to give us an opportunity of reacting the scenes of Washington. To distress an enemy is, no doubt, desirable, but, in the present instance, that distress, even if brought upon the Americans, would cost us dear; whereas, if we failed, it was hardly possible to avoid destruction.

'Such was the reasoning which influenced the council of war to decide that all idea of storming the enemy's lines should be given up. To draw them from their works would require manœuvring, and manœuvring requires time; but delays were all in their favour, and could not possibly advantage us. Every hour brought in reinforcements to their army, whereas ours had no source from which even to recruit its losses; and it was, therefore, deemed prudent, since we could not fight at once, to lose no time in returning to the shipping.

'About three hours after midnight the troops were accordingly formed upon the road, and began their retreat, leaving the piquets to deceive the enemy, and to follow as a rear-guard. The rain, which had continued with little interruption since the night before, now ceased, and the moon shone out bright and clear. We marched along, therefore, not in the same spirits as if we had been advancing, but feeling no debasement at having thus relinquished an enterprise so much beyond our strength.

'When the day broke, our piquets, which had withdrawn about an hour before, rejoined us, and we went on in a body. Marching over the field where the battle of the 12th had been fought, we beheld the dead scattered about, and still unburied; but so far different from those which we had seen at Bladensburg, that they were not stripped, every man lying as he had fallen. One object, however, struck me as curious. I saw several men hanging lifeless among the branches of trees, and learnt that they had been riflemen, who chose, during the battle, to fix themselves in these elevated situations, for the combined purposes of securing a good aim and avoiding danger. Whatever might be their success in the first of these designs, in the last they failed; for our men soon discovered them, and, considering the thing as *unfair*, refused to give them quarter, and shot them on their perches.

'Here we paused for about an hour, that the soldiers might collect their blankets and refresh themselves; when we again moved forward passing the wood where the gallant Ross was killed. It was noon, and as yet all had gone on smoothly without any check or alarm. So little indeed was pursuit dreamt of that the column began to straggle, and to march without much regard to order; when suddenly the bugle sounded from the rear, and immediately after some musket shots were heard. In an instant the men were in their places, and the regiments wheeled into line, facing towards the enemy. The artillery turned round and advanced to the front; indeed I have never seen a manœuvre more coolly or more steadily performed on a parade in England than this rally. The alarm, however, turned out to be groundless, being occasioned only by the sudden appearance of a squadron of horse, which had been sent out by the American General to track our steps. These endeavoured to charge the rear-guard, and succeeded in making two prisoners; but a single shrapnel checked their farther advance, and sent them back at full speed to boast of the brave exploit which they had performed.

'Seeing that no attack was seriously intended, the army broke once more into the line of march, and proceeded to a favourable piece of ground, near the uncompleted position which I have already described, where we passed the night under little tents made with blankets

N

and ramrods. No alarm occurring, nor any cause of delay appearing, at daybreak we again got under arms, and pushed on towards the shipping, which in two hours were distinguishable.

'The infantry now halted upon a narrow neck of land, while the artillery was lifted into boats, and conveyed on board the fleet. As soon as this was done, brigade after brigade fell back to the water's edge and embarked, till finally all, except the light troops, were got off. These being left to cover the embarkation, were extended across the entire space which but a little before contained the whole army; but as no attempt was made to molest them, they had only the honour of being the last to quit the shore.'

The regiment during the engagement on the road to Baltimore was commanded by Brevet-Major Gubbins, in the absence of Colonel Thornton, Lieut.-Colonel Wood and Major Brown, all of whom were still lying wounded at Washington, and two letters from him are here quoted.

Letter from Brevet-Major Gubbins, 85th L.I.

'H.M.S. *Diadem*,
'CHESAPEAKE BAY,
'*September* 16, 1814.

'This letter will, I hope, be in time to accompany the account of another action into which the 85th Reg[t] had again the honour to be the leaders.

'Since the fortune of war has placed me in command of the Reg[t] I have had but little spare time. You will, therefore, forgive me for sending such scraps of letters.

'We disembarked our little army on the 12th and by three o'clock we had a very sharp action with from 8 to 12 thousand Americans. The 85th Reg[t] with the Light Companies of the Army were drawn up in front of the whole. As soon as everything was ready, we sounded the advance and made a grand dash at them, and, notwithstanding a very heavy fire from 6 pieces of cannon and a very close and extensive fire of musketry, we continued to make them fly. One of their guns was taken by Capt[n] De Bathe of my Reg[t]. Our loss in this battle was something above 200. The 85th Reg[t] was very fortunate in only having 32 killed and wounded, including 2 captains and one lieutenant.[1] The

[1] Captains De Bathe and Hicks, and Lieutenant Wellings.

American loss must have been very severe from the number of dead and wounded left on the field of Battle. They carried away a great many and we took numbers in the houses two or three miles off the day after the battle. On the 13th we advanced close to the enemy's entrenched position near Baltimore where after making a reconnaissance it was thought proper to retire. The principal reason for us not attacking the entrenchments in front of Baltimore was the inability of the Navy to assist, without which nothing of consequence could be done. The enemy's works were certainly very strong, and it is said they had 20,000 men.

'Poor Gen[l] Ross was killed by a Rifleman near the road some time before the action commenced—he is a great loss to the 85th Reg[t] which was in high favour with him.'

Another Letter on the same Action from Brevet-Major Gubbins.

'H.M.S. *Diadem*,
'*September* 16, 1814.

'A ten days' rest on board ship put us all in order again for service, and Baltimore being the next point of attack we landed on the 12th, 15 miles from the City, having increased our force with seamen and marines to about 5000. We were again fortunate in not meeting with opposition although the place of landing was favourable for it.

'The plan of operations upon this occasion was, as far as I am able to collect, for several Frigates and Bombs to attack the fort and batteries near the town which, if successful, would have enabled the troops to keep the heights in their vicinity which they easily could have gained. This would have put everything in our power, and have secured a retreat to the shipping.

'However, the Army moved forward as soon as landed and between 3 and 4 o'clock the advance came upon a small party of the enemy's riflemen. I am grieved to say here Major-General Ross met with his death—he was upon all occasions the most forward, and upon the present occasion particularly so. His was the greatest loss that could have befallen our little army; beloved and respected by everyone, he will be long lamented as a good and brave officer.

'Upon driving in the Riflemen we soon perceived the Yankees strongly posted in a thick wood having their flanks pretty well secured by creeks and swamps, a plain open ground in their front. We lost no time in pushing the Light Brigade as close to them as possible and without getting a shot excepting from six field-pieces I had the good fortune to get the 85th Reg^t within 500 yards of their line ; this enabled Col. Brooke who now commanded to bring up the column and form upon advantageous ground.

'The wood in which the enemy was posted was so very thick that we could only see a double line from which we received a heavy fire as soon as an advance was sounded. Much time was not given for firing, we soon got amongst them and put them to flight, one of their guns falling into the hands of the 85th Regt.

'The ground narrowed so much as we advanced that we (the 85th) were obliged to halt and secure ourselves from the fire of our own line in rear which prevented us from taking many more prisoners and I think more guns.

'Our whole loss was little above 200 killed and wounded. The enemy's must have been much greater from the number we saw dead on the ground and from the many wounded prisoners we made the following day two or three miles from the field of action. The strength of the enemy's force was variously stated at from 7 to 12,000 men.

'We moved forward very early the day after the battle and by nine o'clock got a fine view of the entrenched position in front of Baltimore. The Light Brigade was again pushed on and extended, when a grand reconnaissance was made and communication opened with the Frigates and Bombs.[1]

'After much discussion it was made known that Sir A. Cochrane could do nothing to assist the operations of the troops and that he advised the General to retire to the shipping.

'This determined Col. Brooke to retire ; we accordingly moved back to the wood where we halted till the whole of our wounded were sent away and receipts received for those of the enemy, we afterwards marched to within 2 miles of the landing place, and the following day got every man on board ship without the smallest interruption from the Yankeys.

'The failure of our attempt upon Baltimore has caused much talk amongst us, many think that the Navy should, even with the

[1] Gunboats ; or boats with one heavy gun or mortar.

heavy loss they might have sustained, have stormed the Fort—others think the troops should have gone on notwithstanding the Admiral's advice and want of assistance. The entrenchments thrown up by the Americans were very extensive and on commanding ground ; they had also a trench in rear of the position all round the town. There were several redoubts and batteries in all directions with a considerable moveable force of artillery with a force of from 15,000 to 20,000 men, Militia, Volunteers, and Regulars, of the latter very few, the forts, batteries etc. were manned by sailors.

'I am certainly of opinion we could have gained any point we attacked, but whether we could have maintained ourselves with an inferior force, or attempted the destruction of the town without the assistance of our men-of-war I cannot pretend to say.

'The Americans are easily to be made good soldiers, but they have no officers, and have not the smallest idea of ground or movement. At the battle of Bladensburg two companies of the 85th got round their right and decided the battle directly ; on the 12th a company of the 85th got round their left and threw them into great confusion and took a gun. They had no Light Troops in front of their lines and during the 10 days we were in the Washington country and the three days near Baltimore they never harassed our advance or retreat, although the country is the very best I ever saw for small bodies to hang upon the movements of an army, with but little or no risk to themselves.'

Gleig's Diary Continued.

Friday, 16th.—'I have often said to myself who would not be a soldier, and I am sure the insinuation is just. The comfort I enjoy from my last night's sleep, and the real luxury of being clean much more than repays me for all I have suffered these few days back. Employed myself in writing home, and to Alexander, after which went on board the *Tonnant* to see Burrell. Played a rubber of whist in the evening.'

Saturday, 17th.—'A fine fair day which we enjoy after the late rainy weather. A brig came in last night bringing the intelligence of the coming of ten thousand fresh troops. Had they but arrived a few days ago! About eleven o'clock we got under weigh, blowing very hard and stood towards Pine Island. The wind was against us, so it was evening before we got near enough to anchor. Remained at anchor off the Island all night.'

Sunday, 18th.—'Got under weigh at daylight, and stood for the Patuxent river. Passed close by Annapolis and felt not a little vexed that we did not destroy it. It blew very fresh all day—a side wind. Looked out anxiously for the reinforcements of 7 thousand men we hear have arrived within the bay but could not see anything of them. Came to at the mouth of the river.'

Monday, 19th.—'Got under weigh again early in the morning and ran up opposite our old watering places where we anchored. There is a court-martial sitting on board the *Royal Oak* on some seamen for desertion. Sent the boats ashore for water.'

Tuesday, 20th.—'The boats are all watering and every preparation seems to be making for a long voyage. The dispatches sent this morning in a gun brig for England. MacDougall, Hamilton and Thornton go in her. The *Tonnant* has sailed for Halifax and it is supposed we shall wait her return "for war in one" without attempting anything. Two seamen hung on board the *Royal Oak*—Charlton did not go.'

Wednesday, 21st.—'The unhealthy season has at length commenced and dysentery has made great progress through the fleet. I myself have had some slight symptoms of its approach, but have not yet suffered much. Went ashore this evening with Blanchard and Wilkinson, and visited an old negro couple. Got some apples from them. How delightful does a walk ashore appear to a person so long pent up on board of ship as I have been, how sweetly every shrub and flower smells, and how soothing to the ear is the whisper of the wind among the foliage. It spreads a melancholy languor over the mind and recalls more forcibly than anything else could do the recollection of former scenes to the memory.'

Thursday, 22nd.—'Spent the morning in great pain from gripes. The day was raw and unhealthy, but the evening was fine and Blanchard and I went ashore again. We carried as a present for our old friends the Negroes a bottle of rum and some doses of salts for which they were very grateful. Sat under a tree and read a very amusing American publication written in the style of the "Spectator." It is really a work of great merit. Remained here till sunset when we came on board again.'

Friday, 23rd.—' A very rainy bad day, no stirring from the ship. Nothing to be done, no books to read. I was obliged rather than die of ennui to play cards.'

Saturday, 24th.—' Received a short visit from Grey. Got a muster roll made out. Went ashore with an armed party to the house of a Mr. Corral and purchased five sheep, met with no adventures and came safely off again.'

Sunday, 25th.—' Felt very ill this morning. Took a large dose of laudanum which completely overpowered me and I kept my bed all day.

Monday, 26th.—' Went ashore again with ten men armed. Visited several families, and heard of two hundred horse being at a village a mile off. Got some geese and little pigs and remained ashore till dark, when after having paid another visit to Mr. Corral we returned on board.'

Tuesday, 27th.—' Admiral Malcolm with a number of ships sailed from this river and the *Diadem* among the rest. The *Asia* now commands here and it is said we shall follow the rest when the watering is completed. Remained on board all day.'

Wednesday, 28th.—' Wills and the D[r] went fishing but as I did not feel altogether well I did not accompany them. They brought back enough fish to last the whole ship for a week. We continue at anchor doing nothing.'

Thursday, 29th.—' Went ashore again with a party and got some geese. Shot a partridge and remained till quite dark after having gone two or three miles up the country. We were very anxious to procure papers, but the post is not allowed now to come down to the coast. Came on board again in safety.'

Friday, 30th.—' The signal was made to weigh early this morning and we sailed from the Patuxent. This part of the coast is now perfectly familiar to me, so it had not even the charm of novelty to render it interesting. The wind was fair and towards evening we entered the Potomac. The breeze, however, died away before we came up to the shipping so we were obliged to let go and anchor sooner than we wished.'

October, 1814.

Saturday, 1*st*.—' Weighed again about eleven o'clock and got up to the fleet when we brought up close to a little Island in the possession of the British. Some of our Officers came on board from the *Diadem* and told us of the narrow escape of four of them. They had gone ashore with an unarmed party and were attacked by 70 mounted riflemen, who took the whole party except the Officers who escaped into the wood, and burned the boat. The Officers remained in the wood till ten o'clock at night when they got off in a canoe belonging to one of the inhabitants.'

This incident is more fully related in the history as follows :

' Whilst the banks of the rivers continued in our possession, and the interior was left unmolested to the Americans, a rash confidence sprang up in the minds of all, insomuch that parties of pleasure would frequently land without arms, and spend many hours on shore. On one of these occasions, several officers from the 85th regiment agreed to pass a day together at a farm-house, about a quarter of a mile from the stream ; and taking with them ten soldiers, unarmed, to row the boat, a few sailors, and a young midshipman, not more than twelve years of age, they proceeded to put their determination into practice. Leaving the men, under the command of their youthful pilot, to take care of the boat, the officers went on to the house ; but they had not remained there above an hour, when they were alarmed by a shout, which sounded as if it came from the river. Looking out, they beheld their party surrounded by seventy or eighty mounted riflemen ; the boat dragged upon the beach, and set on fire. Giving themselves up for lost, they continued for an instant in a sort of stupor ; but the master of the house, to whom some kindness had been shown by our people, proved himself grateful, and, letting them out by a back door, directed them to hide themselves in the wood, whilst he should endeavour to turn their pursuers on a wrong scent. As they had nothing to trust to except the honour of this American, it cannot be supposed that they felt much at ease ; but, seeing no better course before them, they resigned themselves to his guidance, and plunging into the thicket, concealed themselves as well as they could among the underwood.

In the meantime the American soldiers, having secured all that were left behind, except the young midshipman, who fled into the wood in spite of their fire, divided into two bodies, one of which approached the house, whilst the other endeavoured to overtake the brave boy. It so chanced that the party in pursuit passed close to the officers in concealment, but by the greatest good fortune failed to observe them. They succeeded, however, in catching a glimpse of the midshipman, just as he had gained the water's edge and was pushing off a light canoe which he had loosened from the stump of a tree. The barbarians immediately gave chase, firing at the brave lad, and calling out to surrender; but the gallant youth paid no attention either to their voices or their bullets. Launching his little bark, he put to sea with a single paddle, and, regardless of the showers of balls which fell about him, returned alone and unhurt to the ship.

'Whilst one party was thus employed, the other hastened to the house in full expectation of capturing the British officers. But their host kept his word with great fidelity, and, having directed his countrymen towards another farm-house at some distance from his own, and in an opposite quarter from the spot where his guests lay, he waited till they were out of sight, and then joined his new friends in their lurking-place. Bringing with him such provisions as he could muster, he advised them to keep quiet till dark, when, their pursuers having departed, he conducted them to the river, supplied them with a large canoe, and sent them off in perfect safety to the fleet.

'On reaching their ship, they found the 85th regiment under arms, and preparing to land, for the purpose of either releasing their comrades from captivity, or inflicting exemplary punishment upon the farmer by whose treachery it was supposed that they had suffered. But when the particulars of his behaviour were related, the latter alternative was at once abandoned; and it was determined to force a dismissal of the captives, by advancing up the country, and laying waste everything with fire and sword. The whole of the light brigade was accordingly carried on shore, and halted on the beach, whilst a messenger was sent forward to demand back the prisoners. Such, however, was the effect of his threatening, that the demand was at once complied with, and they returned on board without having committed any ravages, or marched above two miles from the boats.'

Diary Continued.

Sunday, 2nd.—'I this morning went on board the *Asia* to see Grey where I remained till two, and then went to the *Diadem* to dine.'

Monday, 3rd.—'Went to the *Asia* again to spend the day, and returned to the *Golden Fleece* again after dinner, bringing Grey with me. An addition to our party arrived this evening of two Officers and forty men, as the *Diadem* is going home. The *Thames* is now our Head-quarters ship.'

Tuesday, 4th—'Went on shore to the Island for the purpose of shooting, but the day was so insufferably hot there was no moving. Saw some guns and muskets practising and came on board early. An order is issued for the landing of all the troops but the Light Brigade. What the intention is I know not but wish them success. Boats with troops were rowing ashore all night.'

Wednesday, 5th.—'Here we are on board in a state of anxiety for those ashore. We could hear no news nor see anything as the place they landed at is fifteen miles distant. At night they all came back having done simply nothing. Rain, thunder and wind all night.'

Thursday, 6th.—'Remained at anchor all day, no news nor anything going on to amuse.'

Friday, 7th.—'Weighed anchor early this morning, and tried to run down the river, but the wind was so foul we were obliged to anchor again. The *Diadem* is taking in Yankee prisoners for England.'

Saturday, 8th.—'Got under weigh with a fine fair wind and went down the Potomac, and soon got into the Bay again. Cast anchor at night off the mouth of the river.'

Sunday, 9th.—'A beautiful day with a fair wind. Ran down the length of James's river off which we again brought up. A flag of truce followed the Admiral and as we passed by we were delighted to hear ourselves hailed by Dr. Heaven, who had been left with the wounded at Bladensburg, and still more so when he told us he had brought Cols. Thornton and Wood and our wounded men with him.'

Monday, 10th.—'Remained at anchor all day, and though the wind was high and the tide strong, I went on board the *Royal Oak* to see our Field-Officers. Got back with some difficulty bringing Sergeant Higgins with me.'

Tuesday, 11th.—' Ran down with fair wind to the mouth of the Bay where we brought up. The *Menelaus* has gone to Bermuda taking Col. Wood in her. Poor Dr. Williams is dead and in him we lose one of the greatest ornaments of the Corps.'

Wednesday, 12th.—' Got under weigh but the wind proved foul and we anchored again. Towards evening a strange Schooner came in which proved to be the *Whiting* bringing despatches from Sir Alexander Cochrane. She left England on the 13th of August and brings a great deal of news.'

Thursday, 13th.—' A strange Frigate and Brig have come in last night and must have brought orders, for the Fleet this morning put to sea, but we seem to be waiting for something as we only keep beating off and on. Stood out to sea at night.'

Friday, 14th.—' Stood in again this morning, but the wind coming to blow very hard we put about and ran right before it, course S.S.E. We go seven and eight knots under double-reefed topsails.'

Saturday, 15th.—' The gale increases and the sky looks dirty. We carry scarcely any sails and go at an amazing rate. The ship rolls most terribly and is a very uncomfortable residence.'

Sunday, 16th.—' The wind just the same and we stand the same course, where we can be going God knows but they say it is to Jamaica. I hope so.'

Monday, 17th.—' This morning there was not a ship in sight but the Bomb. We lay to and in a couple of hours the Fleet overtaking us, we made weigh again, for sail we have none set. We continue to go eight and nine knots an hour under bare poles.'

Tuesday, 18th.—' The wind much abated this morning, but the swell still continues to render the cabin most uncomfortable. Towards evening there was so little wind that we could scarce make four knots with all sail set. The news of our destination has however transpired. It is Jamaica. This is the pleasantest piece of intelligence I have heard for some time. How delightful will be my meeting with Dr. and Mrs. Watt.'

Wednesday, 19th.—' A dead calm. I have got Burns' works, prose as well as poetical, which I read with great pleasure. His letters, however, except those to Thomson, are in my opinion greatly beneath him, in fact not worth reading. They breath the same sentiments throughout, dressed in almost the same language. His correspondence with Thomson does him honour, and his poetical works it is needless to say delighted me as they do all the world.'

Thursday, 20th.—'A little wind in the morning but it died away towards noon. I have finished "Burns" and have begun "Thaddeus of Warsaw," for, I believe, the third time. It is without exception the most interesting novel I ever read. A boat came on board from the Admiral but brought no intelligence which we did not know.'

Friday, 21st.—'A little breeze again this morning. This is the ever memorable anniversary of the battle of Trafalgar. We did not forget it but drank in the true spirit of patriotism, to the memory of the hero who perished that day. Some Dolphins made their appearance round the ship and the grains were got ready but for no purpose, as they never made their appearance again.'

Saturday, 22nd.—'Quite calm again. The first thing I heard on getting up was the death of one of the Sappers who had died in the night. Poor fellow he was sewed up in a hammock, and after the burial service had been read over him by Blanchard, was committed to the deep. Ball and Wilkinson came on board but had no news to tell, except that the *Ramillies*, the *Thames* and our Ship were to separate from the Fleet and go on to Kingston, while the others followed to Negril Bay. A fair wind in the evening.'

Sunday, 23rd.—'A nice little breeze this morning, but the sky lowers so much, the Fleet seem afraid to carry sail. I never saw the heavens look more threatening, something will happen I daresay before night. As I predicted, a squall came on about two o'clock with a tremendous shower of rain, but did not last long, after which the wind came foul, and we go on with difficulty close hauled. Another squall took us again at midnight, and once more changed the wind to fair.'

Monday, 24th.—'A fine fair breeze blows upon the quarter sending us at the rate of 8 knots. It gradually came round till it blew right aft, which caused a great deal of unpleasant motion. The weather is cooler to-day than it has been for some time. The Companies were mustered to-day.'

Tuesday, 25th.—'The fair wind still continues but the day is immensely hot. We to-day crossed the line expecting every moment to receive a visit from Neptune, but he did not make his appearance all day. At night we lay to, and just as we were thinking of going to bed we were disturbed by a noise upon deck. Upon running up we found that Neptune had just boarded us. He was personated by the Mate dressed in the most ridiculous manner, with his face painted red and black, with a spun-yarn beard and queue, with a rusty cutlass by his side, and his wife a seaman in woman's clothes. It was very badly acted, but he administered the usual oaths and fined each of us a gallon of rum.'

Wednesday, 26th.—'The first thing discernible in the morning was the Caycos Island, and we had got so near the shore that if the ship had missed stays we should certainly have been lost. The Admiral who was just as near fired two guns to alarm the Fleet, and immediately put about. The whole then stood out to sea, and went round by the other side. We did not lose sight of land till late in the evening. I was disappointed in not witnessing any slavery.'

Thursday, 27th.—'The heat is scarcely bearable. Another Island is in sight but we cannot tell the name of her exactly. All this morning taken up with courts-martial. At night the *Volcan* Bomb hailed us, and told us we were to proceed under her convoy direct to Port Royal. Continued with a fair breeze under a good deal of sail all night.'

Friday, 28th.—'There is a dead calm this morning but we have done so well during the night that St. Domingo is opposite to us. The Fleet was a great way astern but near enough for us to discern the Admiral's signal for us to lie-to. While in this situation a very large dolphin made its appearance at the head of the ship. Grey took his grains and after a little delay struck it. It was upwards of four feet long and weighed 26 pounds. The Horse-ship ran foul of us but did us no damage. Four Officers of the 21st have come on board to take a passage to Port Royal. A fair breeze in the evening.'

Saturday, 29th.—'The wind continues fair but light and the weather is excessively warm. A shark made his appearance at the stern this morning. The line and hook were immediately got out, and the fellow was so voracious that he took the bait directly. As soon as he was properly hooked the boat was lowered and a noose thrown over his head. The boat then stood alongside to the gangway when he was drawn up. He was not a large one, only measuring six feet and a half and weighing 90 pounds. He made much less resistance when in the water than I expected, but the moment he was laid on the deck he began lashing with his tail so as very shortly to clear the quarter-deck. They cut his tail off, opened him, and took out his entrails, and after all this was done he still continued to writhe about and bite at every thing. We are still opposite the coast of St. Domingo, which has a very fine appearance, owing to the heights of its mountains and the boldness of its shores.'

Sunday, 30th.—'We leave St. Domingo behind every hour. We have been amused all this morning with water-spouts. They have a very fine appearance, but one came so near us that as there was not time to load a cannon we got five or six muskets ready to fire at it. It dispersed, however, before it reached us, without having properly formed. A fine 8 knot breeze all day.'

Monday, 31st.—'About two o'clock this morning was awoke by the Mate coming in to say there was an American Privateer in chase of us. I started up and the first thing I heard was a shot whistle past us. All hands were immediately called to quarters, the guns loaded and everything prepared for action. Our Commodore the Bomb is a little ugly thing carrying twelve guns, with exactly the appearance of a Merchant-man; so there was every reason to believe he would attack us. We took in a number of our sails and waited for him, but when he saw this he hove to and kept at a distance till daylight. The moment it was light we sent all the men below to deceive him for he was now not very far away from us. At half-past seven he was so near that the Commodore fired a gun at him, and we immediately followed his example. This seemed to enrage the Yankee for he immediately bore down upon the Bomb. All this while the Bomb and we kept up a cross fire of grape and round shot, every one of which went into his decks. He fired a shot at each of us, and when he got near enough gave us a volley of musketry and immediately ran alongside of the Bomb and attempted to board. We now called up all our men and kept up such a hot fire of small arms as very speedily drove every man from his decks. He was now close to the *Volcan*, but seemed to discover his mistake very soon, for he put his helm up and sheered off. In quitting the Commodore he came very near us with his lea side off into which we poured such a shower of musket balls that not a man dared show his nose. We now chased her, pounding away our guns at her, but she sailed so much better than either of us that she succeeded in getting away. She was a large Schooner, I understand, of twelve guns called the *Saucy Jack*. We had not a man touched and the Bomb had only three, while the enemy's loss must have been at least twenty. Had we but been to Windward instead of to Leeward, we should certainly have taken her. We continued our course the rest of the day without any adventure except that the blue mountains of Jamaica were discernible towards the evening.'

November, 1814.

Tuesday, 1st.—' On coming up this morning we had a fine view of the coast of Jamaica, now about five miles from us. The mountains are very bold and rocky and appear to peculiar advantage with their summits concealed as they were in the mist of the morning. A black pilot came on board, but from him we could get no information as he scarcely spoke English. The wind dying away towards evening prevented our getting into the harbour, so we were obliged to anchor about four miles from it. The moon rose beautifully to-night, and shone so bright that every object on the land was discernible, and the smell from the mountains was truly fragrant. I sat alone for some hours on the taffrail enjoying the pleasures of Imagination.'

Wednesday, 2nd.—' The wind for a long time this morning was very uncertain, and it was eleven o'clock before we got into the harbour. The entrance is fine, and the view up the river extremely beautiful. The Bay is defended by several very extensive forts bristling with cannon. We passed close to the shore in coming in, which was crowded with spectators. Immediately on our coming to an anchor which we did close to the small town of Port Royal, I got into a boat with several others and set off for Kingston. It is about nine miles from the anchorage, which distance we performed in something less than two hours, and on landing went immediately to an Hotel.'

Thursday, 3rd.—' Kingston is a place which covers a pretty considerable extent of ground, but composed almost entirely of wooden houses. The streets are not narrow, but excessively dirty and intensely hot. It has completely the appearance of what it really is, a place dependent upon commerce for its very existence, where no attention is paid to beauty of architecture or the improvement of manners, but where every idea seems enveloped with the important object of pounds shillings and pence. I walked about a good deal this morning, but went on board after breakfast. I was pulled to the ship by unfortunate wretches compelled to labour at the oar for the benefit of others under the rays of a burning sun, without a covering to their heads and naked to the waist. Such are the horrors of slavery. Dined and slept on board.'

Sunday, 6th.—'Walked about all day and at four o'clock drove in a gig to the Barracks to dine with the 18th. They gave us a dinner and allowed us to do as we chose, by which means I returned at ten o'clock to my lodgings perfectly sober.'

Monday, 7th.—'After breakfast went out to walk. Met with Duxbury who gave me fifty dollars. Lent West 20 and with the other thirty got myself shirts and other things which I wanted. In walking down one of the streets I was called into a house by some brown ladies who very politely gave me an invitation to a ball this evening, which I of course accepted. Dined with Blanchard, dressed and went to the ball. It was one composed entirely of brown girls. There were not less than forty of this description of woman, whose manners and dress were as unexceptional as those of the modest ladies in England. Danced till four o'clock in the morning when the ball broke up, and we conducted our partners home.'

Here follows a description of Gleig's visit to his uncle, Mr. Watt, who lived 30 miles inland at Belfield Ben. He did not return till the 14th and was apparently absent without leave; for on the 14th the diary continues:—

Monday 14th.—'I arrived at Kingston at nine o'clock in time to breakfast with Grey, from whom I learned that Col. Thornton was so indignant with me for going away without leave that he had put me in arrest. I immediately wrote a long letter to Gubbins in which I explained matters, and called with it myself, but Thornton not being at home he could do nothing. Dined with Mr. Johnstone and called in the evening upon Grey after which I returned to my lodgings to sleep.'

Tuesday, 15th.—'Called after breakfast upon Gubbins but found the Colonel had not yet come home, and as my coin was getting rather short, I was determined not to endeavour any more to see him but to go down to the Ship, which resolution I accordingly put in force.'

Wednesday, 16th.—' Remained on board all day. About mid-day received a note from Gubbins in which he told me I was no longer in arrest. In the evening all the Officers came on board except Blanchard and Fonblanque of the 21st, as we sail to-morrow for Negril. They brought the sea-stock with them the price of which amounts to no less than sixty dollars each.'

Thursday, 17th.—' Got under weigh this morning at daylight, and stood out of the harbour of Port Royal with a fine fresh low breeze. About 8 o'clock the wind died away, and the swell was most disagreeable. The calm lasted till near twelve when the sea breeze came on and we made considerable way. The breeze lasted all day and we continued our course without any adventure.'

Friday, 18th.—' We are now running down to Negril Bay without ever losing sight of the land. Some telegraph passed between our Commodore the *Volcan*, and the *Trave*, but what the subject was I know not. Employed all the morning on courts-martial. Came in sight of the shipping in the Bay that evening, but the wind blowing rather against us we could not get in, a calm towards dark.'

Saturday, 19th.—' The breeze having sprung up again this morning we continued our course and having got well to windward run into the Bay, where we brought up.'

Sunday, 20th.—' Went ashore immediately after breakfast to land our stock. Got a horse from West when he and I set out to take a ride. Passed through an Estate called Providence and went on to another. As we were riding past the door, a gentleman came out and asked us in to dinner. We did not think it was at all necessary to refuse this invitation so we e'en dismounted and amused ourselves with walking about the premises. After dinner we returned to the beach, and on the way met with a countryman of mine a good deal intoxicated, with whom we amused ourselves all the way. Went on board to sleep.'

Tuesday, 22nd.—' Remained in the *Fleece* all day. Sir Alexander Cochrane has at length arrived in Port Royal with the wished for reinforcements, and General Kean ; in consequence of which the horses are all embarked and everything in a state of readiness to set sail the moment he appears off the point.'

Wednesday, 23rd.—' While we were standing on the deck after breakfast some ships of war made their appearance round the point. We were in high hopes it was Sir Alexander Cochrane, but it proved to be only the *Asia, Thames, Trave* and *Weser* from Bluefields.'

Thursday, 24th.—' About 9 o'clock this morning the long wished for *Tonnant* made her appearance and with her came an immense fleet of troop-ships and two transports having the 93rd, six Companies of the 95th and two West India Regiments on board, besides two squadrons of the 14th Dragoons, some Artillery, Rockets, Sappers, etc., and General Kean to command us. We did not as we expected join them on their appearing, but they all entered the Bay and came to an anchor. The detachment was mustered as usual by Grey, and a detachment consisting of five Officers and 70 men has come out to us which makes us once more 400 strong.'

Friday, 25th.—' Sir Alexander weighed anchor this morning and along with the *Ramillies* and two brigs stood to sea. All the rest of the Fleet under Admiral Malcolm remained quiet.'

Saturday, 26th.—' At half past seven o'clock this morning the whole Fleet got under weigh, but the calm was so great that by two o'clock the shores of Jamaica were still discernible. The only thing which attached me to this Island was my friendship with Watt, but even it is not powerful enough to make me regret quitting it. Except the few days I passed at Belfield Ben, I have not spent what I call one happy day since my arrival. It is therefore impossible that I should feel much in bidding an adieu to Jamaica which I hope from my soul may be eternal. The calm continued till towards dark, when a fresh breeze sprung up, but such is the perverseness of human affairs that the arrival of a little Schooner from Port Royal prevented the Admiral from taking advantage of it, and we did nothing but beat off and on all night.'

Sunday, 27th.—' At daylight this morning we began to stand our course in good earnest, and as the wind blew rather fresh we got on at a smart rate. By the time I got upon deck there was no land in view, nothing but one wide waste of water all round. When will this detestable system of marining about end, I am most heartily sick of it.'

Monday, 28th.—'The wind still continues fair and we get over the ground as fast as a few bad-sailing transports will allow. The Fleet has really a fine appearance. It consists of seven two-deckers, and thirty-two sail of Frigates, Sloops of War, Bombs and Transports. The further we get from Jamaica, the pleasanter becomes the climate, though it is still a great deal too hot.'

Tuesday, 29th.—'The breeze still continues fair, but we seem to have got too far to Leeward; for we are now beating up to Windward. About eleven o'clock the cry of land was given, and on looking out it was discernible on both sides of the vessel's head. We steered right towards it, and in a very short time came so close that we could discern houses and trees on the shore. It proved to be the Island of Grand Cayman. Although the sea was running very high two Canoes came off to us, from which we got 538 pounds of Turtle. We remained beating about here for two hours when we again made sail and stood our course. This Island lies very low, and by the account of the Natives is exceedingly healthy, as it is exposed to every breeze that blows.'

Wednesday, 30th.—'Long before the dawn of this morning we had lost sight of the Grand Cayman. We are still favoured by a fair wind, but there is nothing whatever to amuse. I have not even the comfort of a book to read, as I have perused all our Library so often that I have almost got every page in it by heart. We cooked one of the Turtle to-day and like any Alderman I dined on soup and green fat.'

December, 1814.

Thursday, 1st.—'A fine steady breeze continues to blow sending us at the rate of 3 and 4 knots with a calm sea and very little motion. The Admiral this morning made a signal to the Ships of War, that he would change the numbers of the Telegraph flags, but which means we are now left totally in the dark about whatever passes among them. This is no small annoyance to us, as watching signals used to be our principal amusement.'

Friday, 2nd.—'On coming up this morning the land of Cuba was in sight. We continued to sail all day along shore till we came to the utmost extremity of the Island, where the Gulf of Mexico commences, on passing which we were within two miles of the land. The moment we entered the Gulf, there was an almost instantaneous change from the smoothness of the sea to a most disagreeable swell. The wind, however, continued to blow fair and pretty fresh, so we very soon lost sight of Cuba. The shores of Cuba, or at least what we saw of them, are in general low, though there are some places rather rocky, and all overgrown with trees, but of what kind we were not near enough to ascertain.'

Saturday, 3rd.—' The old prospect again meets the eye—sea and air. The swell in this Gulf is more intolerable than anything I ever felt, the Bay of Biscay is a joke to it. It renders the Ship more uncomfortable than ever it was before, as it puts a total stop to walking. The wind continued nearly as it was all day, but towards night it freshened, and the sky had rather a nasty appearance, but it did not turn out anything serious, sending us on our course at the rate of six knots.'

Sunday, 4th.—' The wind continues fair with less swell and weather more moderate. Nothing worth writing occurred all day.'

Monday 5th.—' Smooth sea with fair wind. In the evening for want of something better to do we knocked up a bull-dance to the sound of Grey's fiddle which we kept up till 2 in the morning.'

Tuesday, 6th.—Wednesday, 7th.—' Continuing our course without anything to be seen, except on the last day when a strange brig made its appearance and held a long telegraph with the Admiral.'

Thursday, 8th.—' Towards noon as we were going on with all sail set, the sky suddenly put on a tremendous appearance, and we had u st time to take in all canvas when a very heavy squall took us. It luckily lasted a very short time and we rode it without receiving any injury. We had another dance this evening.'

Friday, 9th.—' The weather has undergone a severe change, and it is now so cold that I can bear a blanket on my bed, and the wind blows against us. We are in sight of land but can only beat off and on without making any progress on our way.'

Saturday, 10th.—' Rainy with cold winds. All that we can distinguish of the land are the trees as the coast lies very low. The Admiral made signal to anchor to-night to which we paid no attention, continuing to reach off and on all night.'

Sunday 11th.—' A fine clear frosty day. The wind has become rather more favourable though it is yet very scant, and we make a little way. Towards mid-day came in sight of Chandelier Island, and ran in so close as to distinguish houses on the shore. It is very low and apparently sandy.'

Monday, 12th.—' Saw Sir Alexander Cochrane at anchor a good way from us, and made all sail to windward. The air is uncommonly clear and frosty, rather too much so. At four o'clock brought up. Went on board the *Ramillies* and found the Regiment had gone to the *Bucephalus*. We are now in.'

Tuesday, 13th.—' Moved further up the lake. 50 men-of-war boats were sent to attack 5 armed cutters. Anchored in the evening, which we spent in dancing to the sound of Grey's fiddle.'

Wednesday, 14th.—' Sailed up and passed the troop-ships but ran aground at dusk. Heard of the success of our brave Tars.'

Thursday, 15th.—' Aground all day. Admiral Malcolm with a number of Engineers and other Officers came on board. Did not get off till night and were crowded with Officers and men who eat up all our stock.'

Friday, 16th.—' Ran a great way further up and again grounded. Bigger and Whitney came on board, and with them Jamie Sawers. A most uncomfortable bustle.'

Saturday, 17th.—' Still on board ship waiting for the boats. A number of the Officers came on board. A heavy night of rain.'

Sunday, 18th.—' At 7 o'clock A.M. boats came alongside, in which we embarked and after having been ten hours on the water exposed to heavy rain, we landed on a place called Pine Island where the troops are bivouacked. It is a complete marsh with scarce a tree on it. In the evening it froze and became very cold. Spent the night on the boat but with very little sleep.'

Monday, 19th.—' Employed ourselves all day in making a hut, but getting an offer from Dr. Baxter of sleeping in his tent we preferred that. More troops arriving all day.'

Tuesday, 20th.—' Turned out an hour before light. All late. To-day General Keane reviewed his little army and expressed his disapprobation of two Regiments in very strong terms. Remained quiet all day and slept in our hut.'

Wednesday, 21st.—' The whole army is now collected and preparations are making for crossing over to the main. [For the proposed attack on New Orleans.] Several American Officers came down deserters, with whom I had some conversation. They all say it will be an easy business.'

Thursday, 22nd.—'At 9 o'clock this morning the advance of the army consisting of the 85th, 95th and 4th Regiments under the command of Col. Thornton got into the boats and commenced crossing. From Pine Island to that part of the main towards which prudence directed us to steer, was a distance of no less than 80 miles. This, of itself, was an obstacle, or at least an inconvenience, of no slight nature; for should the weather prove boisterous, open boats, heavily laden with soldiers, would stand little chance of escaping destruction in the course of so long a voyage. In the next place, and what was of infinitely greater importance, it was found that there were not, throughout the whole fleet, a sufficient number of boats to transport above one-third of the army at a time. But to land in divisions would expose our forces to be attacked in detail, by which means one party might be cut to pieces before the others could arrive to its support. The undertaking was, therefore, on the whole, extremely dangerous, and such as would have been probably abandoned by more timid leaders. Ours, however, were not to be alarmed. They had entered upon a hazardous business, in whatever way it should be prosecuted; and since they could not work miracles, they resolved to lose no time in bringing their army into the field in the best manner which circumstances would permit. The day was wet and cold, so our situation was none of the most comfortable. We continued in the boats all day and at night anchored and spread awnings. But this we enjoyed very short time, for we were soon ordered to go on again. Spent the night in the boats.

'There was a small piquet of the enemy stationed at the entrance of the creek by which it was intended to effect our landing. This it was absolutely necessary to surprise; and whilst the rest lay at anchor, two or three fast-sailing barges were pushed on to execute the service. Nor did they experience much difficulty in accomplishing their object. Nothing, as it appeared, was less dreamt of by the Americans than an attack from this quarter, consequently no persons could be less on their guard than the party here stationed. The officer who conducted the force sent against them, found not so much as a single sentinel posted! But having landed his men at two places, above and below the hut which they inhabited, he extended his ranks so as to surround it, and closing gradually in, took them all fast asleep, without noise or resistance.

'When such time had been allowed as was deemed sufficient for the accomplishment of this undertaking, the flotilla again weighed anchor, and without waiting for intelligence of success, pursued their voyage. Hitherto we had been hurried along at a rapid rate by a fair breeze, which enabled us to carry canvas; but this now left us, and we made way only by rowing. Our progress was therefore considerably retarded, and the risk of discovery heightened by the noise which that labour necessarily occasions; but in spite of these obstacles, we reached the entrance to the creek by dawn.'

Friday, 23rd.—'About ten o'clock this morning the whole of the Advance landed at a place called Baio de Cateline about ten miles below New Orleans, in number 1600. When we were formed we moved forward over very bad roads about two miles, when we came to the main road, which ran parallel with the river and close to the bank all the way to New Orleans. Here we halted and passed the day in peace in a green field where we lit fires and cooked our dinners. We were once a little alarmed by the appearance of some horsemen, but the pickets soon put them to flight. Our dinner was excellent with claret to it, and when it got dark we prepared to lie down. But our rest was soon disturbed by the appearance of a schooner which came stealing up the river and anchored opposite to us and immediately saluted us with a broadside of grape. This of course threw us into no small confusion, and every man ran as fast as he could to get shelter under the bank which happened most providentially to be high. But even this shelter we did not long enjoy, as our pickets were by this time attacked, both in front and on the right. We were now ordered out to oppose the enemy, and Grey's was the first company to go. We moved out amid the shower of grape from the vessel and soon came into a stubble field at the top of which were a great many men, but it was so dark we could not tell friends from foes. They kept firing at us, but we would not return a shot lest they should be our own people. We had now got up to a place where there was good cover, and at my request Grey halted the men while I went on alone to find out who were opposite to us. I got within ten yards of them when I readily distinguished that they were enemies. I returned and told Grey, when we again advanced to within twenty yards of them, when we halted behind some piles of reeds which offered good shelter to the men, as Grey was not yet convinced of their being Americans. Here we agreed to separate, and taking ten or twelve men with me, I went off to the right that I might charge those fellows upon their flank. I left Grey, we never met again. When I had gone a little way to the right I met some of the 95th, and joining them to my party we jumped over a paling and advanced to the attack. The Yankees gave way directly, and we drove them over the field and through a small village of huts, where we released a number of our own men who had been prisoners. In this village we halted, and on mustering my little party I found I had upwards of thirty men and two officers with me. On looking down a field on the left I saw a large line of fellows at the bottom, and leaving my men in the village I went down alone to see who they were. As I walked down the field a man cried out to me not to fire upon them for they were the 2nd Battalion of the 1st American Regiment. I told them that I knew they were and that their first Battalion was upon my right, desiring them at the same time to wait there until I should join them with my party. I then went back, made my men fall in, fix their bayonets, and then follow me down the field. The Yankees allowed us to approach without firing a shot till we were up to them, and their commanding officer coming forward I seized his sword and told him he was surrounded and all his regiment prisoners. At first they began to lay down their arms, but seeing so few they took them up again. I now turned to another officer and demanded his sword, which he refused to give, and we had a regular fight, but he turned to fly when I got one fair cut at the back of his head. We now made a dash into the middle of them, laid about us in all directions and soon put them to flight. There was one rascal had his bayonet in my stock and was fumbling for the trigger when I made a blow at his head which caused him to drop his firelock and run away. The enemy were now repulsed in all directions and we fell back into the village. I was here told of the death of poor Grey. Had he been my brother it would not have shocked me more.

I went to where he lay. I wept over him, but that was all I could do, for we were ordered to fall back to our original position where we formed line, remained under arms the rest of the night, and towards daylight crept under the bank.'

Saturday, 24th.—'The first thing I did this morning was to look for the body of my poor departed friend. I found him lying upon a coat where the soldiers had brought him, shot through the head. When I saw him pale and bloody I thought my heart would have broken. I got a grave dug for him at the bottom of a garden in which I had the melancholy satisfaction of seeing him decently laid. Poor fellow, so ended a friendship which existed without interruption, without one quarrel, for two years. The rest of the day I spent I know not how, but it was a miserable one. When it became well dark we moved off one by one from the main road and by a change of position made our left be in the village.'

We cannot forbear repeating the story of this day as told more fully by Gleig later on in his history :—

'The place where we landed was as wild as it is possible to imagine. Gaze where we might, nothing could be seen except one huge marsh covered with tall reeds ; not a house nor a vestige of human industry could be discovered ; and even of trees there were but a few growing upon the banks of the creek. Yet it was such a spot as, above all others, favoured our operations. No eye could watch us, or report our arrival to the American General. By remaining quietly among the reeds, we might effectually conceal ourselves from notice ; because, from the appearance of all around, it was easy to perceive that the place which we occupied had been seldom, if ever before, marked with a human footstep. Concealment, however, was the thing of all others which we required ; for be it remembered that there were now only sixteen hundred men on the mainland. The rest were still at Pine Island, where they must remain till the boats which had transported us should return for their conveyance, consequently many hours must elapse before this small corps could be either reinforced or supported. if, therefore, we had sought for a point where a descent might be made in secrecy and safety, we could not have found one better calculated for that purpose than the present ; because it afforded every means of concealment to one part of our force, until the others should be able to come up.

'For these reasons, it was confidently expected that no movement would be made previous to the arrival of the other brigades; but, in our expectations of quiet, we were deceived. The deserters who had come in, and accompanied us as guides, assured the General that he had only to show himself, when the whole district would submit. They repeated, that there were not five thousand men in arms throughout the State: that of these, not more than twelve hundred were regular soldiers, and that the whole force was at present several miles on the opposite side of the town, expecting an attack on that quarter, and apprehending no danger on this. These arguments, together with the nature of the ground on which we stood, so ill calculated for a proper distribution of troops in case of attack, and so well calculated to hide the movements of an army acquainted with all the passes and tracks which, for aught we knew, intersected the morass, induced our leader to push forward at once into the open country. As soon, therefore, as the advance was formed, and the boats had departed, we began our march, following an indistinct path along the edge of the ditch or canal. But it was not without many checks that we were able to proceed. Other ditches, similar to that whose course we pursued, frequently stopped us by running in a cross direction, and falling into it at right angles. These were too wide to be leaped, and too deep to be forded; consequently, on all such occasions, the troops were obliged to halt, till bridges were hastily constructed of such materials as could be procured, and thrown across.

'Having advanced in this manner for several hours, we at length found ourselves approaching a more cultivated region. The marsh became gradually less and less continued, being intersected by wider spots of firm ground; the reeds gave place, by degrees, to wood, and the wood to inclosed fields. Upon these, however, nothing grew, harvest having long ago ended. They accordingly presented but a melancholy appearance, being covered with the stubble of sugar-cane, which resembled the reeds which we had just quitted in everything except altitude. Nor as yet was any house or cottage to be seen. Though we knew, therefore, that human habitations could not be far off, it was impossible to guess where they lay, or how numerous they might prove; and as we could not tell whether our guides might not be deceiving us, and whether ambuscades might not be laid for our destruction as soon as we should arrive where troops could conveniently act, our march was insensibly conducted with increased caution and regularity.

'But in a little while some groves of orange-trees presented themselves; on passing which two or three farm-houses appeared. Towards these, our advanced companies immediately hastened, with the hope of surprising the inhabitants, and preventing any alarm from being raised. Hurrying on at double-quick time, they surrounded the buildings, succeeded in securing the inmates, and capturing several horses; but becoming rather careless in watching their prisoners, one man contrived to effect his escape. Now, then, all hope of eluding observation might be laid aside. The rumour of our landing would, we knew, spread faster than we could march; and it only remained to make that rumour as terrible as possible.

'With this view, the column was commanded to widen its files, and to present as formidable an appearance as could be assumed. Changing our order, in obedience to these directions, we marched not in sections of eight or ten abreast, but in pairs, and thus contrived to cover with our small division as large a tract of ground as if we had mustered thrice our present numbers. Our steps were likewise quickened, that we might gain, if possible, some advantageous position, where we might be able to cope with any force that might attack us; and thus hastening on, we soon arrived at the main road which leads directly to New Orleans. Turning to the right, we then advanced in the direction of that town for about a mile; when, having reached a spot where it was considered that we might encamp in comparative safety, our little column halted; the men piled their arms, and a regular bivouac was formed.

'The country where we had now established ourselves was a narrow plain of about a mile in width, bounded on one side by the Mississippi, and on the other by the marsh from which we had just emerged. Towards the open ground this marsh was covered with dwarf wood, having the semblance of a forest rather than of a swamp; but on trying the bottom, it was found that both characters were united, and that it was impossible for a man to make his way among the trees, so boggy was the soil upon which they grew. In no other quarter, however, was there a single hedgerow, or plantation of any kind, excepting a few apple and other fruit trees in the gardens of such houses as were scattered over the plain, the whole being laid out in large fields for the growth of sugar-cane, a plant which seems as abundant in this part of the world as in Jamaica.

'Looking up towards the town, which we at this time faced, the marsh is upon your right, and the river upon your left. Close to the latter runs the main road, following the course of the stream all the way to New Orleans. Between the road and the water is thrown up a lofty and strong embankment, resembling the dykes in Holland, and meant to serve a similar purpose; by means of which the Mississippi is prevented from overflowing its banks, and the entire flat is preserved from inundation. But the attention of a stranger is irresistibly drawn away from every other object, to contemplate the magnificence of this noble river. Pouring along at the prodigious rate of four miles an hour, an immense body of water is spread out before you, measuring a full mile across, and nearly a hundred fathoms in depth. What this mighty stream must be near its mouth, I can hardly imagine for we were here upwards of a hundred miles from the ocean.

'Such was the general aspect of the country which we had entered. Our own position, again, was this. The three regiments turning off from the road into one extensive green field, formed three close columns within pistol-shot of the river. Upon our right, but so much in advance as to be of no service to us, was a large house, surrounded by about twenty wooden huts, probably intended for the accommodation of slaves. Towards this house there was a slight rise in the ground, and between it and the camp was a small pond of no great depth. As far to the rear again as the first was to the front, stood another house, inferior in point of appearance, and skirted by no outbuildings: this was also upon the right; and here General Keane, who accompanied us, fixed his head-quarters; but neither the one nor the other could be employed as a covering redoubt, the flank of the division extending, as it were, between them. A little way in advance, again, where the outposts were stationed, ran a dry ditch and a row of lofty palings; affording some cover to the front of our line, should it be formed diagonally with the main road. The left likewise was well secured by the river ; but the right and the rear were wholly unprotected. Though in occupying this field, therefore, we might have looked very well had the country around us been friendly, it must be confessed that our situation hardly deserved the title of a military position. Noon had just passed, when the word was given to halt, by which means every facility was afforded of posting the piquets with leisure and attention. Nor was this deemed enough to secure tranquillity : parties were sent out in all directions to reconnoitre, who returned with an account that no enemy nor any trace of an enemy could be discerned. The troops were accordingly suffered to light fires, and to make themselves comfortable ; only their accoutrements were not taken off, and the arms were piled in such form as to be within reach at a moment's notice.

'As soon as these agreeable orders were issued, the soldiers proceeded to obey them both in letter and in spirit. Tearing up a number of strong palings, large fires were lighted in a moment ; water was brought from the river, and provisions were cooked. But their bare rations did not content them. Spreading themselves over the country as far as a regard of safety would permit, they entered every house, and brought away quantities of hams, fowls, and wines of various descriptions ; which being divided among them, all fared well, and none received too large a quantity.

'It was now about three o'clock in the afternoon, and all had as yet remained quiet. The troops having finished their meal, lay stretched beside their fires, or refreshed themselves by bathing, for to-day the heat was such as to render this latter employment extremely agreeable, when suddenly a bugle from the advanced posts sounded the alarm, which was echoed back from all in the army. Starting up, we stood to our arms, and prepared for battle, the alarm being now succeeded by some firing ; but we were scarcely in order, when intelligence arrived from the front that there was no danger, only a few horse having made their appearance, who were checked and put to flight at the first discharge. Upon this information, our wonted confidence returned, and we again betook ourselves to our former occupations, remarking that, as the Americans had never yet dared to attack, there was no great probability of their doing so on the present occasion.

'In this manner the day passed without any further alarm ; and darkness having set in, the fires were made to blaze with increased splendour, our evening meal was eaten, and we prepared to sleep. But about half-past seven o'clock, the attention of several individuals was drawn to a large vessel, which seemed to be stealing up the river till she came opposite to our camp; when her anchor was dropped, and her sails leisurely furled. At first we were doubtful whether she might not be one of our own cruisers which had passed the fort unobserved, and had arrived to render her assistance in our future operations. To satisfy this doubt, she was repeatedly hailed; but returning no answer, an alarm immediately spread through the bivouac, and all thought of sleep was laid aside. Several musket-shots were now fired at her with the design of exacting a reply, of which no notice was taken; till at length, having fastened all her sails, and swung her broadside towards us, we could distinctly hear some one cry out in a commanding voice, 'Give them this for the honour of America.' The words were instantly followed by the flashes of her guns, and a deadly shower of grape swept down numbers in the camp.

'Against this destructive fire we had nothing whatever to oppose. The artillery which we had landed was too light to bring into competition with an adversary so powerful; and as she had anchored within a short distance of the opposite bank, no musketry could reach her with any precision or effect. A few rockets were discharged, which made a beautiful appearance in the air; but the rocket is at the best an uncertain weapon, and these deviated too far from their object to produce even terror amongst those against whom they were directed. Under these circumstances, as nothing could be done offensively, our sole object was to shelter the men as much as possible from the iron hail. With this view, they were commanded to leave the fires, and to hasten under the dyke. Thither all accordingly repaired, without much regard to order and regularity, and laying ourselves along wherever we could find room, we listened in painful silence to the pattering of grape-shot among our huts, and to the shrieks and groans of those who lay wounded beside them.

'The night was now dark as pitch, the moon being but young, and totally obscured with clouds. Our fires deserted by us, and beat about by the enemy's shot, began to burn red and dull, and, except when the flashes of those guns which played upon us cast a momentary glare, not an object could be distinguished at the distance of a yard. In this state we lay for nearly an hour, unable to move from our ground, or offer any opposition to those who kept us there; when a straggling fire of musketry called our attention towards the piquets, and warned us to prepare for a closer and more desperate struggle. As yet, however, it was uncertain from what cause this dropping fire arose. It might proceed from the sentinels, who, alarmed by the cannonade from the river, mistook every tree for an American; and till the real state of the case should be ascertained, it would be improper to expose the troops by moving any of them from the shelter which the bank afforded. But these doubts were not permitted to continue long in existence. The dropping fire having paused for a few moments, was succeeded by a fearful yell; and the heavens were illuminated on all sides by a semi-circular blaze of musketry. It was now manifest that we were surrounded, and that by a very superior force; and that no alternative remained, except to surrender at discretion, or to beat back the assailants.

'The first of these plans was never for an instant thought of; the second was immediately put into force. Rushing from under the bank, the 85th and 95th flew to support the piquets, whilst the 4th, stealing to the rear of the encampment, formed close column, and remained as a reserve. And now began a battle of which no language were competent to convey any distinct idea; because it was one to which the annals of modern warfare furnish no parallel. All order, all discipline were lost. Each officer, as he succeeded in collecting twenty or thirty men about him, plunged into the midst of the enemy's ranks, where it was fought hand to hand, bayonet to bayonet, and sabre to sabre.

'I am well aware that he who speaks of his own deeds in the field of battle lies fairly open to the charge of seeking to make a hero of himself in the eyes of the public; and feeling this, it is not without reluctance that I proceed to recount the part which I myself took in the affair of this night. But, in truth, I must either play the egotist awhile, or leave the reader without any details at all; inasmuch as the darkness and general confusion effectually prevented me from observing how others, except my own immediate party, were employed.

'Offering this as my apology for a line of conduct which I should otherwise blush to pursue, and premising that I did nothing, in my own person, which was not done by my comrades at least as effectually, I go on to relate as many of the particulars of this sanguinary conflict as came under the notice of my own senses.

'My friend Grey and myself had been supplied by our soldiers with a couple of fowls taken from a neighbouring hen-roost, and a few bottles of excellent claret, borrowed from the cellar of one of the houses near. We had built ourselves a sort of hut, by piling together, in a conical form, a number of large stakes and broad rails torn up from one of the fences; and a bright wooden fire was blazing at the door of it. In the wantonness of triumph, too, we had lighted some six or eight wax-candles; a vast quantity of which had been found in the store-rooms of the chateaux hard by; and having done ample justice to our luxurious supper, we were sitting in great splendour and in high spirits at the entrance of our hut, when the alarm of the approaching schooner was communicated to us. With the sagacity of a veteran, Grey instantly guessed how matters stood: he was the first to hail the suspicious stranger; and on receiving no answer to his challenge, he was the first to fire a musket in the direction of her anchorage. But he had scarcely done so when she opened her broadside, causing the instantaneous abandonment of fires, viands, and mirth throughout the bivouac.

'As we contrived to get our men tolerably well around us, Grey and myself were among the first who rushed forth to support the piquets and check the advance of the enemy upon the right. Passing as rapidly as might be through the ground of encampment amidst a shower of grape-shot from the vessel, we soon arrived at the pond; which being forded, we found ourselves in front of the farm-house of which I have already spoken as composing the head-quarters of General Keane. Here we were met by a few stragglers from the out-posts, who reported that the advanced companies were all driven in, and that a numerous division of Americans was approaching. Having attached these fugitives to our little corps, we pushed on, and in a few seconds reached the lower extremity of a sloping stubble-field, at the other end of which we could discern a long line of men, but whether they were friends or foes the darkness would not permit us to determine. We called aloud for the purpose of satisfying our doubts;

but the signal being disregarded, we advanced. A heavy fire of musketry instantly opened upon us ; but so fearful was Grey of doing injury to our own troops, that he would not permit it to be returned. We accordingly pressed on, our men dropping by ones and twos on every side of us, till having arrived within twenty or thirty yards of the object of our curiosity, it became to me evident enough that we were in front of the enemy. But Grey's humane caution still prevailed ; he was not convinced, and till he should be convinced it was but natural that he should not alter his plans. There chanced to be near the spot where we were standing a huge dung-heap, or rather a long solid stack of stubble, behind which we directed the men to take shelter whilst one of us should creep forward alone, for the purpose of more completely ascertaining a fact of which all except my brave and noble-minded comrade were satisfied. The event proved that my sight had not deceived me. I approached within sabre's length of the line ; and having ascertained beyond the possibility of doubt that the line was composed of American soldiers, I returned to my friend and again urged him to charge. But there was an infatuation upon him that night for which I have ever been unable to account : he insisted that I must be mistaken ; he spoke of the improbability which existed that any part of the enemy's army should have succeeded in taking up a position in rear of the station of one of our outposts, and he could not be persuaded that the troops now before him were not the 95th Rifle corps. At last it was agreed between us that we should separate ; that Grey with one half of the party should remain where he was, whilst I with the other half should make a short détour to the right, and come down upon the flank of the line from whose fire we had suffered so severely. The plan was carried into immediate execution. Taking with me about a dozen or fourteen men, I quitted Grey, and we never met again.

'How or when he fell I know not ; but, judging from the spot and attitude in which I afterwards found his body, I conceive that my back could have been barely turned upon him when the fatal ball pierced his brain. He was as brave a soldier and as good a man as the British army can boast of ; beloved by his brother officers and adored by his men. To me he was as a brother ; nor have I ceased even now to feel, as often as the 23rd of December returns, that on

P

that night a tie was broken than which the progress of human life will hardly furnish one more tender or more strong. But to my tale.

'Leaving Grey—careless as he ever was in battle of his own person and anxious as far as might be to secure the safety of his followers—I led my little party in the direction agreed upon, and fortunately falling in with about an equal number of English riflemen, I caused them to take post beside my own men, and turned up to the front. Springing over the paling, we found ourselves almost at once upon the left flank of the enemy; and we lost not a moment in attacking it, But one volley was poured in, and then bayonets, musket-butts, sabres, and even fists, came instantly into play. In the whole course of my military career I remember no scene at all resembling this. We fought with the savage ferocity of bull-dogs; and many a blade which till to-night had not drunk blood became in a few minutes crimsoned enough.

'Such a contest could not in the nature of things be of very long continuance. The enemy, astonished at the vigour of our assault, soon began to waver, and their wavering was speedily converted into flight. Nor did we give them a moment's time to recover from their panic. With loud shouts we continued to press upon them; and amidst the most horrible din and desperate carnage drove them over the field and through the little village of huts, of which notice has already been taken as surrounding the mansion on our advanced right. Here we found a number of our own people prisoners, and under a guard of Americans. But the guard fled as we approached, and our countrymen catching up such weapons as came first to hand, joined in the pursuit.

'In this spot I halted my party, increased by the late additions to the number of forty; among whom were two gallant young officers of the 95th. We had not yet been joined, as I expected to be joined, by Grey; and feeling that we were at least far enough in advance of our own line, we determined to attempt nothing further except to keep possession of the village should it be attacked. But whilst placing the men in convenient situations, another dark line was pointed out to us considerably to the left of our position. That we might ascertain at once of what troops it was composed, I left my brother officers to complete the arrangements which we had begun, and walking down the field, demanded in a loud voice to be informed who they were that kept post in so retired a situation. A voice from the throng made answer that they were Americans, and begged of me not to fire upon my friends. Willing to deceive them still further, I asked to what corps they belonged; the speaker replied that they were the second battalion of the first regiment, and inquired what had become of the first battalion. I told him that it was upon my right, and assuming a tone of authority, commanded him not to move from his present situation till I should join him with a party of which I was at the head.

'The conversation ended here, and I returned to the village; when, communicating the result of my inquiries to my comrades, we formed our brave little band into line and determined to attack. The men were cautioned to preserve a strict silence, and not to fire a shot till orders were given; they observed these injunctions, and with fixed bayonets and cautious tread advanced along the field. As we drew near, I called aloud for the commanding officer of the second regiment to step forward, upon which an elderly man, armed with a heavy dragoon sabre, stepped out of the ranks. When he discovered by our dress that we were English, this redoubtable warrior lost all self-command; he resigned his sword to me without a murmur, and consented at once to believe that his battalion was surrounded, and that to offer any resistance would but occasion a needless loss of blood. Nor was he singular in these respects: his followers, placing implicit reliance in our assurances that they were hemmed in on every side by a very superior force, had actually begun to lay down their arms, and would have surrendered, in all probability, at discretion, but for the superior gallantry of one man. An American officer, whose sword I demanded, instead of giving it up as his commander had done, made a cut at my head, which with some difficulty I managed to ward off; and a few soldiers near him, catching ardour from his example, discharged their pieces among our troops. The sound of firing was no sooner heard than it became general, and as all hope of success by stratagem might now be laid aside, we were of necessity compelled to try the effect of violence. Again we rushed into the middle of the throng, and again was the contest that of man to man, in close and desperate strife; till a panic arising among the Americans, they dispersed in all directions and left us masters of the field.

'In giving a detail so minute of my own adventures this night, I beg to repeat what has been stated already, that I have no wish whatever to persuade my readers that I was one whit more cool or more daring than my companions. Like them I was driven to depend from first to last, upon my own energies; and I believe the energies of few men fail them when they are satisfied that on them alone they must depend. Nor was the case different with my comrades. Attacked unexpectedly, and in the dark,—surrounded, too, by a numerous enemy, and one who spoke the same language with ourselves,—it is not to be wondered at if the order and routine of civilised warfare were everywhere set at nought. Each man who felt disposed to command was obeyed by those who stood near him, without any question being asked as to his authority; and more feats of individual gallantry were performed in this single night than many regular campaigns might furnish an opportunity to perform.

'The night was far spent, and the sound of firing had begun to wax faint, when, checking the ardour of our brave followers, we collected them once more together and fell back into the village. Here likewise considerable numbers from other detachments assembled, and here we learned that the Americans were repulsed on every side. The combat had been long and obstinately contested: it began at eight o'clock in the evening and continued till three in the morning—but the victory was ours. True, it was the reverse of a bloodless one, not fewer than two hundred and fifty of our best men having fallen in the struggle: but even at the expense of such a loss, we could not but account ourselves fortunate in escaping from the snare in which we had confessedly been taken.

'To me, however, the announcement of the victory brought no rejoicing, for it was accompanied with the intelligence that my friend was among the killed. I well recollect the circumstances under which these sad news reached me. I was standing with a sword in each hand—my own and that of the officer who had surrendered to me, and, as the reader may imagine, in no bad humour with myself or with the brave fellows about me, when a brother officer stepping forward abruptly told the tale. It came upon me like a thunderbolt; and casting aside my trophy, I thought only of the loss which I had sustained. Regardless of every other matter I ran to the rear, and found Grey lying behind the dung-heap, motionless and cold. A little pool of blood which had coagulated under his head, pointed out the spot where the ball had entered, and the position of his limbs gave proof that he must have died without a struggle. I cannot pretend to describe what were then my sensations, but of whatever nature they might be, little time was given for their indulgence; for the bugle sounding the alarm, I was compelled to leave him as he lay, and join my corps. Though the alarm proved to be a false one, it had the good effect of bringing all the troops together, by which means a regular line was now, for the first time since the commencement of the action, formed. In this order, having defiled considerably to the left, so as to command the highway, we stood in front of our bivouac till dawn began to appear; when, to avoid the fire of the schooner, we once more moved to the river's bank, and lay down. Here, during the whole of the succeeding day, the troops were kept shivering in the cold frosty air, without fires, without provisions, and exhausted with fatigue; nor was it till the return of night that any attempt to extricate them from their comfortless situation could be made.

'Whilst others were thus reposing, I stole away with two or three men for the purpose of performing the last sad act of affection which it was possible for me to perform to my friend Grey. As we had completely changed our ground, it was not possible for me at once to discover the spot where he lay; indeed I traversed a large portion of the field before I hit upon it. Whilst thus wandering over the arena of last night's contest, the most shocking and most disgusting spectacles everywhere met my eyes. I have frequently beheld a greater number of dead bodies within as narrow a compass, though these, to speak the truth, were numerous enough, but wounds more disfiguring or more horrible I certainly never witnessed. A man shot through the head or heart lies as if he were in a deep slumber; insomuch that when you gaze upon him you experience little else than pity. But of these, many had met their deaths from bayonet wounds, sabre cuts, or heavy blows from the butt ends of muskets; and the consequence was, that not only were the wounds themselves exceedingly frightful, but the very countenances of the dead exhibited the most savage and ghastly expressions. Friends and foes lay together in small groups of four or six, nor was it difficult to tell almost the very hand by which some of them had fallen. Nay, such had been the deadly closeness of the strife, that in one or two places an English and American soldier might be seen with the bayonet of each fastened in the other's body.

'Having searched for some time in vain, I at length discovered my friend lying where during the action we had separated, and where, when the action came to a close, I had at first found him, shot through the temples by a rifle bullet so remarkably small as scarcely to leave any trace of its progress. I am well aware this is no fit place to introduce the working of my own personal feelings, but he was my friend, and such a friend as few men are happy enough to possess. We had known and loved each other for years; our regard had been cemented by a long participation in the same hardships and dangers, and it cannot therefore surprise, if even now I pay that tribute to his worth and our friendship which, however unavailing it may be, they both deserve.

'When in the act of looking for him I had flattered myself that I should be able to bear his loss with something like philosophy, but when I beheld him pale and bloody, I found all my resolution evaporate. I threw myself on the ground beside him and wept like a child. But this was no time for the indulgence of useless sorrow. Like the royal bard, I knew that I should go to him, but he could not return to me, and I knew not whether an hour would pass before my summons might arrive. Lifting him therefore upon a cart, I had him carried down to head-quarter house, now converted into an hospital, and having dug for him a grave at the bottom of the garden, I laid him there as a soldier should be laid, arrayed, not in a shroud, but in his uniform. Even the privates whom I brought with me to assist at his funeral mingled their tears with mine, nor are many so fortunate as to return to the parent dust more deeply or more sincerely lamented.'

The loss of the regiment was severe. Captains Grey and Harris, Lieutenant Hickson and 13 rank and file were killed; Captain Knox, Lieutenants Wellings and Maunsell, 4 sergeants, 2 buglers and 59 rank and file wounded.

Another account gives more details of this somewhat curious conflict, and is extracted from the despatch of Major-General Keane dated 26–12–1814 :

'In order to reach the high road, on the left bank of the Mississippi, leading to New Orleans, the Light Brigade under Colonel Thornton of the 85th, consisting of the 85th and 95th with the 4th Regiment in support, was placed in boats, and the 21st, 44th and 93rd in small vessels. The whole sailed at 10 A.M. on the 22nd and landed at daybreak next morning without opposition.

'A way was forced through several fields of reeds intersected by deep muddy ditches and bordered by a low swampy wood. Colonel Thornton with the Light Division then advanced and gained the high road and took up a position with his right resting on the road, and his left on the Mississippi, where he intended to remain until the boats returned with the rest of the troops, the vessels which carried them having grounded at some distance off.

'At 8 o'clock in the evening when the men, much fatigued by the length of time they had been in the boats, were asleep in their bivouac, a heavy flanking fire of round and grape shot was opened upon them by a large schooner and two gun-vessels, which had dropped down the river from the town and anchored abreast of our fires.

'Colonel Thornton in the most prompt and judicious manner placed his brigade under the inward slope of the bank of the river, where it was covered from their guns.

'A most vigorous attack was then made on the advanced front and right flank picquet, the former of the 95th, the latter the 85th, under Captain Schaw. The picquets conducted themselves with firmness and checked the enemy for a considerable time ; but renewing their attack with a large force, Colonel Thornton moved up the remainder of both these corps.

'The 85th Regiment was commanded by Brevet-Major Gubbins whose conduct cannot be too much commended. On the approach of his Regiment to the point of attack, the enemy, favoured by the darkness of the night, concealed themselves under a high fence and calling to the men as friends under pretence of being part of our own force, offered to assist them in getting over, which was no sooner accomplished than the 85th found itself in the midst of very superior numbers, who called on the Regiment immediately to surrender—the answer was an instantaneous attack; a more extraordinary conflict has, perhaps, never occurred, absolutely hand-to-hand, both officers and men. It terminated in the repulse of the enemy with the capture of 30 prisoners.

'The enemy, finding his reiterated attacks were repulsed by Colonel Thornton, advanced a large column against the centre; but a charge by a portion of the 93rd, with the 4th Regiment in support, drove back this attack.

'The enemy now determined on making a last effort, and collecting the whole of his force, formed an extensive line and moved directly against the Light Brigade. All our advanced posts were driven in, but Colonel Thornton, whose noble exertion had guaranteed all former success, was at hand; he rallied his brave comrades round him, and moving forward with a firm determination of charging, appalled the enemy, who, from the lesson he had received on the same ground in the early part of the evening, thought it prudent to retire, and did not again dare to advance.'

Continuation of Lieutenant Gleig's Diary.

Sunday, 25th.—'This day we spent in a state of tolerable quiet, as we had only one man killed by a shot from the schooner. All the officers clubbing together we eat a melancholy Christmas dinner in a small house. I saw poor Nixon to-day, he still lingers.'

Monday, 26th.—'I went on picket this morning a little before day-light. Just as the sun rose a battery of ours which has been constructing for some time opened upon the schooner and in very short time I had the satisfaction of seeing her blown up. We remained unmolested the rest of the day, but after dark they attempted to annoy us by firing at the sentries, by which means they kept us in a state of constant alarm all night.'

In his history Gleig relates how on this night his life was saved by his dog:

'It was on this night, and under these circumstances, that I was indebted to the vigilance of my faithful dog for my life. Amid all the bustle of landing, and throughout the tumult of the nocturnal battle, she never strayed from me; at least if she did lose me for a time, she failed not to trace me out again as soon as order was restored, for I found her by my side when the dawn of the 24th came on, and I never lost sight of her afterwards. It was my fortune on the night of the 26th to be put in charge of an outpost on the left front of the army; on such occasions I seldom experienced the slightest inclination to sleep; and on the present, I made it a point to visit my sentinels at least once in every half-hour. Going my rounds for this purpose, it was necessary that I should pass a little copse of low underwood, just outside the line of our videttes; and I did pass it again and again, without meeting with any adventure. But about an hour after midnight, my dog, which, as usual, trotted a few paces before me, suddenly stopped short at the edge of the thicket, and began to bark violently, and in great apparent anger. I knew the animal well enough to be aware that some cause must exist for such conduct; and I too stopped short, till I should ascertain whether danger were near. It was well for me that I had been thus warned; for, at the instant of my halting, about half a dozen muskets were discharged from the copse, the muzzles of which, had I taken five steps forward, must have touched my body. The balls whizzed harmlessly past my head; and, on my returning the fire with the pistol which I carried in my hand, the ambuscade broke up, and the party composing it took to their heels. I was Quixote enough to dash sword in hand into the thicket after them: but no one waited for me; so I continued my perambulations in peace.'

Tuesday, 27th.—'A little before daylight this morning that part of the picket which was composed of poor Grey's company was withdrawn, and as soon as day broke the whole of the army was put in motion. We advanced in two columns, our brigade forming the left one, proceeded by the main road without opposition for a distance of two miles, the enemy's pickets having been driven in without the firing of a musket. When we had proceeded so far we came to a turning of the road, which we no sooner made than we found ourselves in front of a formidable battery with the Frigate and some gun-boats on our flank, all of which immediately opened upon us and at the first discharge knocked down eleven men together with poor Eden. This, of course, threw us into considerable confusion, but we soon regained our order on being moved into the fields on our right, and advanced in line under a heavy fire of round and grape shot till we got within three or four hundred yards of the enemy's position. Here we halted as we found a deep canal effectually prevented any attack without more preparation than we had made. The place in which we lay down was a wet ditch up to our knees in water, without any cover except what could be got by concealment behind some rushes. We remained here under all their fire for an hour—and then fell back again in as straggling a manner as possible, taking up a position just within range of their guns, and having suffered less than might have been expected.'

Wednesday, 28th ; *Thursday, 29th.*—' During these two days we have remained very quiet except that they continue to fire at our Camp continually but without being able to reach us, and are making batteries on the opposite side of the river to enfilade our position, and during the latter day have opened two or three guns from them, which annoy us much.'

Friday, 30th.—' Went on picket this morning, and found it a most disagreeable one. They continued to fire their great guns at us all day, with one of which they knocked out the window shutter of the house behind which we took shelter, We did not, however, get one man wounded.'

Saturday, 31st.—' Before daylight we were relieved and remained very quiet in camp till it was dark when the 95th and we march to the front to cover the erection of some batteries, which are to open on the American position to-morrow at daylight. Remained during the entire frosty night without so much as a cloak while the rest of the Brigade worked. A pretty way to spend the last night of the year.'

January, 1815.

Sunday, 1st.—' A little before daylight we fell back about two hundred yards, and having formed close column lay down behind some rushes which fortunately concealed us from the observation of the enemy. The morning was foggy, which delayed the opening of our batteries, but as soon as it cleared away the firing commenced.

Ten minutes elapsed before a shot was returned, but the fire then became very smart on both sides. This cannonading continued with little interruption till dusk, when it closed without any effect whatever having been produced. This day cost us only one officer and one man. As soon as the firing was over I was sent on a working party on which I continued till two in the morning, when I returned to the camp and threw myself down in the hut quite worn out for want of sleep.'

Monday, 2nd.—' This day we were left in a state of tolerable quiet, but our pickets had a smart skirmish in which Lieutenant Boyes was wounded.'

Tuesday, 3rd.—' I was this day ordered on command to the rear, a distance of four miles, and returned again without having done anything but procure four geese.'

Wednesday, 4th; Thursday, 5th; Friday, 6th.—' During the 4th I remained in camp all day, and on the 5th went on picket, which was rendered still more disagreeable as we could not light fires. On the 6th I was removed into Lehaw's Division.'

Saturday, 7th.—' And on the 7th again went on picket, but we were relieved again directly. About twelve o'clock the Regiment received an order to march down to head-quarter house for the purpose of crossing the river, together with some marines and seamen in order to carry all the batteries on the opposite side. The difficulty in getting the heavy boats up the narrow canal was so great that it was very late before they were got into the river, by which means the day was just beginning to dawn as we landed.'

Sunday 8th.—' The moment we had landed, which was done without opposition, our division forming the advanced guard, we pushed on directly. We had not gone far when we fell in with a strong party of the enemy posted upon the high road, who, the moment they saw us, formed line behind a deep canal. To dislodge them from this took us only five minutes, and was effected without any loss. We followed them through a wood, and soon saw them again drawn up behind a strong entrenchment, reinforced, and about 1500 in number. Our little party amounted in all to only 450 men, but the inequality of numbers was no obstacle. We attacked immediately, and in a quarter of an hour drove them from their works and completely routed them. We followed them about a mile and a half further and halted behind a canal, when I directly went on picket. The fruits of this victory were 16 pieces of cannon destroyed and two carried off by us. We had remained in this position only half an hour when an order came for us to fall back, which we did, my picket forming the rear-guard, and reimbarked in our boats without further molestation, but completely fatigued. On reaching the opposite bank the first intelligence that met our ears was the unfortunate result of the main attack with the death of Generals Pakenham and Gibbs.'

In order that the reader may follow the general trend of this action it is necessary to amplify the diary and give a general view of the affair.

On Christmas Day, 1814, Major-General Sir Edward Pakenham, C.B., arrived from England and assumed the command. The 85th formed part of the 3rd Brigade under the command of Major-General Keane. The brigade was composed as follows: 85th Light Infantry, 95th Rifle Corps (5 companies), and the 93rd Regiment. The 5th West India Regiment was attached to do duty with the 3rd Brigade.

On December 28, 1814, the army advanced on New Orleans. While the regiment was employed in making a reconnaissance of the enemy's position on this day, a fight took place in which Ensign Sir F. Eden, Bart., 1 bugler, and 5 rank and file were killed; while Lieutenant Ormsby and 13 rank and file were wounded.

'Sir Edward Pakenham ordered the army to be divided, part to be sent across the river to seize the enemy's guns, and to turn them on themselves. The remainder were to make a general assault along the whole entrenchment; but prior to this it was absolutely essential that a canal should be cut across the entire neck of land from the Bayo de Cataline to the river, of sufficient width and depth to admit of boats being brought up from the lake. This was effected by January 6th. The 7th and 43rd Regiments arrived unexpectedly under Major-General Lambert, each 800 strong. Our numbers were a little short of 8000; those of the enemy about 25,000.

'The canal being finished the boats were accordingly ordered up for the transportation of 1400 men; and Colonel Thornton (with the 85th

Regiment, the marines, and a party of sailors) was appointed to cross the river.

'The passage however got blocked, and instead of there being accommodation for 1400 men, only a number of boats sufficient to contain 350 could reach their destination.

'Colonel Thornton's detachment was to cross the river immediately after dark. They were to push forward, so as to carry all the batteries, and point the guns before daylight; when, on the throwing up of a rocket, they were to commence firing upon the enemy's line, which at the same moment was to be attacked by the main body of our army.

'Sir Edward Pakenham divided his troops into three columns:

'General Keane (with the 95th, the Light Companies of the 21st, 4th and 44th and the two black corps) to make a demonstration or sham attack on the right.

'General Gibbs (with the 4th, 21st, 44th and 93rd) to force the enemy's left.

'General Lambert (with the 7th, 43rd and remainder) in reserve.

'Scaling ladders and fascines had been prepared, with which to fill up the ditch and mount the wall. The 44th (being not numerically strong enough, but at the same time accustomed to American warfare) were selected for this service.

'These were the arrangements on the night of the 7th inst., for the 8th was the day on which the fate of New Orleans would be decided. While the rest of the army lay down to sleep till it should be required to fight, Colonel Thornton with the 85th and a corps of marines (amounting to 1400 in all) moved down to the brink of the river.

'As yet no boats had arrived; and when they did, only a few, owing to the explanation given above.

'Still, as they were absolutely necessary in order to put this part of the plan into execution, the Colonel, dismissing the rest of his followers, put himself at the head of his own regiment, with about 50 seamen and as many marines, and with this force (not more than 340 men) pushed off. But the delay was irreparable; for, instead of reaching the opposite bank by midnight, dawn was beginning to appear before the boats quitted the canal.

'It was in vain that they rowed in perfect silence with muffled oars, and gained the point of disembarkation without being observed, when they made good their landing and formed up upon the beach.

'The signal rocket was seen on arrival, while four miles from the batteries.

'Through inadvertence the 44th forgot their ladders, etc., and General Pakenham told Colonel Mullens to return for them. Meanwhile on the left bank of the river our troops were visible and were mowed down in hundreds by a terrific fire. General Pakenham ordered the advance. On the left, detachments of the 95th, the 21st and the 4th Regiments stormed a three-gun battery and took it. On the right, the 21st and 4th were almost cut to pieces, and the 93rd Regiment pushed on and took the lead.

'In vain the ditch was reached—the carnage was frightful, though the most obstinate courage was displayed.

'They fell by the hands of men whom they absolutely did not see; for the Americans, without so much as lifting their faces above the ramparts, swinging their firelocks by one arm over the wall, discharged them directly upon their heads.'

General Pakenham did his utmost to rally his broken troops.

Ordering the 44th with Colonel Mullens to advance, he was even preparing to himself lead them when he received a slight wound in the knee.

Later, he was mortally wounded. Generals Gibbs and Keane were not inactive — in fact, both were finally wounded — but the retreat became a flight, and the ground was quitted in the utmost disorder.

The retreat was covered in the most gallant manner by the reserve.

Making a forward movement, the 7th and 43rd feinted a renewed attack, and the enemy did not venture beyond their lines in pursuit of the fugitives.

While these events were in progress on one side of the river, the party under Colonel Thornton had gained the landing place on the other side, and the first thing they perceived (as has been stated) was the rocket thrown up to show that the battle had begun. This naturally added ' wings to their speed.'

Forming into one little column, and pushing forward a single company as an advanced guard, they reached in half an hour the canal, along the opposite bank of which a detachment of Americans were drawn up.

These were easily dislodged, but this was only an outpost. The main body of the enemy, amounting to no fewer than 1,500 men, were stationed some distance in the rear, and it was not long before these presented themselves. Strongly entrenched behind a thick parapet with a ditch, a battery on their left swept the whole position, and two field-pieces commanded the road.

The British had no artillery, nor had they any means (beyond what nature had bestowed on them) for scaling the rampart.

' The 85th, extending its files, stretched across the entire line of the enemy; the sailors in column prepared to storm the battery; while the marines remained some little distance in rear of the centre as a reserve, On the bugle sounding our troops advanced. The sailors raising a shout rushed forward, but were met by so heavy a discharge of grape and canister that for an instant they paused. Recovering themselves, however, they again pushed on; and the 85th, dashing forward to their aid (though they received a heavy fire of musketry *en route*) endeavoured to charge. A smart musketry fire now continued for a few minutes on both sides, but " our people had no time to waste in distant fighting " and accordingly hurried on to storm the works; upon which a panic seized the Americans, they lost their order and fled, leaving us in possession of their tents and eighteen pieces of cannon.

' Our losses were 3 men killed and 40 wounded, among the latter being Colonel Thornton.'

The enemy were pursued for two miles; but the actual state of affairs becoming known, the men were recalled from pursuit. Eventually the little corps became once more united, and re-embarking on the boats reached the opposite bank again without molestation.

The British had gained a position across the river from which the whole American line was enfiladed, but, alas! it was too late, as the main body on the other side had already retired.

Colonel Thornton in his despatch thus describes the action of his detachment :

Colonel Thornton's despatch to Major-General the Honourable Sir Edward Pakenham is here given *in extenso* :—

'REDOUBT, ON THE RIGHT BANK OF THE MISSISSIPPI,
'*January* 8, 1815.

'SIR,

'I lose no time in reporting to you the success of the troop which you were yesterday pleased to place under my orders, with the view of attacking the enemy's redoubt and position on this side of the river. It is within your own knowledge that the difficulty had been found so extremely great of dragging the boats through the canal which had been lately cut with so much labour to the Mississippi, that notwithstanding every possible exertion for the purpose, we were unable to proceed across the river until eight hours after the time appointed, and even then, with only a third part of the force which you had allotted for the service.

'The current was so strong, and the difficulty, in consequence, of keeping the boats together so great, that we only reached this side of the river at daybreak, and, by the time the troops were disembarked, which was effected without any molestation from the enemy, I perceived by the flashes of the guns that your attack had already commenced. This circumstance made me extremely anxious to move forward, to prevent the destructive enfilading fire, which would, of course, be opened on your columns from the enemy's batteries on this side; and I proceeded with the greatest possible expedition, strengthened and secured on my right flank by three gunboats, under Captain Roberts of the navy, whose zeal and exertions on this occasion were as unremitted as his arrangements in embarking the troops, and in keeping the boats together in crossing the river, were excellent.

'The enemy made no opposition to our advance until we reached a piquet, posted behind a bridge, at about 500 paces from the house in the Orange Grove, and secured by a small work, apparently just thrown up.

'This piquet was very soon forced and driven in by a division of the 85th Regiment under Captain Schaw, of that regiment, forming the advanced guard, and whose mode of attack for the purpose was prompt and judicious to a degree. Upon my arrival at the Orange Grove I had an opportunity of reconnoitring, at about 700 yards, the enemy's position, which I found to be a very formidable redoubt on the bank of the river, with the right flank secured by an entrenchment extending back to a thick wood, and its line protected by an incessant fire of grape. Under such circumstances it seemed to me to afford the best prospect of success, to endeavour to turn his right at the wood; and I accordingly detached two divisions of the 85th under Brevet Lieutenant-Colonel Gubbins to effect that object, which he accomplished with his usual zeal and judgment, whilst 100 sailors, under Captain Money of the Royal Navy, who, I am sorry to say, was severely wounded, but whose conduct was particularly distinguished on the occasion, threatened the enemy's left, supported by the division of the 85th under Captain Schaw.

'When these divisions had gained their proper position, I deployed the column composed of two divisions of the 85th Regiment under Major Deshon, whose conduct I cannot sufficiently commend, and about 100 men of the Royal Marines, under Major Adair, also deserving of much commendation, and moved forward in line, to the attack of the centre of the intrenchment.

'At first the enemy, confident in his own security, shewed a good countenance and kept up a heavy fire, but the determination of the troops which I had the honour to command to overcome all difficulties compelled him to a rapid and disorderly flight, leaving in our possession his redoubts, batteries, and position, with 16 pieces of ordnance, and the colours of the New Orleans Regiment of Militia.

'Of the ordnance taken I enclose the specific return of Major Mitchell of the Royal Artillery, who accompanied and afforded me much assistance by his able directions in the firing of some rockets, it not having been found practicable, in the first instance, to bring over the Artillery attached to his command.

'I shall have the honour of sending you a return of the casualties that have occurred, as soon as it is possible to collect them, but I am happy to say they are extremely inconsiderable when the strength of the position and the number of the enemy are considered, which our prisoners (about 20 in number) agree in stating from 1500 to 2000

Q

men, commanded by General Morgan. I should be extremely wanting both in justice and gratitude were I not to request your particular notice of the officers whose names I have mentioned, as well as of Major Blanchard of the Royal Engineers and Lieutenant Peddie of the 27th Regiment, deputy-assistant-quartermaster-general, whose zeal and intelligence I found of the greatest service. The wounded men are meeting with every degree of attention and humanity by the medical arrangements of Staff-Surgeon Baxter. The enemy's camp is supplied with a great abundance of provisions, and a very large store of all sorts of ammunition.

'On moving to the attack I received a wound which shortly after my reaching the redoubt occasioned me such pain and stiffness that I have been obliged to give over the command of the troops on this side to Lieutenant-Colonel Gubbins, of the 85th Light Infantry; but as he has obtained some reinforcement, since the attack, of sailors and marines, and has taken the best precautions to cover and secure his position, I will be answerable, from my knowledge of his judgment and experience, that he will retain it until your pleasure and further orders shall be communicated to him.

'I have the honour to be, etc.,
'W. THORNTON, *Colonel,*
'*Lieut.-Col. 85th Regt.*

'To Major-General
'The Hon. Sir E. M. Pakenham, K.B., etc.'

Lieutenant-Colonel Gubbins was also in other despatches particularly mentioned for the manner in which he withdrew the picquets upon returning from the lines himself; being the last to enter the boats.

He was subsequently sent to England by the General Commanding with his despatches and the colours taken at New Orleans. Endeavours have been made to trace these colours, but so far without success.

One of the captured guns was a brass 10-inch howitzer inscribed 'Taken at the surrender of York Town, 1781.' The relief of Major-General Jackson on the withdrawal of the British from the position they had captured on his right flank is frankly admitted in his despatch to the American Secretary at War dated January 9, 1815. He also

COLONEL R. GUBBINS, C.B. [*By Constable.*

describes the terms of the armistice and how he speedily re-occupied the redoubts, etc., when Colonel Thornton's force withdrew.

When the army was reunited on the morrow, a two days' truce was agreed upon to bury the dead. Lieutenant Gleig rode out to the scene of the fight, where, within the compass of a few hundred yards, were gathered nearly a thousand bodies—all English.

Out of 5,000 men engaged no fewer than 1,500 were either killed or wounded.

General Gibbs survived but a few hours. General Pakenham was dead and the chief command devolved upon General Lambert. That officer rightly considered that with his slender force the operation of carrying the works could not have been attempted with any reasonable prospect of success. He therefore resolved upon a retreat.

But to retreat it was needful to absolutely construct a road across a quagmire for a considerable distance. Trees there were not; but bundles of reeds were tied together and used as a makeshift to furnish a footing for the men. Occasionally a broad ditch had to be crossed and rough-and-ready bridges were made of large branches carried with infinite labour from distant woods.

Meanwhile the enemy had mounted a six-gun battery on the bank of the river, with which they continually galled the camp. From the front, too, they also kept up a heavy fire by night and day.

There were other troubles also; for desertion was very rife, both sums of money and promises of grants of land being freely offered by the enemy to the men to induce them to desert.

At length the wounded were conveyed to the canal, placed on board boats and sent to the fleet. Then the baggage, stores, civil officers, commissaries, purveyors, etc., followed and also the light artillery. But the ten heavy ship's guns were spiked and abandoned.

On the 17th, all was ready for the withdrawal of the infantry; and these, leaving the camp fires burning, marched silently away on the evening of the 18th, picquets being left behind to furnish a rear-guard.

There were many troubles along the extemporised road; for after the passage of but a few men it practically became non-existent, and the unfortunate troops had to flounder through a mixture of mud, water, and crushed bundles of reeds, during this terrible night.

Two letters from Brevet Lieut.-Colonel R. Gubbins give us some interesting details of the untoward expedition and are here reprinted:

'BANKS OF THE MISSISSIPPI,
'*January* 10, 1815.

'I have an opportunity of sending you a few lines by poor Colonel Thornton who, I am sorry to tell you, is again wounded and obliged to return to England.

'We have had most dreadful work since we came here, the 85th Reg[t] has suffered terribly particularly in officers—all our operations will be before the public, I shall, therefore, say nothing about them at present—we have failed entirely—fortunately the part of the army I was with, and which I commanded after Colonel Thornton was wounded, succeeded completely.

'I mean the 8th, the day of the attack on the American lines, our whole loss on that day is, I believe, about 2000.

'I never saw or heard of such fighting as we have had. At one time I was down in a ditch in the midst of about 200 Yank'ys with a dozen muskets presented at me, luckily, about 20 brave fellows stuck by me and we beat them off, but not without, I am concerned to say, dipping our hands deeply in blood, for we were too close to fire and nothing but swords, bayonets and butts of muskets were used. But this is not a subject to write to you about.

'I have not had my clothes off since the 10th Dec[r].'

'H.M.S. *Seahorse*,
'NEAR THE MOUTH OF THE MISSISSIPPI,
'*January* 25, 1815.

'Once more, thank God, I write to you from on board ship, safe, sound and in good health, which is what very few of six or seven thousand can say. I should rather have said, what was seven thousand six weeks ago.

'The fatal list, as you so well call it, will too soon reach England.

'Never before did troops undergo such hardships, for forty days I never had my clothes off; for several nights together in ditches 2 feet in water unable to lift our heads from the dreadful fire of grape shot—all our attempts failing—it is altogether too shocking to describe.

'You may judge what the loss of the army has been from that of the 85th—out of 40 officers we have only 8 not killed or wounded. In one hour and a quarter we lost two thousand men.

'I am happy to tell you that the 85th still holds its name and on the fatal 8th was the only Reg[t] that succeeded and are reduced to a skeleton and must soon come home.

'I am writing in a most unconnected manner, but I wish this to go with the dispatches.

'I wrote a few lines last week, we had then our retreat to make. I commanded the Picquets the last night and consequently was the last man to retire, we all got away safe, but our anxiety was not small.'

REFERENCES TO MAP OF THE ATTACK ON THE AMERICAN LINES NEAR NEW ORLEANS

December, 1814, *and January*, 1815.

23rd.

a. a. Columns in bivouac on 23rd December.

b. b. British Position towards the close of the action on the night of the 23rd, the outposts being the same both before and for some days after the action, and were furnished by the 85th and 95th Regiments.

E. E. American columns of attack on the 23rd.

12, 13, 14. Schooner and Gun Boats commencing the attack on the 23rd.

24th.

D. British Position taken up on the evening of the 24th.

26th.

No. 1. Battery of 10 Field Pieces or Howitzers which destroyed the schooner on the 26th December.

15. Position of the Louisiana Frigate previous to the destruction of the Schooner.

28th.

F. F. British Position on the 28th after having advanced to reconnoitre as far as the ground afterwards occupied by Batteries 5 and 6.

16. Position of the Frigate, firing on the British Columns on the 28th.

29th.

No. 4. Battery for red-hot shot. 2 9-pounders which compelled the Frigate to move to 17, where she remained.

1st.

No. 5. Battery of 4 long eighteens and 12 Field Pieces or Howitzers.

No. 6. Battery of 6 long eighteens and 4 twenty-four pound carronades, both which fired on the enemy's line on January 1st.

2nd to 7th.

No. 7. Breast Work, 8 and 9 rectangular redoubts for 200 men each, for the security of the British Camp.

No. 3. Battery of 6 eighteens constructed on the 7th.

No. 18. Advanced Work, thrown up by the enemy on the 5th and carried on the 8th by Light Companies.

No. 21. Canal by which British barges passed into the Mississippi.

G. G. British Columns of attack on the 8th.

2. Position of Colonel Thornton's Corps after carrying the works on the right bank on the 8th.

MILITARY DEPOT, Q.M. GENL.'S OFFICE, HORSE GUARDS,
March 16th, 1815.

Continuation of Lieutenant Gleig's Diary.

9th to 17th.—'From this period until the 18th nothing of any importance occurred, and the truth is I had neither pen, ink nor paper to take it down if it had. The army has been employed in preparations for a retreat. The stores, wounded and horse have been embarking every day and finally all the light Artillery withdrawn. The heavy 18 prs. we have been obliged to leave behind. The enemy have

Facsimile of Siborne's Map in the Royal United Service Institution. [*By kind permission.*]

Landing Place, 9 miles from the mouth of the Bayoue. 72 from the anchorage of the Frigates and 87 from that of the Line of Battle Ships.

Marshy — covered with Reeds

Marshy and nearly impracticable Wood

Mississippi River

American Lines

Marshy and nearly impenetrable Wood

Scale

NEW ORLEANS.

contrived by giving their guns great elevation to throw shot and shell into our camp, which however has done very little execution.'

Wednesday 18th.—'Everything having now been got off but the Infantry, the 93rd Regiment began its retreat at half-past seven o'clock; we followed at half-past eight, the 95th an hour after, the 43rd after them, and the 7th bringing up the rear. The 21st, 44th, Marines, Seamen and West Indian Regts. had been sent off some time before. Our pickets did not withdraw till three o'clock, and our sentries till just before daylight. The road we passed over was without exception the worst I ever travelled over. Every step was up to the knees in mud, and several places actually up to the breast. We contrived to get through it however with the loss of only one man smothered in the mud, and halted an hour after daylight at a place called the Fishermen's huts.'

Thursday, 19th.—'This halt arrived about as seasonably as any I ever enjoyed, for the fag of last night's march was beyond everything. When we were across the river I brought off with me a Yankee tent which I pitched immediately, took off my muddy clothes, and lay down to sleep. The place where we are is a marsh overgrown with reeds, without so much as one tree to make a fire of. The reeds are the only things we have to burn. But there is a still more disagreeable circumstance attached to our present encampment which is the want of provisions. We can get nothing but biscuit and rum. Spent all the day in my tent.'

Friday, 20th.—'Immediately after the turn out I went to shoot something for dinner, and came back to breakfast with a wild duck and water rail. On this we entirely depended. The troops are embarking as fast as possible, some for Cat Island, and some for the shipping at once. The 21st and part of the 44th are already gone with all the Blacks, &c.'

Saturday, 21st.—'Went out again to get a dinner, but with very bad success, only one water hen out of three.'

Sunday, 22nd.—'The Fishermen's huts are a parcel of miserable straw huts built on the bank of the Baio del Cateline.'

Monday, 23rd.—'The Americans had a picket there which was surprised the night before we first landed. There were no inhabitants in them now but a parcel of miserable Negroes. From this period we remained in the same miserable state until the 24th.'

Tuesday, 24th.—'About one o'clock to-day we embarked on board a launch to go down the length of the gun-boats, but had not got 400 yards from the mouth of the creek when we ran aground, and stuck till past three, when a boat came and took some of our men out, by which means we got off again. It was dark when we reached the gun-boat, and from her we were ordered to a transport brig, where we found several of our officers, and a great many men before us. In this dirty little vessel we passed the night.'

Wednesday, 25th.—'This morning boat after boat full of men came to us, till at last we had 400 men and 17 officers on the brig. We are so crowded we have literally scarce room to turn. Deshon came on board, and we got off with some difficulty but ran aground again. Spent a still more disagreeable night.'

Thursday, 26th.—'We got off this morning in real earnest and got down to within 15 miles of the shipping, when we grounded once more, and a Schooner going up went ashore also, and seeing no prospect of getting off with so many men on board, Major Deshon went on board the schooner and prevailed on the midshipman to take the 95th on board. He proving to be Sawers I went with him and slept on board the schooner.'

Friday, 27th, to Tuesday, 31st.—'After having run foul of several ships I at last left the schooner and to my no small satisfaction found myself once more on board the *Golden Fleece.* The satisfaction was a melancholy one, for in this ship I had spent many happy hours with the best of friends. Poor fellow—I went on board the *Thames* and read his will, in which he has left me his pistols, spyglass and a dog, besides some books. Dined with Blanchard. Here we have remained the rest of the month without any adventure occurring.'

February, 1815.

Wednesday, 1st, to 3rd.—'During these three days we have remained perfectly quiet at anchor, without being able to move on account of foul winds, and the weather to add to it has been very bad. On the last day however the signal was made to follow the Captain's Agent who weighed, and stood down towards the other anchorage, but the wind dying away no one could do it.'

Saturday, 4th.—'We prepared to sail this morning but could not from contrary winds. I went on shore to Cat Island, a place totally uninhabited except by one family, to shoot along with little Boyes. We found no game, but as the day was beautiful we continued our walk through a pine wood, till we came to the only cottage on the island. Here entirely separated from the rest of the world dwells an old Spaniard with his wife, two lovely daughters grown up, a son, and two little children. This hut, which is made of wood, is surrounded with trees, and stands about one hundred paces from the margin of the lake. It is a most romantic spot, and the beauty of it is not a little increased by the small cultivated garden which is behind the house. I remained here about an hour, and having purchased a cow returned, and after considerable difficulty got it on board.'

Sunday, 5th.—'Again we prepared to weigh anchor but there was no wind. I went on board the *Thames* to settle some disagreeable business with Dunbar, and got back to the ship just as she was under weigh. Wills and Mudge dined with us. We ran down the length of Ship Island and again dropped anchor. Poor Mudge left us this evening. The army is now divided and about to act separately. We, the 93rd, 43rd and 44th go by ourselves, while the 4th, 7th, 21st and 40th, together with the two West Indian Regiments, remain here.'

Monday, 6th.—'Got under weigh with a fair wind a little after daylight, and about nine o'clock again brought up at the Admiral's anchorage. Weighed again about two o'clock with a fresh breeze and anchored at eight not far from Mobile Bay. Weighed at seven and by the time we came on deck found the ship in the entrance of the Bay. A Yankee fort with its flag proudly floating in the air is discernible, and a schooner lying at anchor near it. We cast anchor again near an uninhabited island about eleven. At twelve Leban came on board with an order for us to disembark immediately. We had very little time for preparation as at one the boats came alongside for us, and by three o'clock we were landed upon the island, the name of which is Dauphin Island. We came expecting and prepared for opposition, but met with none, and after having marched to the eastern extremity of the place bivouacked for the night.'

Wednesday, 8th.—'As the report was we were to remain here some time I sent for canteens and Boyes went out to forage. Dauphin

Island is a place twelve miles long and two broad. It has always been uninhabited except by two families, and of these only one now remains. It is well stocked with cattle, but they are exceedingly wild and difficult to catch. It was esteemed so healthy by the Americans that some time ago they made it a sort of depot for the sick of their army. The soil is sandy and dry, except upon one side, which is marshy, and it is covered with pine, laurel and oak trees. Boyes succeeded in getting two fine fat young cows which we sent on board. Got the canteen on shore and had a comfortable dinner. The 43rd, 95th, 93rd, 40th and 7th are all landing here, while the 4th, 44th and 21st are besieging Fort Bozeos which commands the entrance of Mobile Bay.'

Thursday, 9th.—'Went on board the *Fleece* for more baggage, and brought Wills off with me. Spent the rest of the day in forming plans for our future comfort, but most luckily put none of them in execution, for as we sat after dinner the order came for us to cross the water at daylight. This put an end to all preparations. The baggage was packed and sent on board, and everything made ready once more to rough it.'

Friday, 10th.—'Rose two hours before daylight and struck the tent expecting to march every moment, but it was six o'clock before the boats were ready for us, and even then there were only enough for three divisions, so ours and Gubbins' returned to their old ground. It was not till after we had dined that they returned for us, by which means we got over barely in time to have everything settled before dark. A deserter, it seems, has come in with the news that we are to be attacked to-night.'

Saturday, 11th.—'The night has passed in quietness. I am this morning on a very quiet picket. Our Regiment is employed not in the siege of the Fort but in covering the besieging army. The place where we are encamped is in front of the 4th, 44th and 21st upon a narrow neck of land, not more than a quarter of a mile across, to the natural strength of which we have added a number of breastworks. The Fort has this morning surrendered, but owing to the disorderly state of the garrison they are allowed to remain inside till to-morrow.'

Sunday, 12th.—'At daylight this morning I was relieved, and very soon after we received an order to re-embark and cross over to the island again. As there were not boats for all, our division and Captain Ball's marched along the beach and piled arms near the fort until they should return for us. I went into the fort along with two other officers, and such a miserable little place I never beheld. I could with ease throw a biscuit from one side to the other of it, and towards the land side the walls are barely grape-proof. It is entirely built of sand, and faced with wood, so that hot shot would set it on fire as soon as a ship. At twelve o'clock the Garrison consisting of 400 of the 2nd Regiment fell in, and having first paraded in the ditch marched out with the honours of war and laid down their arms upon the glacis, while their colours were given to the 4th Regiment, who were drawn up to receive them. Very soon after, the boats having arrived, we embarked and landed once more on Dauphin Island.'

Monday, 13th.—'The 1st and our Brigade were this morning inspected by General Keane, and to my no small astonishment and delight a ship arrived with the Official Communication of Peace. Thank God all our toils are nearly at an end, and I may once more anticipate the comforts of home.'

Tuesday, 14th.—'The Americans it appeared came down last night upon the 21st, but were immediately repulsed. There is nothing new to-day. Everything depends upon Mr. Maddison. Should he refuse his signature to the Treaty, the war goes on, but should he not, then for Old England.'

Wednesday, 15th, to Saturday, 18th.—'During these four days nothing particular has occurred. All the army, except part of the 4th Regiment which is left in the fort, have assembled here, and nothing seems intended, until Maddison's determination is known.'

Sunday, 19th.—'The Chaplain for the army to-day read prayers and preached a sermon to the Brigade. This is the second time I have heard service read by a clergyman since I left England. I heard him with more awe and reverence than I almost ever did before. This evening the whole army was inspected by General Lambert.'

Monday, 20th.—'Went out shooting and killed a wild duck and three snipe.'

Tuesday, 21st.—'Went out again and killed a curlew. These birds are most acceptable when we have nothing but salt meat issued out.'

Wednesday, 22nd.—'Shaw and Halton of the 43rd dined with us. I am now in a mess with the two Brigades. Shaw has got leave of absence, and I to-morrow enter upon the command and payment of his Company. A great deal of thunder during the night.'

Thursday, 23rd.—'A very rainy bad day, but as we have all got tents we do not feel it so much. The change is sudden from intense heat to intense cold.'

Friday, 24th.—'The Regiment as usual was mustered, but as the day was bad I stayed in my tent.'

Saturday, 25th; *Sunday, 26th.*—'Nothing particular during these days except that on the latter the Chaplain again read prayers to the Brigade.'

Monday, 27th.—'Remained in my tent all day, and during the evening Boyes left me for England. There is something in saying farewell to a man with whom you have lived even for a short time in intimacy that I cannot bear. Felt particularly low after his departure.'

Tuesday, 28th.—'West dined with me, and I went home and drank porter with him. What a miserable life this is. I shall in future leave my journal blank until something worthy of notice occurs.'

March, 1815.

'During this month nothing occurred worthy of notice until the 5th when I went upon picket. While I was wandering along the beach I discerned a schooner bearing a white flag approaching, and very soon saw her anchor and a boat shove off from her. On running to the place where the boat landed I found that an American General and Colonel had come ashore and proceeded up to head-quarters. I then addressed myself to one of the boat's crew who was a soldier in their artillery, and learned from him that the intelligence they brought was that of the confirmation of peace. The rest of the day I spent in delightful reflections on the pleasures of home. I had almost forgot to mention the occurrences of the 4th and 3rd.'

Friday, 3rd.—'We have lately for want of something else been amusing ourselves by throwing fir tops at each other like a parcel of schoolboys. The officers of the 43rd and 7th, who are just such another set of fine young men as our own, to-day took it in their heads to declare war against us, and attacking us when we were totally unprepared took from us one hut and its enclosures, after an action of two hours. Seeing that they were too strong for us we formed an alliance with the 93rd, the 95th were originally with us, and set about to meet them next day in good style. Captain Travers of the 95th is voted Commander-in-Chief of our forces, I am second in command, and Captain Hart of the 93rd the next.'

Saturday, 4th.—'Having assembled all our forces, and formed them into three columns under their respective chiefs, we this morning after parade retook the redoubt we lost yesterday, and waited the attack of the enemy. About twelve o'clock they came on our column, and drove in our outposts, our centre also according to the preconcerted plan giving way. Just at this moment when they thought themselves sure of success, I put myself at the head of the 93rd, which had till now remained in ambush, and stealing quietly through the trees, attacked their flank and completely routed them. We drove them within their fort which it was agreed should be stormed directly. The attack was immediately made, and being one of the first to get in I was overpowered and made prisoner. However, the rest pressing on, the outwork was soon carried and all their army except four taken. They then offered to surrender, which was agreed to and we marched in. We drank some wine with them and then returned home with flying colours. But this is not the only amusement we have. A theatre has been erected in which those Officers who like to exhibit themselves perform, and it being open to-night I went to it, and was highly pleased both with the house and acting.'

Monday, 6th.—'Came off picket this morning and found Travers' tent converted into a fortification, as the 43rd refusing afterwards to own themselves beat the war is renewed. Everything was ready for the renewal of hostilities, but it rained so hard that a truce was agreed upon till next day at twelve. Continued working all day and greatly improved our fort. An exchange of prisoners took place and I am free.'

Tuesday, 7th.—'Finding this morning that we could not bring more than twelve into the field, we agreed that the best plan would be to shut ourselves up in our fort and wait a siege. They soon attacked us with twenty-seven combatants, but after an action of three hours they found all attempts at carrying our fort useless, and began to retire, having lost one prisoner. We had recruited our force up to the number of 25, and making a sally we attacked them so furiously and unexpectedly that we drove them once more within their own works where we made them sign a treaty of peace. Went to the play again this evening.'

Thursday, 9th.—'Nothing else occurred until the 9th, when Wills came ashore and dined with me, and having drank rather freely we agreed to walk out. I had wandered among the huts of the 40th and was talking to a person who pretended to be angry, when some one passing by took notice of what I was doing in what I conceived to be an impertinent manner. I followed him, and finding him to be an Officer of that Regiment whom I did not know, high words ensued and we exchanged names. I did not however look upon the matter as of such consequence as to require any further discussion, and therefore went on board with Wills, as I had previously intended to do.'

Friday, 10th.—'This day I spent on board with Wills, and in getting those things I wanted for our little mess.'

Saturday, 11th.—'Having returned on shore this morning I found that an Officer had called upon me several times on the part of Mr. Foulkes to require satisfaction. An apology on my side was out of the question, and I therefore agreed to meet him next morning at daylight. I procured my good friend West for my second, got my pistols put in order, and went to the play in the evening.'

Sunday 12th.—'A little before daybreak this morning I rose to keep my appointment. I certainly felt a little queer when I reflected that I was going out either to be shot myself or to shoot a man against whom I bore not the slightest malice, and with whom moreover I was totally unacquainted. Such are the laws of honour. Our seconds having joined each other, the ground was measured and we took our station twelve paces asunder, with our backs towards each other. While I was waiting for the signal to face about and fire, I heard our seconds in earnest conversation, and all at once mine asked Mr. F. if in the speech he made the other night he had intended to insult me, to which he replied No. His second then came up and said he presumed

I had no objection to make it up and say I was sorry I had spoken so warmly. This I refused to do unless my opponent would come forward and say to me that he had not intended to hurt my feelings, and had not spoken to nor at me at all. Having met half way he affirmed the above three distinct times, when I told him I was satisfied and sorry the circumstance had happened at all. We then shook hands and thus was the unpleasant affair concluded. I returned home, eat a good breakfast, and spent the day at West's.'

Monday, 13th.—'The weather has now become so hot that I this day commenced bathing. The day passed without any occurrence worthy of mention.'

Tuesday, 14th.—'Got up at rather an early hour, and strolling out overheard Admiral Malcolm telling Major Deshon that our baggage was to embark to-day, and ourselves to-morrow. I had barely time to finish my breakfast when the baggage was packed and sent off, mine on board H.M. Frigate *Thames*. As soon as it was gone I took a firelock and went out to look for alligators, as a large one had been seen lately near the camp of the 7th. As I approached the pool in which he had been I met some people dragging him along in triumph. He had been shot a moment before by De Bathe, and an Officer of the 7th, and measured nearly eight feet from head to tail. Saw another about three feet, and a number of smaller ones, and killed five snakes, the largest of which was six feet long. This island swarms with vermin. Spent the evening with Dempster.'

Wednesday, 15th.—'Rose at five o'clock and struck my tent at half-past five, and at six o'clock the Regiment fell in, marched to the beach and embarked immediately. I found the *Thames* in a horrid state of filth and confusion, but having got a small cabin to myself I shall make myself very comfortable. Received in the evening a letter from my dear Father.'

Thursday, 16th.—'Weighed anchor early this morning, but the wind coming foul were obliged to anchor again. I am Orderly Officer for the day. Remained at anchor all the day in a state of great impatience.'

Friday, 17th.—'A signal was again made to weigh, but the Admiral would not allow us to go until a man had been flogged round the Fleet by which time the wind died away and we once more brought up. St. Patrick's Day, but nobody drunk.'

Saturday, 18th.—' Weighed anchor this morning with a fresh breeze, and at last stood our course towards the Havannah. Towards night it came on to blow hard, so much so that there was no comfort either in bed or out of it.'

Sunday, 19th.—' The wind is so high and the ship rolls so much that everything is in the greatest confusion. Although I am now a pretty old sailor I could not help being a little sick this morning. Towards evening it began to moderate, and by night settled in a nice steady fair breeze.'

Monday, 20th.—' Nothing particular to-day. Going at the rate of 7, 8, and 9 knots.'

Tuesday, 21st.—' About ten o'clock this morning the Commodore made the signal for land, and about eleven we could distinguish the high mountains of Cuba, like in the horizon. The wind has now fallen off very much so that we make very little way, but at night it freshened again considerably.'

Wednesday 22nd.—' We this morning found ourselves about thirty miles from the land, but upwards of eighty from the harbour. The breeze was fresh and fair and we carried all sail.'

Thursday, 23rd.—' This morning the town of Havannah was in sight. Crowding all sail we were by ten o'clock near enough to admire the beauty of the place. The harbour of Havannah is formed by a little bay that runs in shore, and is certainly better protected against both bad weather and an enemy than any I ever saw. It is defended upon the left as you enter by two forts of considerable extent, and apparently of amazing strength, and on the right by a small fort and the walls of the town. The appearance of the town is particularly fine, owing to its numerous spires, and magnificent buildings. The flags when we came in were all half-mast high, which was a cause of surprise at first, but I soon found out the reason, this is Holy Week. It was pretty late when we came in, so I did not go ashore. We struck coming in and got off again before dark with difficulty.'

Friday, 24th.—' We were this morning mustered at a very early hour, after which I went ashore. The town of Havannah is a large, magnificent place, like all other Spanish towns full of religious houses, but considerably cleaner than the generality of them. This being Holy Week there are no public amusements going on, but the grand procession of the burial of our Saviour took place to-day. It was certainly very fine, but it——

[*Here four leaves, equivalent to eight pages, have been torn out some time, and only a few fragments of words remain from March 25th to April 15th.*]

because a scene took place this morning to which I never before was witness, and which left a melancholy impression upon my mind, the committing of a dead body to the deep. The bell was tolled as if for a funeral, the crew collected on the deck and the body was brought to the gangway on a grating and covered by a flag. The Captain read the burial service, and at the solemn words Earth to Earth, etc., the flag was removed and all that remained of poor mortality was consigned to the waves. We were going very fast at the time, and the sea was pretty rough, so I could scarcely catch one short glimpse of the body as it sunk, but that was enough to recall to my recollection Lord Byron's beautiful poem. In the evening a thunderstorm came on, which lasted for an hour, and effectually changed our fair wind into a foul one.'

Monday, 17th.—'It has been blowing a gale of wind in our favour these some days, by which means we this evening, after a great deal of trouble and anxiety discerned the shores of Bermuda. The wind however blew so strong and it was besides so late that we did not attempt to go in, but beat off and on all night.'

Tuesday, 18th.—'A pilot coming on board this morning we stood our course towards the anchorage, but the wind blowing right against us it was evening before we brought up. We were amused all day watching whales that played about in great numbers. This is a place I sincerely hoped never to have seen again. What painful ideas it raises up in my mind. How different was the Regiment when we were here before. My poor Grey was then alive. But such is the fortune of war.'

Wednesday, 19th.—'All the Officers went on shore except myself, and one or two others. For my own part I have not the slightest desire ever to put my foot on the desolate place.'

R

Thursday, 20th.—This day's entry is produced in facsimile.

Friday, 21st.—'Two Companies of our Regiment have gone on board the *Menelaus* with De Bathe, Greene and Boyes, for the purpose of going home in her. The report now is that all the ships sail independently. We remained at anchor all day, but at night weighed and put to sea. The breeze was very light, so we made but little way, and at midnight the low shores of Bermuda were still visible.'

Saturday, 22nd.—'Towards morning a fresh breeze sprang up, and by the time I came on deck we were nearly sixty miles from land. Thank God the next we shall see will be Old England. We are all alone carrying every stitch of sail and going eight knots an hour. I shall now pause till something uncommon occurs.'

May, 1815.

Monday, 1st.—'The wind with very little interruption has continued to blow fair and fresh all along, by which means we are now something more than half-way home. This month has set in in rather a melancholy way. A poor little boy fell off the cross-jack yard, and though a boat was lowered immediately and he was seen for some time, they did not contrive to save him. It is strange that a single incident of this kind has more effect upon the mind than the death of hundreds in battle.'

Sunday, 6th.—'After many fruitless tries we this day came into soundings, in one hundred and four fathoms, with white sand and shells. The wind continues to blow fresh and fair, and we confidently expect to see Ushant this evening or to-morrow morning.'

Monday, 7th.—'At about ten o'clock this morning breakers were distinguished from the poop, and in a short time more we saw land which proved to be the Saints. As we were not exactly sure about our situation we stood in towards land, and were not a little astonished to see the tri-colour flag flying. Many were the opinions sported upon the occasion, but we at last concluded that it must be a signal of some kind. We stood in very close to the coast and had a fine view of Brest and the rocky shores all along.'

Tuesday, 8th.—'Sailing up channel with a fine fresh breeze. Hailed a Schooner but could not make out what she said. At three o'clock

Most Gracious God, receive my humble
thanks for the great mercies thou hast
heretofore vouchsafed to grant me. By
thy goodness alone have I been preserved
from the dangers of war and the horrors
of shipwreck to add this day another
year to the span of my existence. May
the manifold errors of my past life be
forgiven by thee, and may they prove
such a warning to me that in future
I may so conduct myself that when
the last solemn hour shall arrive, my
death may be as that of the righteous
man. Grant this almighty Father
for the sake of thy well beloved son
Jesus Christ our Lord and Savior — Amen

FACSIMILE OF GLEIG'S DIARY.

we came in sight of the Isle of Wight. What a sensation to view the shores of one's native land after so long an absence. How my heart panted for one kiss of the beloved soil. The wind lasting we stood our course well, and by seven o'clock p.m. came to anchor off Spithead. We had previously been boarded by the Lieutenant of the Guard-ship from whom we learnt the astonishing particulars of Buonaparte's re-entrance into France. This piece of intelligence hurled me from the height of joy to the depth of misery, for the prospect of going out again without seeing my relations was of course the first thing that struck me.'

Wednesday, 9th; and Thursday, 10th.—'Landed early and the first thing I did was to kiss the ground.'

Friday, 11th.—'The Regiment was this day moved out of the *Thames* on board of the *Amelia*, which immediately sailed, but for want of wind was obliged to anchor at night.'

Saturday, 12th.—'A fresh breeze springing up we weighed and stood our course towards the Downs. The wind was so fair that before dark we were at an anchor off Deal. The wind however was so high and the sea so rough that no boats were allowed to go ashore.'

Sunday, 13th.—'This morning the head-quarters of the 85th Regiment disembarked after an absence of nearly two years, in which time it had been most actively employed, and took up its quarters in Deal Barracks.

'I shall now therefore, after having given a correct Journal of all that has befallen me since I quitted England with that Regiment, until my return along with it, put a stop to my writing until my country shall again require my absence from her shores.'

[*Here ends the Diary.*]

CHAPTER IX

PEACE, 1815—1834.
HOME SERVICE, 1815—MALTA, 1821—GIBRALTAR, 1827—MALTA, 1828—ENGLAND, 1831—IRELAND, 1834

ON arrival in the Downs the regiment disembarked at Deal on May 13.

Two days later it marched for Hythe, arriving on May 16.

On May 29 the regiment was ordered to proceed to Chatham in two divisions and reached that place on May 31 and June 1st.

Here it remained until August 7, when it again returned to Hythe by divisions, arriving there on the 9th and 10th of the month.

On August 12, 1815, the following letter was received granting the regiment a new denomination.

'HORSE GUARDS,
'10 *Aug.* 1815.

'SIR,

I have the honour to acquaint you, by direction of the Commander-in-Chief, that in consideration of the discipline and good conduct evinced by the 85th Regiment, on various important services on which it has been employed in Europe and America, since the year 1813, His Royal Highness the Prince Regent in the name and on behalf of His Majesty, has been pleased to approve of the 85th Regiment being in future styled the 85th (or Duke of York's Own) Regiment of Light Infantry, and also to permit the Regiment to bear on its colours and appointments, in addition to any other badges or devices which may have been heretofore granted to the Regiment, the motto " Aucto Splendore Resurgo."

(Signed) 'HENRY CALVERT, *A. General.*

'General Stanwix or Officer
'Commanding 85th Light Infantry Regiment.'

The aptitude of this motto 'Aucto Splendore Resurgo' is obvious.

Its meaning is 'I rise again with augmented splendour (or glory).'

By whom it was selected is not recorded. To resume, on August 28, 1815, the regiment marched from Hythe to Canterbury, leaving Canterbury on September 20 and 21, and arriving at Chatham on the two following days.

On February 1, 1816, the establishment of the regiment was ordered to consist of the following numbers as from December 25, 1815: 10 Companies, 1 Colonel, 1 Lieut.-Colonel, 2 Majors, 10 Captains, 22 Lieutenants, 8 Ensigns, 1 Paymaster, 1 Adjutant, 1 Quartermaster, 1 Surgeon, 2 Assistant-Surgeons, 1 Quartermaster Sergeant, 1 Paymaster-Sergeant, 1 Schoolmaster-Sergeant, 1 Armourer-Sergeant, 50 Sergeants, 50 Corporals, 22 Buglers and 760 Privates; total, 936 officers, non-commissioned officers and men.

While mentioning the new establishment of the regiment it may be well to note that:

'During the Half Year ending 24 Sept. 1816, the Regiment received 64 recruits, 26 of whom were enlisted at the Head Quarters of the Corps, 14 raised by the recruiting parties in England, 10 by the party in Ireland, 5 Boys from the Royal Military Asylum and 9 German Musicians.'

The enlistment of German musicians is interesting.

'The Regiment also received the following Transfers during the above period, 38 from the 72nd Regt., 1 from the 39th, 1 from the 89th, and 2 from the Isle of Wight, with 9 Deserters returned.'

The regiment left Chatham in three divisions on January 18, 1816, and marched to Winchester. On May 27 it marched to Northampton, arriving there on June 4 and 5. On June 17 seven companies of the regiment marched from Northampton to Weedon Barracks, leaving three companies at the former place.

At its half-yearly inspection at Weeden on October 17, Major-General Sir Henry Fane, K.C.B., issued the following district order:

'WEEDON,
'18 *Oct.* 1816.

'*General District Order.*

'Major-General Sir Henry Fane has much satisfaction in expressing to Col. Thornton and the 85th Reg[t] under his command, his perfect approbation of their general appearance, steadiness in the field, and interior economy, and he will not fail to make a favourable report of their state to H.R.H., the Commander-in-Chief.

(Signed) 'E. C. RADCLIFFE,
'*Lt.-Col. and Major-Genl. of Brigade.*'

October 30, 1816. Six companies of the regiment, under the command of Lieut.-Colonel Warburton, marched from Weedon and Northampton to Coventry, there to await orders from Major-General Sir H. Fane.

The regiment was now about to change quarters and commenced its march for Liverpool in three divisions—1st Division from Weedon on November 6, 1816, the 2nd and 3rd Divisions from Coventry the 9th and 10th of the same month.

November 20 and 21. Six companies of the regiment arrived at Liverpool, which was established as headquarters, while three companies were detached to Chester and one to the Isle of Man.

December 20, 1816. Headquarters being moved to Chester.

The reports of the inspections for 1817 are here given, as also is a complimentary letter on the removal of the regiment to Gloucester.

'PONTEFRACT,
'10 *May*, 1817.

'*District Orders.*

'Major-General Sir John Byng has derived much satisfaction from his inspection of the 85th Reg[t]. and from hearing from several of the most respectable inhabitants of Chester and Liverpool of the excellent conduct of the officers and men, which has occasioned them to be much respected at both places. The Major-General will consider it his duty to endeavour to do justice to the Reg[t]. in his report of it to the Comd[r]. in Chief.

(Signed) 'JOHN BYNG,
'*Major-General.*'

'CHESTER,
'18 *Sept.* 1817.

'*District Orders.*

'Major-General Sir John Byng has again to express the perfect satisfaction he has derived from his inspection of the 85th Regt., he has equally noticed their clean uniform and soldierlike appearance under arms, the regularity and precision with which they exercised and their good behaviour in Quarters which has been communicated to him from most respectable authority which he will not fail to report to H.R.H. the Comd[r]. in Chief.

(Signed) 'JOHN BYNG,
'*Major-General.*'

'HEADQUARTERS,
'PONTEFRACT,
'26 *Decr.* 1817.

'SIR,

'The Quartermaster-General of the Forces having notified to me the necessity (in furtherance of general arrangements) for the removal of the 85th Reg[t] from the District under my orders and that they are to commence their march towards Gloucester on the 29th inst., as personal feeling must always give way to the general convenience of the Service, I can only express my regret in losing so distinguished a corps from my command, and have to request you will accept yourself and convey to Lieut.-Col. Warburton, the Field officers and officers of the Reg[t], my best thanks for their uniform and great attention to the best interests of the service, in their strict observance of Discipline and good conduct.

'I have derived much satisfaction in making their acquaintance which I shall be happy to renew when and wherever I may again have the pleasure to meet them. You will also be pleased to notify to the Regiment that my high opinion of them has increased with each inspection, that I have witnessed with much satisfaction and have reported with equal pleasure to H.R.H. the Comd^{r} in Chief in terms of high approval my opinion of their excellent appearance and high state of discipline, and with every good wish for the prosperity and happiness of yourself and the 85th Reg^{t}, I have, etc.,

(Signed) 'JOHN BYNG,
'*Major-General.*

'Colonel Thornton,
'Commd^{g} 85th Reg^{t},
'Chester.'

During the month of March, 1817, the regiment was called upon, for the first time, to aid the civil power.

Serious rioting was anticipated in Manchester, and thither on March 9 four companies proceeded under the command of Colonel Thornton. Three of the companies returned to their quarters on the 13th, leaving one company at Manchester till the 17th. During the month, however, a second application to aid the civil power was made from Manchester; and on March 29 six companies, again under the command of Colonel Thornton, proceeded thither. They returned to quarters on April 3.

The recruiting at this time was seemingly very successful; 143 recruits joining the regiment, of which 65 were obtained at headquarters, 28 in England by recruiting parties and the remaining 50 in Ireland; 47 men were also obtained by transfer as follows: 2nd Battalion, 67th Regiment, 11; do. 87th Regiment, 3; do. 89th, 24; and from the 101st Regiment, 9.

On December 29, 1817, the following changes of quarters took place. Captain Williams' company marched from Chester to Carmarthen and Milford; Lord Churchill's Company from the same place to Pontypool and Tredegar; Brevet-Major Ball's from Liverpool to Brecon, as also did the company commanded by Captain Schaw.

On January 7, 1818, five companies of the regiment marched for Gloucester; Captain the Hon. Charles Gore's, Captain Charleton's and Captain Johnson's from headquarters, Chester; Brevet-Major De Bathe's and Captain Fairfax's from Liverpool. The three former companies arrived at their destination on the 16th and the two latter

LIEUT.-GENERAL SIR WILLIAM THORNTON, K.C.B.
Commanded the Regiment during the Peninsula and American Campaigns; 1813-1814.

on January 17, 1818. On the following day the companies of Captains Fairfax and Charleton proceeded to Bristol.

The regiment was now stationed thus : at Gloucester 3 companies, Bristol 2, Brecon 2, Carmarthen 1 (with 30 men detached to Milford), Pontypool 1 (with 30 men detached to Tredegar), and one company was stationed in the Isle of Man.

By April, 1818, the regiment was found to be above its strength and a notification was received on the 11th from the Adjutant-General ordering the recruiting parties to be withdrawn.

During the next few months a few changes of quarters took place.

The regiment marched to Plymouth on April 20, leaving only a subaltern's detachment at Milford. The regiment reached its destination on April 29, April 30, and May 1.

Meanwhile recruits were still coming in, many re-enlisting from regiments that were disbanded.

On October 26, 1818, a detachment of the regiment, consisting of 1 Subaltern, 1 Assistant-Surgeon, 2 Sergeants, 1 Bugler, and 32 Rank and File, marched to Pendennis Castle, which place it occupied till April 20, 1819, on which date it returned to headquarters.

On October 24, 1818, the establishment of the regiment was ordered to be reduced; the new establishment consisted of 746 officers and non-commissioned officers and men. The only change, apparently, being this: that instead of 30 Sergeants we find 10 Colour-Sergeants and 20 Sergeants, 30 Corporals, 21 Buglers, and 620 Privates. Previously there had been 50 Sergeants and 50 Corporals.

November 20, 1818. Thirteen Sergeants and 144 Rank and File supernumerary to the above establishment were discharged and invalided from the regiment.

On Christmas Day, 1818, in consequence of the above reduction, the following officers were placed on half-pay:

Lieut.-Colonel Warburton, Lieutenants Ashton, Barlow, Macdougal, R. Boyes, Brooks, Levinge, Peel, Thompson, Dutton, Dixon, and Assistant-Surgeon Tedlie.

On February 15, 1 Sergeant and 30 Rank and File, volunteers from the late 97th Regiment, joined the 85th, and a similar number were in consequence discharged from it.

July 26, 1819. The regiment embarked at Plymouth Dock for Dover.

August 21, 1819. Marched for Chatham.

September 21, 1819. Marched for Northampton.

September 29, 1819. Four companies marched for Leicester and one for Lutterworth.

September 30, 1819. Lieut.-Colonel Warburton from half-pay was appointed to command *vice* Colonel William Thornton, appointed Deputy Adjutant-General to the Forces in Ireland.

October 16, 1819. New colours were presented by the Duke of Buckingham. These were in use till 1827.

One of these colours is preserved in a case in the Officers' Mess, with a letter.

On October 27, 1819, and following days, marched for Knutsford, Northwich, Middlewich, and Altrincham.

December 3 and 4. Three companies to Bury and Oldham.

December 11. Five companies to Manchester.

December 24. Three companies to Huddersfield.

June, 1820, to Lichfield, Derby and Burton.

July 25, 1820, to Chesham, Amersham, Watford, High and West Wycombe.

September 11, 1820, to Brighton.

On September 25 the regiment was reviewed at Brighton by H.R.H. the Duke of York, Commander-in-Chief, when His Royal Highness was pleased to express, in a very flattering and condescending manner, his entire and unqualified approbation, both as to the general and soldier-like appearance of the regiment, their appointments, and the steadiness of the men under arms.

On November 27 the half-yearly inspection of the regiment took place by Major-General Sir Henry Torrens, K.C.B., who was pleased to express his very high opinion of the regiment in the most flattering manner. At this period the regiment was finding the guards over the Royal Pavilion at Brighton, where King George IV was in residence. The King was seeking a divorce from his wife at the time. The story goes that a hostile demonstration was made against His Majesty by some of the audience at the theatre. The 85th officers, who

THE KING'S.

Colours presented at Northampton by the Duke of Buckingham, October 1819. They were in use till 1827. They were returned to the Regiment by Admiral Marra of the Royal Italian Navy in 1894. (*See pages 332 and 333.*)

THE REGIMENTAL.
1819-1827.

were also present, turned the rioters out, at which His Majesty was pleased to express his approbation.

'Horse Guards,
'11th April 1821.

'Sir,

'I have the honour to acquaint you, by direction of the Commander-in-Chief, that his Majesty has been pleased to command that the 85th Light Infantry Regiment shall bear, in addition to its present County title, the title of the 85th or King's Light Infantry Regiment, instead of the Duke of York's Own Regiment of Light Infantry, and that the Uniform of the Regiment shall be faced with Blue and laced with Silver.

'I have the honour to be, etc., etc.,
'H. Torrens, *A. General.*

'Officer Commandg,
'85th Light Infantry.'

Extract from a Letter to the Editor, 'U.S. Gazette,' o/d Birr, April 26 (?)

'The facings of the Regt were changed from yellow to blue April 11, 1821, at Brighton by particular desire of his Late Majesty King George IV, who was (two months previously) pleased to counter-order the Regt being removed from Brighton to Malta, where, however, they went on June 16, 1821.'

With regard to this the Regimental Digest tells us that on February 6, 1821, 'A detachment of the Regt marched this day from Brighton on its route to Portsmouth there to embark for Malta, but was countermanded at Chichester by the command of His Majesty and rejoined Head Quarters at Brighton.'

June 10, 1821. To Portsmouth and embarked for the Mediterranean.

Arrived at Malta July 11 and occupied Floriana Barracks and Fort Manuel.

The following letter was received from Major-General Sir J. W. Gordon, Bart., the Colonel of the regiment, on his being removed from the 85th to the 23rd Regiment of Foot:

'HORSE GUARDS,
'*May* 6, 1823.

'MY DEAR SIR,

'In thus announcing to you my appointment to the command of the 23rd Reg^t I trust that you will do me the justice to believe that I feel a more than common pain and regret in taking leave of the 85th Reg^t.

'The King, in conferring this honour upon me, was graciously pleased to say that it was a National Regiment—and while I cannot but receive this fresh mark of favour and gratitude I at the same time think it due to you to declare that no other consideration could have induced me voluntarily to resign the command of the 85th, a Reg^t which was dear to me from many happy circumstances, and latterly is none more so than in the conviction that it is officered and commanded in the best and most efficient manner and by gentlemen of whose friendship and regard it is my pride to boast.

'I must now beg of you to convey to the Reg^t and to each officer respectively my best thanks for their services and support to me as their Colonel, my best wishes for their welfare, and my assurance that it will afford me great pleasure to be in any manner conducive to it.

'I remain, etc.,
(Signed) 'J. W. GORDON.

'To. Lt.-Col. Warburton,
'85th L. I.'

The regiment remained at Malta till Janary 25, 1827.

The following letter was received from the Horse Guards:

'HORSE GUARDS,
'25 *July* 1826.

'SIR,

'I have the honour to inform you by direction of His Royal Highness the Commander-in-Chief that His Majesty has been pleased to approve of the 85th Reg^t having on its colours and appointments, in addition to any other Badges or Devices which may have heretofore been granted to the Regiment, the words "Fuentes D'Onor" and "Nive," in commemoration of the distinguished services of the regiment at Fuentes D'Onor on the 5th May 1811, and in the Passage of the Nive, in the month of December 1813.

I have etc.,

(Signed) 'H. TORRENS, *A. General.*

'Lt.-Genl. Sir Herbert Taylor,

'Colonel 85th Reg[t].'

'HORSE GUARDS,

'6 *Septr.* 1826.

'SIR,

'I have the honour to acquaint you by direction of His Royal Highness the Commander-in-Chief, that His Majesty has been pleased to approve of the 85th or King's Light Infantry Reg[t] bearing on its colours and appointments, in addition to any other Badges or Devices which may have heretofore been granted to the Regiment, the word "Bladensburg" in commemoration of the distinguished conduct of the Regiment in the action on the heights above Bladensburg on the 24th August 1814.

(Signed) 'H. TORRENS, *A. General.*

'Officer commanding,

'85th King's Light Infantry.'

The following General Order was issued when the regiment left Malta.

'HEAD-QUARTERS,

'VALETTA

'16 *Jany.* 1827.

'*General Orders.*

'Major-General Woodford cannot allow the 85th Light Infantry to quit Malta without bearing testimony to the high character the Reg[t] has maintained during the period it has been in this command.

'The Major-General has the satisfaction of seeing it embark in the highest state of drill, discipline and equipment and ready for any service it may be called to perform.

'He requests Major Fairfax will convey to Colonel Warburton and accept for himself his acknowledgment of their attention to the good of the service, and to the interests of the corps which has contributed so much to the soldierlike habits and distinguished appearance of the 85th on every occasion.

'The Major-General desires to assure the Reg. of his best wishes and he will always feel a warm interest in its welfare and success.

(Signed) 'GEORGE FLUDYER, *Capt.*,
'*Act*[g] *Military Secretary.*'

'HORSE GUARDS,
'28 *Aug.* 1827.

'*Memorandum.*

'His Majesty has been pleased to permit the 85th Reg[t] of Foot to discontinue the motto "Aucto splendore Resurgo" as well as the words Bucks Volunteers and to retain the designation of "The King's Light Infantry" only, upon its colours and appointments.'

January 15, 1827. The regiment proceeded to Gibraltar, occupying Windmill Hill barracks.

On November 20, 1827, a pair of new colours were presented to the regiment.

[NOTE.—These Colours were in use till 1844, when they became the property of General Frederick Maunsell, then Colonel of the regiment. On this officer's death, in 1875, they passed into the possession of his son, Major Robert Maunsell, late 85th King's Light Infantry. They now hang over the memorial to the former in Limerick Cathedral.]

'HEAD-QUARTERS,
GIBRALTAR,
'24 *March* 1828.

'*General Orders.*

'His Excellency the Lieut.-Governor cannot allow the 85th King's Light Infantry to embark for Malta without expressing his strongest approbation of the regular, orderly and soldierlike conduct they have uniformly maintained during the period they have been under his command, and he desires to offer his thanks to the Officers, Non-commissioned officers and Privates of the Regiment, whom he is very sorry to lose from this garrison, but more particularly he begs to acknowledge his sense of the able and unremitted attention of Colonel Warburton which has so eminently contributed to the high state of discipline of his corps, and which he shall have great satisfaction in reporting to General Lord Hill.

(Signed) 'D. FALLA,
'*Town Major.*'

THE KING'S.

THE REGIMENTAL.
1827-1844.

March 25, 1828. The regiment returned to Malta and occupied St. Elmo Barracks till October, 1831, when it returned to England, landed at Portsmouth and marched to Winchester; and, in December of the same year, to Dudley, Wolverhampton, Worcester, and Stourbridge.

About this period (November 1831) the silver laurel-leaf embroidery worn in place of lace on the officers' full dress, or 'wing jackets,' was replaced by gold embroidery of the same pattern, in accordance with an order restricting the use of silver lace to the Militia and Yeomanry. Gold embroidery of the above pattern was worn on the collars, cuffs, and wings by officers of the regiment until the introduction of the tunic in 1856.

January, 1832, to Haydock Lodge, Lancashire; and Wigan, Liverpool, Eccles, Altrincham, and Leigh.

March, 1832, to Bolton, Oldham, Preston, Rochdale, and Blackburn.

July, 1832, to Manchester.

'Haydock Lodge,
'24 *Decr.* 1833.'

The following letter received by Colonel Warburton from Major-General Sir H. Bouverie:

'Bantry,
'22 *Decr.* 1833.

'As the 24th is the day fixed for the 1st Div[n] of the 28th Reg[t] to arrive in Dublin on their way to Liverpool it will not be many days before the 85th Reg[t] will embark.

'I cannot allow you to go from this district without requesting you to accept for yourself and your whole Regiment my warmest acknowledgments of the exemplary manner in which the duty of your regiment has been conducted throughout the whole period of their being under my command—I am quite aware that such acknowledgments as this are nothing new to the 85th Reg[t], but it is gratifying to me to have this opportunity of adding my mite to the general voice, and I flatter myself that it will not be unacceptable to you.

(Signed) 'H. Bouverie.

'Col. Warburton.'

August, 1833, to Haydock Lodge, Liverpool, Wigan, Chester, and Hollings Green.

December, 1833. Arrived at Dublin and marched to Limerick with detachments at Killaloe, Tipperary, Newcastle, etc.

May, 1834, to Galway with detachments at Castlebar, Loughrea, Oughterard, and Westport.

May, 1835, to Dublin from Galway.

January, 1836, to Cork from Dublin.

On May 23, 1836, Major Frederick Maunsell was appointed Lieutenant-Colonel of the 85th King's Light Infantry without purchase *vice* Colonel Warburton, deceased.

The regiment was now about to proceed on foreign service again, this time to Halifax, Nova Scotia. The history of the regiment in Nova Scotia, Canada, and the West Indies will form the subject of the next chapter.

THE OLD CHATEAU AT UROGNE.
See page 108.

CHAPTER X

Foreign Service 1836–1846—Nova Scotia, 1836—New Brunswick, 1837—Canada, 1838—West Indies, 1843–1846

On June 21, 1836, the 1st Division of the regiment embarked at Cork on board the *Stakesby* transport under the command of Major H. J. French for Halifax, Nova Scotia.

The Division consisted of three companies, viz. those of Captains Hunt, Power, and the Hon. J. Stuart; its strength being 12 Sergeants, 8 Corporals, 12 Buglers, and 216 Rank and File. The Division arrived at its destination on August 8, 1836, disembarked, and marched into the South Barracks, there to await further orders.

The 2nd Division, with the headquarters, consisting of the companies commanded by Captains Taylor, Brockman, and St. Quintin, embarked on July 5, 1836, at Cork on board the *Katherine Stewart Forbes* transport, Lieut.-Colonel Maunsell being in command. The Division arrived safely on August 12, disembarked the next day, and joined the 1st Division at the South Barracks, the strength of the 2nd Division being 18 Sergeants, 12 Corporals, 2 Buglers, and 238 Rank and File.

The total strength of the regiment, therefore, at Halifax was: 30 Sergeants, 20 Corporals, 14 Buglers, and 454 Rank and File.

The Regimental Records do not state the numbers of the Majors, Lieutenants, Ensigns or Staff that went to Nova Scotia, but the last table of effective strength gives the establishment of the regiment as 1 Colonel, 1 Lieut.-Colonel, 2 Majors, 10 Captains, 10 Lieutenants, 10 Ensigns, 3 Staff, 3 Medical Officers, 7 Staff Sergeants, 36 Sergeants, 36 Corporals, 14 Buglers, and 616 Rank and File.

It will therefore be easily seen that the Reserve or Depot Companies left at home were a strong body of officers and men, though their exact numbers cannot be stated. We know, however, that there were

616 effectives and that 87 men were lacking to complete. Hence the total strength of the regiment should have been 703 Rank and File.

On September 23, 1836, Sergeant-Major Patterson, of the 85th King's Light Infantry, was promoted to an ensigncy in the 41st Foot without purchase.

Ensign Patterson was transferred back to his old regiment on the 30th of the same month, *vice* Crofton, promoted.

The only other event of importance during the year was the inspection of the regiment on October 24 by Major-General Sir Colin Campbell, K.C.B.

The same officer also inspected the regiment again on May 22, 1837. During the year from June to November a number of small detachments were sent out from headquarters to garrison different places as follows:

June 7, 1837. Lieutenant Garnier with 1 Sergeant and 24 Rank and File of Captain Taylor's Company marched to York redoubt; and Captain Brockman and Ensign Dering, with 2 Sergeants and 50 Rank and File, proceeded to Cape Breton.

On the following day, Captain St. Quintin and Lieutenant Tennant with 2 Sergeants and 60 Rank and File proceeded to Prince Edward Island; while Lieutenant Hamilton and Ensign Lord John Butler, with 2 Sergeants and 34 Rank and File of Captain Taylor's company, went to Annapolis.

On June 9 the headquarters, with the companies commanded by Captains Hunt, Stuart, and Power, removed from the South to the North Barracks.

On August 3 a draft consisting of 1 Subaltern, 1 Sergeant-Major, and 40 Rank and File arrived at headquarters.

On September 16 the detachment from York redoubt rejoined headquarters. Two days later, Ensign Domville with 2 Sergeants and 27 Rank and File marched from headquarters to St. John, New Brunswick.

September 19, 1837. The detachment at Annapolis embarked for St. John, to remain there awaiting further orders.

September 21, 1837. Ensign Lord Butler, with 1 Sergeant and 10 Rank and File, rejoined headquarters from Annapolis.

October 9, 1837. Lieutenant Hamilton, with 1 Sergeant and 24 Rank and File, sailed from St. John and arrived at Annapolis.

October 14, 1837. Ensign Domville, with 2 Sergeants and 24 Rank and File joined headquarters from New Brunswick.

On October 30 the half-yearly inspection of the regiment took place, the inspecting officer being Major-General Sir Colin Campbell, K.C.B.

The regiment was now about to leave Halifax, and before leaving received the following address from the members of Her Majesty's Council, the clergy, magistrates, and other inhabitants of the place :

'HALIFAX, N.S.,
'11 *Novr.* 1837.

'SIR,

'We the members of Her Majesty's Council, Clergy, Magistrates, and other inhabitants of Halifax, beg to express our sincere regret at the intended removal of the excellent Regiment under your command from this province. We would fail in our duty were we to omit the opportunity of bearing witness to the order and good conduct of the 85th Regt. during the short time they have formed a part of this garrison.

'The social intercourse which has existed between yourself and Brother officers and the inhabitants of Halifax has grown into a feeling of sincere regard and good will on our part and has been fully reciprocated by them.

'The N.C.O.'s and Privates have always cheerfully seconded the example of their officers, and we beg you will be good enough to express to them the thanks of the inhabitants of Halifax for their active zealous and efficient services during the calamitous visitations from which this town has suffered since your Reg^{t}. has been stationed here. Our best wishes will most cordially accompany your well disciplined Corps wherever Duty and the high behest of our Sovereign may call them and we feel confident that wherever stationed they will win the best feelings and regard of the people among whom they may be quartered.

'We are, Sir, etc.'

(Here follow 148 signatures.)

During November, 1837, the regiment left Halifax for St. John, New Brunswick, marching as far as Windsor and travelling the remainder of the distance by ship.

On December 2, in consequence of the rebellion in Lower Canada, the regiment received orders to prepare to move to the assistance of the troops stationed in that province. The men were provided with two pairs of moccasins each, two blankets, warm mitts, etc.: and each company with a proportion of camp kettles, felling axes, etc.

The first division of the regiment started for Quebec on December 16.

The *Regimental Record* gives the following account of the adventurous winter journey:

'December 16, 1837. The first division (Captain Power's company) proceeded *en route* to Quebec.

'December 18, 1837. The headquarters, consisting of Captains Taylor's and Stuart's companies, marched to Poverty Hall (seven miles), where sleighs with two horses and accommodation for 8 men each were waiting on the ice on the St. John River for our reception and in which we proceeded on the river to Gouldings and Watsons (33 and 36 miles) where the division halted for the night.

'December 19. Proceeded on the river to Fredericton (40 miles), where the men were placed in barracks for the night.

'December 20, 1837. The headquarters with Captain Taylor's company (Captain Stuart's forming a separate division) proceeded in sleighs on the land to Lewis Hewster at Queenstown, 25 miles.

'December 21, 1837. To Jones, opposite Eal River, 25 miles.

'December 22, 1837. To Dingle Settlement, 30 miles.

'December 23, 1837. To Tobingare Settlement, 30 miles.

'December 24, 1837. To the great falls on the river of St. John (the men, sleighs, and horses having crossed the River Restock in a boat, the ice not being strong enough to bear a man), 22 miles.

'December 25, 1837. On the River St. John to Madawska, 35 miles.

'December 26, 1837. On the River Madawska (except the last six miles) to the Deycli (?) Camp, 26 miles.

'December 27, 1837. On the Lake Temisquata to the Camp at the entrance of the St. Laurence portage (?), 16½ miles.

'Here the H^d^Q^rs^ found Cap^n^ Power's Co^y^ when the whole proceeded as a grand division. In consequence of the roads being too narrow for the New Brunswick sleighs we were supplied with carioles, one for every two men and drawn by one horse, in which we proceeded on :

'December 28, 1837, to the camp half way across the Portage (?), 16 miles.

'December 29, 1837. Village of St. André on the St. Laurence, 18 miles.

'December 30, 1837. Riviere Onilb, 31½ miles.

'December 31, 1837. L'Istell, 32 miles.

'January 1, 1838, to St. Michael, 34 miles.

'January 2, 1838. Point Lovi, 18 miles ; and across the St. Lawrence in canoes to Quebec and quartered in the Jesuit Barracks.

'January 3, 1838. Captain Brockman's and Stuart's co^ys^ arrived.

'January 5, 1838. Capt^ns^ Hunt's and St. Quintin's Co^ys^ under Major French arrived.

'January 15, 1838. H^d^ Q^rs^ with Captains Brockman's and St. Quintin's Companies left Quebec in carioles for Sovel by the following routes :

'Cape Santi, 30 miles.

'January 16, 1838, to St. Anne, 27 miles.

'January 17, 1838, to Three Rivers, 28 miles.

'January 18, 1838, to Rivière de Loup, 24 miles.

'January 19, 1838, to Sovel, 27 miles.

'January 20, 1838. Captains Taylor's and Power's Co^ys^ arrived.

'January 22, 1838. H^d^ Q^rs^ with Captain Brockman's and St. Quintin's co^ys^. left Sovel in carioles to Montreal, 45 miles.

'The men quartered in Quebec Square Barracks, and Captn Power's Co^y^ (who arrived following day) in Gaol Barracks, and Captains Hunt's and Stuart's (who arrived under Major French on January 27) in St. Paul Street Barracks.

'February 1, 1838. Captains Brockman's and Power's Co^ys^ under Major French proceeded in carioles to L'Acadie (?).

'February 28, 1838. In consequence of the rebels having assembled on the frontiers the Reg^t^ moved to St. John.'

March 1st. The 43rd and 85th Regiments with two guns moved to

Henryville, where having learned that the rebels had been disarmed and dispersed by the American General Nool, the troops returned to St. John and on March 3rd to Montreal.

'June 4, 1838. The regiment marched to Lachine Canal, embarked in *bateaux*, and proceeded by canal to Kingstown, Upper Canada.

'June 22, the regiment embarked on board the "Cobourg" steamer and proceeded to Toronto.

'November 24, 1838. Embarked for Hamilton *en route* to London, U.C., disembarked same day and marched seven miles to Ancester where men quartered for the night in the church and chapels.

'November 25. Regiment proceeded in wagons 18 miles to Brentford.

'November 26. Right wing proceeded in wagons to Tugorsall.

'Left wing proceeded in wagons to Woodstock.

'Arrived London, U.C., on 5 Dec^r, remained in billets there till 19th Decr when the left wing proceeded to Sandwich in sleighs and the right wing to St. Thomas's.

'April 9, 1839. Lieut.-General Sir William Thornton, K.C.B., appointed Colonel *vice* Sir Herbert Taylor, K.C.B., G.C.B., deceased.'

The following letter was here received from the War Office:

7 August 1839.

'SIR,

'We have the honour to acquaint you the following memo^m. appeared in the *Gazette* of the 6th inst.

"WAR OFFICE,
"16 *May*, 1839.

"Her Majesty has been pleased to permit the 85th or the King's Light Inf^y. Reg^t. to resume on its colours and appointments the motto "Aucto splendore Resurgo" which was authorized to be borne by that Reg^t. on the 3rd August, 1815."

(Signed) 'Cox & Co.

'Officer Comd^g.
'85th Foot.,
'St. Thomas.'

The overland march of the 85th Regiment from New Brunswick to Quebec in December, 1837, is most graphically described in a letter which was published in the *United Service Journal* for April 1838.

The letter is long and cannot be reprinted *in extenso*; still, the most interesting passages may be well inserted here:

'On the 16th of December, Captain Power's company left St. John. Each man was provided with two blankets, ear-covers, and moccasins. They marched seven miles, in consequence of the snow not being sufficiently deep for sleighing. On reaching St. John River they found the sleighs were waiting for them. Each vehicle took eight men; and the officers had a sleigh to themselves, and one for their baggage, which consisted of one portmanteau and carpet-bag to every officer.

'On the 18th, Colonel Maunsell and the headquarters, consisting of two companies, left in the same way, and arrived on the following day at Fredericton; 19th, Captain Brockman's company left; 20th, Hunt's; 21st, St. Quintin's, the last division, left St. John. We travelled rapidly, and were very little inconvenienced by the cold, which, though 18° below zero, seemed scarcely below freezing.

'At Fredericton we were entertained by Sir John Hervey.

'On the morning of the 23rd of December, having changed our sleighs, we were again *en route*. The weather beautiful, and sleighing excellent. We kept upon the road, as the ice would no longer admit of travelling on the river. The appearance of the latter was most singular; the ice, which had been broken and again united, presented a surface like a plain which had been blown up by an earthquake. About eighteen miles from Fredericton we crossed the river between two walls of ice and frozen snow, and following the Skirts for about three miles returned again to the road. I thought I had never travelled over so precipitous a road; and it required considerable experience in sleigh-driving to prevent slewing over the precipices on either side. This, however, was only a prelude to what we were to meet within the sequel. The day's journey was twenty-five miles. We were put up at three houses, and had very tolerable accommodation and fare.

'December 24. We resumed our journey at eight A.M. The

road much the same, as well as the state of the ice—which latter, in many places, looked like a sea of frozen ice.

'The banks of the river were exceedingly high and precipitous; the cleared country was thickly studded with cottages of no mean appearance, and there were many tolerable farms; the interminable forest bounded the horizon. We reached our destination, twenty-five miles, at half-past 3 P.M. Our fare and accommodation much like yesterday's.

'December 25. A beautiful morning; thermometer at zero.'

Here follows a description of the appearance of the inhabitants, which we omit.

'We reached Dingee settlement at 4 P.M., having performed thirty miles. Our quarters here were certainly not equal to an English hotel, but to one who had seen service before, they were duly appreciated. Two of us got tolerable beds, and the rest made it out on the floor in their buffalo-skins.

'December 26. The roads were awfully precipitous; and to add to our discomfort, one of our horses was as unmanageable as a wild beast, and constantly put our lives in jeopardy. Thrice we were on the brink of a precipice of unmeasured depth—and at length over we *did* go, but luckily the fall was not great. I sprang out in time to save myself but saw the horses roll over, followed by the sleigh, containing only the driver and one officer, who narrowly escaped having his head dashed against the trunk of a tree. After this accident we always took the precaution of jumping out when near a dangerous place; and two of us applying ourselves to the back part of the vehicle, were able to steer it in whichever way we pleased, in spite of our unruly horse. Before we reached Tobique, our halting place, we had to descend on to the river, a place so steep that I could hardly have imagined any man bold enough to ride down, much less drive. The track was not more than thirty yards, and then turned off at a right angle, at the point of which was a large hole which invited any inexperienced driver to *slew* into; nevertheless, the whole party got down without accident.'

Here the writer mentions the manifestations of loyalty on the part of the inhabitants.

'December 27. We commenced our journey by crossing the Erastic, a small river which runs into the St. John; its breadth was about 150 yards.' The ice would not bear. The horses were taken out and led separately across and the sleighs pushed over by hand All crossed without accident in about an hour; 'though the driver of my sleigh certainly tempted fate to the utmost verge, driving over alone at a furious rate, whilst the whole sheet of ice undulated like the representation of the sea on a stage.'

He continues:

'We now entered the portage (a track of land which connects two waters) which led to the Grand Falls. Never in my life did I see such a road; it was a succession of precipices flanked by a dark, gloomy, boundless forest, where the arm of a single lumberman might in one night have baffled all our attempts to march onward, and where fifty men, nay twenty, well armed and with snow-shoes, could have destroyed every man of us without the slightest risk. You have seen an American wood; I therefore need not describe what an avalanche of trees is, and how utterly impassable those woods are to any one except an Indian or a woodsman.'

The writer then states that his driver (an expert lumberman) could have felled more trees in one night than the expedition could have removed in a fortnight. Also that a single expert axeman will cut down the largest tree in the forest in twenty minutes and six of the best lumbermen would be needed to remove it in an hour.

'We reached Grand Falls about half-past four—twenty-five miles. Here we met Sir J. Caldwell, a gentleman of considerable property here, who has extensive saw-mills which supply the provinces with timber. He took us to see the Falls, which were near his house, and then entertained us with a good dinner etc.

'December 28. We left the Grand Falls at 8 o'clock, and proceeded on the river. The thermometer was 10° above zero, which was unusually mild, considering that a few days before it had been 27° below zero. The snow was deep, and consequently made our progress rather slow. We nevertheless reached the Madewaska, thirty-five miles, before 6 P.M. Here we put up at the first house where French was spoken, and our entertainment was not bad.

'December 29. This morning considerable delay occurred in consequence of our having to change sleighs, and a fresh supply of rations being issued; it was half-past 9 before we got off. Our route was along the Madewaska . . . the ice was excellent for upwards of twenty miles, when we were obliged to take to a portage which connects this river with the Tamasquatha lake.

'Here we had a specimen of what the people here call *cahos*.

'These are a succession of deep holes, which are formed when the snow is on the ground by the bad construction of the carioles, the shafts of which are fastened on the very runners, and having a broad board to connect them, sloping at an angle of 45° ; the snow is thereby scraped up into mounds, between three and four feet high, so that really the motion of our sleigh was precisely that of a boat in a heavy sea, only its effects were ten times more violent; and this suggested to me the name I gave the portage, viz., *Passage des ondes glacées*.

'It was dark when we got into camp—a number of large log huts erected on purpose for the troops. We passed a very uncomfortable night owing to the smoke from our fire, which was at times large enough to roast an ox by, and obliged us to rouse out and put snow on it, when shortly after it would get so low that we were in danger of freezing. The thermometer was 4° below zero.'

December 30. The distance covered on the previous day was twenty-six miles. A stage of eight miles had now to be covered on the Tamasquatha Lake to reach the camps at the outlet. The ice was very dangerous, being full of holes, but the party arrived safely at noon. Here they were at the entrance of the Grand Portage which extends for thirty-six miles and terminates at St. André. The writer of the letter here determined to go on ahead with the Commissary in order to avoid two more nights in camp, and also to give his horse a much needed rest before the Division arrived. The poor beast, however, was quite done up and could not proceed over the *cahos*; so it had to be left in camp and the sleigh abandoned in the snow. Having gone about a league with the Commissary the horses, unaccustomed to the violent concussions caused by the sleigh pitching into the holes, floundered and fell into the deep snow and broke the shafts, the travellers being thus left stranded in the forest.

There was nothing for it but to walk back, and to walk back in travelling dress was no joke. However they at last, about 4 P.M., regained the camp, where they luckily found plenty of carioles in readiness. They obtained two, hired two *traineaus* for their luggage, and again started.

They had twelve miles to go; but such was the state of the road that they expected not to arrive before midnight. It was far worse travelling in a cariole than in a sleigh, as the latter is a long vehicle compared with the former and glides over the *cahos* with comparative smoothness to the way these shorter vehicles plunged into them. A little before 10 P.M. they saw a solitary light in a hovel, the only one on the portage. Here they halted. The hut was 'full of artillery-men and cariole-drivers; and smelt foully. A dish of fried pork and onions was provided for us; but my stomach was too sick to encounter it, and I was too happy to throw myself on the floor when I soon got into a sound sleep.'

'December 31. At 4 A.M. we were up, and prepared to start. The bustle which ensued awakened the other inmates of our small room who instantly unkennelled from their bed, and, to my infinite surprise, proved to be two young damsels of elephantine dimensions. Not the least abashed by our presence, they began and completed their toilette with wonderful expedition and perfect *naïveté*.'

The road proved rather better, the ponies were able to trot at times and the motion thus engendered resembled a game of leap-frog. About 1 P.M. the first view of the St. Lawrence burst on the travellers. The river is thirty miles across at this place. They remained two days at St. André, during which Captain Hunt and Lieutenant St. Quintin arrived with their company and the Royal Artillery also.

On January 2 they started for Quebec in a line of some three miles in length and composed of 130 carioles. They proceeded thirty miles that day and reached Oreille. The country was fine, level, well-cultivated, thickly populated and full of capital farms. At Oreille they were 'handsomely entertained by Mons. Cosgraie and his brother,' who had previously shown similar hospitality to the officers who had gone ahead. The men were also comfortably put up, ample fuel and all necessaries being provided.

On January 3 they reached Islet, a distance of thirty miles, passing through the villages of St. Ann's, St. Roch, and St. John. Bad travelling was anticipated owing to heavy rain the preceding night, but the going proved not so bad after all; and they reached Islet before dark. Here the *curé* and other gentlemen of the place entertained them and provided beds.

'January 4. This day's stage was to St. Michele, thirty-three miles, through the villages of St. Ignace, St. Thomas, Berthier, and St. Vallier; of which we could see but little, owing to a snow-storm which assailed us *en chemin*, and nearly stopped us. At St. Michele the men were admirably quartered, and the *curé* (a most gentlemanlike man) received all the officers, and treated us *en prince*.'

January 5. The snow having drifted during the night, it was needful to get the assistance of the peasantry to clear the road; otherwise it would have been impossible to have reached Point Levi, fifteen miles on, in time to cross the St. Lawrence that day.

'We all embarked in canoes, which we found waiting in the street, and as the carioles drove up, their contents of men or baggage were removed into the canoes assigned for them. This done, at a given signal they were shoved over the ice into the water and paddled with great dexterity through the floating masses of ice.'

This concludes the account of the journey. The writer states that he is about to begin a second one to Sorel with the last division of the regiment.

The last few paragraphs of the letter contain certain uncomplimentary remarks about Lord Gosford, who, it would seem, was universally unpopular. The political situation appeared to be very critical and fears are expressed that unless vigorous action is taken to support the authorities the Colonies will be lost. The letter is dated Quebec, January 22, 1838.

On August 9, 1839, a draft consisting of 2 Subalterns (Lieutenant Grant and Ensign Knox), 1 Sergeant, and 63 Privates, joined headquarters from the reserve companies in the depot at home.

This Ensign Knox was afterwards Lieut.-General and Hon. General

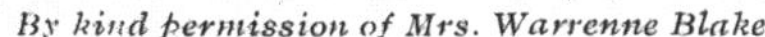
By kind permission of Mrs. Warrenne Blake.

LIEUT. THOMAS EDMOND KNOX.

Thomas Edmond Knox, C.B. He remained in the regiment from February 23, 1838, to August 17, 1852.

The following notes are extracted from his diary, some of which furnish amusing reading:

'April 22, 1838.

'Put my name down for the Army and Navy Club then just formed.

'Apr. 24. Attended Lord Fitzroy's levée to thank him for removing me to the 85th.

'May 3. Attended the C.-in-C. (Lord Hill's) levée and not knowing better I asked him when he thought our depot would come over from Ireland upon which he bowed me out.[1]

'May 11. Elected member of the A. and N. Club.

'June 17. The 85th Depot landed about 3 o'clock at Devonport after a very stormy passage of 48 hours from Cork in the "Messenger" Gov^t^ steamer and marched into Cumberland Barracks, Major Jackson commanding. Cap^ns^ Pipon and Browne, Lieut. Dickson, Adj^t^ Jackson, Ensign Maunsell, 8 Serg^ts^, 4 Buglers and 139 Rank and File. We all dined at Elliott's Hotel.

'June 18. Mounted my blue coat and horrified old Jackson by carrying an umbrella.

'June 20. Began drill under Serg^ts^ Watson and Barnet and Porter, Serg^t^-Major—got rather more drill than if I had learnt none at R.M.C.[2]

'June 23. John James, late 85th, came on a visit.

'June 28, 1838. The coronation—garrison reviewed, *feu de joie*, Royal salutes, illuminations and grand ball etc.

'Lethbridge joined on appointment half in uniform and half in plain clothes.

'July 23. Commenced fire-lock drill.

'Aug. 24. I was reported to and blown up by the General, having been seen in bad company.

'Aug. 27. The "Atholl" troopship for Canada arrived in the Sound.

'Aug. 28. Maunsell and 46 of our depot sailed in the "Atholl" for Canada.

[1] The depot was at Tralee.

[2] Royal Military College, Sandhurst.

'Sept. 9. Supernumary on the Main Guard with Grant, a lively place on Sundays.

'Sept. 12. Supernumary on the Dockyard Guard with Captn Pipon . . . there were 17 sentries to visit and the rounds, usually performed once by the Captain and twice by the subaltern, took ¾ hour going.

'Sept. 13. The "Atholl" returned minus a foremast. Maunsel came back and slept on shore. Lent him £20 which he repaid.

'Sept. 17. Supernumary on a C. Martial for the first time.

'Oct. 11. Out at ball practice between Stonehouse and the main guard—we were armed with the Brown Bess with flint and steel. Shots rarely went over 100 yards.

'Oct. 18. Our depot marched into the Citadel, a change I was glad of as the habits of the Picquet Barracks from their public situation were rather debauched.

'Nov. 10. A fight between the 46th and 85th against the 29th.

'Dec. 21. My allowance from my father was £120 a year.

'Our uniform in full dress consists of a heavy broad-topped chaco, black ball-tuft and flat scale chinstrap. Double-breasted swallow-tailed coatee with gold embroidery on cuffs and collars, gilt wings, embroidered blue cuffs and collars. Embroidery on pocket flaps, crimson sash with tassels tied round the waist, sword worn with buff shoulder-belt and with breastplate, chain and whistle, leather scabbard. In winter Oxford mixture, in summer white trousers.

'1839.

'Jany. 31. Ensign Tynte joined a nice quiet good-looking fellow.

[*Note added.*—'In these days songs used to be the order after mess, little smoking.']

'Mch. 20. Sir Herbert Taylor (our Colonel since 1823) died at Rome.

'June 7th. Embarked on the "Atholl" for Cork *en route* to Canada.

'The "Atholl" was built of timber from the Duke of Atholl's property and once carried 28 guns—now, being a troopship, she only carried two.

'June 29. A man on board tried and got 100 lashes for insubordination. The first corporal punishment I ever witnessed which made me feel squeamish.

'July 17. Reached Quebec.

'Aug. 8. Marched into St. Thomas' about 1 P.M. Much struck by the large size and soldierlike appearance of the head-quarter men.

'Reg[t] weak in consequence of constant desertions.

'Shell jackets are worn by the officers on parade.'

From the Regimental Digest of Services we learn that on September 5, 1839, the headquarters marched from St. Thomas to London and quartered in the Log-Hut Barracks.

The Diary continues:

'Sept. 5. Two men flogged for drunkenness and insubordination.

'Broughton, one of our draft, tried by G.C.M. for abusive language towards Grant and myself and got 200 lashes.

'Septr. 21. The motto "Aucto splendore resurgo" restored at the request of Sir William Thornton.

'Octr. 10. Garrison flogging parade. Broughton got 200 lashes.'

Regimental Digest of Services.

'October 22, 1839. One company under the command of Ensign Horrocks marched from Sandwich *en route* to London where it joined headquarters on October 28th, 1839.'

From the Diary.

'SANDWICH, CANADA.

'Octr. 23. P[te] Hughes N[o] 6 co[y] received 150 lashes for stealing some cabbage.

'Nov. 7. The first fall of snow.

'Nov. 11. Tried two men for attempted desertion. Transported one for life, the other for fourteen years.

'Bought some racoon skins to make the men's winter caps.'

From the Digest we learn that on October 22, 1839, a letter was received from the War Office dated August 20, 1839, augmenting the strength of the regiment as from the 12th instant. The establishment was now as follows: 10 Companies, 1 Colonel, 1 Lieut.-Colonel, 2 Majors, 10 Captains, 10 Lieutenants, 10 Ensigns, 1 Paymaster, 1

Adjutant, 1 Quartermaster, 1 Surgeon, 2 Assistant-Surgeons, 1 Sergeant-Major, 1 Quartermaster-Sergeant, 1 Paymaster-Sergeant, 1 Armourer-Sergeant, 1 Schoolmaster-Sergeant, 1 Hospital-Sergeant, 1 Orderly Room Clerk, 10 Colour-Sergeants, 30 Sergeants, 40 Corporals, 1 Drum-Major, 13 Drummers and Fifers, 760 Privates.—Total 901.

On November 20 the following Circular was received from the Adjutant-General:

'HORSE GUARDS,
'20 *Nov.* 1839.

'*Circular Memo.*

'The following is to be the distribution of the Service and Depot Companies of the Regiments of Infantry serving on Foreign Stations excepting those in the East Indies and New South Wales, and the General Commanding in Chief directs that the numbers under stated be considered as the Establishment of the respective portions of the Regiments abroad and accounted for accordingly in the Regimental Returns, viz.

Six Service Companies at 100 Rank and File each,	600 R. and F.	4 Depot Companies at 50 Rank and File each,	200 R. and F.
7 Staff-Sergeants 24 Sergeants, i.e. 4 per Company including Colour-Sergeants.		16 Sergeants, i.e. 4 per Company, including Colour-Sergeants.	
10 Drummers, 24 Corporals, 576 Privates.	600 R. and F.	4 Drummers, 16 Corporals, 184 Privates.	200 R. and F.

Total Establishment.

	Staff-Sergts.	Sergts.	Drummers.	Corps.	Pts.
6 Service Companies	7	24	10	24	576
4 Depot Companies		16	4	16	184
10 Companies	7	40	14	40	760
				800	

From the Diary.

'Dec. 4. Colonel Prince burnt in effigy at Detroit, it being the anniversary of him shooting the sympathisers here last year.

'Dec. 11. Sleighing began.

'Dec. 30. Hughes received 150 lashes for stealing.

'1840.

'Jan. 14. Walked across the river nearly to Grope Island on the U.S. side and fell into an air hole, in which I remained about 5 min. unable to get out. Three of my fingers were frost bitten and I had a very narrow escape with my life.

'Jan. 17. About ten American soldiers deserted and came over to our side wanting to enlist, we of course refused to take them.

'Feb. 2. Major Deedes' groom deserted to Detroit with his horse and saddle. The authorities sent them back and imprisoned the man, but refused to give him up.

'Feb. 22. The Washington anniversary ball at Detroit to which we were invited by the American garrison of Detroit.

'Went in full uniform. Introduced to Major-General Brady and several other American officers.

'The ball room was ornamented with English and American flags.

'The band of the Brady Guards, a local volunteer corps dressed in grey with black facings and epaulets, played during the evening.

'The Yankees were politeness itself and gave up their claims to any lady we expressed a wish to dance with.

'March 9. Four of the United States Art^y^ came over and lunched with us.

'Their men are of a very bad class, Germans, Irish and deserters from our service, the discipline is very severe and desertion prevails to an immense extent.

'March 17. Dined with the 34th. Large dinner party, being St. Patrick's day. Sat up till 4 A.M. One of the Light Company killed himself by drinking off a pint of raw whisky. I saw his body next day—his face quite blue and a strong smell of whisky about him.

'The 34th a very nice Reg^t.^

'Ap^r^ 25. Head Quarters left London to-day—intensely hot. The Armourer Serg^t^ (Newman) died from the effects of heat.'

From the Regimental Digest.

'April 25, 1840. Headquarters with four companies marched from London, Upper Canada, to proceed to Montreal, Lower Canada, in two divisions; first division under the command of Lieut.-Colonel Maunsell, the second under the command of Major Hunt.'

At this time news was received of the death of Sir William Thornton, the Colonel of the regiment.

His successor in the colonelcy was Major-General Sir John Fitzgerald, K.C.B.

On leaving London the following letter was received by the regiment:

'To LIEUT.-COL. MAUNSELL,

'Commanding 85th Regiment.

'We, the undersigned members of the Board of Police and inhabitants of the town of London, avail ourselves of the present occasion of expressing to yourself and the officers and men under your command the feelings we entertain towards the 85th Rgt. upon its departing from this place.

'The kind and urbane conduct of yourself and the officers of that Regt. together with the general good conduct of the men in all their intercourse with the inhabitants of this town during the period of their being quartered here, have left an impression which will be long remembered by the inhabitants with sentiments of esteem, and we cannot but write with heartfelt wishes for the welfare of that corps whatever may be its future destination.

(Signed) 'G. J. GOODHUE (?),
(and 30 Gentlemen). '*President, Board of Police.*

'London,
'April 24, 1840.'

April 28, 1840. The 2nd Division joined headquarters at Brantford. The whole, under the command of Lieut.-Colonel Maunsell, arrived on April 30 and embarked in the *William IV* steamer the same day for Kingston, at which place it arrived the following evening, and quartered in Fort Henry for the night.

May 2, 1840. Embarked for Montreal by the Ridian Canal, reached that place May 8 and quartered in the Quebec Gate Barracks, there to be stationed until further orders.

From the Diary.

'May 4. Murray of N° 5 received 150 lashes for striking Serg[t] Enright.

'May 5. Saw Phillip Huffman hanged here for committing incest with his own daughter and destroying the child. Hundreds of Yankees came over to see it. The Rev. Mr. Johnson turned him off, the hangman not understanding his business properly.'

[This last is a most curious fact, and one would be inclined to think as an execution its circumstances were unparalleled.—ED.]

'May 17. Proceeded to Montreal, met by the band and marched into Quebec barracks, where the H[d] Q[rs] and four companies had arrived a few days previously.'

On May 19 the detachment of two companies stationed at Sandwich, Upper Canada, under the command of Captain Taylor joined headquarters at Montreal, having left Sandwich on May 8.

June 1, 1840. Ten privates of the 11th regiment, 5 sergeants, 3 corporals, and 76 privates of the 15th regiment were transferred to the 85th, having volunteered to serve in Canada. The Mess house of the 85th was, we learn, the centre house in Dalhousie Square. While in Montreal the regiment entertained very extensively and the Mess used to be an open house to all comers.

June 9, 1840. A draft under the command of Captain Dickson, consisting of 2 subalterns (Ensigns Ross and Lethbridge), 1 sergeant-major, 1 sergeant, 3 corporals, and 60 privates, joined headquarters from the reserve companies.

The Diary continues:

'June 10. The 85th trooped on the *Champs-de-Mars.*

'June 11. Trooping on the *Champs-de-Mars.* Our reg[t] is very strong now. Total in the service co[ys] 615 Rank and File.

'June 18. Carried the colours on "Waterloo day." Review on the *Champs-de-Mars.*

'June 29. Glass 97° in the shade.

'July 9. Half-yearly inspection by M. Gen[l] Clitherow. Dinner party at mess, hot, stiff and uncomfortable, like all inspection dinners.'

T 2

The writer of the Diary now went on a month's leave, and his next entry concerning the regiment is as follows:

'July 25. To Bladensburg where our Reg[t] was engaged in Aug. 1814. It is a small town on an extensive plain covered here and there with underwood

'Aug. 21. Inspected by Gen[l] Clitherow, a silly old Guardsman.

'Aug 22. Colonel Maunsell gave a soup tureen to the mess.'

From the Digest.

'December 1, 1840. Sergeant Peter McGowan of the 23rd regiment was transferred to the 85th having volunteered to serve permanently in Canada.'

From the Diary.

'Apr. 27, 1841. New accoutrements and breastplates taken into wear. The Sergeants discontinued chains and whistles.

'May 6. Horrocks got a year's leave—his twin brother Charles in the 15th so like him that once they changed uniforms and paraded with each other's regiments!'

On May 28, 1841, Bugle-Major Samuel Lambert joined headquarters from the reserve companies.

June 1, 1841. 1 sergeant and 13 privates from the 73rd regiment; 1 corporal and 8 privates from the 32nd regiment; 13 privates from the 24th regiment; 6 privates from the 65th and 3 privates from the 34th regiment joined the 85th, having volunteered to serve permanently in Canada.

June 10. A draft under the command of Capt. J. K. Pipon, consisting of Captain H. S. Brown, Ensign E. L. Parrott, Armourer-Sergeant Reardon, 1 sergeant, 1 corporal, and 21 privates, joined headquarters from the reserve companies.

November 23, 1841. Major-General Sir J. F. Fitzgerald, K.C.B., the Colonel of the 85th, was promoted to be a Lieut.-General.

The Diary now continues:

'1842.

'Jan[y] 16. The Reg[t] took black pack straps into use. The Colonel put red collars on the men's shell jackets with white bugles.

'Feby 3. Our sergeants gave a grand ball at Nelson's Hotel to which many of us went and enjoyed it very much. One of my partners Mrs. Sergt. McGowan asked me for my washing!

'March 17. The usual festivities on St. Patrick's Day. An Irish party of 36 at Mess who consumed 86 bottles of wine.'

The regiment was now about to leave Montreal.

On April 27 the right wing under the command of Major Hunt proceeded from Montreal to St. John's per steamboat *Vittoria* and railway cars.

The following day, headquarters and the left wing under the command of Lieut.-Colonel Maunsell travelled in the same manner and for the same destination. This is mentioned in the Diary as follows:

'Ap^{r} 28. H^{d} Q^{rs} and left wing left for St. John's.—Grant and Lethbridge late for parade and put under close arrest by the colonel, who was furious at having no subalterns to carry the Colours out of Montreal.

'May 20. The colonel employed the Reg^{t} by wings in transplanting trees into the barracks and improving the place.

'May 25. Took white trousers into wear.

'May 30. The Colonel (Maunsell) allows no officers to drill the Batt^{n}. He has a splendid word of command.

'June 17. The "Chackot" (Shako) with curb-chain chin-strap taken into wear. We now gave up the old broad-topped "chacots."

'July 23. The buglers sent to the ranks for some time for bad behaviour.

'Octr 26. Mrs. Maunsell delivered of a son and heir, to the great joy of the colonel.

[Viz. Major Robert Maunsell of 85th K.L.I.]

'Novr 16. Left St. John's for New York (for a year's leave home). The Reg^{t} turned out and cheered me as I passed the barracks, nearly all the officers came to see me embark.

'Novr 20. Called on Miss Rodgers who limited her civilities to asking me to go to church with her which I did.'

A reduction in the establishment of the regiments of Infantry of the Line was now ordered by which the strength of the 85th became 740. A hundred supernumeraries were henceforth to be allowed to regiments serving in India and China. In all others serving abroad the number of supernumeraries was limited to 30.

The standard of height for the Infantry of the Line, with the exception of those regiments serving in India, China, and New South Wales, was ordered to be raised to 5 feet 6½ inches for men not exceeding 25, and 5 feet 6 inches for growing lads under eighteen years of age. Regiments serving in India, China, and New South Wales were allowed to recruit men at 5 feet 6 inches, but they were not to be under eighteen years of age. During this year there was a serious fire at St. John's, to help extinguish which the regiment performed good service. A letter of thanks for what they had done was received by the regiment on May 19, 1843, bearing the signatures of four magistrates.

On May 23 an augmentation in the establishment of the regiment took place and the total of the rank and file now again numbered 800. The regiment was now about to leave Canada and to proceed to the West Indies. The following address was received by Major French, commanding the regiment, from the inhabitants of the town of Dorchester (St. John's).

'ST. JOHN'S,
'*September* 29, 1843.

'To Major French and the Officers, Non-commissioned Officers and Privates of Her Majesty's 85th Regt. of Light Infantry.

'We, the undersigned Magistrates and Citizens of the Town of Dorchester and its vicinity beg thus publickly to express the sincere regret with which we view the approaching departure of this gallant and distinguished corps for the West Indies.

'A sojourn of 18 months in this garrison which has been marked by the most friendly intercourse on all sides, would of itself be sufficient to enlist our best wishes for the future success of the 85th Regt. but to this is added the grateful remembrance that in the perilous times which have now happily passed away the 85th was one of the first gallant Regts. who at the close of the eventful season of 1837, so promptly accomplished a toilsome winter's march to support Her

Majesty's authority in this Colony and maintain that connection with the British Empiere which every loyal subject in Canada is so desirous to perpetuate.

'To that excellent and gallant officer Lieut.-Colonel Maunsell and also to Major French is due the greatest praise for their unceasing exertions to improve and beautify the grounds which adjoin the garrison, and which have been so literally thrown open to the Public for amusement and recreation—in these improvements so well commenced the 85th have left a lasting proof, not only of their good taste but also of their liberal feelings which prompted them to undertake the work and which must serve as a bright example to their successors.'

(Here follow a number of signatures.)

To this, as to all other similar addresses, a suitable reply was of course returned by Major French.

On October 1, 10 sergeants, 4 corporals, and 142 privates, who had volunteered for permanent service in Canada and North America, were transferred from the 85th to various other regiments.

October 2, 1843. The left wing of the regiment, consisting of Captains Tennant's, Taylor's and Blackburn's companies under the command of Captain Tennant, left St. John's for Barbadoes in the *Princess Royal* transport. They reached Barbadoes November 14, and thence proceeded to Antigua, where they landed in English Harbour.

On November 25, Captain Blackburn's company (No. 6) marched for St. John's and the other two companies occupied barracks on the Ridge.

October 11, 1843. The headquarters of the regiment (consisting of Captains Coape's, Grey's and Taylor's companies) left St. John's, Canada, under the command of Major French. They arrived at Barbadoes on November 22, and anchored in Carlisle Bay, sailing on the morrow for Antigua.

A draft from England had come out in the *Java* consisting of Captain Grey, Lieutenant Day, Ensign Mainwaring, 3 sergeants, and 171 rank and file to join the service companies. They were taken on the strength from October 13, having arrived on October 11. The draft was then embarked on board the *Boyne* transport at Quebec and sailed two days later for Barbadoes.

The strength of the regiment on its arrival in the West Indies was 546 officers, non-commissioned officers and men. Its full strength should have been 581. It appears that 2 ensigns and 35 privates were wanting to complete.

The writer of the Diary that has been so freely quoted (Lieutenant T. E. Knox) now on October 13, 1843, became Adjutant *vice* Patterson, who resigned.

Continuing from the Diary:

'Nov[r] 25. Joined the Reg[t] on board the "Boyne."

'Nov[r] 27. Making but little progress the band played on deck.

'Nov[r] 28. Sighted Guadaloupe, Descada (?) and Montserrat, and our skipper, Capt[n]. Hammond, not being exactly aware of the English Harbour was running into Willoughby Bay when the bugles of our left wing on Shirley Heights warned us off in time and we tacked and stood out.

'Nov[r] 29. Antigua. Anchored at the mouth of English harbour and the ship was warped in. Landed the men in the afternoon in the dockyard, a most unhealthy spot, and marched to the Ridge.'

The Colonel of the regiment (Lieut.-General Sir John F. Fitzgerald. K.C.B.) having been removed to the colonelcy of the 62nd Foot, he was succeeded by Lieut.-General Sir Thomas Pearson; K.C.B., as Colonel of the 85th; November 21, 1843.

'Nov[r] 30. The men allowed to wear smock frocks off parade.

'Dec[r] 9. Our detachment for St. Kitts [also called St. Christopher's—Ed.] consisting of Captains Coape and Dering, Lieuts. Lethbridge and Lord S. Compton, Ensigns Keyt and Massy, 8 Sergts., 186 R. and F. sailed in the *Boyne* for St. Kitts, where they proved very unfortunate.'

'Lethbridge and 33 men with several women and children died of yellow fever, one man shot himself, 7 were discharged for very serious crimes, a great number were tried for drunkenness and other offences and all the officers quarrelled with Coape.'

'1844.

'Jan[y] 1. Rode over to a ball at Government house, St. John's. Very slow affair and no supper, rode home half-starved alone.

'Jan[y] 13. Our parades used to be at 5 P.M. The officers in the unusual dress of open shell jackets, white waistcoats and trousers, swords and sashes.

'Antigua not a lively quarter.

'Feb. 20. The *General Palmer* transport arrived quite unexpectedly to convey the H[d] Q[rs] and 2 co[ys] to Barbadoes, to our great disgust.

'Feb[y] 23. Weighed anchor in the afternoon with the wind right in our teeth. All very sick and uncomfortable on board. The Captain and Agent uncivil and unaccommodating.'

From the Regimental Digest we find:

'February 23rd, 1844. The headquarters of the regiment with Nos. 2 and 3 companies embarked on the *General Palmer* Transport for Barbadoes and sailed next day, the strength being 1 Field Officer, 2 Captains, 2 Lieutenants, 1 Ensign, 4 Staff, 16 Sergeants, 7 Corporals, 9 Buglers and 169 Privates. They arrived in Carlisle Bay, Barbadoes, on March 7th and landed the same day, occupying part of the Brick Barracks at St. Ann's. One Private died on the voyage.'

Lieutenant Knox gives these further details:

'Mch. 7. After a fortnight of foul winds and bad seamanship we anchored at Carlisle Bay. Dined with our old friends the 23rd.

'Mch. 20. Dined at McChlerg's, capital dinner. Land-crabs, turtle and iced punch in perfection.'

The following letter was now received from the War Office:

'War Office,
'4 *May* 1844.

'Sir,

'I have the honour to signify to you the Queen's pleasure that the establishment of Drummers and Fifers of the 85th Reg[t] of Foot under your command shall from 1st April 1843 consist of the undermentioned Nos. viz.

'One drum major and 16 Drummers or Fifers.

'I have, etc.,
(Signed) 'H. Hardinge.

'Officer comd[g],
'85th Reg[t] of Foot.'

From this the buglers seem to have officially vanished.

Lieutenant Knox continues:

'July 15. The hurricane season began. All homeward bound ships left.

'Aug. 1. Kennedy and Farnifar flogged and subsequently drummed out for disgraceful conduct at St. Kitts.'

On August 25 the headquarters of the regiment arrived from St. Kitts under the command of Captain Coape. Its strength was 8 sergeants, 6 corporals, 2 buglers, and 146 rank and file. Lieutenant Knox here describes a somewhat comic field day:

'Sept[r] 16. Campbell had the whole garrison out on the Savannah at Roseau including even the military labourers, little wretches in grey not 5 ft. high, and had a most absurd field day with 10 rounds of ammunition and ended by suddenly charging the spectators with the bayonet who fled for their lives.

'Sept[r] 7. The moves fixed. Our reg[t] for St. Vincent, St. Lucia and Berbice, St. Vincent.'

The movements of the various detachments into which the regiment was at this time split up were as follows:

November 1, 1844. Captain Coape, Lieutenant Keyt, and Ensign Ogilvy with No. 3 Company embarked on board the *Flora Macdonald* for Berbice, there to be stationed until further orders. Strength: 1 captain, 1 lieutenant, 1 ensign, 3 sergeants, 4 corporals, 1 bugler, and 76 privates.

November 9, 1844. No. 5 Company (Captain Dering's) consisting of 3 sergeants, 2 corporals, and 51 privates embarked for St. Lucia in H.M.S. *Hermes* under the command of Lieutenant Lord S. Compton.

November 13, 1844. The headquarters with No. 1 and No. 2 Companies, consisting of 3 subalterns, 4 staff, 14 sergeants, 7 corporals, 11 buglers, and 146 privates under the command of Major French sailed from St. Vincent in the schooner *Maid of Erin* for Bermudiaima (St. Vincent), and on arrival there on the next day occupied the barracks in Fort Charlotte, relieving the 46th Regiment which had been stationed there.

Lieutenant Knox continues:

'Nov[r] 27. Lt.-Col. Maunsell rejoined alone from leave of absence. The men lined the walls of the Fort and gave him three hearty cheers.

'Dec[r] 28. The Colonel presented the new colours in the citadel square without much ceremony, he in his evening dress coat.'

On the same day No. 4 Company (Captain Tennant's) and No. 6 Company (Captain Taylor's), under the command of Captain Tennant, proceeded from Antigua to St. Lucia, relieving two companies of the 46th stationed there. Their strength was: 3 subalterns, 8 sergeants, 1 bugler, and 135 privates. They arrived on December 27. A draft also arrived at Barbadoes on the *Princess Royal* on January 12, 1845, consisting of Captain Todd, Ensigns Walters and Filder, 1 sergeant-major, and 90 privates. Captain Todd and 10 rank and file landed next day, the remainder proceeding to St. Vincent and St. Lucia. Ensign Filder, 1 sergeant-major, and 36 rank and file to the former to join headquarters; while Ensign Walters and 49 rank and file joined the left wing at the latter island.

Dated January 26, 1845.

How scattered the regiment was at this time is shown by the following table.

	Sergeants	Corporals	Buglers	Privates
Headquarters	15	8	11	174
St. Lucia	11	11	1	229
Berbice	3	4	1	76
Barbadoes				11
Antigua		1		1
Total Effectives	29	24	13	491

On January 26 Lieutenant Knox tells us 'The Kettledrums for the Bugles arrived.'

The regiment now received orders to hold itself in readiness to proceed from St. Vincent to Barbadoes.

The following address was received on the regiment being ordered

NOTE.—Lt.-Col. Maunsell was at the time the Acting Governor of the island. Both pairs of colours eventually became his property (for old pair see previous Note in 1827). The new colours are now in possession of Major Robert Maunsell Amberd, Dean Park, Bournemouth.

to hold itself in readiness to embark for Barbadoes from St. Vincent—which order was afterwards cancelled.

'COMMITTEE ROOMS,
'4 *June* 1845.

'SIR,

'It is with sincere regret that the Legislative Bodies of St. Vincent have heard that the Comd[r] of the Forces intends within a short period of time to remove your gallant corps from this Island.

'While they are assembled in Sessions the Legislative Bodies cannot separate without expressing their opinions of the orderly and peaceful conduct of the Privates and non-com[d]. Officers of the 85th Kings' Light Inf[y] and of the gentlemanly bearing and kind hospitality of the officers under your command, the manner in which they have mixed and associated with the inhabitants has been duly appreciated and has won for them the good opinion of all.

'We, sir, presiding over the Legislative Bodies are desirous to convey to you and your non-comm[d] Officers and Privates, their best wishes for your future happiness and welfare and to assure you and the gallant 85th Reg[t] they will long be remembered by the inhabitants of this Island.

'We have, etc., Sir,
(Signed) 'G. C. GRANT,
'*President of the Board of Council.*
'N. STENTH (?),
'*Speaker of the House of Assembly.*'

To this Lieut.-Colonel Maunsell returned a suitable reply.

Lieutenant Knox's last entries for this year (1845) are brief but interesting:

'Aug. 25. Charles Thomas called me a variety of abusive names.

'Sept[r]. 27. Charles Thomas received 200 lashes.

'Dec[r]. 26. The Duke's order against "smoking" in or near a mess room issued.'

His first entry for 1846 is a sad one:

'Jan[y]. 12. Coape died at 4 o'clock this afternoon.

'Jan[y]. 13. Saw his body laid out and reduced to a mere skeleton and his face hardly to be recognised. He was buried in Kingston churchyard with military honours.'

The next is hardly more lively.

'Jany. 25. P^{te} J. Donohear (?) of No. 1 shot Parker of No. 2 dead in the barrack room without any assignable cause, for which he was tried by the civil power and hanged in company with a black man in front of Kingston gaol on 11th Feby.'

A note on uniform follows:

'Jany. 26. White trousers discontinued by the army at home and lavenders came in.'

On February 2 the headquarters sailed for Barbadoes, where they landed on the 6th and occupied the Stone Barracks, their effective strength being 2 field officers, 3 captains, 7 subalterns, 5 staff, 30 sergeants, 10 buglers, and 479 rank and file. Lieutenant Knox thus records the event:

'Feby 6. Anchored in Carlisle Bay (Barbadoes). The reg^{t} all together for the first time since they left St. John's, Canada, in 1843.'

On February 9 the regiment was inspected by Lieut.-General G. Middleman, C.B. (Commanding the Forces in the Windward and Leeward Islands).

In the following month the 85th again embarked for home.

CHAPTER XI

HOME SERVICE, 1846–1853

WE left the regiment under orders for Home Service.

Accordingly, on March 25, 1846, the headquarters of the 85th under the command of Lieut.-Colonel Maunsell, consisting of the companies of Captains Patterson, Grey, and Taylor, embarked on board the *Cressy*, freight ship, and sailed the following day. The *Cressy* arrived at the Cove of Cork on April 22. The troops landed next day and marched to Buttevant, where they arrived on April 25. The losses on the voyage were two rank and file.

At Buttevant they were joined by the depot companies under Major French.

The second division of the service companies under the command of Major M. Power, and consisting of the companies of Captains Todd, Dering, and Tennant, embarked at Barbadoes on March 28, arrived at the Cove of Cork on May 1, landed on May 3, and joined headquarters on May 5. Losses on the voyage: 1 rank and file.

The total effective strength of the regiment was 46 sergeants, 15 buglers, and 802 rank and file.

The losses of the service companies during their voyages out and home and while in the West Indies amounted to 2 captains, 1 lieutenant, 4 sergeants, and 77 rank and file.

When the regiment was once more united at Buttevant it immediately marched to Limerick in six divisions, arriving at that place on May 18. Here it was quartered in the New Barracks.

From the Knox Diary we gather:

'March 25. Embarked in the *Cressy* for Cork.

'Apr. 22. Beat into Cork harbour—a Custom House officer came on board to whom we gave a heavy lunch and some bundles of cigars to pass our baggage without search.

'Landed at the Custom House quay, Cork, and marched up to Barracks, where Major-Gen[l] Turner walked down the ranks and expressed his approval of the appearance of the men.

'Our men got dreadfully drunk and riotous in barracks.

'Apr. 24. Having collected the men with some trouble, many of them very drunk, we started for Mallow where we arrived after a harassing march of 21 miles which it took us 11 hours to accomplish.

'Apr. 25. March into Buttevant. Flogged 5 or 6 men.

'Apr. 29. The dreadful drunkenness on the march was caused by our men selling their tobacco at Cork.

'May 3. A cockfight in barracks got up by E. Y. Peel, 85th.

'The Reg[t] brought home 512 N.C.O.'s and men and the Depot co[ys] were 352 strong. The average height of the service co[ys] was 5 ft. 8¼ in.

'May 18. Marched into Limerick, played in by the 83rd Band, 868 N.C.O.'s and R. and F.

'June 18. The Colonel (Maunsell) made the first announcement of his exchange with Sir Gaspard Le Marchant.'

On June 19 Sir Gaspard Le Marchant was appointed Lieut.-Colonel of the 85th from being Inspecting Field Officer of a Recruiting District.

The half-yearly inspection of the regiment was held the same day by Colonel R. E. Maunsell, Assistant Adjutant-General, Limerick.

The Diary continues:

'June 19. The old colonel published a farewell order whch was read on parade by French, 26 officers and 693 men on parade to hear it!!

'June 23. Sir Gaspard came over to see the Reg[t] and brought his standing orders for the 99th Regt. which are to be adopted by us.

'Sept[r] 4. (Returned from leave). Found Sir Gaspard in command who has introduced many thorough and beneficial changes in the interior economy of the Reg[t] which improved wonderfully during the short time he remained with us.'

September 9. The regiment was inspected by Major-General His Royal Highness Prince George of Cambridge.

Diary.

'Septr 24. Sir Gaspard mad on drill.

Octr 9. 300 of our Regt. and the Trp. of Horse Art^y^ sent out to Doonas to protect Sir Hugh M——'s cattle. They returned the same day. Major-Genl. Prince George of Cambridge arrived to take command of the district.

'Dering's company went out to Bohew (?) to protect some property.

'The country beginning to be very disturbed in consequence of the potato famine (which brought on the great famine of 1846).

'Octr 13. Prince George dined at mess.'

On October 27 the headquarters and five companies of the regiment marched in three divisions from Limerick to Birr, where they arrived on November 18 and occupied barracks. Sir Gaspard Le Marchant now exchanged and was succeeded by Brevet Lieut.-Colonel Benjamin Chapman Browne from half-pay unattached.

From the Diary.

'Novr 10. Marched to Birr.

'Decr 1. Sir Gaspard offered the Lieutenant-Governorship of Newfoundland.

'Decr 4. The famine now raging all over Ireland and deaths from starvation very frequent in the Counties of Cork, Limerick and Clare.

'The Board of Works commenced cutting a quantity of useless new roads to employ the people. A great deal of distress and misery about this place. Soup kitchens set up.'

Brevet Lieut.-Colonel Browne, however, retired on December 29, and was succeeded by Lieut.-Colonel Brook Taylor by purchase. For the remainder of 1846 there is no other event to record.

'March 10, 1847. New officer's breastplate taken into wear.'

'April 2. The half-yearly inspection of the regiment was held by Major-General Sir Guy Campbell.

'May 21. Sir Thomas Pearson our Colonel died and was succeeded by Lieut.-General Sir John Guise an ex-Guardsman.

'June 1. Half-yearly inspection by Sir Guy Campbell which went off very well indeed. The Reg[t] worked beautifully in six divisions of 24 file. Ball in our mess-room in the evening.'

Lieut.-General Sir J. W. Guise was a baronet and a K.C.B.
Lieut.-General Sir Thomas Pearson was a C.B., and K.C.H.

'June 8th. Out ball-firing on a " boy " (buoy) for several hours.'

On September 15 Captain Daniell's company (No. 4), which had been detached at Loughrea, marched to Dublin, arriving on September 18.

On September 21 Captain Todd's company (No. 1) from Portumna marched to the same place, arriving on September 23.

On the same date Captain Thurloe's company (No. 6) and Captain Dering's company (No. 9) marched from Banagher and Shannon Harbour, arriving at Dublin on September 27.

Captain Grey's (No. 2), Captain Parrott's (No. 3) and Captain Day's (No. 8) with the bugle-major and bugles under the command of Major Blackburn marched from Birr on September 23, arriving at Dublin next day. The headquarters, consisting of the companies commanded by Captain Grant and Captain Lord S. S. Compton, the staff, colour and band, under the command of Major Tennant, marched for Dublin from Birr on September 28. They arrived next day and occupied the Royal Barracks, Royal Square. The half-yearly inspection of the regiment by Major-General H.R.H. Prince George of Cambridge was held on October 23.

From the Diary.

'November 15. Dined at mess (he was married at this time) to meet the Buffs and lost 12 guineas at hazard. Old Sir J. Dennis commands the Buffs. He has been 51 years on full pay!'

The next entries of interest are as follows, and give a very interesting though brief account of the condition of affairs at the time.

'Dublin, 1848.

'Mch. 16. Out practising street firing in the Park under Prince George with 600 of the 75th and 85th and four Troops of Enniskillens and Lancers and Art[y].

'Mch. 17. St. Patrick's Day, under arms and in readiness to turn out all day. Great excitement in Dublin and preparations making for an outbreak.

'The guards doubled at the Bank and Castle and 200 men quartered for some weeks in Trinity College. Half the officers of each Reg[t] only allowed to go to the Castle ball.

'Apr. 11. A picquet of 2 Capt[ns], 3 Subalterns and 132 Ca[y] and Inf[y] mount every evening at the Viceregal Lodge.

'Apr. 15. The garrison all confined to barracks. The marines and sailors marched up from Kingston, an outbreak being expected.

'Apr. 28. Inspected by the Prince.

'May 5. Light drill in the Park.

'May 13. John Mitchel again arrested for felonious writing to the *United Irishman.*

'May 26. Mitchel convicted, great excitement.

'May 27. Mitchel sentenced to 14 years transportation and quietly escorted to the steamer by squadron of the Carabiniers, whom he had a few days previously accused in his paper of being ready to join him. He sailed in the *Shearwater* the same night in irons.

'July 1. The Great Southern and Western Railway opened as far as Dundrum in state by the Lord Lieu[t]. On the guard of honour.

'July 24. Rumours of an outbreak in Waterford, all officers on leave recalled.

'July 25. The garrison confined to barracks by wings daily.

'Commencement of the search for arms.

'July 26. The Habeas Corpus act suspended till 1st March next. Great excitement in Dublin and reports of risings in the provinces.

'July 30. The eight companies and H[d] Q[rs] left at 7 for Thurles with Lt.-Colonel Taylor, Captains Grant, Todd, Grey, Thurlow, Daniel, Parratt and Compton.

'Lieuts. Seymour, Massy, Williamson, Warburton, Maunsell and Knox.

'Ensigns Floyd, Gubbins, Sitwell, Thistlewayte, and Brown.

'Lieut. and Adjt. Thompson, Pay[mr] Thompson and Asst. Surgeon Thompson and about 530 all ranks.'

The author of Diary did not accompany the Regiment.

From a Sketch by R. Ebsworth, 16/10/1852

BAND. Undress Uniform. BUGLE BAND.

From the Regimental Digest we read:

'July 30, 1848. The Regiment (Sick, Recruits, Women and Baggage left in Dublin) proceeded by Railway, *en route* for Templemore, ordered to proceed on to Thurles, encamped near the railway station at the latter place for the night.

'July 31, 1848. Moved from encampment, ordered to take up a position between Thurles and Littleton so as to hold both towns. Encamped at Turtulla Park.

'August 2nd. Marched to New Birmingham and encamped near that place.

'August 3rd. Received orders to return to former encampment at Turtulla Park.

'August 5th. At 9 P.M. one wing of the regiment with a Division of the 2nd Batt[n] 60th Rifles and 1 Wing of the 74th Reg[t] were ordered into Thurles to keep the Town and escort Wm. S. O'Brien to the Railway Sation. They returned to camp at 11.30 P.M.

'August 6th. Struck tents and marched into Thurles where the regiment was accommodated in the Police Barracks, gaol, etc., for the night. Early next morning they proceeded by rail to Limerick and returned the same day to their former station at Turtulla Park and remained there in Camp.'

From the Diary.

'Sept. 5. The camp broken up. Three companies marched for Waterford. A most flattering order issued by Major-General McDonald to the Regt.'

The order in question is here given:

'Field Force Order.

'THURLES,
'3 *Sept.* 1848.

'Major-General McDonald having been directed to hold the troops under his command in readiness to break up their encampments and be placed in quarters,

'Takes this opportunity of returning his very warmest thanks to the officers comdg. the Regiments and detachments of which the moveable column has been composed, and requests they will have the goodness to convey the same to those under their command.

'He has not failed to bring to the notice of his superiors, nor is he himself likely to forget, the absence of crime, and the admirable discipline which has been maintained, the zeal and cheerful alacrity with which every order and movement has been executed through one continued series of wet and inclement weather.

'Although the Major-General rejoices that on the present occasion, the services of the troops were not required in actual collision, still their readiness and Loyalty were very manifest and he never could wish for better men or officers, should he be called upon to meet the enemy of his Queen and Country.

'By order,
(Signed) 'KENNETH MCKENZIE,
'Major of Brigade.'

The regiment remained at Turtulla Park in camp until September 5, when headquarters marched to Waterford and detachments to New Ross, Duncannon Fort, Piltown, Wexford, and Dungarvan.

The Diary continues:

'Sept[r]. 9. Marched into Waterford at 1, the H[d] Q[rs] arrived immediately afterwards. No. 2 co[y] at New Ross, No. 3 at Duncannon Fort.

'Sept[r] 12. A dispatch having arrived from Portlaw stating that the police barracks had been attacked, the Trp. of the 4th Light Dragoons and 20 officers and 290 men of ours marched at nine for Curraghmore.

'Passing through Portlaw we perceived signs of the fight and plenty of blood on the road but no rebels to be found.

'Halted in the Park at Curraghmore and we were joined by about 80 of the Buffs and a Squadron of the 4th Dragoons under Sir Jas. Dennis.

'The Buffs and Police set off in cars to scour the woods and having told off Todd and a strong co[y] to remain and had a good heavy lunch at Lord Waterford's, whose house was well fortified and victualled for a siege, we marched back to Waterford bringing five or six prisoners and arrived at 9 P.M. having had a march of twenty-four miles.

'Sept[r] 14. Todd's party returned having had a few shots at the rebels in the woods of Curraghmore, the subalterns had to patrol at night.

'Sept[r] 20. Basquit our new German bandmaster arrived, unable to speak a word of English.'

The half yearly inspection was held on October 20, 1848, at Waterford by Major-General J. McDonald, C.B.

The next extracts from the Diary are as follows:

'Dec. 31. The blue frock coats were now discontinued and shell jackets with sling belts were taken into wear. The coatee, open, with a blue embroidered waistcoat, is worn at mess, as shell jackets are not permissable at mess.

'We still wear the lavender trousers in summer which do not look bad but are difficult to keep alike in hue.

'We took soldiering easy. Brook Taylor used to be away hunting perhaps half the week, then come back late in the evening and tell off the accumulation of prisoners in his hunting clothes.

'His great delight was to be with one of the three Marquises in these parts—Waterford, Ormond, or Drogheda.'

There is no other entry for the year 1848 which need be recorded. The events for the year 1849 are of little interest.

From the Diary we have:

'Feb[y] 1. Reduction of the Army by 10,000 men ordered. Our companies reduced to 75 R. and F. All bad characters and men unfit of service discharged.

'Aug. 3. Went into Waterford for the garrison theatricals. Saw the "Jacobite" and "The Illustrious Stranger." Keyt and Gubbins far the best actors.

'Aug. 4. The Queen and Royal Squadron anchored at Passage nearly opposite Duncannon Fort. Prince Albert and the Prince of Wales came up to Waterford in the "Fairy" . . . Crowds of people went to Passage to see the Queen. The colonel went on board with a "state."'

For some reason neither the entry for February 1 nor that for August 3 appears in the Regimental Digest.

The usual half-yearly inspection was held at Waterford on May 7 by Major-General McDonald.

A few changes of quarters took place during the year, companies relieving one another in various places.

In March 1850 the regiment was ordered to Mullingar. Headquarters arrived there on April 18 under Lieut.-Colonel Brook Taylor, the baggage being forwarded by steamer from Waterford to Dublin and thence by canal boats to Mullingar.

Prior to leaving Waterford the following letter was received by the regiment from the Mayor and Corporation of that City, and a suitable reply was returned by Colonel Brook Taylor:

'*April* 4, 1850—

'At a meeting of the Town Council of the Borough of Waterford held on Thursday the 4th April 1850.—

'Richard Cooke Esq., Mayor, in the chair. Moved by Henry Denny, Esq., seconded by —— Mackay, Esq. Resolved unanimously:

'"That the marked and special thanks of the Mayor and Town Council of this Borough are justly due, and hereby given to Colonel Brook Taylor, and the officers of the 85th Light Inf[y]; not only for their liberal and munificent support of the various charitable Institutions of the City since their sojourn in Waterford (now nearly two years) but also for their soldierly and honorable bearing towards all classes of our Fellow Citizens, which are the characteristics of this excellent corps on all occasions, both in these countries and Her Majesty's North American Colonies.

'"The thanks of the Council are also bestowed on the Non-Commissioned officers and men of this Gallant Regiment for their excellent, praiseworthy and general good conduct since their arrival here."

(Signed) 'Richard Cooke,

'*Mayor of Waterford.*'

The half-yearly inspection was held at Mullingar on May 13, 1850, by Major-General H.R.H. Prince George of Cambridge.

The period of service in Ireland had now expired.

On May 24 the headquarters of the regiment (consisting of Nos. 6, 8 and 9 companies with staff, colours, and band, under the command of Lieut.-Colonel Brook Taylor) marched from Mullingar and proceeded by rail to Dublin. There they embarked at the North Wall on board the *Princess Alice* (Glasgow Company's steamer) the same evening. They disembarked at Fleetwood at 6 A.M. next day and proceeded by rail to Preston Barracks in Lancashire. The second division similarly followed on May 25 and the third division on May 27.

From the Diary we get:

'June 11. Went to Liverpool races with nearly all the Reg^{t}—had to return on the roof of a railway carriage, there being such a scramble for seats.

'June 31. Amateur theatricals. Parratt, Keyt, Warde, Gubbins and Young acted in the "Jacobite" and "Your life's in danger"—Keyt and Gubbins acted capitally.

'Aug. 2. Our colonel, Sir John Guise, inspected the Reg^{t}. dressed as Colonel of the 85th. We skirmished on the moor and he dined at mess and gave us a haunch of venison.'

September 20. The half-yearly inspection of the regiment was held at the headquarters, Preston, by Lieut.-General Earl Cathcart, C.B.

October 29. A theatrical performance was held. A programme printed on pink silk is still in existence and is here reproduced as a curiosity.

On April 15, 1851, orders were received removing the regiment from Preston to Hull citadel, with detachments at Leeds, Bradford, Halifax and Scarborough. The various detachments, according to orders proceeded forthwith to those places.

From the Diary.

'Feby 9. Wrote for the *United Service Magazine* a historical sketch of the services of Reg^{ts} disbanded and raised since 1756.[1]

'Apr. 2. Under orders for Hull, Scarborough, Halifax, Leeds, Bradford, Chester.'

The last entry from this interesting diary which we shall give is this:

'1852.

'Feby 21. Heard of sudden order of the Reg^{t} to Portsmouth.'

[1] This paper does not appear to have been published.—ED.

THEATRE ROYAL, PRESTON.

GARRISON AMATEUR

THEATRICAL PERFORMANCE.

FOR THE

BENEFIT OF THE LADIES' CHARITY.

ON TUESDAY EVENING, OCTOBER 29th, 1850,

Will be performed,

BY THE OFFICERS OF THE PRESTON GARRISON,

The petite Comedy, in 2 acts, of

FLYING COLOURS;

Or, CROSSING THE FRONTIER.

The Duke de Sombreuse (Governor of Verdun)..... Mr. C. PEDDER, 3rd Dragoon Guards
Col. Amadon (Commandant of the Garrison)......Mr. J. GUBBINS, 85th Lt. Inf.
Count Alfred de Belleville......Mr. WARDE, 85th Lt. Inf.
Captain Sans-souci .. .. Mr. HAMILTON AIDE, 85th Lt. Inf.

Margaret (Duchess de Sombreuse)...................................Miss CRUPPENDEN
Helen de Monterean ....Miss FEILDING
Madame Thibaut (Hostess of the White Horse) .. ..Mrs. MILLS
Justine (her Neice, and Maid to the Duchess) ....Miss MILLS

To be followed by the Farce of

STATE SECRETS.

Master Hugh Neville ....Mr. GUBBINS, 85th Lt. Inf.
Calverton Hal (a Cavalier).. ...Capt. HODGSON, 79th Highlanders
Humphrey Hedgehog (Landlord of the Black Bull) ....Mr. KEYT, 85th Lt. Inf.
Gregory Thimblewell (Tailor of Tamworth)...Major FERGUSON, 79th Highlanders
Robert (his Son) ....Mr. ADDERLEY, 79th Highlanders

Maud Thimblewell (the Tailor's Wife) .... Mrs. MILLS
Letty (Daughter of Hedgehog)............ ..Miss MILLS

To conclude with an incident in Married Life, entitled

DOMESTIC ECONOMY.

John Grumley.................... .....Mr. KEYT, 85th Lt. Inf.
Joey (his Son)................... ....Mr. ADDERLEY, 79th Highlanders
Sergeant Tom Brown ....Capt. PARRATT, 85th Lt. Inf.

Mrs. Grumley ....Miss FEILDING
Mrs. Shackles... ....Mrs. MILLS
Mrs. Knaggles ....Miss MILLS
Sally............ .. .. Miss BROWN
Peggy................. ....Miss C. BROWN

Doors to be opened at Seven o'clock. The Performance to commence at half-past Seven o'clock.

Box. 4s. Up. Box. 2s. Pit 2s. Gal. 1s.

NO HALF-PRICE.

Tickets may be had and Places secured at Messrs. Addison's.

ADDISONS, CHURCH STREET, PRESTON.

On February 23, 1852, the regiment moved from Hull to Portsmouth, going first into the Anglesea barracks and later to the Clarence barracks. The regiment travelled by train and consumed no less than three days to cover the distance. On the first day proceeding from Hull to Derby, where they went into billets for the night. The second day, from Derby to Oxford, where they again went into billets. The third day completed the journey from Oxford to Portsmouth.

While at Portsmouth a small number of Minie rifles was issued to each company. On April 13, Lieut.-Colonel Manley Power was appointed to command *vice* Brook Taylor who exchanged.

On November 1, 1852, the Commanding Officer received an order from the Horse Guards to hold the regiment in readiness to embark for Mauritius in the course of January 1853. The 85th had been at home for more than seven years, a rather unusually protracted period of service.

CHAPTER XII

MAURITIUS, 1853—CAPE OF GOOD HOPE, 1856–1863.

AT Mauritius, the destination of the regiment, the 85th were to replace the reserve battalion of the 12th Foot, which had been transferred unexpectedly from that station to the Cape frontier some months previously. Prior to sailing, the Minie rifles lately issued to part of the regiment were recalled to store.

The service companies of the regiment therefore embarked on January 27 and 28, 1853, in two divisions.

The 1st division under Major Grey with Captains Warde, Keyt and G. Maunsell, Lieutenants Bond, Armitage and Lord T. H. Taylour, Ensigns Hogge, Barton, Mytton and Bayley, Assistant-Surgeon Bartley, 13 sergeants, 10 buglers, 275 rank and file, 19 women and 37 children, embarked on the *Marion* at Portsmouth, January 27.

The 2nd division with headquarters, under Lieut.-Colonel Manley Power, with Captains Thompson, Williamson and Massey, Lieutenants Sitwell and the Hon. E. T. Boyle, Ensign Glyn, Paymaster Pechell, Adjutant Rooper, Surgeon Clerihew, Quartermaster Rouse, 18 sergeants, 1 bugler, 285 rank and file, 19 women and 24 children, embarked on the 28th on board the *Roman Emperor*. The remaining four companies were formed into a depot and sent to Parkhurst, Isle of Wight, whence they were subsequently moved to Horfield barracks, near Bristol, and eventually, in succession, to Bury, Preston, Sunderland and Newcastle-on-Tyne. The regiment, on board the two transports, duly sailed for Mauritius, where it arrived in the middle of April after a voyage of 73 days.

For those times the voyage was a comparatively quick one, and on arrival the Commanding Officer, in an order, congratulated the regiment on such a short and prosperous voyage.

On the voyage the headquarters division (2nd) had no casualties. The 1st division, however, lost by death five privates and two children.

The following letter was received from the Adjutant-General:

'HORSE GUARDS,
'2nd. *Feby*, 1853.

'SIR,

'I have the honour by direction of the Comd^{r}-in-Chief to express to you His Lordship's high gratification at having received a report from the General Officer Com^{dg} at Portsmouth of the state of order and discipline in which the 85th Reg^{t} embarked from thence for the Mauritius, every man being present, without one instance of drunkenness or irregularity having occurred, and the men appearing in excellent spirits, circumstances which Viscount Hardinge considers very creditable both to the officers and their men.

'I have, etc.,
'E. BROWN,
'*Adjutant-General.*

'Officer Com^{dg},
'85th Reg^{t},
'Mauritius.'

The regiment remained at Mauritius for three years, during which time but little occurred of interest in a military point of view. The Russian war, although productive of great excitement amongst the French population of the colony, was too remote to exert any material influence over the monotonous tenor of garrison life, of which the most important event was the annual interchange of quarters between the troops at Port Louis and those stationed at Mahebourg, Flacq, Grande Rivière, Sud-Est and Rivière Noir, the posts on the windward side of the island.

At this time the 5th Fusiliers were also stationed in the island.

The only representative of the regiment in the Crimea was Captain James Gubbins who was aide-de-camp to Sir de Lacy Evans. Captain Gubbins was promoted to the rank of brevet-major December 12, 1854, 'for distinguished service in the field.'

The regiment disembarked, on arrival at Mauritius, at Port Louis, and marched to the infantry barracks there.

But although, from a military point of view, there was no history to record, the stay of the 85th in Mauritius was not without its periods of anxiety and stress. Early in 1854 a terrible outbreak of Asiatic cholera occurred in the island.

For over thirty years this scourge had not attacked the population. But, notwithstanding its remote station in mid-ocean and the rigour of the quarantine regulations, the ravages of this frightful disease were appalling; the native population, without exaggeration, being decimated, over eleven thousand dying. By good fortune and by the prompt and energetic measures taken by the military authorities to isolate the troops, the garrison on this occasion as far as the 85th was concerned suffered no loss. Eighteen months afterwards a second outbreak took place and the regiment then lost nineteen men in one month.

Meanwhile, the withdrawal of troops from the Cape, consequent on the Russian war, had led to the usual uneasiness and rumours of disturbance on the frontiers, and on May 15, 1856, H.M.S. *Penelope* arrived from Table Bay with an urgent request that a regiment might be despatched forthwith to East London. On May 18, within 36 hours of her arrival, the 85th, under Colonel Manley Power, had embarked on board her in light marching order, leaving the women and baggage at Port Louis.

The officers who went to the Cape were: Captains Thompson, Boyle, Armitage and Lord J. H. Taylour; Lieutenants Hallowes, Fitzgerald, Chichester, Mathews, Beadon, Galbraith and Noyes; Ensigns Grant and Wilson, Paymaster Pechell, Adjutant Ash, Assistant-Surgeon Bartly and Staff-Assistant-Surgeon Kirwan.

The strength in non-commissioned officers, buglers, and rank and file was as follows: 33 sergeants, 13 buglers, and 544 privates.

On reaching East London, the bar proved too rough to admit of any attempt being made at a landing. The *Penelope* accordingly steamed down to Algoa Bay. But the weather was unfavourable and in place of ten days, as anticipated, the voyage lasted three weeks, provisions ran short, and from the crowded state of the ship the men suffered severely. Their conduct throughout was admirable, and formed the subject of a special communication from the Governor of the Cape.

Some time after the arrival of the regiment in South Africa, a letter was received from the Horse Guards, expressive of the gratification with which H.R.H. the Commander-in-Chief had received reports so creditable to the conduct and discipline of the 85th regiment:

COLONEL MANLEY POWER.

From a Miniature in the possession of K. M. Manley Power, Esq., dated 1832.

Commanded the Regiment from 1852 to 1857.

'HORSE GUARDS,
'*20th Novr.* 1856.

'SIR,

'I am directed by the General Comd[g]-in-Chief to transmit you the accompanying copy of a letter from the War Department with its enclosure from Governor Sir George Grey in which he represented the admirable conduct exhibited by the men of the 85th Reg[t] while on their passage from Mauritius to the Cape of Good Hope under very trying circumstances and to inform you that His Royal Highness has received with great satisfaction a report so creditable to the 85th Reg[t].

(Signed) 'C. YORKE.

'Lt.-Genl.
'Sir J. Jackson, K.C.B.'

On February 18, 1857, the following extract from No. 593 General Orders, dated Headquarters, Grahamstown, February 5, 1857, was received:

'Para. 1. The Commander of the Forces has much pleasure in publishing the following enconium upon the 85th Light Inf[y] elicited by a report from His Excellency the Governor to the Right Hon[ble] The Secretary of State for the Colonies, detailing the hardships experienced by that Reg[t] under very trying circumstances when on board Her Majesty's steam frigate *Penelope* on the passage from Mauritius to Algoa Bay on which occasion they were at sea 21 days, during 18 of which almost incessant rain prevailed accompanied by stormy gales of wind. When, out of 600 men, 400 were from want of space obliged to remain constantly on deck; but, aware that they were going on a special and pressing service, they bore their hardships cheerfully, not a single complaint was heard from them, and their entire conduct was excellent.'

The regiment disembarked on June 7, 1856, at Algoa Bay and encamped near the town till June 26, when it moved to Grahamstown, leaving Captain Thompson's company to escort a quantity of ammunition to that place. The regiment halted at Grahamstown from July 4 to July 8, when four companies marched to Fort Beaufort, arriving there on July 10.

On August 7, 1856, Major Williamson took over command

vice Colonel Manley Power on sick leave. The Colonel did not rejoin, dying at Bath, April 27, 1857.

On August 15, 1856, Captain Armitage's company marched to Fort Brown and Koonap to relieve the 6th Regiment.

After a short stay at Port Elizabeth, and at Port Beaufort, where on August 10 the women and baggage under Quartermaster Rouse rejoined from Mauritius, the regiment was ordered up the country to the Queenstown district. Here, upon the grassy 'flats' which still swarmed with antelope, the regiment lay in camp for nearly a twelvemonth, passive spectators of the events which rendered the years 1856–7 memorable in the annals of the frontier, and which ended in the disintegration and final exodus of the Amakosa tribes, whose proximity had so long imperilled the peace of the Colony.

On May 29, 1857, Major J. W. Grey, with two companies under Captains Coussmaker and Baker, Lieutenant Reeves, Ensigns Urquhart, Henderson, Drage, Assistant-Surgeon Leet, 12 sergeants, 8 buglers, 239 rank and file, 17 women, 243 children, arrived at East London. The former took over command of the regiment.

In announcing the lamented death of Colonel Manley Power the following order was published by Major Grey:

'Colonel Power served upward of 30 years in the 85th, during which time he devoted his best energies to the welfare and efficiency of the Regiment and Major Grey is sure that his memory will be respected by all ranks, who must feel that they have lost a kind friend and a just and impartial commanding officer.'

In July, 1857, Major J. W. Grey was appointed Lieut.-Colonel, *vice* Brevet-Colonel Manley Power, as of date April 28.

July 1, 1858. Snap caps were issued to the regiment.

At Queenstown camp the regiment underwent its first course of instruction in the use of the Enfield rifle—under great difficulties, as not only a range, but, with the exception of rifles and ammunition, all the requisite appliances, targets included, had to be improvised for the purpose.

When the frontier had become more tranquillised, the wing of the 85th was brought down to Fort England, Grahamstown, to replace the regiments which, from their great contiguity to the coast, had been forwarded to India as reinforcements during the mutiny. From Grahamstown this wing was moved to Natal in June 1858 to replace an equal number of the 45th Regiment returning to England after 16 years' service in South Africa.

On March 26, 1858, an order came out ordering the companies of all regiments to be numbered consecutively throughout, from 1 to 12, including the depot companies.

March 30, 1858. The Light Infantry plumes (green horsehair) authorised by the dress regulations of 1857 were issued to the regiment.

In Natal, the 85th right wing and headquarters remained between four and five years—the headquarters and three companies at Fort Napier, Pietermaritzburg, and one company detached at the Port of Durban. The left wing remained at the posts on the Kei until they were abandoned, and was then removed to the Keiskamma Hoek, British Kaffraria. The depot had, meanwhile, been attached to the depot battalion at Preston, and subsequently transferred to that at Pembroke Dock.

In the course of the year 1860, Natal was honoured by a visit from H.R.H. the Duke of Edinburgh—then Prince Alfred—who was making a tour through South Africa in company with His Excellency Sir George Grey. New colours for the regiment had shortly before been issued by the War Department (in accordance with a regulation which had come into force not long previous) and the Prince was pleased to accede to the request of the commanding officer, Major Williamson—at that time Acting Lieut.-Governor of the Colony—that His Royal Highness would undertake the task of presenting them to the corps. The ceremony was accordingly performed in the market square of Pietermaritzburg with all the éclat which the limited resources of the little garrison could command, on September 4, 1860.

On presenting the new colours, H.R.H. Prince Alfred, addressing Captain S. J. Boyle, who handed them to him, said:

'It is with great pleasure that I present these new colours to so distinguished a Regiment. I do so in the full confidence that, if I live to see old age, I shall, if necessity arise, see the colours which

[NOTE.—The old colours becoming, as previously stated, the property of General Frederick Maunsell, then Colonel of the regiment.]

I now give in my youth covered with names of other victories in addition to those which already adorn them.'

On January 21, 1861, the new blue serge summer trousers were issued to the regiment

On January 29, 1861, the 85th received orders to hold itself in readiness to embark for New Zealand, where additional troops were urgently needed. This movement was never carried into effect.

In July, 1861, the Colony of Natal was thrown into a state of commotion and alarm by rumours of an impending invasion from Zululand. Umpanda, the great chief of the Zulu nation, had always maintained friendly relations with the English, but his power was on the wane; he was aged and infirm, his sons rebellious, his exchequer oft-times empty. Five years previously a great battle had taken place near the borders of the Colony between the adherents of two rival sons, when, according to native accounts, the slain were counted by thousands, and the broad waters of the Tugela ran red with blood for three days. Since the battle, which was decided in Zulu fashion (a close hand-to-hand fight), the fortunes of a son named Cetewayo had been in the ascendant; he had become 'the feet' of the people of whom his father was still 'the head,' and 'the feet' were wont not unseldom to assume a defiant attitude towards the Colonial authorities.

Upon the present occasion he had collected an army—20,000 braves, it was said, one estimate being 30,000—ostensibly for the purpose of the chase, but in reality with a view to a sudden raid upon the Colony. Prompt measures appeared necessary, and the handful of troops composing the garrison of Fort Napier, consisting of three companies of the 85th, a troop of Cape Mounted Rifles, and some artillery, were pushed forward to the banks of the Tugela, together with a few mounted Colonial Volunteers and 2,000 Natal Kaffirs under a feudatory native chieftain. Application was made to the Cape and Mauritius for reinforcements at the same time. But the impending invasion had no further results and the Volunteers and native contingent were sent back to their homes. The necessity of posts of observation along the Tugela having been clearly shown, two small outposts under Lieutenants Henderson and Reeves, were established on the right bank, the upper of the two receiving the name of 'Fort Buckingham';

THE QUEEN'S.

THE REGIMENTAL.
1860-1877.

These Colours now hang in the private chapel at Stowe, and are the property of the Baroness Kinross. (*See page 332.*)

and at these a portion of the right wing remained until the embarkation of the regiment for England.

The details of the movements of detachments of the regiment during this year are as follows:

July 14, 1861. Lieutenant Reeves and Ensign Ramsbottom and 40 rank and file marched to D'Urban to reinforce Lieutenant Beadon's detachment at that station. Lieutenant Beadon marched to the Tugela mouth.

Major Williamson took the command of this force with a detachment of the Cape Mounted Rifles and 100 volunteer cavalry.

July 15, 1861. The regiment with one 6-pounder and 30 Cape Mounted Rifles marched from Fort Napier under Lieut.-Colonel Grey (with Lieutenant Taylor, Captain Boyle, Lieutenants Henderson and Stace, Ensign Cooper, Lieutenant and Adjutant Drage, Assistant-Surgeon Norris and Quartermaster Watts and 160 non-commissioned officers and men), and halted on the 17th at Greytown (50 miles from Fort Napier), where they were joined by 50 Natal Carabiniers and a small burgher force under Chief Justice Harding.

The inhabitants of the neighbourhood had congregated in considerable numbers in the Laager. Lieutenant-Governor Scott, the Colonial Secretary, and the Secretary for Native Affairs, arrived in the evening, when it was determined to send on the detachment of Cape Mounted Rifles and Mounted Volunteers towards the Tugela River to open communication with Major Williamson from Van Steadden Farm by the right bank of the river and that the 85th should move on to-morrow to Kranz Kop overlooking the valley of the Tugela.

July 20, 1861. Encamped on the heights above the Tugela, distant about 30 miles north-east of Greytown. In the evening the Governor reported the native spies had brought intelligence that we were to be attacked in force by Cetewayo at midnight—preparations were made accordingly, but the night passed quietly.

July 21, 1861. Colonel Grey, the Lieutenant-Governor, and Captain Grantham, R.E., reconnoitred the neighbourhood to fix a site for an entrenched post.

July 22, 1861. About 3000 Natal Kaffirs arrived.

July 23, 1861. The regiment began to entrench the post which was called 'Fort Buckingham.'

July 25. Colonel Grey and Lieutenant Taylor left for Fort Napier.

August 2. Regiment marched to Fort Napier, leaving a detachment under Lieutenant Henderson to complete Fort Buckingham. A similar detachment was left at an entrenched post on the Lower Tugela under Lieutenant Reeves.

In October, 1861, white trousers were abolished.

The following general order was received on the departure of the regiment from the Cape of Good Hope for England.

'*General Order, No.* 514.

'Para. 1. The Lieut.-General comd[s] the Forces cannot permit the 85th Light Inf[y] to quit this command after a long period of colonial service without conveying to Colonel Grey, the officers, non-commissioned officers and men of this fine old corps his unqualified approval of the conduct of the Reg[t] while serving in South Africa. Whether in the Cape Colony, Kaffraria, or Natal, this has been equally conspicuous and has in every instance secured the marked approval of the authorities civil as well as military.

'The well-known discipline of this gallant Reg[t] has not, by its long colonial service, suffered in any degree whatever while under the judicious command of the late Colonel Power or the present commanding officer Colonel Grey, and in leaving the shores the 85th Light Infantry carries with it a character for discipline, orderly conduct, and regularity, that must ever place it high in the estimation of those under whose orders it may hereafter be placed, as well as the Lieut.-General's best and warmest wishes for a successful future.

'By command,
(Signed) 'F. V. BISSEL, *Colonel,*
'*Deputy Quartermaster-General.*'

In April, 1863, on the arrival of H.M.S. *Himalaya* from Mauritius with the 2nd Battalion 5th Fusiliers, the right wing of the Fusiliers having been landed at Durban, the right wing of the 85th was embarked; the *Himalaya* left the same night for East London, and landed the left wing of the Fusiliers and received the left wing of the 85th. From East London she proceeded to Simon's Bay, and thence to Spithead, calling at St. Helena and St. Vincent. From Spithead she steamed round to Dover, where the 85th landed on June 8, taking over quarters in the Citadel and South Front barracks, the latter of which were then occupied for the first time. Shortly afterwards the regiment received new rifles, muskets pattern 1853, and new clothing, the new infantry cap and pouches, sergeants' pouches, belts, whistles and chains, waist-belts, and slings. The rank and file new ditto and ball-bags.

The following is an account of the doings of the 85th King's Light Infantry from May, 1856, to August, 1857, compiled from the diary of Ensign H. K. Wilson, then serving :

'On 14th May 1856 the 85th, then stationed in Mauritius, received orders to embark in H.M.S. *Penelope* on the 18th for Cape Colony where a Kaffir rising was anticipated. The Regt., consisting of 6 Coys. only, reached East London on June 4th, having been considerably delayed by the weak state of the ship's boilers, but was unable to land owing to the surf : the *Penelope* therefore proceeded to Port Elizabeth, where she finally disembarked her troops on June 7th. Owing to lack of transport the 85th was further delayed here till June 19th when Nos. 2 and 5 Coys. left for Grahamstown, the headquarters with Nos. 1, 3 and 6 Coys. following on June 26th and No. 4 Coy., Capt. Thompson's, on the 27th. Grahamstown was reached on July 4th and the Regt. hospitably entertained by the officers of the 6th Royals and Cape Mounted Rifles during its stay of four days. On July 8th the march was resumed via the Koonap Pass and through the Fish River bush to Fort Beaufort, which was reached on the 10th and where the Regt. was accommodated in very scattered quarters.

'On the 17th the Colonel was laid low by a paralytic stroke just before going on parade and left the Regt. on sick leave on August 7th never to return, the command devolving upon Major Williamson. At this time in spite of persistent rumours of the outbreak of hostilities with the Kaffirs, the officers enjoyed very fair shooting, Thompson and Boyle being the most successful Nimrods.

'On August 15th No. 2 Coy., Captain Armitage, was ordered to Fort Brown, two marches distant from Beaufort, and on Sept. 20th the remainder of the Battn. on being relieved by the 80th Regt. left Beaufort for Queenstown, No. 2 Coy. eventually rejoining Headqrs. on Oct. 8th after a severe struggle with mud and cold.

'At Queenstown the 85th lived under canvas and suffered considerably from the weather, which was persistently cold and stormy. The officers, however, enjoyed good sport with spring buck and pan (bustard), the former being especially numerous round the station, and Capt. Lord John Taylour performed the feat of killing a buck at a range estimated at 800 yds. The men also built a regimental theatre of wattle and daub, in which a series of very successful entertainments were given.

'An interesting personality at Queenstown at this time was that of a Swiss veteran who had served in, he stated, over 300 engagements with Napoleon's Old Guard and who still wore the remnants of the uniform of that immortal Corps.

'The weather continued its vagaries till the end of the first week in 1857, when intense heat replaced the rain and wind, a temperature of 108° being registered in the officers' tents on Jan. 12th. The extremes of temperature tried the health of the Regt. considerably and several deaths occurred among the N.C.O.'s and men. The monotony of existence at Queenstown was temporarily relieved by the arrival of a large draft on May 29th of 260 men and 7 officers, viz. Major Grey with Leet, Baker, Coussmaker, Henderson, Reeves and Drage. Bad weather set in again with gales of wind and rain, while on July 17th 1 foot of snow was registered. On July 6th it is interesting to learn that the new Enfield rifles for the re-arming of the Regt. arrived and that their introduction was much appreciated, the new weapon being lighter by 2½ lbs. than the old Brown Bess.

'Two days later came the sad news of the death in England of Col. Power (the C.O.) and gloom settled more heavily in consequence upon the Regt. with whom their late C.O. had been extremely popular. Even news of the outbreak of the Indian Mutiny which arrived at this juncture failed to rouse the Regt. from its lethargy. The 85th was finally made up to full strength by the arrival on July 28th of a small draft from Mauritius of 20 men and 3 officers, viz. Doughty, Bruce and Athorpe.'

[The diary ends abruptly on August 21, 1857, with no further item of interest to chronicle].

The diary of Ensign Wilson from which the above account has

been compiled is so interesting that it would indeed be a pity not to give some absolute extracts therefrom here. Moreover, as we have drawn upon the supply of sketches with which its pages are here and there studded the context may well add to their interest.

He states that the 85th was selected to go from Mauritius to the Cape for two reasons:

'1st. Because it was then stationed at Port Louis while the 5th Fusiliers were scattered about on different outposts or detachments.

'2nd. Because Gov. Higginson wished to keep the stronger force in the Island, the 85th having only six service companies out.

'One Portemanteau, a Carpet Bag and Bedding—each—were all we had to take—the *Penelope* not having room enough to stow away more.

'Revolvers and Rifles were in great request—"Colts" and "Dean and Adams" were most sought for—but a great number of us were obliged to be contented with the famous "Patent," which did not look, or afterwards prove, a very safe weapon.

'At 6 A.M. on the 18th of May, the Right Wing and Head Quarters marched through the town amidst the shouts and hurrahs of the 5th.'

The officers had to sleep in hammocks on board, and were much crowded and most uncomfortable.

An account of the pranks played upon the staff Assistant-Surgeon is related in a most amusing mannner, a sketch or two of that officer being given. He had been for a short time in the Crimea and 'he wore about two inches of blue ribbon, sewn on the breast of every one of his coats.'

A facsimile page of the diary describing the ship's officers and the weather towards the conclusion of the voyage is given.

At East London they could not land. Provisions and water were running short and Sir William Wiseman determined to run for Algoa Bay, despite the fact that on shore tents had been pitched and waggons were seen to be in readiness.

He gives a graphic description of the landing in the surf-boats.

These boats 'are generally manned by six or seven Fingoes with long broad oars, who pull them when laden to another one, anchored a quarter of a mile from the shore and with which a large thick rope is

(8) June

gentlemanly set of fellows & Sir William Wiseman was a general favorite - He was exceedingly kind to us, & placed his library, which was a pretty good one at our disposal

We had tolerably fine weather till we got within a hundred Miles off the Coast of Natal - when the Wind suddently shifted to the S.W. from which quarter it blew pretty stiffly This continued for four days alternately blowing very fresh with thunder & lighting or else raining in torrents - - during the whole of this weather four hundred of our men were obliged to remain on deck, exposed to the rain wind & storm - as 200 alone who get below: The Hardships they had to undergo were very great - however not a murmur or complaint was heard - their conduct was beyond all praise —

On the 4th June the wind abated & the sea fell considerably - About 8 Am got up

FACSIMILE PAGE OF ENSIGN WILSON'S DIARY.

connected. The Fingoes then pass this rope through two holes cut in the boat, one at the stern and the other at the bow, and keep hauling on it until it grounds when they jump out and carry the cargo to shore.' How the Colonel (who was the last to leave the ship) and the two Ensigns (Wilson and Grant) reached the shore the illustration shows. The diarist remarks, 'It was rather a comical sight to see our old chief between his two Ensigns hurrahing away—till we were landed safely on shore.' He then goes on to describe Port Elizabeth and to contrast it with Grahamstown. They landed on June 7, on which day there were peace rejoicings in the place, 'races by day and "Bond Fires" by night.' The races were not much: 'small raw boned poneys' being ridden by uncouth-looking savages or by Dutchmen.

Of the looks of the Dutch girls he writes in terms of admiration. While under canvas here the wind caused much trouble and anything in the shape of a hat was not easily recovered before it had been annexed by some unclean native.

The people at Port Elizabeth were very hospitable and among other entertainments gave them a big ball. The room was decorated with the flags of the *Penelope* and the colours of the 85th—these last guarded by two colour-sergeants. What might have been a bad accident occurred during the festivities. The chandeliers had been decorated with bayonets and wreathed in artificial flowers. One of them caught fire, and shrieks and hysterics on the part of the girls ensued. One of the lieutenants, however, secured a broom with which he extinguished the conflagration. Fresh candles were obtained and the ball was soon again in full swing, the floor being much improved by the wax from the candles of the chandeliers.

On this occasion somebody unfastened the pouch of the Assistant-Surgeon and down it rolled at his back, displaying a long 'row of lancets, scissors, pill-boxes, etc., which extended to the skirt of his tunic. With this flying about he continued to dance, till—making a false step—he tripped up and fell down, dragging his unfortunate partner with him.'

It was at this time that the 'German Legion' raised during the Crimean War was being so expensively carted off to the Cape to settle there. It cost no less than £200 per man.

His adventures on a journey by mail-cart with a brother officer for the purpose of purchasing two bouquets of camelias are told with many amusing touches. At one stopping-place he met a Mr. and Mrs. Anderson. The conversation turned on shooting and 'we were astonished to hear the lady joining in, and doubly so when she told us that she liked to shoot a springbok.' So interested did they become in the 'stories this Amazon' was telling them that they narrowly escaped missing the coach. As it was, the inside was full and they had to clamber to the top somehow and cling hold like grim death to an iron rail. The heat and dust and jolting of the ride were intense.

At Port Elizabeth there had been some good private theatricals,

GRAVESEND, PORT ELIZABETH, FROM THE CAMP.

and fired by their success the regiment in its turn got up a piece. The title is not stated; but Private Harrington of No. 5 Company 'was excellent, who together with a ridiculously ugly face, was endowed with a good deal of wit and comical humour.'

Several times Ensign Wilson in his early days at the Cape endeavoured to get some shooting, but if he went with a rifle after buck he generally saw partridges and *vice versa*. The mimosa thorns too, appear to have been particularly fond of rending his garments.

After a stay of three weeks in Port Elizabeth the regiment marched for Grahamstown. A good description of the salt-pans is given, and how they crossed the river the illustration shows. Ensign Wilson writes:

'At 11 A.M. we got to the ford, which we had to wade over. However I was spared the necessity of taking off my shoes and socks, by

From a Sketch by Ensign Wilson.

CROSSING THE DRIFT ON THE MARCH TO GRAHAMSTOWN.
June 1856.

Ashe carrying me across on his back, much to the amusement of the Colonel and the men.'

A little further on, the ferry was crossed in flat-bottomed boats which, secured by two ropes, were pulled and hauled over by a few Fingoes.

The diarist then describes the ox-waggons in which the sick and baggage were conveyed. The teams numbered from fourteen to sixteen animals each. The method employed for stopping a team was somewhat original. 'They begin by roaring out "Wo"; at the same time, picking up a stone as large as their heads, they throw it with great force at any refractory bullock, whom it invariably hits either over the eyes or else on the horns. The effect of the blow is always the means, of stopping the animal.' Halting at a place called Adcock's, tents were pitched. The only three beds obtainable were secured for senior officers; the remainder slept outside. Towards night it was cold. A large fire was lighted and round this they sat in a large circle. Tents were struck on the morrow, but some of the bullocks had strayed and one of the officers (Noyes) who was on baggage-guard had perforce to remain behind till the beasts were recovered.

Shortly after starting, a party of the Cape Mounted Rifles met them.[1] Their uniform was then dark green, with rifles slung at their backs. It was the advance escort of Sir George Grey and his staff, who were on their way to Port Elizabeth.

On arrival at Sunday River the regiment had to halt till the waggons had crossed. When over the river they pitched their tents in front of an inn, where they were lucky enough to find plenty of beds. On this page is a sketch, presumably of the halting-place.

The regiment left Sunday River on June 29. The next halting-place was Vogel's Farm, where tents were pitched and which appears, from the description, to have been a very beautiful spot. Quail abounded in the district and wild turkeys were also reported, but several officers who attempted shooting failed to get bags. No bucks were seen though 'their footprints were all about.' That night was very cold. A large fire was lighted and 'blazed away so gloriously that we felt as comfortable as possible seated round it, each with a good

[1] This corps, it may be remarked, ceased this year (1912) to exist as a distinct corps.

plateful of beefsteaks, which he had cooked himself. After which we infused some tea or coffee—lighted our pipes, and felt as happy as any *bon vivant.*' The regiment had now reached the other side of Algoa Bay and could see Port Elizabeth in the far distance. Bad weather now came on and they reached the next halting-place, a small inn, 'wet, tired and benumbed with cold.' The place was known as Larter's Inn. Here the food was plentiful, but except bread and cheese was very bad.

A body of the Field Mounted Police was there, also on escort duty.

'They were all as drunk as they possibly could be, swaggering about the place and proving a great nuisance. They are a fine body

THE INN AT SUNDAY RIVER.
June, 1856. From a sketch by Ensign Wilson.

of men, dressed in their moleskin suit, leather cap, and rifle slung across their shoulders. They are a very efficient force and exceedingly useful in catching thieves and following the spoor of stolen cattle.'

Here Sir George Grey with Majors Hoffman and Grant arrived, the two last riding while Sir George drove in a waggon. Major Hoffman had belonged to the German Legion.

A sketch here given shows the character of the country at this part of the march. The next camping ground was at Nassau River, nineteen miles from Larter's Inn. Owing to the overcharges at the inn some of the officers now determined to put up together in tents and mess on their rations. Ensign Wilson describes the arrangements in the case of himself and two brother officers. Tea, coffee, chocolate and 'Cape Smoke' were the only drinkables obtainable. The last-named was then, as now, a vile concoction and his opinion of it is fairly and fully expressed.

The regiment proceeded on its long march, sometimes having the luck to get a partridge or a hare but more often not.

Two days later they reached the mouth of Howitzon's Poorte after passing through some very beautiful scenery, and there encamped. On this day's march the General (Sir James Jackson) and his staff met them. Sir James and the Colonel of the 85th were old brother officers. Later they traversed a pass or ravine some four miles long and distant six from their destination, Grahamstown. In this ravine baboons were seen scampering up and over the rocks, 'chattering and grimacing at us.' The 85th marched into Grahamstown at 1 P.M. and found not a single living person, save a few Hottenots, in the streets. The fact was that on that day a 'General Thanksgiving' was being held in the church on the conclusion of peace with Russia. The diarist describes the town as he saw it, and rather wonders at the very large number of 'Churches and Meeting Houses in all directions.'

At Grahamstown the 6th regiment was quartered and showed the 85th much hospitality.

An inspection of the regiment was held here by Sir James Jackson. His aide-de-camp was an officer of the regiment, a Captain Orme, who had lost an arm from a cannon-shot in India while serving in the 3rd Dragoons. That night Ensign Wilson dined at the mess of the Cape Mounted Rifles. The mess-house was about three miles from Fort England, on the west end of the town. Great difficulty was experienced in finding the place. An interesting account of the former and present uniform of that corps here follows:

'The Cape Corps officers are a very nice set of fellows and have an exceedingly good mess. Their uniform used to be a very pretty one, viz. dark green pelisse, all over black lace and embroidery, worn over one arm—with another jacket with lots of braid and stuff on the chest. A red and gold sash round the waist, green trousers, and a shako with cock's feathers. Now they wear the tunic, which does not look half so well.'

After mess they adjourned to the quarters of an Engineer officer who gave them his account 'of his engagement with the Kaffirs on the Koonap Hill. This took him some time, as he had to fight every waggon in succession.' His story lasted from 12 midnight until 2 A.M. When they did at length depart, the difficulty was to find the road. One of the party volunteered to show the way, but soon came to grief. After many twists and turns and several falls into holes, ditches and similar delectable spots, they at length reached Fort England, where 'we could not help laughing at the different appearances we each wore, encased in a thick coating of clay—our faces all bedaubed with mud and our hair sticking out like "quills from the fretful porpentine."'

Next morning they found that all the rest had similar adventures to relate, the Colonel included. That officer had completely lost his way in a wheat field.

On Monday, July 7, the regiment received orders to march to Fort Beaufort, a place about 46 miles due north.

Ensign Wilson now describes and contrasts the Cape horse and the 'Winkler,' a type of pony usually ridden by the Dutch. He appears to prefer the latter as being more hardy and with less vice.

His adventures at a local ball follow in which he was first carted round the room by the hostess, a lady who had, it turned out, visited the refreshment table somewhat frequently and who ultimately subsided on to a sofa. The entertainment does not appear to have been a success, the ladies were few and 'no bonnie.'

The march to Beaufort began on July 8. It appears to have been the custom of the General to escort any large body of men out of the town with all the military force in Fort England and the other barrack. Accordingly the 85th were accompanied by the 6th Royals, who preceded them by half an hour, and were followed by the Cape Mounted Rifles. On reaching the top of a hill they found themselves on a very extensive plain. Here the 6th were drawn up and presented

From a Sketch by Ensign Wilson.

ON THE MARCH; July 1856.
In the Ecca Pass.

arms to the 85th as the latter marched by. There was a march past the General and his staff, who were accompanied by a few ladies. The Cape Mounted Rifles and the 6th then had a field day. Some artillery also took part in the proceedings. It was, however, a bitterly cold day.

On reaching the Ecca Pass the 6th halted while their band played the 85th a short way down. The General and his staff halted on a ridge overlooking the pass, while the artillery blazed away over the heads of the regiment.

'Ecca Pass is composed of a road cut round the side of a precipitous and thickly wooded hill, and winds round and round it. On looking about you it appears as if you were in one dense forest. At your feet lies a deep chasm or kloof, full of aloes, mimosa

FORT BROWN, 17 MILES FROM GRAHAMSTOWN.
26 August 1856. From a sketch by Ensign Wilson.

and other bushes; while above are mountains covered with the same brushwood. The place abounds in all kinds of game, baboons and rock-rabbits (not unlike a large rat) running about the rocks overhanging the road, the former grinning and chattering as we marched along. Wild jessamine and geraniums grew in great luxuriance, and filled the air with their sweet perfume.'

The Ecca Pass is about three miles in length and descends into a large kloof where there is a little water and then winds up another hill at the top of which begins the Fish River bush.

At about 2 P.M. the regiment sighted Fort Brown, distant therefrom about one mile. It was situated at the base of a small conical hill called Vereker's Kopf, and was built on the banks of the Fish River.

This bush is mainly composed of mimosa, a very thorny plant which resembles the sensitive plant and whose spikes are sometimes as much as three inches in length.

That day the 85th covered about eighteen miles and all were very tired as the road was most fatiguing. The regiment was most hospitably received by a detachment of the 6th then quartered there. Having crossed the river on the next morning by a bridge, said to be one of the best then existing in the colony, but really ' big clumsy and wooden,' the country improved. Six miles from Fort Brown the regiment passed under Blue Krantz, 'an immense precipice about 500 feet high,' where baboons as usual abounded. With regard to this place Ensign Wilson expresses wonder that ' during the late wars no body of men have been attacked by Kaffirs in passing this defile, where the rocks and bush growing about the precipice would form a splendid cover for the assailant.' Having crossed the Koonap River by means of one boat capable of containing nine persons at most and rowed by two soldiers of the 6th, the regiment at length reached Koonap Post, though, as may be imagined, the operation of crossing consumed a considerable amount of time.

At Koonap Post they visited the hut of the officer in charge, who had hospitably obtained luncheon in readiness for them. ' I never saw ' writes Wilson, ' such miserable looking quarters in my life, you could not compare it to anything better than the poorest Irish hovel.' However, other and better quarters were in process of building and the men's barracks were half finished. When the regiment came up they traversed the famous Koonap Pass, ' the ugliest and most dangerous road ' you can imagine, especially during war time. They saw the spot where the Engineer officer before mentioned had had his fight and where he had behaved with great gallantry. He was at the time escorting a number of waggons to Grahamstown, loaded with Minie rifles. With a handful of Engineers he fought his way to Koonap Post and then made a stand until he was able to retire to the inn, where he obtained assistance. He, however, lost every waggon, as the Kaffirs shot every mule at the beginning of the attack. The Minie rifles were of course captured, but being unprovided with locks were useless to the enemy.

That night they camped at Liewfontein, a roadside inn about

seventeen miles from Fort Brown. Their dinner here cost them 15*s*. a head, and as there were sixteen officers there the landlord (a drunken old fellow named Pye) did not do badly. Next day Wilson was on baggage-guard and had to remain behind until the last waggon had started. He rode on one of the rear waggons, the driver of which was a one-eyed Englishman of most forbidding appearance, but who notwithstanding turned out by no means a bad sort. The old fellow regaled his passenger with blood-curdling stories of the Kaffir War, notably one of the murder of the bandmaster of the 74th, the full and terrible details of which he related at length. About 2 P.M. the column came in sight of the town of Beaufort, which lay beneath them in the valley. Before reaching the town, however, the Kat River had to be crossed. However, as there was a tolerable bridge, not much time was lost, and early in the afternoon the whole regiment had safely arrived.

'DESOLATION COTTAGE.'

The principal barrack was at the top of the town and consisted of one long building capable of holding about 300 men. Several smaller barracks, each able to contain about 100 men, were scattered in different parts of the town. 'For instance the Officers' Quarters, the Main Barracks, and the Mess House form three points of a triangle, being about half a mile from one another, and the Officers' Quarters about three-quarters of a mile from the Mess House. This is certainly a great nuisance during the wet season, when the whole square is one immense pool of water, through which one has to wade every night in order to get his dinner.'

The messhouse appears to have been 'about the finest building in the town,' with a large messroom, anteroom and verandah.

There were not sufficient officers' quarters in Beaufort. So Wilson and a brother officer took a small hut which they named 'Desolation Cottage'—it consisted of two rooms, a pantry, and a kitchen.

Here follow a number of descriptions of the inhabitants of the neighbourhood, both civil and military; but they need not be quoted.

On Monday, July 14, 1856, the Colonel had a stroke of paralysis. Ensign Wilson thus records it:

'We were all very much shocked, as we were waiting for the Colonel on Parade, to see him come out of the Orderly Room, led by Thompson and Bartley. He had received a stroke of Palsy. The plucky old man endeavoured to get on his horse, but after several vain efforts, he was obliged to take another fellow's arm. He would not even allow himself to be carried in a chair, but walked as well as he could to his quarters, which luckily werc not far off. Towards evening he got better, but all his left side was powerless. However, the Doctors are in hopes of his recovering.'

These hopes, however, were not to be realised. On Thursday, August 7, we read:

'The Colonel accompanied by Bartley started for Graham's Town in a small spring cart on his way to England. We were all exceedingly sorry to say good-bye to the chief as he always had been very much liked.'

Wilson gives a long and interesting account of a great hunt one day in this month when hundreds of naked Fingoes beat the bush, but beyond one hare and one buck bagged by Boyle, one of the officers, the remainder of the bag, viz. 12 buck, 3 boars, and 2 wild dogs, fell to the assegais of the beaters.

On August 13 orders arrived for 'No. 2 Company with two efficient subalterns to march on the 15th to Fort Brown leaving a subaltern and 50 men at Koonap Post and to relieve the 6th stationed at Koonap and Fort Brown. We were also to march with the greatest precaution.' Accordingly Captain Armitage with Lieutenant Lord Taylour and Ensign Wilson proceeded thither, starting as above. No. 2 Company safely reached its destination, Lieutenant Taylour being left with 50 men at Koonap Post. At Fort Brown nothing of importance occurred.

On August 28 a notification came that the Governor, General, and staff would visit the post on the morrow. Accordingly a luncheon was prepared and a bugler stationed on the high ground to blow the 'assembly' immediately the party hove in sight, or, as Wilson puts it, 'saw the swells.' The lunch consisted of '4 roast fowls, 2 ducks, a ham and a leg of mutton, Vegetables, &c., Beer, Wine, and with Tea not so dusty, eh!!'

'At about half past 1 we heard the "assembly" go—we formed up outside of the fort and presented arms as the Governor, etc., rode up, while our bugles sounded the "Point of War." The General then rode down the ranks.' An amusing incident took place, an assistant-surgeon who was in rear of the line with his professional paraphernalia drew his sword and 'flourished it about—never having learnt to salute—while the horse began to plunge and kick about so much that he soon sent groom, paniers and all, flying about him.' This officer was a most eccentric man. His portrait—no doubt a caricature to some extent—is on one page of the diary. In it he wears 'an enormous jackall's tail round his wideawake, he had let his red beard grow, which was now not unlike a lot of hog's bristles, and he looked anything but handsome.'

The 85th was now relieved by the 80th and was ordered to Queenstown. From this point the character of the diary quite changes, and we shall in consequence conclude our extracts with a final excerpt. Ensign Wilson was evidently known as a caricaturist, and on one occasion we find the General requesting a sight of his sketch-book in which sundry personages appeared. Wilson tells us that one of the subjects was present and that he was not certain whether he relished the contents of the book as much as the General did.

One more entry in this most interesting diary must be quoted. It is the last entry but one in the book and runs as follows:

'When returning, little M—— astonished Urquhart and me by his stories. He first of all had been in the Stock Exchange—then Editor of *Bell's Life,* after which he had got a cornetcy in the 11th Hussars and was now a Purveyor's Clerk!! He told us that when last in Scotland he had shot six Red deer, and his father 4—at a distance of 700 yards!!!'

The name of this gentleman is given. Reference, however, to the 'History of the 11th Hussars' fails to substantiate the cornetcy. The editorship of *Bell's Life* is similarly not to be traced. For those days the 700 yards' range was an impossibility, but the purveyor's clerkship was an undoubted fact.

With this we close our quotations from Ensign Wilson's manuscript book and cannot but think our readers will be as much amused by them as we were ourselves.

CHAPTER XIII

ENGLAND, 1863–66—IRELAND, 1866–68—INDIA, 1868–78

ON March 30, 1863, orders were received in Natal for the regiment to hold itself in readiness to embark for Portsmouth in H.M.S. *Himalaya* upon the arrival of that ship with the 2nd battalion 5th Fusiliers from Mauritius.

Early in the month a party of invalids had already been sent home under the command of Captain Barton.

Between these dates a wing of the regiment had been sent from East London to King William's Town to relieve the 2nd battalion 13th Light Infantry, which was ordered to Mauritius. The wing of the 85th returned to East London on April 7, having been relieved by the 96th Regiment from England.

On April 21 a detachment of the regiment then stationed at Fort Buckingham on the Lower Tugela having been previously concentrated at Durban, where the remainder of the regiment then lay, marched from Fort Napier leaving Captain Herberden, R.A., in command. Durban was reached the next day, where the regiment encamped at the post. The *Himalaya* had arrived and was anchored outside the bar.

On April 24 at 4 A.M. (not a man being missing) the regiment marched down the railway track to the point. The 5th Fusiliers were disembarked and the 85th with its baggage was embarked during the afternoon.

The *Himalaya* weighed anchor early next morning.

The strength was as follows (*Headquarters*): Colonel Grey; Captains Boyle, Harford and Hallowes; Lieutenants Chichester, Doughty and Henderson; Ensigns Ramsbottom, Davison, Lewin and Drage; Lieutenant and Adjutant Drage, Quartermaster Watts, and Dr. Johnson;

with 24 sergeants, 16 corporals, 11 buglers, 293 privates, 35 women and 80 children. These belonged to Companies Nos. 2, 5, 9 and 10.

Next day the *Himalaya* anchored off Buffalo River mouth and on the 27th embarked the wing from East London, consisting of Major Appleyard, Captains Hogge, Mytton and Mathews, Lieutenant White (Acting Adjutant) and Assistant-Surgeon Herbert; with 20 sergeants, 10 buglers, 273 privates, 12 women and 20 children of Nos. 1, 3, 4, 6 and 7 Companies.

Two officers of the regiment remained at the Cape, namely Lieutenant Wilson, who had been appointed aide-de-camp to General Wynnyard, and Ensign Cooper, aide-de-camp to Colonel Staunton. At the last moment Lieutenant Grant also obtained six months' leave of absence to remain at the Cape and therefore did not sail with the regiment.

On April 29 the ship anchored in Simons Bay and was detained there until May 6.

The dates of the remainder of the voyage were as follows: St. Helena, May 24; St. Vincent, where the vessel coaled, on the 26th; Spithead, June 6; Dover, June 8. Here the regiment disembarked and took up quarters in the Citadel and South Front barracks.

On June 11 the regiment was inspected by Major-General Sutton, then commanding the garrison. Strength as follows: 2 field officers, 6 captains, 12 lieutenants and ensigns, 4 staff, 44 sergeants, 21 drummers, 34 corporals and 547 privates. The only casualty on the voyage was one man who died at Simons Bay.

On June 24 a draft from the depot joined the regiment under the command of Captain Thompson with Ensigns Garnett, Wilmer, Hodgson and Campbell and 76 rank and file.

The same date in compliance with an order (Horse Guards April 1, 1863) the establishment was fixed as under: 10 service companies, 3 field officers, 10 captains, 11 lieutenants, 9 ensigns, 5 staff, 48 sergeants exclusive of schoolmaster, 21 buglers, 40 corporals and 640 privates; 2 depot companies, 2 captains, 3 lieutenants, 1 ensign, 10 sergeants exclusive of schoolmaster, 4 buglers, 10 corporals and 110 privates.

The strength, therefore, of the non-commissioned officers, buglers and privates thus amounted to 58 sergeants, 25 buglers, 50 corporals and 750 privates.

On July 20, 1863, the regimental dinner was held in Willis's Rooms, the following officers being present:

Major-General Brook Taylor; Colonels Pipon, F. E. Knox and Charles Knox; Lieut.-Colonel G. Warde; Majors Tennant, Stuart Knox, R. Maunsell, W. Maunsell, Blackburn and Williamson; Captains Rooper, Thompson, Boyle Orme, Hogge, Mytton, Mathews and Aytoun; Lieutenants L. H. Warde, Sir George Osborn, Nelson Rycroft, Sir H. Edwards and Taylor; Ensign Lewin and Mr. Neil.

The 85th remained at Dover until April, 1864, when it moved to Shorncliffe Camp, and thence in September of the same year to Aldershot, where the depot companies from Pembroke joined. In June, 1865, the several companies of the regiment were ordered to be designated by letters (A to M, less J) instead of by numbers.[1]

In the autumn of 1865 the regiment proceeded to Manchester under command of Lieut.-Colonel Williamson, giving detachments to Chester, Carlisle and Newcastle-upon-Tyne; but the Fenian disturbances in Ireland necessitating the presence of additional troops there, the detachments were withdrawn again, and the regiment transferred to Dublin.

Major-General Grey, who had been so many years with the regiment, now retired, and was succeeded on October 27, 1865, by Lieut.-Colonel Williamson.

The following letter was received:

'SIR,

'I have the honour by desire of the Major-General Comd[g] to acquaint you that H.R.H. The Field Marshal Comd[g]-in-Chief is much gratified at receiving from Lieut.-Gen[l] Sir James Scarlett a satisfactory report of the Reg[t] under your command which is considered by that General Officer to be a most excellent and efficient corps.

(Signed) 'G. BROUGHAM, *Col.*,
'*A. A. Genl.*'

[1] Under this new arrangement the numbered companies in the order named—2, 10, 5, 7, 11, 1, 8, 9, 12, 3, 4, and 6—were now designated A, B, C, D, E, F, G, H, I, K, L, and M respectively. On parade it was ordered that the companies 'will be numbered from right to left, the company actually on the right being No 1.' Also, 'In the event of any reductions, the company will not be re-lettered but each will retain the above letter.'

The transfer of the regiment from Manchester to Dublin was thus. Two detachments, of two companies each, were lying at Chester and Newcastle-on-Tyne respectively. These were recalled, as also were another two companies quite recently detached to Tynemouth. On February 19 the orders to proceed to Ireland arrived. On February 21 the whole regiment went by train to Liverpool, embarked on board two steamers, and reaching Dublin the next morning proceeded at once by rail to the Curragh Camp.

March 24, 1866. The establishment of the regiment was by order of the Horse Guards reduced from April 1 to ten companies.

'A and D Companies were therefore broken up and distributed between the remaining ten companies. Captains Noyes (D) and White (A), with Lieutenants Hodgson and Smith and Ensigns Anstruther and Robinson, together with the sergeants and corporals of these companies being borne as supernumeraries. The excess over the reduced Establishment at this date being 58 Officers, non-commissioned Officers and Privates.'

During April, 1866, eight companies of the regiment proceeded to Dublin for duty and were followed, on July 3, by the headquarters and remainder of the regiment. For the next five months the regiment was broken up into detachments, occupying Ship Street, Linen Hall, Portobello and Pigeon House Fort barracks; but on December 6 the whole regiment took up their quarters in the Royal barracks.

On December 5, 1866, the regiment received the first issue of the Snider converted rifle, the muzzle-loading Enfield being returned into store and steel scabbards taken into use.

February 1, 1867. The breech-loading rifles converted on the Snider principle were taken into use, and on the 4th, 5th and 6th of the month the muzzle-loading Enfield (2nd class) arms, cap, pouch, etc., were returned into store.

March 6, 1867. Major F. E. Appleyard was promoted Lieutenant-Colonel.

March 7, 1867. The officers were recalled from leave in consequence of the attempted rising of the Fenians on Shrove Tuesday, March 5.

In April, 1867, a new blue infantry patrol-jacket came into wear, the frock coat being abolished.

In May the establishment was raised to twelve companies.

On July 16 the regiment proceeded to the Curragh, sending one company to Carlow for three months.

The guards and duties in Dublin had been very heavy on account of the Fenians. During this year 333 recruits joined; while the wastage from deaths, desertions, transfers, discharges, etc., amounted to 143.

The regiment was now about to proceed on its first term of service in India.

Orders being received to embark for India, the married families and heavy baggage left for Queenstown on January 28, 1868, and embarked on H.M.S. *Serapis* next morning. The regiment followed next day and the ship sailed on the evening of the 30th.

F.O.	Capts.	Subns.	Staff.	Staff-Sergts.	Sergts.	Buglers.	Privates.	Women.	Children.
2	10	19	6	8	36	21	725	98	85

The following officers formed the service companies: Lieut.-Colonel F. E. Appleyard, Majors Thompson and Lord John Taylour (on leave till July 21); Captains Barton, Hallowes, Taylor, Galbraith, Noyes, White, Hancocks, Oldfield, Cooper, Illingworth; Lieutenants Davison, Lewin (Staff College), Jebb (Instructor of Musketry), Hon. C. Dutton, Spencer-Smith, Hon. E. à Court, Burland, Ravenhill, Knox, Robinson, Purdon, Rudkin; Ensigns Burrell, Ives, Kettlewell, Smythe, Stevenson, Welman, Seton, Vivian; Lieutenant and Adjutant Drage; Paymaster Hughes, Quartermaster Humphreys, Surgeon Skeen, Assistant-Surgeon Campbell, Assistant-Surgeon Kerans.

After a stormy passage as far as Gibraltar, which was passed February 4, the *Serapis* arrived at Malta on the 9th, and that evening Captain Soady and the officers of the ship were entertained by the regiment to dinner at the Malta Club.

Leaving next day, Alexandria was reached on the afternoon of the 13th, and while mooring the ship in the harbour, a strong wind blowing at the time, one of the hawsers gave way and the vessel began to swing, causing much excitement on the Khedival yacht and another vessel moored close by; but by skilful management a collision was avoided. On the 14th the heavy baggage was unloaded and sent by rail to Suez under a guard in charge of Captain Galbraith with

Lieutenant Robinson; Lieutenants Jebb and Rudkin, having volunteered for the transport service in Abyssinia, also accompanied this party. (Later, both these officers were invalided home.)

The regiment reached Suez on the 16th and embarked on H.M.S. *Jumna*, which sailed on the 21st (being detained for repair of boilers), and arrived at Bombay on March 10, the voyage having lasted forty-one days. White clothing and helmets were issued at Bombay and next day the regiment transhipped to the P. and O. steamer *Ellora* and transports *Corona* and *Camperdown*. The headquarters and seven companies, in the *Ellora* and *Corona*, reached Karrachee on the 15th, disembarked on the 17th, went on by rail to Kotree, arriving next morning, and at once proceeded up the Indus in the Indus Steam Flotilla ship *Havelock* and two large flats lashed to it, one on either side. 'I' Company embarked on the Indus Flotilla steamer *Outram*, which also had two flats lashed to it, and on the 19th was joined by the three companies from the *Camperdown*. The *Outram* started at noon. Owing to the swift current and numerous mud-banks, which were always altering the position of the channel, pilots had to be changed every five or six miles, and every day before dark a halt was made for the night at one of the wood stations—wood being used for fuel instead of coal. Eventually, after twice running on to a bank, the *Outram* reached Rajghat on the evening of April 4. A few hours later the headquarters and the whole regiment, with its baggage, proceeded in one train *via* Mooltan to Mian Meer, which was reached early on April 5. Only one casualty occurred during the entire journey of sixty-seven days—Private Cox, who was drowned in the Indus, falling from the gangway between the *Outram* and one of the flats. The native at the wheel saw him fall and jumped into the river after him; but Cox never rose to the surface, and the native was carried more than a mile down the river before he was picked up by a boat.

On April 6 it was found that a box containing over twenty silver promotion cups was missing. A few days later the box was found in a well, but the cups were never recovered.

On April 20, 'H' Company, under Captain Taylor, proceeded on detachment to Fort Lahore, and this detachment was relieved monthly. During the hot weather the regiment suffered much from fever and lost Assistant-Surgeon Campbell and twenty-three non-commissioned officers and men.

On March 13, 1869, 'A' (Captain Cooper's) Company reinforced the detachment at Fort Lahore during the residence there of Shere Ali Khan, Amir of Afghanistan, and rejoined on March 22.

On March 15, the regiment with the rest of the garrison marched at 3 A.M. to Lahore and lined the road from the Ravee Bridge to Fort Lahore on the arrival of the Amir, the detachment providing the guard of honour at the Fort. A public durbar was held on the 19th, and the next evening the regiment with the rest of the garrison proceeded to the Shalimar Gardens, which were brilliantly illuminated with native lamps for the Amir.

In April, 150 men under Captains Taylor and Galbraith and four subalterns marched to Banikhet near Dalhousie in the Himalayas for road-making and remained there till the autumn. Detachments also went to convalescent depots for the hot weather.

On July 16, fines for drunkenness came into force in India, and all men in confinement for habitual drunkenness had the unexpired portion of the imprisonment remitted.

On February 9, 1870, the regiment with the 39th, 92nd, and five native infantry regiments, 20th Hussars, 9th Bengal Cavalry, Royal Horse Artillery and Royal Artillery, lined the road approaching to Government House, Anarcullie, on the arrival of H.R.H. the Duke of Edinburgh, and on subsequent days found guards of honour at Fort Lahore, etc. On the 11th, H.R.H. the Duke of Edinburgh lunched with the officers of the regiment and was received by a guard of honour under Major Thompson. He afterwards proceeded to open the Soldiers' Exhibition.

Over 200 men were again sent to various hill depots for the summer, the regiment suffering a great deal from fever.

From April 1 the service companies were reduced to eight companies :

Lt.-Col.	Majors	Capts.	Lts.	Subns.	Paymr.	Qrmr.
1	2	8	10	6	1	1
Adjt.	**St.-Sergts.**	**L.-Sergts.**	**Sergts.**	**Corpls.**	**Buglers**	**Ptes.**
1	9	8	32	40	16	780

and the depot was reduced by two ensigns.

On October 21 two companies under Captain Galbraith proceeded to Dugshai and on December 9 the regiment moved from Mian Meer to Umballa and on the 17th reached Dugshai, two companies proceeding to Solon on the 18th.

On November 20, 1871, three companies under Captain Beadon marched to Umballa for duty, rejoining headquarters on February 18, 1872. Purchase had been abolished on November 1, 1871, and also the rank of ensign—all ensigns being gazetted lieutenants October 28, 1871. F. Vice was the last ensign. Shortly afterwards the rank of sub-lieutenant was introduced.

Sub-Lieutenant Spens was the first appointed to the regiment.

An annuity of £15 per annum with a silver medal was granted to Sergeant-Major Rock, late 85th Light Infantry, for his long and meritorious services in the regiment, dated April 30, 1872.

On June 1 the regiment was inspected by Lord Napier of Magdala, Commander-in-Chief.

On July 29 cholera broke out in the military prison, and a few days later in barracks; five cholera camps were formed.

The regiment lost 19 non-commissioned officers and men, 4 women, and 4 children from this epidemic, which lasted till early in October. Altogether, 35 men, 6 women and 5 children in the regiment caught the disease. Lord Napier of Magdala and staff arrived on October 13, and the following day Lord Northbrook, Viceroy and Governor-General of India, arrived from Simla, the regiment lining the approach to the officers' mess, where he was received by a guard of honour.

During the day, accompanied by the Commander-in-Chief, he inspected the station and barracks, and dined with the officers in the evening. On the 15th he left for the plains. A few days afterwards cholera again broke out and in the same cell as the first case, and it was the very cell the Commander-in-Chief entered when the Viceroy inspected. The prisoners went into camp again, but luckily there were no more cases.

On November 1 the regiment left for Umballa *en route* for Meerut, arriving at the latter place on November 13.

On December 3 three companies proceeded on detachment to Fatehgurh.

On February 6, 1873, a general order was published by His

Excellency the Commander-in-Chief and was forwarded to the regiment for publication in which His Excellency was pleased to mention with approbation the behaviour of Surgeon Skeen, Assistant-Surgeons Bourne and Richards, and certain men of the regiment during the epidemic of cholera that visited the regiment at Dugshai (G.O.C.C., 19. 12. 1872).

A letter dated August 9, conveying the satisfaction expressed by His Excellency the Commander-in-Chief at the progress made by the regiment in its musketry practice for 1872–3, was also received.

The headquarters and five companies from Meerut took part in a camp of exercise near Roorkee from November 22 to January 8, 1874.

On January 15, 1874, scarlet patrol-jackets were taken into wear by officers and on March 17 a new pattern forage-cap and ornament.

On May 29, 1875, Colonel F. E. Appleyard was appointed ordinary member of the 3rd Class, or Companion, of the Bath.

November 2. Lieutenant-General Campbell appointed colonel of the regiment *vice* General F. Maunsell, deceased.

On December 7 the regiment (one company remaining at Fatehgurh) proceeded to Delhi under Major Hallowes for the camp of exercise with the 8th and 32nd Native Infantry and formed the 1st brigade, 3rd division, under the Hon. General A. Hardinge.

The whole force was inspected by His Royal Highness the Prince of Wales on January 12, 1876, outside Delhi. He also attended a ball given by the officers of the force at the palace in the Fort.

At the termination of the manœuvres the regiment proceeded to Lucknow, arriving there on January 25 and 26, to relieve the 40th Regiment.

On June 7, 1876, the 42nd Brigade Depot (composed of the 52nd and 85th Light Infantry) was formed at Oxford.

On December 23, Lieutenant-General Arnold Charles Errington was appointed Colonel of the regiment *vice* Lieutenant-General George Campbell, C.B., deceased.

On January 9, 1877, new colours were presented to the regiment at Lucknow by the Duke of Buckingham and Chandos, G.C.S.I.

His Grace and staff arrived from Delhi the previous day and dined with the regiment, a dance being given in the evening. He inspected the regiment on the 9th, and presented the colours, afterwards addressing the regiment as follows :

'Colonel Appleyard and soldiers of the 85th, I have accepted with pleasure the duty of presenting these colours ; for I have always thought that it was for the good of the soldier and of the army that every tie which tended to encourage the maintenance of the connection of the regiments with their districts, the soldier with his home, the regiment with its county, should be maintained. Hence I have a peculiar satisfaction as Lord-Lieutenant of Buckinghamshire in placing the new colours in the hands of the 85th (The Bucks Volunteers), a regiment in the raising of which my ancestors took an active part.

'We stand here, far from home, in Lucknow, where once the bravery and discipline, the unflinching energy and determined perseverance of the British soldier were most sorely tried but conquered and triumphed.

'More than 80 years have passed since the regiment was raised in Buckinghamshire. The names blazoned on these colours show that it has, during that time, taken part in some of England's hardest fought fields—in those fights it fought gallantly and well and earned credit and honour.

'Since the formation of this regiment the periods and conditions of a soldier's enlistment have been much changed. A young man now enters with the prospect of being able to return to his home and native pursuits at an early age. But you, young soldiers, do not let this make you less eager to maintain the character and the credit of the regiment of which you are a part. Young and old, let it be your constant thought to maintain by your conduct as good soldiers and as good men the character of your regiment. Think upon those who are at home, "The girls you've left behind you," and when they see any reference to the 85th let them always find it spoken of and welcomed as a highly efficient regiment of good soldiers and well conducted men.

'I now leave these colours in your hands, confident that I entrust them to worthy hands, that wherever these colours are unfurled, whether in the stately pomp of ceremonial procession, or on active service, they will herald the approach of a gallant and well disciplined regiment, and that when borne to action they will be carried bravely to the field and will leave it with victory.

'In conclusion I would say, remembering the regimental motto, "Aucto splendore resurgo"—"I rise again with renewed splendour"—let every duty bring additional credit and honour.'

The strength of the regiment on parade was 16 officers and 594 non-commissioned officers and men.

The following officers were present on parade: Colonel F. E. Appleyard, C.B.; Major E. M. Beadon; Captains Grant, Knox and Robinson; Lieutenants Purdon, Langford, Welman, Vivian, Burrell, Spens and Campbell; Sub-Lieutenants Capper and Rivett-Carnac; Staff-Lieutenant and Adjutant Collete and Quartermaster Hawkins.

The following officers were present, but not on parade: Captain Galbraith; Sub-Lieutenants Lyle, Godfray, Galloway, Bulman and Stirling.

After the conclusion of the parade his Grace expressed to Colonel Appleyard his high approval of the appearance of the regiment and of the manner in which the parade had been carried out. He was also pleased to accept the old colours as a memento of the occasion and of his connection with the regiment as Lord-Lieutenant of Bucks. In the evening his Grace attended a ball in the sergeants' mess. These colours now hang in the private chapel at Stowe Park, Buckinghamshire, and are the property of Baroness Kinross.

The following letters give an account of the means by which these colours were restored to the regiment (see illustrations facing page 304):

R. AMBASCIATA D'ITALIA
IN LONDRA,
ADDETTO NAVALE.

'17 REDCLIFFE SQUARE,
'SOUTH KENSINGTON,
'LONDON, 25.11.1894.

'SIR,

'His Excellency the Italian Ambassador, Count Tornielli, has received a letter from one of our Naval Officers at Naples, Rear Admiral Marra, to say that he has found among his effects there the old Regimental Colours of the 85th Duke of York's Light Infantry Regiment, which in 1877 was given to the Duke of Buckingham on the presentation of the new colours at Lucknow. It appears that they were afterwards given by his Grace to General Brook Taylor, whose adopted daughter was married to Admiral Marra, and hence he is in possession of the colours, his wife being now dead.

'The Admiral thinks that so interesting a souvenir of many a hard fought field of battle is likely to be dear to its old Regiment, and that it should find a home there rather than with strangers. Under this impression His Excellency has asked me to write to you as the Colonel commanding the old 85th Regt. to communicate this statement by Admiral Marra, and to add that if you and the officers of the Regiment would like to become the guardians of this brave old flag, it shall be sent to you or to anyone whom you may indicate.

'I am,
'Sir,
'Yours truly,
'A. PERSICO,
'*Captain R. I. Navy.*'

'Lieut.-Colonel P. H. Murray,
'Commandant 85th Foot,
'Portland.'

This most courteous offer was, as may be supposed, gratefully accepted by the regiment, and in due course the colours arrived, accompanied by the following:

'NAPLES, *the 19th Xbre* '94,
'36 GIOVANNI BANSEN (?)

'DEAR SIR,

'I am sending you by the means of Col. Slade in Rome, in two wooden boxes the old flag of the 85th Regiment English Light Infantry, which has belonged to me for several years, and was left to late my wife by General Brook Taylor. Our Italian Ambassador Count Tornielli, has been so kind to make the offer to you, has advised me to address it to you. I am happy to express my best thanks to the officers of your Regiment, for having gladly accepted this precious relic. With the assurance of my esteem,

'Yours most obliged
'ADMIRAL MARRA.'

January 1, 1877. The new pattern mess-jacket and waistcoat were taken into wear.

May, 1877. Martini-Henry rifles issued in place of Sniders.

On December 22, 1877, the regiment was inspected at Lucknow by His Excellency Sir F. Paul Haines, G.C.B., Commander-in-Chief in India, who expressed his satisfaction at the appearance and general good conduct of the regiment. The regiment was also thanked in General Orders by His Excellency the Commander-in-Chief for the great efficiency shown in the musketry course for the years 1876–7.

April 10, 1878. In consequence of the threatening condition of affairs in the East of Europe the regiment was confidentially warned by telegram from army headquarters to hold itself in readiness for service.

This order was afterwards countermanded.

This year the regiment was first on the list of regiments in India in the annual course of musketry and second in the army for 1877–8.

War was now about to begin, and the regiment received orders to hold itself in readiness for active service in Afghanistan.

The history of this campaign, as far as the regiment is concerned, will be dealt with in the next chapter.

THE PRESENT COLOURS.
Presented by the Duke of Buckingham at Lucknow, 9th January, 1877.

CHAPTER XIV

AFGHANISTAN, 1879–1880—NATAL, 1881

THE political position in Afghanistan at this time was as follows. Negotiations for an alliance with the Amir had been in progress. These were, however, broken off in 1877 on the refusal of that potentate to permit an English Agent to reside at Cabul. When, however, in the autumn of 1878, the Indian Government received intelligence—intelligence which subsequently proved to be perfectly correct—the situation changed.

The fact was that, while a British Envoy had been refused permission to proceed to Cabul, a Russian Mission, the head of which had full powers vested in him to conclude a treaty with the Amir, had been received at Cabul. Obviously this was very detrimental both to British interest and British prestige. The first step taken by the Government was to notify the Amir that a British Mission would at once proceed to Cabul. The Mission started, but was turned back at Ali Musjid. In November, war was declared by the Viceroy against Afghanistan, and three separate columns were ordered to advance on Kandahar, Jelallabad and Kurrum.

This then was the position of affairs when, on October 8, a telegram was received from army headquarters directing the regiment to hold itself in readiness to proceed at once on active service on receipt of orders. Therefore on October 29 the headquarters under Colonel Appleyard (strength 20 officers, 720 non-commissioned officers and men) left by rail for Umballa, halting at Bareilly and Meerut, and arrived at Umballa on November 1. A depot under Captain Welman, consisting of families, invalids and heavy baggage (strength 1 officer, 101 men) remained at Lucknow.

The following farewell order was published on the departure of the regiment from Lucknow:

'On the departure of the 85th King's Light Infantry the Lieut.-General has much pleasure in conveying to Colonel Appleyard, to his officers, N.C.O.'s and men the sense he entertains of the high spirit and feeling which pervades the Regiment.

'He also expresses his regret at so fine a Regiment leaving his command, for during its 2½ years' stay in garrison its conduct has been exemplary in all respects, and no one knows better than the Lieut.-General how thoroughly duty is done, how all ranks work with a will and how excellent is the system, which under Colonel Appleyard's vigilant care and watchfulness makes the Regiment efficient in every respect.

'It leaves the Oudh Division in a high state of Discipline which is the sure guarantee for efficient service in the field, and in saying Farewell, the Lieut.-General hopes that fresh honors may be added to the glorious names now inscribed on the colours.

'By order,
(Signed) 'T. C. MINTO, *Major,*
'*Brigade-Major.*'

Colonel Appleyard, C.B., having been appointed to the command of the 3rd Brigade, 1st Division, Peshawar Field Force, left Umballa for Peshawar; and on November 6 the regiment (under command of Major Beadon) proceeded by rail to Loodhiana, thence by route march, via Ferozepore, to Mooltan, where it arrived on December 5 and camped near the barracks, one company occupying the fort in relief of a detachment of the 1st Battalion of the 18th Regiment.

On December 7 Major Galbraith was appointed Deputy Assistant Adjutant-General to Kurrum Field Force under Major-General Sir F. Roberts, V.C., and Captain (local Major) G. W. Smith, Deputy Assistant Adjutant-General to 1st Division, Peshawar Field Force, under Lieut.-General Sir Samuel Browne, V.C. On March 23, 1879, Major Beadon having proceeded to England on sick leave, Captain Grant assumed command. On April 6 the depot and families rejoined from Lucknow and the regiment settled down in cantonments, leave being opened shortly afterwards.

By the treaty of Gundamuk, signed May 24, Amir Yakub Khan besides giving up the Khyber, Kurrum, and Pishin Valleys, consented to the establishment of a British Envoy at Cabul, and shortly afterwards Sir Louis Cavagnari, with a small staff and escort of the 'Guides,' under Lieutenant Hamilton of the 'Guides,' proceeded to Cabul and resided there until September 3, when they were massacred.

On receipt of the news in India the war was renewed and the force in the Kurrum Valley was at once despatched under Sir F. Roberts, V.C., across the Shutargardan pass against Cabul.

The regiment again received orders to hold itself in readiness to proceed on immediate service in Afghanistan and all officers were recalled from leave. 'D' Company rejoined from the Fort, and on September 23 and 24 the regiment, under command of Captain Grant, left Mooltan in two wings by train for Jhelum, halting a day at Mian Meer, and reached Jhelum on 25th and 26th, where it received over field-camp equipage and transport. Strength: 25 officers, 805 non-commissioned officers and men.

A depot under Lieutenant Vivian of 57 non-commissioned officers and men, with families and heavy baggage, remained at Mooltan until November 14, when it moved to Jullundur on being relieved by the 88th Regiment.

September 28, the regiment marched from Jhelum via Rawul Pindi to Kohat, where it arrived on October 13, and was posted to the 3rd Infantry Brigade, Kurrum Division, Cabul Field Force, under Brigadier-General Fraser-Tytler, V.C., C.B.

Leaving Kohat on the 15th, it reached Thull on the 20th and went on to Kurrum, where it remained encamped for a month, while preparations were being made for a punitive expedition against the Zaimukht tribes, who inhabit the valleys north of the road from Thull to Kurrum, as they had recently committed several murders, including those of Lieutenant Kinloch, 92nd Regiment, and Dr. Smyth.

On October 29 three companies of the 85th, with Captains the Hon. E. à Court, Robinson, Ives, Lieutenants Bulman, Thompson, and Surgeon-Major Boyd, left Kurrum for Balish Kheyl to strengthen the force at that place, Captain Plowden (political officer) arriving the same day. On the 31st a large number of the tribesmen assembled on the hills and came down on to the plateau. In consequence a force of all arms at once moved out and the guns opened fire. The tribesmen then fell back to the hills.

On November 20 the headquarters of the 85th and five companies with four guns of the 1/8th Mountain Battery, arrived at Balish Kheyl from Kurrum, and were followed the next day by Brigadier-General Fraser-Tytler, V.C., C.B., and his staff.

On the 28th, 80 men of the 13th Bengal Lancers with 150 of the 85th King's Light Infantry under General Fraser-Tytler reconnoitred along the plateau for about six miles; and while the infantry escorted Mr. Scott (Indian Survey) to the top of Kulgarda (5700 ft.) for survey work, the cavalry pushed on and found several deserted villages; but others, across the river called the Gowaki, were occupied, the only resistance being that a few shots were fired by the tribesmen.

On December 1 two more reconnaissances were sent out.

The first, under Colonel Gordon, 29th Native Infantry, consisting of 2 mountain guns, 150 of the 85th King's Light Infantry, 50 of the 1st battalion of the 8th Regiment and 300 of the 29th Native Infantry, started at 4 A.M. for the Drab Mountains, seven miles off, and then ascended to the crest (7300 ft.). The guns and mules with a part of the infantry stopped half-way up, an escort of the 85th going to the top with Mı. Scott for survey work. This party got back to camp at 7.30, not being interfered with.

The second, under Colonel Low, 13th Bengal Cavalry (strength—100 cavalry, 2 guns, 200 of the 85th King's Light Infantry and 200 of the 20th Native Infantry), proceeded as far as the Krun defile, north of the Gowaki River, and got back about 5.30, having seen no enemy.

On December 3 a force consisting of 2 guns, 50 cavalry, 200 of the 85th and 200 of the 20th Native Infantry under Colonel Rogers (20th Native Infantry), with Majors Plowden and Connolly (political officers) and Mr. Scott, left camp at 4 A.M. to reconnoitre the Abbansikor Pass and reached the village of Tatung (5700 ft.) at 8.30 A.M., nine and a half miles from camp. Leaving the cavalry, artillery, and part of the infantry here, and occupying the heights with picquets, the rest of the force began to ascend the pass for two and a half miles.

The political officers and Mr. Scott then went on with an escort of forty of the 85th King's Light Infantry, and after a steep and rocky ascent for one and a half miles reached the top (7700 ft.), whence extensive views of the Massozai country and Lankai Pass were obtained. Just as

Mr. Scott finished his survey work, the Massozai began to assemble and opened fire at long range; so the party withdrew, and on reaching Tatung the whole force returned unmolested to camp, arriving at 9 P.M., having covered over thirty miles and successfully carried out the work required.

By December 7, sufficient supplies and transport having been collected, the Brigadier-General issued orders for the expedition to start.

The force was made up of 4 guns 1st Mountain Battery 8th Brigade R.A.; 2 guns No. 1 Kohat Mountain Battery; 85th King's Light Infantry (strength—21 officers and 702 men); 250 native cavalry, and 4 native infantry regiments (20th Native Infantry, 4th Punjab Infantry, 29th Native Infantry and 13th Native Infantry), and one company of Native Sappers and Miners.

Starting early on the 8th, the force reached Gowaki, and next day proceeded to Manatoo, without opposition.

Early on the 10th three separate columns moved out to punish the Watizai Zaimukhts.

Colonel Low, 13th Bengal Cavalry, with detachments of the 1st and 13th Bengal cavalry, 1/8 Mountain Battery, 85th King's Light Infantry, and some native infantry, reconnoitred as far as Kundela village, which was found to be deserted. The cavalry then went on and found that the enemy were occupying the rocks on their left. The infantry and guns were then brought up; a guard was left at Kundela, while the 85th men and Royal Artillery guns advanced on Tura village and opened fire on the Zaimukhts, who had fallen back and occupied rocky ground beyond the village; meanwhile a party went forward and destroyed the stores of grain, rice, bhoosa, etc., and set fire to the houses. Cuttack Mela was also destroyed. The troops bivouacked at Kundela for the night and destroyed it together with quantities of grain, etc., next morning, also capturing a number of cattle. They then marched back to Zaitunak, to which place General Fraser-Tytler had moved.

Colonel Gordon's force (2 guns No. 1 Mountain Battery, 2 companies 85th and part of 4th Punjab Infantry) started at 6.30 A.M. up a defile which got so steep that the transport mules had to be sent back, the blankets, etc., being sent on later by kahars under an escort. The battery mules got to the crest with difficulty, though the ground looked impossible for mules. A covering party under Lieutenant Renny

(4th Punjab Infantry) now occupied Dresola Peak (7700 ft.) during the advance on the Zawaki village, which was found to be deserted. The troops bivouacked there for the night, and next morning, after destroying the village and stores of grain, etc., in it, proceeded to Mela, which had been occupied the previous night by Colonel Rogers' force without opposition. This village also was destroyed, and the troops marched back to Zaitunak with several head men of the villages as prisoners. The headquarters camp was reached in the evening, the last three miles over boulders in the bed of a stream in a nullah being done in the dark.

On the 12th the whole force marched to Chenarak, destroying two villages *en route*. During this night the camp was fired into.

Before daylight on the 13th General Fraser-Tytler moved out of camp against Zawa, the Zaimukht stronghold, which was situated in a mountain pass, the rocky precipitous sides of which were intersected with ravines.

The force consisted of 50 Bengal Lancers, 4 mountain guns, and 1350 infantry, viz. 14 officers and 457 men of the 85th King's Light Infantry and detachments of the 4th Punjab Infantry, 13th, 20th, and 29th Native Infantry.

Advancing for about three miles along an open valley, it reached Raga village, a little beyond which the valley narrowed into a defile with precipitous hills on either side.

The force was now divided into two columns. The right column under Colonel Gordon with 700 infantry (viz. four companies of the 85th under Captain Grant and a detachment of the 29th Native Infantry with two guns of the Kohat Mountain Battery) were ordered to ascend the ridge and clear the heights of the enemy on the right of the advance.

General Fraser-Tytler then sent three companies of the 4th Punjab Infantry to occupy a high hill on the left which commanded the defile, up which he then advanced slowly along the rocky bed of the stream.

The detachment of the 85th with this column was with the rest of the force commanded by Captain the Hon. E. à Court.

The two guns 1/8 Royal Artillery took up a position on a ridge to the right of Raga and opened fire on the enemy, which held the crest of the next ridge, while the 85th under Captain Grant advanced in skirmishing order up the steep ascent. On reaching the crest, the Zaimukhts were seen falling back along the ridge to a second position on a high spur which had been strengthened by stone breastworks (*sungars*). From this they opened fire on the 85th who, unchecked, followed them up rapidly and soon dislodged them, compelling them to fall back on the next ridge, the only frontal approach to which was up a precipitous rocky cliff 150 feet high, down which the enemy hurled rocks and kept up a fire with their *jezails* (long native rifles).

The 85th halted here to give the 29th time to move round for a flank attack.

Going down a steep descent and crossing a gorge, the 29th Native Infantry ascended the spur held by the enemy. Rocks were rolled down on them from a sungar; but, pushing on, this was captured, and approaching the main position, fire was opened at close range.

At the same time the No. 1 Mountain Battery guns opened fire from a lower ridge, having gone round by the defile, and simultaneously the 85th succeeded in getting to the top of the cliff and the Zaimukhts fled in all directions, pursued by the 85th. This strong position was captured with no loss except one jemadar, 29th Native Infantry, severely wounded. The enemy's loss was about twenty.

As this crest commanded the whole defile the 85th occupied it, bivouacking for the night, and suffered greatly from thirst. They could get no water as the bhisties (water carriers) sent from Bagh emptied their mussochs on the way up the mountain owing to the steep ascent. The 29th Native Infantry and the Mountain Battery held a lower ridge.

The left column reached Bagh in the afternoon, and on arriving there was fired on by the enemy from a hill on the left. This was shelled by the guns 1/8 Royal Artillery; and a company of the 20th Native Infantry advancing the enemy retired and the hill was occupied for the night. General Fraser-Tytler now bivouacked at Bagh, picquets being posted.

Early on the 14th General Fraser-Tytler continued his advance up the defile, the advance guard being 'F' Company of the 85th, under Captain Robinson, with Lieutenant Thompson, and three companies of the 4th Punjab Infantry, Colonel Close commanding.

On the right Colonel Gordon's force held the heights gained on the previous day and sent forward the two mountain guns, which had now reached the crest, escorted by the 85th. The ascent was by means of a deep gorge. The guns opened an effective fire on a large body of the enemy moving on Zawa and checked their advance. The other two companies of the 85th were sent down to join the force in the defile. The heights on the left were crowned by two companies of the 20th Native Infantry, while the guns of the 1/8 Royal Artillery shelled the enemy, who kept up a sharp fire as they fell back from ridge to ridge. After advancing about two miles the 4th Punjab Infantry, except a half company under Lieutenant Renny, were ordered up the hill on the left. They advanced along the slopes, and shortly afterwards a company of the 20th Native Infantry was sent to the right to dislodge some of the enemy there. The firing was now brisk in front and on both flanks. A little later the men of the 4th Punjab Infantry under Lieutenant Renny, leaving the river bed, ascended a narrow path in single file under a heavy fire and showers of rocks, hurled down from the heights above.

Here they occupied a small rocky spur at the bend of the path. 'F' Company of the 85th followed them. It was while fighting at this corner that Lieutenant Renny and two of his men were dangerously wounded. Here too Captain Burrell of the 85th (the Deputy Assistant Quartermaster-General) had a narrow escape, a bullet passing through all his clothes and shirt, beneath his armpit, without cutting the skin. Captain the Hon. E. à Court now came up with an order from General Fraser-Tytler. In consequence, 'F' Company of the 85th rushed forward followed by Renny's men across a short ravine, and reaching the next ridge, which sloped down to Zawa, came on a large party of Zaimukhts at close quarters and opened fire as these fled. Killing several, the company followed for some distance across the fields and then formed up, being joined by another company of the 85th, and held some rising ground while the villages were destroyed by parties of other corps.

The troops were then ordered to retire, passing along the stream through a very narrow gorge, the rocky cliffs on either side almost meeting above. Bagh was reached about 5 P.M., the force being unmolested, while villages were destroyed on the way. Here the troops bivouacked for the night, Colonel Gordon's force and the 20th Native Infantry holding the ridges above as on the 13th.

Next morning the troops returned to Chenarak, Gordon's column rejoining at Raga. Lieutenant Renny died at Chenarak the same day.

The casualties were Lieutenant Renny and one sepoy mortally

wounded, one havildar seriously wounded. The enemy lost about 150 men.

On the 16th Major Galbraith rejoined the staff at Nawakila from Cabul and assumed command, and on the 23rd the force returned to Thull; three Zaimukts being shot at Sangroba in the morning for the murders of Lieutenant Kinloch and a native Syce. The Government programme had been successfully carried out and the Zaimukhts severely punished. Besides having twenty villages burnt, and their defensive towers destroyed in the villages that were spared, they had to pay a fine of 25,000 rupees and give up over 1000 stand of arms.

Their country, never before traversed by a European, had been thoroughly surveyed, and the effect on the border tribes was most salutary.

The operations, it may be added, were throughout conducted with consummate skill by Brigadier-General Fraser-Tytler. As a termination to the expedition, on January 1, 1880, a durbar was held in three large tents. The whole of the troops were drawn up with all the Jirghas, Durbarees and inhabitants of Thull and the neighbouring villages seated in a semicircle in rear of them. On the arrival of General Fraser-Tytler and his staff they sat in the centre of the durbar tents with the officers and all the friendly khans on either side. The latter were then presented and their services mentioned.

The Jirghas then came forward and paid their fines, laying them in their shields in a row in front of General Fraser-Tytler, who afterwards addressed the friendly khans, saying that their services would be rewarded. At night the dâk bungalow and surrounding fortifications were illuminated with native lamps and around a large fire about forty Cuttacks, wearing loose short-skirted coats and very full pyjamas, danced their war-dance with swords in one hand and scabbards in the other and flowing puggarees. They rushed round the fire, span round, cutting and pointing at it or at each other, then crouching down and springing into their places in the circle. The Afridi pipes and tom-toms were meanwhile playing a very noisy accompaniment. It was altogether a wonderful and weird performance.

On returning to Thull a large number of the men were practically bootless owing to the rough work they had done. Many too went into hospital, there being several fatal cases of pneumonia. The officers all occupied one sepoy pāl during the expedition, and ate their Christmas dinner in it, sitting on the ground; Captain Ravenhill, the mess president, providing an excellent repast and champagne. This had been left at Thull when the regiment first came up from India.

On December 26, 1879, four companies under Captain Grant proceeded to Balish Kheyl, and on January 7, 1880, returned to Chuppri for road making and were joined on January 14 by the headquarters and three companies. 'F' Company followed on March 31, and next day the regiment left for Kurrum, arriving there on April 5. Here they were attached to the Upper Kurrum Brigade under Brigadier-General Gordon. At the end of February, to the regret of all who had served under him, Brigadier-General Fraser-Tytler, V.C., C.B., died of pneumonia at Thull, brought on by the exposure and hardships of the late expedition.

From Kurrum the regiment went to Shalozan, forming part of a flying column, and while here Major Beadon rejoined from sick leave and assumed command.

Shortly afterwards the headquarters and five companies proceeded to the Peiwar Kotal, where they encamped in a pine forest above the Kotal Pass and left detachments at Thull, Shalozan and Kurrum.

While here large quantities of commissariat stores had been collected for the troops returning from Cabul. After the battle of Maiwand, however, plans were changed and all stores sent down by degrees to Thull, and eventually the Peiwar Kotal was handed over to Afghan officials and the troops returned to Thull, the regiment arriving there on October 21.

Ever since the commencement of hostilities the country round Thull had been unsafe through the constant attacks of the Waziri and, to punish them, on the night of October 27 (the orders only being issued two hours before), a force 1000 strong, composed of cavalry and of the 85th (about 400), 13th and 20th Native Infantry and the 8th Brigade R.A. (two guns) under General Gordon, with the political officer Major Plowden, made a raid by a night march through Cabul Kheyl in the Waziri country, a distance of some seventeen miles. They crossed the Kurrum River three times. Their objective was the Mauk Shahee settlements, which were simultaneously surrounded and surprised at dawn on the 28th at three different points, the forces being divided into three parties; 126 prisoners and 1500 head of cattle and 700 loads of fodder were brought. One Waziri was killed and two wounded in self-defence. The troops were on the move twenty-three hours. This successful incursion brought about the prompt payment of a heavy fine imposed on the Mahsud Waziris for offences committed during the last two years round Thull.

Shortly after this the headquarters and four companies moved to Doaba, one company to Hangu and thence to Kohat. During 1880 there were thirty-three deaths. The undermentioned promotions were made in recognition of services in the Afghan Campaign, 1879–1880.

To be Brevet-Lieut.-Colonel: Major W. Galbraith, November 22, 1879.

To be Brevet-Major: Captain G. W. Smith, November 22, 1879.

To be Brevet-Major: Captain the Hon. C. Dutton, March 2, 1881.

To be Brevet-Major: Captain D. A. Grant, March 2, 1881.

It may here be mentioned that in 1883 a memorial was erected in St. Mary's Church, Shrewsbury, to the memory of the men who lost their lives in this campaign, the cost being defrayed by voluntary subscriptions from the officers, non-commissioned officers and men of the regiment.

In 1881 the regiment was authorised to bear 'Afghanistan, 1879–80' on its colours and appointments.

On June 21, 1880, Major E. M. Beadon was promoted to command the regiment *vice* Brevet-Colonel F. L. Appleyard, C.B.

February 3, 1881. The regiment arrived at Rawal Pindi and thence proceeded to Jellundur.

Shortly after orders were received to be in readiness to proceed to the Cape.

On March 1 orders were definitely received for service in Natal.

March 9. The regiment embarked on H.M.S. *Euphrates* (20 officers, 721 non-commissioned officers and men), arrived at Durban April 1 and proceeded by train in open trucks to Pinetown, where it encamped with the 7th Hussars and 41st Regiment. Later, one company was sent on detachment to Pietermaritzburg. The regiment camped in a hayfield, which was gradually cut by the men with knives, and several cricket matches were played during the months the regiment was in Natal.

A draft of the 52nd arrived from England on April 8, consisting of four non-commissioned officers and eighty privates, accompanied by Majors White and Smith, Captains Knox, Robinson, and Welman, and Lieutenant Campbell.

On July 1, 1881, the regiment was transferred from the 42nd Brigade to the 53rd Regimental District and with the 53rd Regiment and Shropshire and Herefordshire Militia formed 'The Shropshire Regiment, The King's Light Infantry,' the 85th being the 2nd Battalion.

This title was changed on March 10, 1882, to 'The King's (Shropshire Light Infantry),' the official title of the battalion to be '2nd Battalion, Shropshire Light Infantry.'

In June, 1881, valise equipment was issued in place of knapsacks.

November 7, 1881, the regiment marched for Durban and proceeded to England on the transport *Balmoral Castle.*

The embarkation was a somewhat lengthy business owing to the fact that there was such a heavy swell. In consequence it became needful to haul the men three at a time only on board in a basket.

The *Balmoral Castle* touched on the voyage at Port Elizabeth, Cape Town and Madeira.

The regiment landed at Cowes, Isle of Wight, on December 6, 1881, and marched to Parkhurst Barracks, after thirteen years and eleven months' foreign service.

The strength of the regiment on landing was 24 officers and 724 non-commissioned officers and men.

CHAPTER XV

HOME SERVICE, 1881–1899

ON December 7, 1881, the regiment was inspected by Major-General His Serene Highness Prince Edward of Saxe-Weimar, who expressed great satisfaction at the appearance of the regiment.

December 16, 1881. A detachment of the regiment under Captain G. N. Atkinson marched to Osborne to form a guard during the stay of Her Majesty Queen Victoria, a similar detachment under Major D. A. Grant performing this service in July, 1882.

For the year 1881 there are no other events to record. The strength on December 31 was as follows: 28 officers, 42 sergeants, 40 corporals, 16 buglers, and 587 privates.

On March 11, 1882, the battalion proceeded to Southsea, and, with other battalions in the command who had served in Afghanistan 1879–80, was inspected by His Royal Highness the Duke of Cambridge.

The other battalions present were the first battalion of the 17th Regiment, the second battalion of the 60th, and the 92nd.

June 9, 1882. Mounted buttons were approved in place of stamped ones for officers.

Gold bugles to be worn on collar of mess-jacket

Badges. Title on helmet plate and sword belt to be changed from 'Shropshire Regiment' to 'Shropshire Light Infantry.'

The following movements of the regiment took place:

On September 13, 1882, the regiment moved to Aldershot and on May 15, 1884, to Woolwich.

June 21, 1885, Lieut.-Colonel and Colonel N. Ximmines Gwynne was appointed to command the battalion *vice* Beadon.

December 8, 1885. The following officers were promoted in

recognition of their services, viz.: Major Charles Edmond Knox to be Lieutenant-Colonel for services in Bechunaland Expedition and Captain William Baume Capper to be Major for services in the Soudan.

October 7, 1886. The battalion proceeded in H.M.S. *Assistance* from Woolwich to Ireland and was quartered at the Curragh.

July 1, 1887. Lieut.-Colonel Guy N. Atkinson succeeded to the command of the battalion.

January 19, 1888. The battalion proceeded to Kilkenny, finding detachments at Waterford, Wexford, and Duncannon Fort.

March 7, 1890, an annuity of £10 together with a silver medal was granted by Her Majesty to Sergeant John McMahon as a reward for his long and highly meritorious services, including the Afghan war of 1879 and 1880 and with the Natal Field Force 1881.

February 10, 1890. Lieut.-Colonel G. N. Atkinson died while in command of the battalion. The following orders were published on the occasion:

'It is with deep sorrow and regret that the 2nd in Command has to announce to the Batt[n] the death of their Commanding Officer Lieut.-Col. G. N. Atkinson. In him the Reg[t] has lost the kindest and best of Commanding Officers, always anxious to see those under him, Officer or Private, happy and contented.

'Serving as he did for 26 years in the Reg[t], he loved the old corps, and his greatest ambition was attained when he was appointed to command it.

'Colonel Knox feels sure all ranks will join with him in deep sympathy for his widow and family.'

'*District Order:*

CORK, 12 *February*, 1890.

'It is with great regret that Major-General Davies Comd[g] Cork Dist. has to announce the death of Lieut.-Col. G. N. Atkinson, late Comd[g] 2 Batt[n] King's Shropshire Light Infantry, which occurred on Monday last.

'By Lt.-Col. Atkinson's death Her Majesty loses a good officer and the Shropshire Light Inf[y] an excellent Comd[g] Officer who was respected and beloved by all ranks.

'The Major-General feels sure that he is only expressing the feelings of all under his command when he offers his condolence to the Shropshire Light Inf[y] on the great loss the Reg[t] has sustained and tenders the assurance of his sincerest sympathy with Mrs. Atkinson in her deep affliction.'

On February 11, 1890, Colonel C. E. Knox succeeded to the command.

January 26, 1891. The battalion moved from Kilkenny to Cork, finding a detachment at Charles Fort, Kinsale, and later at Fort Carlisle and Youghal.

February 11, 1894. Lieut.-Colonel P. H. Murray succeeded Colonel C. E. Knox.

May 26, 1894. The battalion moved, in H.M.S. *Tyne*, from Cork to Portland.

November 27, 1895. A party of 25 non-commissioned officers and men under Captain R. N. R. Reade proceeded to join a Special Service Corps for service on the Gold Coast (Ashanti Expedition).

The following letter was received from Lieut.-Colonel F. W. Stopford commanding the Special Service Corps during the expedition :

'S.S. *Coromandel*.

'SIR,

'I have the honour to bring to your notice the very satisfactory way in which the Detachment from the Batt[n] under your command performed its duties whilst forming a portion of the Special Service Corps for service in Ashanti.

'The conduct of the men was excellent throughout and the greatest credit is due to all ranks for the soldierlike spirit shewn during a trying march in an unhealthy climate. When Serg[t] Smith was taken ill I made Corporal Connor a Lance-Serg[t] and he carried out his duties as senior N.C.O. of the Detachment very satisfactorily. 3160 P[te] Morris and Perkins whom I appointed Lance-Corp[ls] performed their duties very well. I have, etc.

(Signed) 'FRED W. STOPFORD, *Lt.-Col.*,
'*Commanding Special Service Corps.*'

The following formed the detachment :

Captain R. N. R. Reade
* 1791 Sergeant H. Smith
2778 Corporal W. H. Martin
1136 ,, J. Connor
2767 Private G. Morris
2743 ,, J. Lowe
4108 ,, R. Blannin
3164 ,, A. Landon
1467 ,, D. Griffiths
3148 ,, W. Perkins
1801 ,, A. Williams
* 2999 ,, W. Anthony
* 2982 ,, F. Page
* 2970 Private A. Hodges
3073 ,, J. Parker
3117 ,, D. Davies
1505 ,, W. Meech
2138 ,, J. Wheatley
3272 ,, W. Reynolds
3160 ,, J. Morris
3240 ,, E. Cook
3255 ,, H. Hanson
3109 ,, J. W. Cobb
2500 ,, C. Ward
3344 ,, E. Pearce
1176 ,, S. Lewis

The above (with the exception of those marked with an *) received the Ashanti Star, 1896.

October 3, 1896. The battalion moved from Portland to the Portsdown Hill Forts.

June, 1897. For active and distinguished service before the enemy during the Dongola Expedition Captain James Ross O'Connell received the 4th class of the Imperial Order of the Medjidie.

June 22, 1897. The battalion was present in London and took part in Her Majesty Queen Victoria's Diamond Jubilee celebrations, lining the Westminster Bridge Road. The following officers were present on the occasion :

Lieut.-Colonel P. H. Murray (*in command*)
Major J. Spens
Captain R. N. R. Reade
,, A. H. J. Doyle
,, A. R. Austen
,, S. J. Judge, D.S.O.
,, R. C. Mounsey-Heysham
,, J. H. Hicks
,, J. C. Forbes
,, C. P. Higginson
Lieutenant J. J. White
Lieutenant F. H. Mackenzie
,, R. R. Gubbins
,, C. Marshall
,, H. M. Smith
,, W. C. Wright
,, E. M. Sprot
2nd Lieutenant J. H. Bailey
,, C. W. Battye
,, H. G. Bryant
,, C. E. Atcheson
,, E. R. M. English

The regiment camped for the occasion in the grounds of Lambeth Palace and after the ceremony returned there for refreshments.

The following special army order was issued June 23, 1897:

'The Queen desires to express her gratification at the appearance upon this occasion of all the forces, including the Blue Jackets, Marines, Militia, Yeomanry and Volunteers as well as of the Colonial and Indian contingents whose presence was an additional source of pride and satisfaction to the Queen Empress.'

From August 28 to September 6, 1897. The battalion took part in the South-Eastern District manœuvres under Major-General Sir William Butler. Only one man fell out during this period, during which the weather was most trying, and a special order was published on the occasion complimenting the troops on their admirable spirit.

September 11, 1897. The battalion moved into Victoria Barracks, Portsmouth.

December 14, 1897. The battalion was inspected by the Colonel of the regiment, General Sir Henry H. P. de Bathe, Bart., on which occasion he presented a gold watch to 4429 Private L. Taylor, the best shot of the year.

February 11, 1898. The battalion furnished a guard of honour on the occasion of Her Majesty the Queen's visit to Netley Hospital under Captain O. H. E. Marescaux with Lieutenant R. R. Gubbins and Second Lieutenant J. H. Bailey, a complimentary letter being afterwards received from the General Officer Commanding on the smart appearance and good physique of the guard.

February 11, 1898. Lieutenant-Colonel James Spens succeeded to the command.

November 27, 1898. Royal Humane Society certificates were presented to 4059 Private P. Ashdown and 4953 Private H. Busby on a garrison parade by Major-General R. M. Stewart, C.B., for having on August 16 gone to the rescue of Private J. Hamer, who was drowning.

August 5, 1899. By special command the band of the regiment under Mr. James Forrest played a selection of music during and after the Queen's dinner party at Osborne, at which she was pleased to express her gratification through Sir John McNeil.

While at Portsmouth the regiment was ordered to hold itself in readiness for South Africa.

A move to Aldershot had already been ordered and it was expected that this would be cancelled.

The move was, however, carried out on September 14, 1899, and the regiment on arrival occupied the Corunna barracks, Stanhope Lines, and formed part of the 1st Infantry Brigade.

CHAPTER XVI

THE SOUTH AFRICAN WAR TO THE CAPTURE OF PRETORIA

THE South African War now broke out and the regiment was detailed as one of the battalions to hold the lines of communication. Accordingly the reserves were called up and considerable enthusiasm took place in Shrewsbury, where they assembled before being forwarded to Aldershot.

They turned up extremely well, 408 joining out of the 409 summoned to rejoin the colours, and of these only twenty-one were medically unfit. The reservists took the place of about 300 men of the battalion who were ineligible either from youth or length of service. The latter were formed into a provisional battalion and later on supplied drafts to replace casualties. On October 31, 1899, orders were received to embark on the *Arawa* on November 7. The strength of the regiment on embarkation was as follows: 29 officers, 1 warrant officer, 50 sergeants, 54 corporals, 786 privates, 16 buglers—a total of 936.

The following officers proceeded with the regiment to South Africa:

Lieut.-Colonel J. Spens (*in command*)
Major P. Bulman (*second in command*)
,, C. T. Dawkins
Captain F. L. Banon
,, E. W. K. Money
,, R. A. Smith
,, T. G. Forbes
,, J. J. White
,, R. R. Gubbins

Lieutenant G. R. Sowray
,, W. C. Wright
,, H. M. Smith
,, H. G. Bryant
,, C. E. Atcheson
,, E. A. Underwood
,, H. W. Kettlewell
2nd Lieutenant P. F. Fitzgerald
,, E. P. Dorrien Smith
,, R. C. Middleton
,, T. T. Simpson

2nd Lieutenant R. P. Miles	*Adjutant:*	Captain C. P. Higginson
,, T. M. Carter	*Quartermaster:*	Hon. Lieutenant J. Forrest
,, P. L. Hanbury		
,, J. C. Hooper	*Medical officer:*	Lieutenant E. P. Connelly (attached)
,, P. R. C. Groves		

At the last moment two companies under Captain R. R. Gubbins were ordered to proceed to the Royal Albert Docks, London, and embark on the s.s. *Chicago*, the accommodation on the *Arawa* not being sufficient.

The departure of the *Arawa* was delayed by the inferiority of the meat, which was condemned, and also by the breakdown of the electric lighting plant on board. However, on December 1, 1899, both ships arrived in Table Bay, and the regiment disembarked at Capetown the following day, and proceeded by train to Orange River Bridge, leaving Second-Lieutenant T. T. Simpson and a small party at Cape Town to form a base. It arrived at Orange River Bridge on December 4, the journey from Cape Town taking fifty hours. The train stopped periodically for the men to cook their meals and at times went so slowly up the steep inclines that the men could get out and stretch their legs on the veldt and amuse themselves by catching lizards and beetles. The inhabitants on the Karroo were most hospitable and gave the men baskets of fruit and other provisions and a hearty welcome.

Orange River Bridge, where the regiment was destined to remain for several weeks, was a most important post on the lines of communication. The bridge carried across the river the only railway which was in our possession and by which entry could be gained into the Orange Free State and the Transvaal. This bridge was eventually used by Lord Roberts for the passage of the whole of his army and at the time was the only means of communication with Lord Methuen's force which had just fought the battle at Modder River.

The outpost duty in consequence was very heavy; this was taken over from the 1st Gordon Highlanders, who proceeded to join Lord Methuen's force at Modder River on December 8.

The tedium of this duty was, however, considerably relieved by the opportunities for sport; partridges, guinea-fowl, k'oran (a small species of bustard), hares and small buck being fairly numerous. Cartridges, however, were scarce. Fishing in the Orange River was also a source of amusement, if not productive of much addition to the larder.

On December 6 our transport animals arrived from De Aar, five ponies and 125 mules, the latter tied together by the neck in bunches of five—and most of them being only half trained, their education by their Kaffir drivers was a source of constant amusement.

The men not employed on the outposts were fully occupied in fatigue work and in making extra railway sidings. These additional sidings greatly relieved the pressure on the railway. Lifting the sun-baked rails, however, was not the pleasantest of duties, but it had to be done. On December 10 a heavy downpour of rain practically swept away our camp, which was pitched on a level sandy spot on the Karroo surrounded by rough scrub, ant heaps and ant-bear holes; but, alas! it proved in wet weather to be the bed of a river!

The drinking water, drawn from the Orange River, was of a dark yellow colour, and required stirring with alum to make the sand sink to the bottom.

The officers on arriving in the country were armed with a sword and revolver and wore a Sam Browne belt but these were soon withdrawn and they were armed with a rifle and equipment the same as the men.

Supply trains were forwarded daily to Lord Methuen's force, and Orange River station was a busy place with reinforcements including Australian and Canadian contingents continually arriving, and wounded, sick and prisoners trickling to the rear.

On December 8 the line connecting us with the Modder River force was blown up by the Boers, but soon repaired; and on the 11th the attack by Lord Methuen on the Magersfontein position took place. But news was always hard to get, the passengers on the trains of wounded being for the most part our only informants, and it was necessary to wait for the English mails before accurate information was obtained, although Lord Methuen's force was only a few miles north of us.

On Christmas Day the following message from Her Majesty the Queen to the Commander-in-Chief was communicated to the troops:

'I wish you and all my brave soldiers a happy Christmas. God protect and bless you all.

(Signed) 'VICTORIA R. & I.'

And on New Year's Day :

'I wish you all a bright and happy New Year. God bless you all.
(Signed) 'VICTORIA R. & I.'

On January 6 three companies of the regiment under Lieutenant-Colonel T. Spens with two companies of the Duke of Cornwall's Light Infantry, a squadron of the Greys, a party of mounted infantry and the 1st Battery Royal Horse Artillery and 2nd Howitzers, marched at 3.45 A.M. and occupied Zoutpans Drift to safeguard our right flank. This was the first force to actually occupy the enemy's country.

Our left flank had been previously safeguarded by blowing up the road bridge over the Orange River at Hopetown, at which place Major C. T. Dawkins proclaimed martial law on January 17.

On January 20 a draft of 150 men joined from England with Lieutenant G. R. M. English and Second-Lieutenant Delmé-Murray.

On January 24 a New Year's gift from Her Majesty was presented to all ranks consisting of a tin box of chocolate with Her Majesty's portrait and cypher and 'I wish you a happy New Year' in red, blue and gold picked out on the lid. These boxes were highly valued, and for the most part were sent home by the men to their relations for safe keeping, after the contents had been consumed by them.

On the same day instructions were received for the battalion to furnish 103 specially selected men to form a company of Mounted Infantry. 'D' Company was in the main selected for this; and under Captain J. J. White with Lieutenant H. M. Smith and Second-Lieutenants P. F. Fitzgerald and J. C. Hooper they proceeded to join the 4th battalion Mounted Infantry at De Aar. This company was augmented on February 9 by Second-Lieutenant E. A. Underwood and twenty-six non-commissioned officers and men.

Their proceedings during the war are chronicled later.

Towards the end of January, Orange River station became a very busy spot. Large reinforcements arrived on the scene and generals with their staffs succeeded one another in rapid succession. Major-General C. E. Knox and his aide-de-camp (Captain O. H. E. Marescaux) receiving a warm welcome and send-off from the regiment as they passed through with their brigade. Lieut.-General Kelly-Kenny, Lieut.-General

Charles Tucker, and later Lord Roberts and Lord Kitchener came and went.

The traffic over the bridge became very congested. The bridge, be it understood, was a single-line railway bridge, 445 yards long ; and the whole of Lord Roberts' army with its miles of transport, sixteen oxen or twelve mules to each waggon, together with cavalry, artillery, and mounted infantry, had to pass across this narrow defile. The approaches on both sides were bad. Heavy sand, overloaded waggons, untrained mules, occasional trains passing across, continually delayed matters. There was nothing between the passengers crossing over the bridge and a drop of many feet into the Orange River but a few inches of board, temporarily placed for lighter traffic between the iron girders and rails. However, assisted by the cheery voice of General Charles Tucker, all were safely passed over with but few accidents. A few broken poles and a waggon with its team of sixteen oxen run down by a train to the accompaniment of a certain amount of expressive language summed up the mishaps, and Lord Roberts' army was safely on its way to Paardeburg.

The 85th looked with somewhat jealous eyes at regiments who had only lately landed in the country moving up to the front ; but a welcome order was received on February 5, 1900, that we, together with the 2nd Duke of Cornwall's Light Infantry, the 1st battalion of Gordon Highlanders and the Royal Canadian Regiment, were to be formed into the 19th Brigade under Brigadier-General H. L. Smith-Dorrien, who up to then had commanded the Sherwood Foresters, and that our brigade, with the Highland Brigade, was to form the 9th Division, commanded by Lieut.-General Sir H. Colville, K.C.B.

Amongst the troops who passed across Orange River at this time were our company of Mounted Infantry under Captain J. J. White, who had joined the 4th Mounted Infantry Battalion, and also a party of Mounted Infantry under Lieutenant C. W. Battye, who had joined the 1st Mounted Infantry at Aldershot under Colonel Alderson and had landed in South Africa before the regiment and taken part in the unfortunate affair at Stormberg under General Gatacre.

At last, on February 11, our outposts were withdrawn and taken over by Nesbitt's Horse, a newly raised Colonial corps, and we started at 2 A.M. February 12 by train to Graspan, being followed by our three companies from Zoutpans Drift. On arrival at Graspan we found the remainder of our brigade had already started for Ramdam, and at 4 P.M. on 13th we followed and joined them at the latter place at 10 P.M. The following day (February 13, 1900) we marched at 5 A.M. with the division to Watervall Drift on the Riet River, an exceedingly hot and trying march, the regiments marching in brigade and the battalions in column. This formation would be suitable on a flat and level plain with no change of direction; but marching, as we did, over rolling country and swinging to the right and left, with the weather excessively hot and water scarce, the troops suffered considerably.

We reached Riet River at 5 P.M., where Lord Roberts visited our camp, and at 4 A.M., February 15, marched and bivouacked 3 miles west of Jacobstal, at Wegdraai. From here a column was sent back to our last camp to assist a convoy of 200 waggons which had been captured by the Boers, but time being precious the convoy was abandoned and we were placed on half rations.

On the 16th we marched to Jacobsdal, where we heard the news of General Cronje's abandonment of his position at Magersfontein and his flight across our front—all speed had then to be made to capture his force. We marched all that night as rearguard to the division, and at dawn on February 17 arrived at Klip Drift, where we bivouacked in a sheep-kraal.

Two companies (Major A. R. Austen's and Captain F. L. Banon's) left our last bivouac for Modder River as escort to a telegraph company of Royal Engineers. The remainder of the regiment continued its march at 4 P.M. and arrived at Paardeberg Drift on the Modder River about 4 A.M. on the 18th, where General Cronje with the greater part of his force was headed up in the bed of the river.

Since the loss of the convoy at Riet River rations had been short, and after a hasty and frugal breakfast we were again astir. We were marched behind Signal Hill, whence a good view was obtained of the enemy's laager in the river bed; but we did not remain there long before orders were received for the brigade to cross the Modder River and envelop the enemy's position from the north bank.

Two companies ('H' and 'F') under Captain Gubbins with Lieutenants Sowray and Groves were ordered to remain on the south bank and clear the bushes on that side, while the remainder of the battalion

PAARDEBERG.
February, 1900.

waded across the river which was breast high with a strong current, but by means of a rope stretched from bank to bank all got safely across.

The brigade extended and advanced to within 1000 yards of the Boer lines, where they took up a position whence fire could be directed on the Boer laager.

Here they remained for the day (February 18, 1900) under heavy fire, the losses being eight non-commissioned officers and men killed, four officers, thirty-two non-commissioned officers and men wounded.

On the 19th there was a temporary cessation of hostilities. 'B' and 'C' companies rejoined the battalion from Modder River, bringing with them a waggon full of supplies which they succeeded in getting across the drift to the bivouac of the battalion, and at dusk trenches were dug and occupied by the battalion.

On the succeeding days the regiment held its ground, losing under a galling fire one man killed and five wounded.

On February 21 after nightfall the regiment rushed 200 yards nearer the Boer trenches.

The incident is thus described by Reuter :

'The scene of the last five days' fighting is one of the prettiest spots in South Africa. The river where General Cronje is surrounded and fighting for his life resembles parts of the Thames, the ground all round slopes towards the river. All the higher ground is covered with our artillery so that General Cronje is faced front and rear on both sides by our men, while General French's cavalry operating far from our flanks prevents any sudden inrush of Boer reinforcements. The 2nd King's Shropshire Light Infantry occupied the bed of the river from Sunday to to-day (February 22).

'Yesterday our artillery continued firing till dark. During the night, after the last gun had been fired, the Shropshire Light Infantry rushed forward and seized nearly 200 yards of new ground where they spent the remainder of the night in entrenching themselves. When the morning came General Cronje thus found himself with 200 yards less space. As the Shropshire Light Infantry, ever since Sunday, had been under a galling fire and had done good work, they were relieved by the Gordon Highlanders. The manner in which the relief was effected was amusing in spite of the danger. The Highlanders crept up to the trenches on their stomachs and the Shropshire men crawled over the bodies of the relieving force.'

On February 21, 'B' company under Major Austen paraded at 6.30 P.M. and crossed the river by boats. They advanced towards the laager after dark and dug a trench in the hard rocky ground. When the moon rose about 1 A.M the trench was complete and occupied, sentries being posted on the flanks. At daylight on the 22nd the company advanced about 150 yards and endeavoured to make another trench. The ground here was almost solid rock and as the fire of the Boers became too hot the company fell back to its former trench, which was held until the Gordon Highlanders relieved the company at 12 noon, when they recrossed the river.

From the 22nd to the 25th the regiment was on constant outpost duty.

On February 26, the anniversary of Majuba, the 19th Brigade made a night attack on the laager—the battalion being ordered to cover the attack with its fire. The attack was gallantly carried out and a trench dug by a party of Royal Engineers about eighty yards from the Boer trenches and enfilading them.

This trench was later occupied by the Royal Canadian Regiment.

At daybreak General Cronje with his force of about 4000 men and six guns surrendered.

On the last day Lieutenant C. E. Atcheson was dangerously wounded.

The following were killed in action at Paardeberg:

1481 Sergeant F. Robson	5914 Private J. Medcalfe
3827 ,, J. Machell	5794 ,, T. Bradbury
2570 Private W. R. Jones	1704 ,, S. Gittoes
3371 Lance-Corporal E. Howarth	5820 ,, E. Lees

Died of wounds:

3532 Private J. Ruscoe

Wounded:

Captain R. A. Smith	Lieutenant E. R. M. English
,, R. R. Gubbins	,, H. W. Kettlewell

Lieutenant C. E. Atcheson
3041 Colour-Sergeant S. Lea
3114 Corporal W. Knowles
3758 " C. Williams
5423 " A. Peel
5244 Lance-Corporal T. Avery
4347 Musician L. Harding
5569 Private T. Gavin
5961 " E. Wilson
1997 " E. Riley
1635 " A. Davies
3229 " T. Stroud
3142 " W. Fuge
5800 " R. Chinnery
3459 " E. Williams
2488 " T. Thomas
3126 " J. Stanley
5513 " W. Olner
3054 " A. Clayton
4246 Private R. Meredith
3240 " E. Cook
3137 " A. Dingley
5782 " W. Moore
2223 " C. Mills
3029 " E. Jones
1827 " A. E. Gilbert
2787 " E. Morgan
5970 " W. Smith
1467 " D. Griffiths
2997 " J. Tague
2710 " J. Moss
4202 " J. Pugh
5815 " J. Johnson
1165 " C. Morris
6456 " W. Lowe
3675 " E. Roberts
5763 " J. Ingram

The following were mentioned in despatches on the occasion :

Lieut.-Colonel J. Spens
Major P. Bulman
Captain C. P. Higginson
Lieutenant H. G. Bryant (special act of gallantry)
2294 Colour-Sergeant Bartram
3041 Colour-Sergeant Lea
4246 Private R. Meredith

In addition, two very gallant acts were performed by 5244 Lance-Corporal T. Avery and 5817 Private T. Hill, who were with the two companies operating in the river bed.

After the surrender of General Cronje the 19th Brigade advanced and bivouacked near the Boer laager and when all the prisoners had been removed was engaged in clearing away arms, etc.

The following order was published on February 28, 1900 :

'Lord Roberts has received the following telegram d/d 27th inst. from the Most Honble. The Secretary of State for War :

'"Her Majesty's Government sincerely congratulate you and the Force under your command on your great and opportune achievement."

'In publishing it for information The Field Marshal Com[dg]-in-Chief desires to express to the troops under his command his high appreciation of their conduct during the recent operations; by the endurance they have shewn through long and tiring marches and the gallantry they have displayed when engaged with the enemy they have worthily upheld the traditions of Her Majesty's army. He has every reason to rely on the spirit and resolution of British soldiers and he confidently trusts to their devotion to their Queen and country to bring to a successful close "the operations so auspiciously begun."'

Owing to the insanitary condition of the laager on March 2nd the brigade moved to Osfontein and remained there until March 6th, when it moved on to Bank's Drift.

The following telegrams were published:

From Her Majesty the Queen, d/d Windsor 27.2.00:

'Accept for yourself and all your command my warmest congratulations on the splendid news.'

From His Royal Highness the Prince of Wales:

'Sincerest congratulations.'

From Field-Marshal Viscount Wolseley:

'Well done. I congratulate you and every soldier under your command with all my heart.'

From His Royal Highness The Duke of Connaught:

'Our heartiest congratulations to you and your gallant troops.'

On March 7, 1900, we marched twenty miles to Poplar Grove where the enemy were in position. The regiment acted on the flank of the enemy, but was not closely engaged. However, 'G' company under Captain Forbes captured a gun which the enemy in his haste to retire had left on a kopje.

The following brigade order was published the next day:

'The Major-General Comd[g] the Brigade wishes all ranks of the Brigade he has the honour to command to understand how thoroughly he appreciates the spirit and zeal shewn by them since the Brigade assembled at Graspan on 12th Feby. All have been called upon for extraordinary exertions and have had to undergo forced marches, short rations, frequent wettings and want of water and sleep and also severe and trying fighting, concluding with an extremely arduous flank march yesterday of some 20 miles.

'It will be gratifying for them to know that yesterday's march turned the Boer position, threatened their rear and caused them to retreat in haste, making them cease firing on our naval guns and abandon their own gun (which the King's Shropshire Light Inf^{y} captured). It also enabled the Highland B^{de} to advance direct on the enemy's trenches without opposition and further caused the retirement of a large force of mounted men which had held the Mounted Inf^{y} (on the left flank) in check throughout the morning.

'It will be gratifying to all to know that, thanks to the untiring energy shewn by everyone, the 19th Brigade has established a high name for itself, which the Major-General feels confident all ranks will do their utmost to maintain.

'He regrets the B^{de} has suffered the loss of many brave officers and men and especially deplores the loss of Lt.-Col. Aldworth, D.C.L.I.'

March 9. The brigade crossed to the south bank of the Modder River and on the 10th encountered the enemy at Driefontein. Here the brunt of the fighting was borne by the 7th Division. No opposition was afterwards met with and the 6th, 7th and 9th Divisions marched on Bloemfontein, the capital of the Orange Free State, viz:

March 11, to Assvogel Kop; March 12, Venters Vallei; March 13, to Ferreira siding and bivouacked close to the railway.

Bloemfontein was occupied on March 13, 1900.

March 15. Marched at 5.30 A.M. to Bloemfontein and bivouacked about three-quarters of a mile from the town on the rifle range. The following impression from the pen of Dr. A. Conan Doyle was published in *The Friend* at Bloemfontein on April 6, 1900:

'A FIRST IMPRESSION.

'It was only Smith-Dorrien's Brigade marching into Bloemfontein, but if it could have passed, just as it was, down Piccadilly and the Strand, it would have driven London crazy. I got down from the truck which we were unloading, and watched them, the ragged, bearded, fierce-eyed infantry, straggling along under their cloud of dust. Who could conceive, who has seen the prim soldier of peace, that he could so quickly transform himself into this grim virile barbarian? Bull-dog faces, hawk faces, hungry-wolf faces, every sort of face except a weak one. Here and there a reeking pipe, here and there a man who smiled, but the most have their swarthy faces leaned a little forward, their eyes steadfast, their features impassive but resolute. Baggage wagons were passing, the mules all skin and ribs with the escort tramping beside the wheels. Here are a clump of Highlanders, their workmanlike aprons in front, their keen faces burnt black with months of the veldt. It is an honoured name they bear on their shoulder-straps. " Good old Gordons " I cried as they passed me. Here are a clump of Mounted Infantry, a grizzled fellow like a fierce old eagle at the head of them. Some are maned like lions, some have young keen faces, but all leave an impression of familiarity upon me. And yet I have not seen irregular British cavalry before. Why should I be so familiar with this loose-limbed, head erect, swaggering type? Of course, it is the American cowboy over again! Strange that a few months of the veldt has produced exactly the same man that springs from the western prairie. But these men are warriors in the midst of war. Their eyes are hard and quick. They have the gaunt intent look of men who live always under the shadow of danger. What splendid fellows there are among them! Here is one who hails me; the last time I saw him we put on seventy runs together when they were rather badly needed, and here we are, partners again in quite another game. Here is a man of fortune, young, handsome, the world at his feet, he comes out and throws himself into the thick of it. He is a great heavy-game shot, and has brought two other " dangerous men " out with him. Next him is an East London farmer, next him a fighting tea-planter of Ceylon, next him a sporting baronet, next him a journalist, next him a cricketer whose name is a household word—those are the men who press into the skirmish line of England's battle.

'And here are other men again, taller and sturdier than infantry of the line, grim solid men, as straight as poplars. There is a leaf-maple, I think, upon their shoulder-straps, and a British brigade is glad enough to have those maples beside them. For these are the Canadians, the men of Paardeberg, and there behind them are their comrades in glory, the Shropshire Light Infantry, slinging along with a touch of the spirit of their grand sporting Colonel, the man who at 45 is still the racquet champion of the British army. . . .'

March 18. The following army order (d/d 14.3.00) was published :

'It affords the Field-Marshal Comdg-in-Chief the greatest pleasure in congratulating the Army in S. Africa on the various events that have taken place during the past few weeks, and he would especially offer his sincere thanks to that portion of the army under his immediate command who have taken part in the operations resulting yesterday in the capture of Bloemfontein. On 12th Feby last this force crossed the boundary which divided the Orange Free State from British territory. Three days later Kimberley was relieved, on the 15th day the bulk of the Boer army in this State under one of their most trusted Generals were made prisoners. On the 17th day the news of the relief of Ladysmith was received, on the 13th March, 29 days after the commencement of operations, the capital of the Orange Free State was occupied.

'This is a record of which any army may well be proud ; a record which could not have been achieved except by earnest well disciplined men, determined to do their duty and to surmount whatever difficulties or dangers might be encountered ; exposed to extreme heat by day, bivouacking under heavy rain, marching long distances, not infrequently with reduced rations, the endurance, cheerfulness and gallantry displayed by all ranks are beyond praise and Lord Roberts feels sure that neither Her Majesty the Queen nor the British Nation will be unmindful of the efforts of this force to uphold the honour of their country.

'The Field-Marshal desires specially to refer to the fortitude and heroic spirit with which the wounded have borne their sufferings.

'Owing to the great extent of country over which modern battles have to be fought, it is not always possible to afford immediate aid to those that are struck down. Many hours have indeed at times elapsed before the wounded could be attended to, but not a murmur or word of complaint has been uttered. The anxiety of all when succour came was that their comrades should be attended to first.

'In assuring every officer and man how much he appreciates their efforts in the past Lord Roberts is confident that in the future they will continue to show the same resolution and soldierly qualities to lay down their lives, if need be (as so many brave men have already done), in order to ensure that the war in S. Africa may be brought to a satisfactory conclusion.'

On March 25, 1900, Major A. H. J. Doyle joined.

A considerable time was spent at Bloemfontein while the railway to the south over Norvals Pont was restored and supplies were accumulated for a further advance. The troops were reclothed and sadly they needed it, some of the men being hardly decent and all of them in rags.

Tents, not being an urgent necessity, were not brought up for some time, and owing to the very inclement weather the troops suffered considerably ; added to this a bad outbreak of enteric broke out which was probably intensified by the cutting off of the Bloemfontein water supply. The result was rather a feeling of dissatisfaction at doing nothing after the active experiences undergone previous to the occupation of the Free State capital. The officers had the use of a good club in the town, but the frequent funerals that passed were very depressing.

A sudden move was made on March 31.

The 9th Division under Lieutenant-General Sir H. Colvile marched at 5.30 A.M. to Waterval Drift on the River Modder—a long and trying march of over twenty miles—to cover the retirement of General Broadwood's cavalry brigade from Thabanchu, who were found to have been surprised and defeated at Sanna's Post; unfortunately we were not in time to be of any practical assistance to this force; but the following day the 19th Brigade under General Smith-Dorrien, supporting General Porter's cavalry brigade, were successful in recapturing eighty-seven of our prisoners from the enemy.

With these we retired on Bushman's Kop, but the waterworks were left in the hands of the enemy.

April 2. March to Springfontein, and April 3 back to Bloemfontein.

The following day (April 4) we were again suddenly ordered to march, this time in a south-east direction. Starting at 2 P.M. we marched twelve miles and bivouacked at Rietfontein, and the following morning marched to Leeuw Kop; but the enemy had eluded us and on April 6 through much slush and mud we marched back to Bloemfontein.

On April 6 our tents arrived and we were in luxury again.

April 17. A draft of 100 men under Lieutenant C. Marshall arrived.

On April 20 orders were received to move on 21st to Springfontein; we were informed that we should be back in three days, but after parading at 8 A.M. and eventually moving off at 1 P.M. we never saw Bloemfontein again.

Lieutenant G. R. Sowray was left in 'temporary' charge of the camp, which was left standing and we did not see him again as both he and Lieutenant W. C. Wright accepted companies in other regiments. Lord Roberts' advance on Pretoria had commenced.

We remained at Springfontein till the 23rd, hearing heavy firing to south-east of our position.

April 24. Marched at 11 A.M., stopping two or three hours at Bushman's Kop and arriving at 5 P.M. at Klip Kraal where we bivouacked, marching at 6 A.M. on the following morning to the waterworks.

These were recaptured with but little opposition. Before reaching them we crossed the Koorn Spruit which the Boers had lined when they surprised Broadwood's force, and passed over the ground where the action at Sanna's Post had been fought. The ground bore many signs of the disaster, not the least pathetic of which were the bodies of three of our men, still unburied. Owing to lack of spades we were forced to leave them where they lay. Moreover the dead horses, cattle and broken down waggons told their own tale.

On April 25 the 19th Brigade marched at 9 A.M. for Thabanchu, leaving the Highland brigade with General Sir H. Colvile at the waterworks. The 9th Division was never reformed. After the first few miles from the waterworks the road to Thabanchu leads into a shallow valley five or six miles wide, but very gradually the features become more and more accentuated until at last the rising ground on either flank becomes mountainous so as to be almost impracticable for mounted troops. About 11 A.M. the enemy under Grobler was discovered across our front, holding a position at Israel's Poort, a series of kopjes stretching across the entrance to this valley. The battalion was sent to clear the left flank of the brigade and had two men wounded

(Sergeant A. Crowden and Private C. Rowe). On the enemy retiring, the brigade bivouacked on the scene of action, and two very fine pigs here found their way into our mess-cart somehow. We marched the following day without opposition to Thabanchu, where we halted outside the village. Here much bargaining took place with the natives for Kruger coins, chickens, eggs and bread. Our parson caused us considerable amusement by appearing with two large fowls, one under each arm, both crowing lustily, followed by the mess president with two more chickens and a haversack full of eggs, while the rear was brought up by a Kaffir with five more struggling birds.

After collecting these delicacies we marched through the town and bivouacked on the east side of it, the Boer outposts covering a laager of De Wet and his brother, Olivier, Grobler, Lemmer and Fourie, and situated about six miles to the east of Thabanchu, being plainly visible a few miles further on.

Here we enjoyed a well-earned lunch off pork chops from the pigs found at Israel's Poort.

On April 27 after an easy morning, employed in watching small parties of Boers and our men sniping one another on the neighbouring hills, we got a sudden order to fall in at 1.45 P.M. and were marched towards a kopje about four miles off, on the right flank of the Boer position. After marching for about a quarter of an hour we were vigorously shelled by two Boers guns on our right flank. Our own guns were unable to reach them, but no damage was done.

The kopje in question (Schumann's Kop) was attacked by Kitchener's Horse and we had a good view of the engagement, both sides being plainly silhouetted against the skyline with the setting sun behind them.

We then retired under cover of the darkness to our bivouac at Thabanchu, but had considerable difficulty in finding a way for the guns over the innumerable gullies.

We afterwards heard that Kitchener's Horse were in difficulties and at 11 P.M. the 1st Gordon Highlanders were sent off to assist them.

The following morning (April 28) we paraded at 4.45 A.M. and returned to Schumann's Kop to relieve the Gordons and Kitchener's Horse. When we arrived there we discovered Kitchener's Horse had retired in the night and the Gordons had gone astray and were not to be found, while the Boers were again in possession of the kopje.

The battalion, however, drove them off without much difficulty and occupied the kopje for the remainder of the day. We covered the bodies of one of Kitchener's Horse and a Boer with stones, the ground being too hard to dig a grave. Both were old men and both shot through the head. The Boer had a large blue bottle of snuff by his side and wore long leather gaiters, reaching to his thigh and racing spurs.

We watched the cavalry manœuvring to the north and east of us all day, and towards evening were given to understand that we had succeeded in surrounding 500 Boers and their two guns.

But as usual the report was a false one.

The object of these operations had been to close in the Boer laager behind the mountain. Under the orders of General French, Gordon's cavalry brigade was sent round the south and Dickson's round the north of Thabanchu mountain, and the two brigades were to join hands. But this was found to be impossible and the cavalry had to retire, Dickson being so hard pressed by Olivier that if it had not been for a diversion of a company of the regiment under Major Austen, who supported his left rear, his casualties would probably have been serious.

At dusk our whole force retired to their former bivouac at Thabanchu and remained there during the whole of the following day (Sunday, April 29), an occasional shell dropping into our lines during that afternoon.

On April 30 the march north for Pretoria began. Hamilton's force had up to now been getting into position as the right flanking column of Lord Roberts' general advance.

The Commander-in-Chief's scheme at this period was to advance up the railway to Pretoria, in one great line extending from Kimberley on the west to Ladysmith on the east.

Hunter (with 10,000 men and 18 guns) was on the extreme west with Methuen next to him. In the centre was Lord Roberts himself, with two infantry divisions and French's cavalry guarding his left flank, and on the extreme right came General Ian Hamilton with the 19th Infantry Brigade and 4th Mounted Infantry corps under General Ridley. This latter force was to be afterwards increased to 14,000 men by the addition of Broadwood's cavalry brigade and the 21st Infantry Brigade under Major-General Bruce Hamilton.

Lord Roberts hoped that Sir Redvers Buller would be able to push up through Natal and guard the right flank of this general advance; but this was not done, so there was a considerable gap on Ian Hamilton's right and the whole Free State army was later on able to slip round it.

This flank to which the regiment belonged was considerably exposed during the whole march through the Free State and saw most of what fighting there was.

On April 30 we paraded in the dark and marched north towards Winburg, which was to be reached by May 3, Hamilton's force being composed of Ridley's Mounted Infantry Brigade, the 19th Brigade, 'P' Battery and the 74th Battery.

For ten miles the road north of Thabanchu is flanked on the east by a line of ridges; these ridges then curve round westward and end abruptly in the imposing mass of Toba mountain. The road we were on passed over the ridges at Houtnek.

We proceeded without molestation till 9 A.M., when the enemy under Philip Botha was discovered to be holding the nek in considerable force, also Toba mountain on our left flank and the kopjes on our right flank. The 85th were this day the leading battalion and extended in rear of the mounted infantry facing the nek, while two companies ('B' and 'E') under Major Dawkins and Major Austen and Lieutenant C. Marshall were, with the 1st Gordon Highlanders, sent to reinforce Kitchener's Horse on Toba mountain, which appeared to be the key of the position.

The Canadian Regiment was held in reserve and the Duke of Cornwall's Light Infantry were left as guard to the baggage.

Owing to the lack of cover no frontal attack could be successful without heavy loss; and we had to remain in our positions under a dropping fire until the left flank attack was successful. The lower slopes of Toba mountain were gained without difficulty; but on the crest the enemy were strongly posted behind the rocks and could not be dislodged. They were later reinforced by 150 foreigners under Colonel Maximoff, a Russian, and further the enemy's left flank was protected by the Boer guns in the pass. At 6 P.M. Major Dawkins was wounded and the command of 'B' and 'E' companies devolved on Major Austen.

The situation did not alter during the day and we remained in position that night, and stood to arms at 4.15 A.M. on the following morning (May 1). At daybreak fighting was resumed, General Smith-Dorrien himself proceeding to the top of Toba mountain to organise the attack there; and some of the Gordon Highlanders on the summit together with the two companies of the regiment charged the enemy and cleared the hill by 1 P.M., half a company of the regiment ('B') under Colour-Sergeant Scouse distinguishing themselves by seizing a nek and holding it for some hours under a heavy cross fire where they had ten casualties. At the same time the 8th Hussars appeared round the flank and the appearance of Broadwood's cavalry and the 21st Infantry Brigade, who were joining Hamilton's force from the west, hastened the departure of the Boers, who, foreseeing they would have to retire, had sent their convoy away during the night.

The remaining companies of the regiment then advanced over the nek and as far as Jacobsrust, from whence a strong force of the enemy could be seen retiring on Winburg. Here we bivouacked for the night.

Our casualties on these two days were:

Killed

5799 Private J. Smith

Died of Wounds

2497 Private A. Morris	5618 Private H. Fryer
5636 Private A. Eldridge	

Severely wounded

Major C. T. Dawkins

Wounded

1356 Sergeant F. Haskins	5756 Private B. Field
3654 Private A. Moore	5762 ,, J. Bedwell
3670 ,, T. Browne	5755 ,, A. Winterbottom
1545 ,, E. Mabby	3088 ,, A. Rudge
1630 Colour-Sergeant W. Herbert	2734 ,, J. Cale
1249 Sergeant R. Talbot	3778 Bugler A. Brewer

Another account of the events of this action is here given:

'On 30th April, after a march of 10 miles, the action commenced. "B" and "E" Cos. were the two rear companies in the Battn. "B" Coy. had had a pretty hard day on 28th covering General Dickson's cavalry on their retirement on Tabanchu, and expected to be in reserve on this occasion. However about 9 A.M. General Smith-Dorrien rode up and directed these two companies together with the 1st Gordon Highlanders under Colonel Macbean to go at once to Toba Mountain (then about 2 miles on our left front) to reinforce Kitchener's Horse. We went as hard as we could go and got to the lower slopes without being fired on. Just before the crest of the hill is reached the slope is very steep and the men had to use their hands to climb up the rocks. It was hard work and the men had just halted to get their breath, when a nine-pounder shell burst just behind us, wounding Pte. Mabby. Seeing where the Boer gun was orders were given to get up over the crest where they would be safe. The men who had been puffing for want of breath a moment before were up over the crest in no time, not, however, before two more shells screamed over us. Once on the hill we were under cover from the gun as it was some way below us and the range of the crest was not accurately taken by the Boers. Several more shots were fired, and then we began to advance. Almost immediately we came under rifle fire from the Boers in front. "B" Coy. then advanced to the edge of a small rise from where they could fire on the ridge where the enemy was supposed to be. We never saw any of them. This position was held till the next morning. Major C. T. Dawkins having been shot through the foot the command of these two Companies devolved on Major A. R. Austen. The night was very cold and we had no blankets and nothing to eat. Some men were sent with a lot of water-bottles to get water at a spring some way to our left rear and they were successful in finding it and returning with a good supply.

'The next morning firing commenced as soon as it was light. The Canadians were now on our left, the Gordons on our right, and we began an advance at 6 A.M., the Canadians making a flanking movement. "B" company advanced by Sections, and No. 1 Section under Colour-Sergt. Scouse succeeded in getting on to a ridge between two positions held by the Boers. This action together with the flanking movement of the Canadians made the Boers retreat from a position on their right and our force gradually swung round to the right, driving in the Boers to the left of their position. About 12 noon Genl. Smith-Dorrien arrived and directed "B" and "E" companies to advance across an open space on to the Boer position. As was usually the case, directly we began to move steadily forward the fire of the Boers became wild and then ceased. A Highlander on our left saw some of them retiring and gave a cheer and rushed forward. Our men took it up and fixed bayonets and charged to the top of the position, only to find the Boers in full retreat a long way off. We emptied our pouches and magazines at them but did not, I am afraid, do them much harm. Shortly after this some food arrived for us—somewhat welcome after about 30 hours' hard work. We buried seven of the enemy and one of our men, and some of Kitchener's Horse, and marched on to rejoin the Battn. at Jacobs-Ruste.

'After the engagement at Hout Nek on 30th April–1st May 1900 General Ian Hamilton came down to the camp of the Battalion and congratulated Col. Spens on the success of the two companies (" B " and " E ") who charged the enemy on Toba Mountain.'

At Jacobsrust on May 2 the regiment was reinforced by a draft of 100 non-commissioned officers and men under Captain W. S. R. Radcliffe and the 1st Volunteer Service Company consisting of 113 non-commissioned officers and men from the Volunteer battalions of Shropshire and Herefordshire under Captain W. H. Trow with Lieutenants B. Head and J. C. Cutler, who were heartily welcomed. These drafts had marched from Bushman's Kop during the day and did not arrive till 11 P.M. With the arrival of the 2nd Cavalry Brigade and the 21st Infantry Brigade, Ian Hamilton's force was now complete. General Smith-Dorrien took command of the Infantry Division and Colonel Spens assumed that of the 19th Brigade. The headquarters of the 9th Division, with the Highland Brigade, received orders to march five to ten miles in rear of Hamilton's column so as to be available for either flank if required.

On May 3 we marched fifteen miles to Isabellafontein (Verkeerde Vley), where there was a capital store and a regular poultry yard of ducks, chickens and pigeons, which we secured.

On May 4 we reached the River Vet at dusk. A force under Philip Botha had taken up a position on a long kopje on our right flank called Babiaansberg to oppose our advance, and from which we were shelled in the flank by a Creusot gun.

Hamilton had left this kopje alone and pressed on to capture the drifts over the River Vet, but Sir H. Colvile coming on in our rear had attacked the southern end of it and the regiment was drawn up ready to attack the northern end when the Highlanders appeared coming along the top and the Boers retreated. We bivouacked at Welkom Drift and marched next day to Winburg, crossing the Klein Vet without opposition, a flag of truce being sent on to demand the surrender of the town. We halted outside and heard General Botha had left as we arrived.

Winburg is a picturesque little town, and the inhabitants, who included a large number of the fair sex, gave us quite a hearty reception. The town was well stocked with provisions and a welcome addition was made to our supplies, not the least of which was four dozen bottles of beer! Winburg being situated on a branch of the railway, a number of sick men under Captain J. G. Forbes were left behind here when we next advanced, as hard marching was expected and orders were given for all weakly men to be left behind.

Our next objective was Kroonstadt. Starting at 5 P.M. on May 6 we marched about eight miles north of Winburg and bivouacked by a deserted farmhouse called Dankbaarfontein. We remained there during the whole of the 7th and 8th. Everything had been cleared out of the farmhouse except a bedstead with an old dog lying beneath it. The contents of the feather mattress were strewn a foot deep over the floor. There was a beautiful grove of lemon trees, some of which had been cut down for firewood; but a guard was placed over the remainder and close by was the family graveyard and amongst the ancient tombs was one only freshly filled in. A beautiful tree covered with yellow flowers shaded the grave and some of the flowers from this tree graced our dinner table that night, if a waterproof sheet spread on the ground can be dignified with such a name.

Hamilton had wished to press on and harass the Boers before they had time to take up a position along the Zand River, but Lord Roberts would not sanction this as the enemy were expected to make a firm stand at this spot and he was not ready himself to co-operate in the centre. So it was not till 6.30 A.M. on May 9 that we continued our march.

After proceeding twelve miles we bivouacked at Bloemplatz, a fine farm about four miles south of the River Zand.

During the march a fat ant-bear was put up. It had somewhat the appearance of a pig. To the great joy of the whole column it was caught after an exciting chase by two dogs and quickly despatched by the hungry soldiers and a ham procured for our mess.

Bloemplatz was one of the finest farms we had seen so far. It was inhabited by an old Dutchman who was much struck at the strength of our column. His son was fighting against us and he had covered his house with white flags. General Ian Hamilton took possession of it for his headquarters for the time being. Large herds of springbuck and besbuck gladdened our eyes, and sportsmen were soon afoot and bullets were flying about in all directions as if it were a general action, until some unlucky trooper was shot through the stomach and the sport was stopped—but not before Captain R. A. Smith had managed to down a springbuck and wounded a besbuck, which, coming right through the bivouac, was caught by the men and added to the pot.

We dined off springbuck chops that night. The reader may think I refer somewhat frequently to this question of food, but with us it was a serious matter. It must be remembered that the whole army of over 100,000 men was dependent for its food on one line of railway, which slender means of communication was daily growing longer. All bridges and culverts had been destroyed by the Boers and the Royal Engineers had their hands full repairing the line and making deviations at the bridges before the trains could pass. The food question had kept Lord Roberts six weeks at Bloemfontein. He would not advance until thirty days' supplies for 100,000 men had been accumulated. Since leaving Bloemfontein his own column on the railway had been comparatively easy to feed, but Hamilton's column marching thirty to sixty miles from the railway and requiring fifty waggon loads of food and forage a day was not so easy to supply, so that the troops were seldom on full rations, and more often on half—nor were supplies to be drawn from the country except on rare occasions—fowls and pigs were for the mounted troops, who were scattered about on the front and flanks—the infantry marching in close order could seldom obtain these delicacies and the strict orders against looting, threatening death to the offender, were faithfully adhered to—at any rate if there was a chance of being found out. But few titbits were left by the cavalry and mounted infantry, who would come into camp laden with ducks and chickens and with large gateposts or railings balanced across the saddle to cook them with, and I am afraid we poor infantry cast envious eyes on them and wondered how they had paid for their supplies and fuel!

A reconnaissance was made of the Boer position that afternoon and a battalion from the 21st Brigade thrown across the river that evening to cover the passage of Junction Drift which had been most unaccountably left unguarded by the Boers.

At 5.30 A.M. on May 10 the action began by a shot from our five-inch gun. The enemy opposite us consisted of Lemmer's, Olivier's, Fourie's and De Wet's commandoes and they considerably overlapped our right flank—so much so that General Tucker's Division was ordered to co-operate with Hamilton.

The enemy occupied the rising ground north of the river and a considerable body were posted in the river bed.

Lord Roberts' idea was to turn both flanks, sending French round their right and Broadwood with his cavalry brigade and the mounted infantry round their left. We were to force a passage for Broadwood.

This latter was successfully performed by the 21st Brigade, the 19th Brigade being in support.

The enemy's pompom posted to the north of the river was very active this day until stopped by a magnificent shot from our 4·7.

This was the first occasion our Volunteers were under fire, and very well they behaved. They were selected as escort to a battery and subjected to a considerable amount of shelling. Luckily the shells did not burst properly, being fitted with percussion fuses which, bursting in the soft ground, did no damage.

The turning movements were unfortunately not successful in cutting off the retreat of the Boers, who as usual, as soon as their flanks were turned, were off. The battalion was not seriously engaged this day, but 'H' company (who were posted in the river bed covering Hamilton's artillery on the south bank) were able to fire into the flank of 200 or 300 Boers who were holding up the 7th Mounted Infantry and threatening to surround our right flank by creeping up a donga. Our men soon caused them to beat a hasty retreat up the river bed to the east.

The battalion then advanced over the Boer position and bivouacked about four miles north of the river at Boschkop.

The enemy on the hill-side had sheltered themselves in holes blasted out of the rock to get stone for posts for their wire fences. These holes afforded capital cover.

The capture of the Zand River position meant the fall of Kroonstadt which was the second largest town in the Free State and the capital of that country since the middle of March, and large supplies had been accumulated there. The Boers did not hold their prepared position at Boschrand; but, having burnt the stores at Kroonstadt, hastened off, the Boers towards Rhenoster River and President Steyn and the Free Staters to set up a new capital at Heilbron.

On May 11 we marched about fifteen miles as rear-guard to about four miles of transport and reached our bivouac at Twistniet about 10 P.M., marching the following day to Kroonspruit, a point four miles south of Kroonstadt.

Here it was like returning to civilisation, with the railway in view again, and we had hopes of receiving a mail or two, but in this we were disappointed.

We remained here till May 15, having a parade service on the 13th and an inspection by Lord Roberts on the 14th, when we heard we were to march to Lindley, a town afterwards the scene of much fighting, about fifty miles to the east, and thence to Heilbron, Hamilton's orders being if possible to capture President Steyn and his government. But the expedition was an abortive one on the whole. Still, though it had the effect of making Botha give up the idea of opposing the passage of the Rhenoster River, yet it had no moral effect on the Free Staters, who harassed our column considerably during its passage through this difficult country.

Starting at 8 A.M. on May 15 we marched 200 yards and halted for one and a half hours. Then, starting again, we managed to cover another 200 yards and then rested for one and a half hours. After marching six miles we bivouacked and advanced the following day fourteen miles to Tweepoort.

May 17. Marched as rear-guard to many miles of transport waggons. We had a number of bad drifts to cross and did not arrive in camp at Bankfontein till 11.30 P.M., and one company not till 4 A.M. Many carts stuck and some were overturned in the drifts, but luckily the River Valsch was crossed by a good bridge. Broadwood's cavalry this day captured Lindley.

On May 18 the 21st Brigade pressed on to Lindley, and we remained at Bankfontein this day to escort an expected convoy. We were ordered to come up on the left of the column when it turned north for Heilbron.

The day was very hot and ice was on our buckets at night.

On May 19 we were ordered to move at 6 A.M., but this was changed to 11 A.M., being altered again to 9 A.M. We fell in at this hour but did not march till 10.30 A.M., when we marched two miles and then halted for three hours, then moved on six miles, and eventually arrived at our camp at Quaggerfontein in pitch darkness.

Captain Trow, who commanded the Volunteer Service Company, was sent back sick from here and died of enteric at Kroonstadt. He was a great loss.

On May 20 we marched at 6 A.M., being informed that we were going twelve miles. We arrived at our camp at 12.30 and after halting there for half an hour were fallen in and moved on about 200 yards, where we remained for one and a half hours, and were again moved 200 yards in another direction. Brigade markers were then put out and we marched on them, but, being ordered to stand by, stood by for half an hour and were then ordered to march on two miles further; after advancing about three miles we eventually camped. It was a bitterly cold night and a strong wind was blowing. Our baggage was miles behind and with a bad drift to cross. Truly the soldier's life is a happy one, but an incompetent or thoughtless staff officer may make a regiment very uncomfortable.

On May 21 we were ordered to parade at 6.50 A.M.; at 6 A.M. ordered to fall in as soon as possible. We fell in and stood by for an hour and were then told to fall out till 10 A.M. Eventually we moved off at 11 A.M. as rear-guard—400 Boers were reported on our right. We marched for about two miles when we were informed that the enemy was advancing from the south-east in considerable force; so we took up a position on the kopjes on our right and watched the burghers. Some came over the neighbouring hills and we put the pom-pom on to them. They retired—and so did we. Camped ten miles south of Heilbron.

May 22. We arrived at Heilbron, Mr. Steyn and the Free Staters having left shortly before. De Wet's convoy could be seen trekking towards Frankfort; but the bulk of the Free Staters broke away round our right flank. It was at this point that the co-operation of Buller's army was missed. We marched through the town, which was full of women and children. Here we got two days' supplies for the whole force and the men had a ration of flour issued to them instead of biscuit. Large quantities of eggs were to be purchased and we procured a dozen bottles of port for £6; so we were in luxury again. The stock of boots in the town was bought up and some of the men whose boots were without soles were fitted up in patent leather boots and fancy shoes.

On May 23 we marched to Elandspruit.

On May 24 we marched at daybreak to the main line at Vreedeport siding through miles of burnt veld and then marched three miles up the line to Marseilles. Here we joined Lord Roberts' main body—we camped at Erste Geluk near Prospect and it being the Queen's birthday an issue of rum was made and there was much singing and rejoicing, which was intensified later on when we heard that Mafeking had been relieved and that French had crossed the Vaal.

May 25. The order of march was now to be changed, the Free Staters were behind and the left flank became the most important, with the Witwatersrand and Johannesburg to occupy. For this reason Hamilton's column was to cross over the line in rear of Lord Roberts' column and be on his left flank, Hamilton's place on the right flank being given over to Colvile and the Highland Brigade and Spragge's yeomanry.

This somewhat awkward manœuvre was carried out without confusion on May 25, and we halted and watched Lord Roberts' divisions pass by. On the 26th, after a tiring march of fourteen miles through heavy sand, we crossed the Vaal about 3.30 P.M. at Wonderwasser Drift, where the water only came up to our knees. The Volunteer Company were the leading company and the first to step on Transvaal soil.

On May 27 we marched about fifteen miles north to Wildebeestefontein and about 3.30 P.M. came in sight of the Boer position south of Johannesburg. Lord Roberts' plan was for French with his cavalry to describe a circuit to the north of the town and Hamilton was to advance direct on Florida.

On May 28 we had orders to proceed to a point west and south of the Boer position and while French turned the enemy's right flank and Roberts his left flank we were to demonstrate in front, or force a passage for French if he could not get round; but these plans were unnecessary, Botha had retired to his main position on the Kliprivers-berg which forms an outwork to the Witwatersrand.

The front of his position was protected by the Klip River, which forms a series of swamps and marshes and which was impassable except at the crossings. His right rested on Doornkop, celebrated by Jameson's raid, and the greater part of his force was on this flank.

On the night of the 28th we camped at Cyferfontein and could see the tall chimneys of the mines in the distance.

The following day it was the turn of the 21st Brigade to lead the division and the regiment's turn for rear-guard. It was the custom for the brigades to lead on alternate days and each regiment in the brigade to lead its brigade in rotation, so that each regiment should have its turn. We were thus, through the fortune of war, for the most part spectators of the action that followed — in which the Gordons so much distinguished themselves—but Colonel Spens had command of the brigade and was mentioned in despatches for his skilful manœuvring of it; while Captain Higginson, his acting staff-officer, was recommended for the 'V.C.' for displaying great gallantry and coolness in taking a message from Smith-Dorrien to the Gordon Highlanders, cantering along their line under heavy fire until he found their colonel to give him the order.

We were on very short rations at this time and it was necessary for this reason to force the position in spite of heavy loss. This was done in gallant style by the 21st and 19th Brigades. The regiment, acting as baggage- and rear-guard, was not so fortunate, as, French's baggage getting mixed with ours, we followed the former towards Doornkop and by the time the mistake was discovered darkness had descended and our own transport was out of sight; so, after wandering rather aimlessly about, we bivouacked and waited for daylight. We spent a hungry and cold night, without food or covering, and were glad to rejoin the column the following day at Florida; but rations were still short and mealie flour about all we could get. However, the following orders which were published made up for any little hardships:

From Lord Roberts to General Ian Hamilton

'Well done, but much grieved to hear your men are without rations, will send you a train load at once—so sorry you cannot be here with us. Tell the Gordons I am proud to have a Highlander as one of the supporters of my coat of arms.'

From Lord Roberts to General Ian Hamilton

'I am delighted at your repeated successes and grieved beyond measure at your poor fellows being without food, a trainful shall go to you to-day.

'I expect to get the notice that Johannesburg surrenders this morning and we shall then march into the town.

'I wish your column which has done so much to gain possession of it could be with us.'

But the promised train never came, and on June 1 we marched through a pleasantly wooded country to Braamfontein, seven miles nearer Johannesburg, where supplies could be obtained, and a proportion of the troops were allowed to go into Johannesburg. The return to civilisation was much appreciated; for to smoke 1*s.* 6*d.* cigars, drink beer at 4*s.* a bottle, and dine off a table-cloth, were unwonted luxuries.

But on June 3 we were again on the move, marching at 6.30 A.M. fifteen miles towards Pretoria and camping at Diepsloot on the Yekeskei River.

With the idea that Pretoria would be held, Lord Roberts ordered French to advance round the Boer right and cut the line to Pietersburg, Hamilton supporting him in rear. But on June 14, hearing that the Boers would not seriously defend their capital, he drew Ian Hamilton's column in towards his own main body, and that night we camped on the river Hennop.

On June 5, parading at 6 A.M., we marched straight for Pretoria through a narrow rough pass. On reaching a small hill about a mile from the town, the whole of Hamilton's force was drawn up and we were informed that General Botha was to be received and Lord Roberts was to demand the surrender of the town and the Boer army. After waiting some time, and the Boer general not appearing, we entered into the town and camped close to the race-course, and later in the afternoon proceeded through the town past President Kruger's house and marched past Lord Roberts, who with his staff and a large concourse, including the released prisoners, were drawn up in the Market Square.

Thus ended a march of about 425 miles since we left Bloemfontein. The result was the occupation of the enemy's capital and the safety of the gold industry at Johannesburg, but the Boer forces had not been beaten. They had been merely swept aside, and the long single line of communication in our rear was very vulnerable, as we were soon to discover; for instead of having, as we expected, a rest at Pretoria, we on the following day (June 6) suddenly received orders at 8.30 A.M. to march south again at 9.30 A.M.!

Lord Roberts' advance had been so rapid that he had left the line in his rear practically undefended from Johannesburg to Kroonstadt and the reconstruction of the railway could not keep pace with him. De Wet had discovered this weak spot and was beginning to take advantage of it. General Smith-Dorrien came to see us off and was loudly cheered by the men, who were unable to restrain their feelings towards this popular commander. The General, in response, made a short speech in which he said:

'The first time I saw you under fire was at Paardeberg and I was much struck by the magnificent way in which the leading companies opened out for the attack. The next time was at Poplar Grove and I asked Colonel Spens if he could advance and take up a certain position. This was over an open piece of ground commanded on both sides by kopjes occupied by Boers with guns. This required a great deal of dash and determination, but it was done—the Boers were driven out and a gun was captured by you—and whenever we met the enemy and I knew the Shropshires were in front I was certain of the best results.

'You are now going to Vereeniging, which the Boers are reported to be threatening. You have been specially selected for this on account of your good work during the campaign, but I hope you will soon join me again; and lastly I must commend you for your good conduct and smartness both in the field and in camp. When you marched past yesterday, Lord Roberts said to me, "I am given to understand that is a very fine regiment," to which I replied, "Sir, it is one of the best in the service." With regard to my own feelings, I may add that whatever praise or credit I get for this campaign I shall always remember a great deal of it is due to Colonel Spens and the Shropshires.'

On the same day an order was published by General Smith-Dorrien to the 19th Brigade, viz.

'The 19th Brigade has achieved a record of which any infantry might be proud. Since the date it was formed, viz. 12th Feby. 1900, it has marched 620 miles, often on half rations and seldom on full. It has taken part in the capture of 10 towns, fought in ten general actions and on 27 other days. In one period of 30 days it fought 21 of them and marched 327 miles. Casualties between 400 and 500. Defeats Nil.'

The following order was published by Lord Roberts to the army in South Africa at a later date:

'(1) In congratulating the British Army in South Africa on the occupation of Johannesburg and Pretoria, one being the principal town and the other the capital of the Transvaal, and also on the relief of Mafeking after an heroic defence of over 200 days, the Field-Marshal C.-in-C. desires to place on record his high appreciation of the gallantry and endurance displayed by the troops, both those who have taken part in the advance and those who have been employed in the less arduous duty of protecting the lines of communication through the Orange River Colony.

'After the force reached Bloemfontein on the 13th March, it was necessary to halt there for a certain period. Through railway communication had to be restored with Cape Colony before supplies and necessaries of all kinds could be got up from the Base. The rapid advance from Modder River and the want of forage *en route* had told hardly on the horses of the cavalry, artillery and mounted infantry, and the transport mules and oxen, and to replace the casualties a considerable number of animals had to be provided.

'Throughout the six weeks the army remained halted at Bloemfontein the enemy showed considerable activity, especially in the south-east portion of the Orange River Colony, but by the beginning of May everything was in readiness for a further advance into the enemy's country, and on the 2nd of that month active operations were again commenced.

'(2) On the 12th of May, Kroonstadt, where Mr. Steyn had established the so-called government of the Orange Free State, was entered; on the 17th of May Mafeking was relieved; on the 31st May Johannesburg was occupied, and on the 5th June the British Flag waved over Pretoria. During this 35 days the main body of the force marched 300 miles, including 15 days' halt, and engaged the enemy on six different occasions.

'The column under Lt.-General Ian Hamilton marched 400 miles in 45 days, including 10 days' halt, and was engaged with the enemy 28 times. During the recent operations the sudden variations in temperature between the warm sun in the daytime and the bitter cold at night have been peculiarly trying to the troops, and owing to the necessity for rapid movement the soldiers have frequently had to bivouac after long and trying marches without blankets or firewood and with scanty rations. The cheerful spirit with which difficulties have been overcome and hardships disregarded is deserving of the highest praise, and in thanking all ranks for their successful efforts to attain the objects in view, Lord Roberts is proud to think that the soldiers under his command have worthily upheld the traditions of Her Majesty's army in fighting, in marching and in the admirable discipline which has been maintained throughout a period of no ordinary trial and difficulty.'

OUR MESS.

SOUTH AFRICA

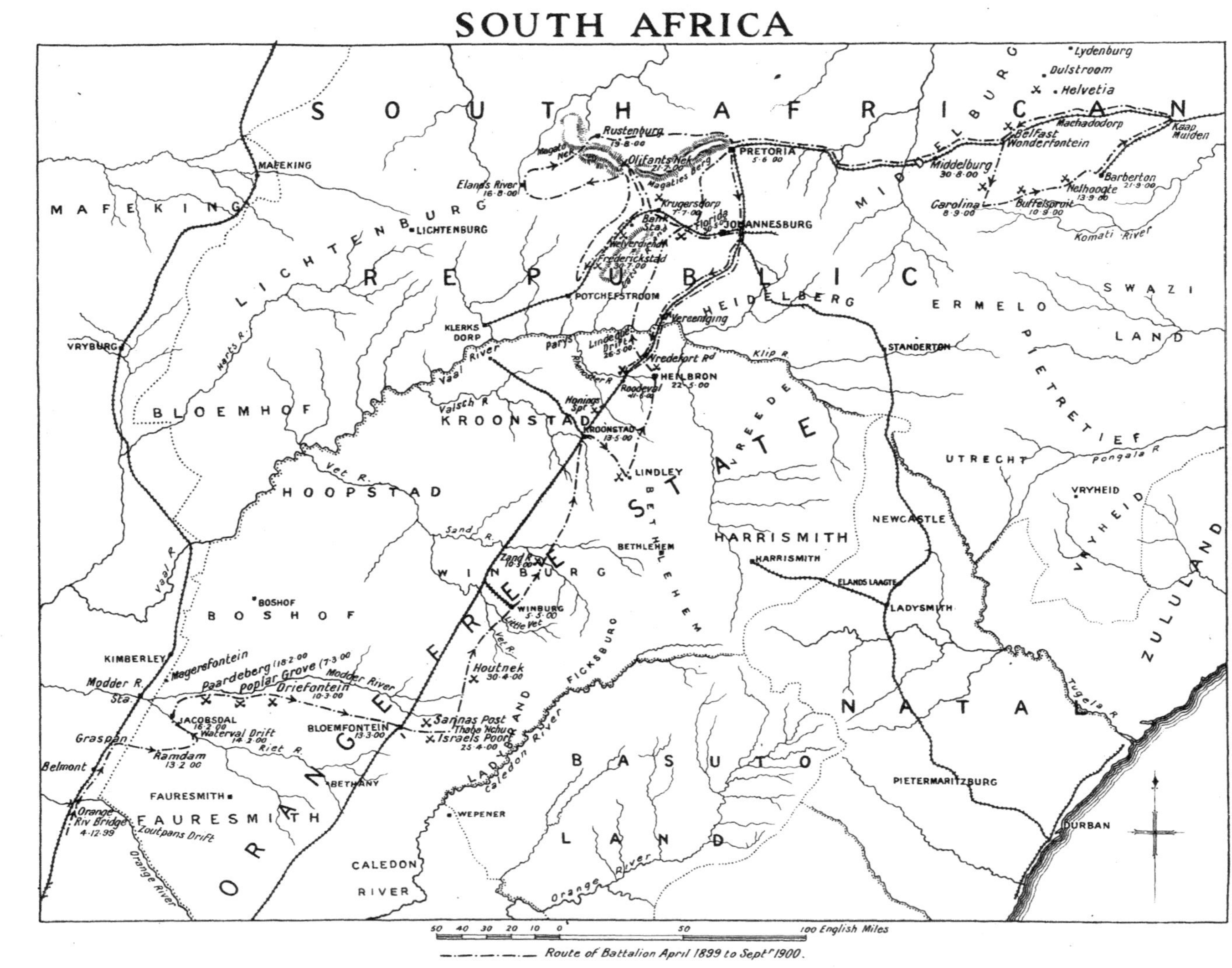

Route of Battalion April 1899 to Septr 1900.

CHAPTER XVII

THE SOUTH AFRICAN WAR FROM THE CAPTURE OF PRETORIA TO 1902

IT is convenient for the purposes of this history to open here a new chapter. The style of the campaign was quite changed.

The position of affairs was totally different from that in 1899, both from the political and military point of view. The situation, as far as the regiment was concerned, was this. Lord Roberts had recognised the insecurity of his line of communications between Kroonstadt and Pretoria, and on June 6 we started south again; but De Wet had also discovered our weak spot, and on June 7, dividing his force into three columns, he attacked Vredefort Road station, the Derby Militia at Rhenoster River, and a party at Roodeval guarding a huge supply of stores, clothing, and ammunition together with seven or eight weeks' mails destined for the army at Pretoria. These three attacks were entirely successful.

But this was not of course discovered by us till later. Leaving Pretoria at 9.30 A.M. we had a hot and dusty march to Irene, whence the line was open as far as the Vaal. Six companies under Colonel Spens left here at 6 P.M., the remaining two under Captains Smith and Gubbins being left to follow later, together with two guns of the 74th Battery by a train expected in an hour's time; but the train did not come that night. Instead, a heavy water tank came bounding along the line, having broken loose from a station twelve miles south, and dashing into some empty carriages on the line smashed itself to bits, and spilt all the water meant for the engine. At daybreak next morning the expected train did arrive, full of supplies which had first to be unloaded; then (after shipping the guns, ammunition and transport, together with forty horses and fifty mules) our belated train with much whistling and creaking started about noon with very little water in the boilers, caused by the loss of the water tank, and with the brakes all out of order; however, we managed to reach the next station by the skin of our teeth, the boilers practically empty, and luckily found a water tank, but no means of filling the engine—eventually by means of half a dozen buckets and two hours' hard work we were enabled to proceed as far as Zuurfontein, where we met another engine, which had been sent to look for us and which took us on to Elandsfontein. Here to our surprise we were met by the whole of the local staff, who were expecting Lord Kitchener and were in a great state of excitement. Lord Kitchener had been hurriedly sent south by Lord Roberts, who, subsequent to our departure, had probably heard of De Wet's activity in his rear. Our next stop was Vereeniging, where we arrived about midnight, being greeted by a volley from the outposts, who were firing at their own patrols; here we spent a peaceful night on the platform, being sent out on outpost duty the following day (June 8).

Lord Kitchener arrived in the afternoon, followed later by the South Wales Borderers, and at 8.30 A.M. on 9th we crossed the Vaal by a drift and marched eight or nine miles to Tybosch, then the railhead, where we again entrained and proceeded to Vredefor Road. On the morning of June 10 we were aroused at sunrise by the sound of the enemy's guns, who were attacking our picquets; but they were soon driven off, one man of the Volunteer Service Company being wounded by a shrapnel bullet.

We were here joined by Lord Methuen's column, consisting for the most part of yeomanry. The following day (June 11) the combined column moved south, down the railway line, and attacked the Boers who, under Christian De Wet, had taken up a position between the latter's own farm, Roodepoort, and Honing's Kopjes.

The column on the east of the line operated under Lord Methuen, that on the west under Colonel Spens.

The yeomanry turned the Boer right, and the regiment his left, and advancing drove the Boers off in a westerly direction and captured De Wet's farm, the floors of which were covered a foot deep with our letters; cigarettes, chocolates, shirts, registered letters, photographic films lying about everywhere in profusion. The house was a poor one, but nicely situated near a large lake or 'dam' with swarms of game—quail, hares, duck and geese.

We bivouacked at Rhenoster River on the spot where three days previously the Derby Militia had met with disaster and who, after losing 40 killed and over 100 wounded, had been forced to surrender. The ground was strewn with their helmets and blood-stained clothing and littered with defaulter sheets, mess accounts, etc.; and the empty tents, riddled with bullet-holes, were strewn with boots, the whole telling a sad tale.

The railway line too had all been blown up again and the bridge destroyed; and every now and then pealed forth the loud explosion of a bursting shell and the crackle of small-arm ammunition, a large accumulation of which was burnt, together with the blankets and warm clothing so badly wanted by the troops. But the loss of our mails affected us chiefly, several truck-loads lay scattered about—the nine-inch shells that were bursting round us were not now required—they were to have silenced the forts round Pretoria—but the mails we had been looking forward to for weeks.

On June 12 Lord Methuen's column left us and at 6 A.M. marched southwards to relieve Kroonstadt, which was reported as likely to be attacked by De Wet. In the afternoon we moved our camp to the top of the kopje and were now in a very strong position.

Lord Kitchener remained with our force but took up his residence at Kopje station two miles north; on June 14 he was nearly captured by De Wet, who, recrossing the line near this spot, attacked a construction train which had come to repair the line.

Lord Kitchener rode into our camp at 3 A.M. shouting 'Colonel Spens! Colonel Spens! turn out your men at once!' In a very few minutes two of our companies together with all our mounted men and two guns started off in the moonlight and were successful in driving the Boers off. The remainder of us stood to arms—a bitterly cold night and freezing hard, a quarter of an inch of ice was on the buckets. The Boers hung round us all day, shelling us from three directions, but did no damage; and towards evening, on the approach of Lord Methuen's column, they retired.

On the 16th De Wet's farm was burnt and the following proclamation issued:

'V. R.

'NOTICE.

'Under instructions from the Field-Marshal Commanding-in-Chief all Commandants of Stations will make every effort to make known to the inhabitants through their Mounted Troops, that interference with our Railway and Telegraph Lines cannot be tolerated, and that in all future cases the nearest farm to the break will be burnt to the ground. The burning will be carried out immediately under the orders of the nearest Station Commander.

'In this connection the Farm of Christian de Wet has been burnt because a bridge close to it was destroyed by the Boers two days ago.

(Signed) 'H. L. SMITH-DORRIEN,
'Major-General,
'Commanding Lines of Communication,
'*June* 15, 1900. *'Pretoria to Kroonstadt.'*

(The above order was then repeated in Dutch.)

On June 18 'E' company and the Volunteer Service Company under Captain W. S. R. Radcliffe and Captain Head proceeded to Honings Spruit to guard the line; they only just had time to entrench themselves when they were attacked by the Boers under Olivier; but, owing to the excellence of the entrenchments, very little damage was done in a fight of several hours, during which they were freely shelled.

About this period a number of boxes and bales of warm clothing arrived, generously provided by the inhabitants of Shropshire and Herefordshire. The contents were much appreciated. The nights were now bitterly cold and the men had nothing but their much worn khaki to wear. Owing to all the warm clothing having been captured and burnt by De Wet at Roodeval station there was little chance of it being renewed at this period, therefore these presents were particularly welcome.

We remained in the neighbourhood of Rhenoster till July 7, nothing of great importance happening. A number of farms were burnt during the period and Lord Roberts' lines of communication were considered now so safe that Lady Roberts passed through with a train-load of nurses, after a specially watchful night on our part. The arrival of these ladies (who distributed papers, cigarettes and bread to the men) was a welcome interlude to us, who for some months had not gazed upon the fair sex.

At 3 P.M. on July 7, 1900, we left Rhenoster, and arrived at Viljoens Drift on the Vaal the following morning. Our engine made several attempts to surmount the steep ascent from the temporary bridge over the Vaal, but eventually had to give it up and take us up piecemeal. At 3 A.M. on 9th we arrived at Elandsfontein, where we boarded another train and with the men sitting on some coal trucks we started gaily off; but we only got a mile when we stopped and went back again, the driver saying he could not pull the load. So, after leaving three coal trucks behind, we started again and eventually arrived at Irene station at 11 A.M. Here we met our old brigade with the exception of the Canadians, whose place was taken by the Suffolk Regiment.

Before proceeding any further it may be as well to briefly explain the general situation in the western Transvaal.

The Boers in this part of the country had offered no resistance to the advance of our forces westwards and General Barton had been left behind as military governor of the whole district from Klerksdorp to Krugersdorp with only half his Fusilier brigade, a battery, and a few mounted troops. But the Boers (encouraged by De Wet's successes, and under the leadership of De la Rey) were being quietly re-organised. After having peacefully returned to their farms and given Lord Roberts reason to believe that this part of the country, at any rate, was tranquil, they were preparing to rise and once more take the field.

Lord Roberts' energies were chiefly directed against the main Boer army under Botha on the east of Pretoria and to the capture of the railway line to Delagoa Bay. The preparations of the Boers in the western Transvaal were not looked upon as very serious. Several small garrisons had been dotted about with the idea of holding the country, which was supposed to be pacified, but which would be a source of weakness in a hostile country.

Lord Roberts, thinking the western Transvaal quiet, ordered General Baden-Powell, who was occupying Rustenburg, to advance to the north of Pretoria. Preparatory to this move General Baden-Powell retired from Rustenburg with the idea of occupying Commando and Zilikats Neks. He had no sooner done so than the Boers under Lemmer advanced on Rustenburg and occupied Olifants Nek, thus obtaining communication with Liebenberg in the Hekpoort Valley. Baden-Powell had then to hurry back to Rustenburg. Lord Roberts met the situation by sending the Scots Greys and four guns to hold Commando and Zilikats Neks and by bringing up Smith-Dorrien with the Gordon Highlanders and King's Shropshire Light Infantry and a few mounted men and two guns to clear out the Boers south of the Magaliesberg.

We left the regiment encamped (half at Irene, half at Elandsfontein) and under orders to proceed to Pretoria to join in the general advance towards Delagoa Bay.

However, late on the night of July 9 we received orders to proceed, together with the Gordon Highlanders, to Krugersdorp, where we arrived about 4 A.M. the following day and were most hospitably treated by the Royal Welch Fusiliers.

At 7 A.M. on July 11 our small force, consisting of the Gordon Highlanders, two guns of the 78th Battery under Lieutenant Turner, half the 50th company of Imperial Yeomanry under Sir James Miller, and ourselves—the whole under General Smith-Dorrien—started north against the Boers in the Hekpoort Valley. We were to be supported by the Scots Greys and four guns of 'O' Battery, advancing from the north. Having marched about ten miles, the enemy under Sard Oosthuizen were discovered occupying a position at Dwarsvley, which position closes in the Hekpoort Valley on the south. The guns were sent forward to a position between two knolls facing the ridge; but, going beyond the knolls, which were to be held by the Gordons, soon found themselves exposed to a deadly fire from a party of Boers who had galloped to a hill only 800 yards away. The gunners displayed great bravery, but all the guns were soon out of action. In the meantime the Gordons had occupied the two knolls in question and the Boers were unable to capture the guns. While this was taking place in the front the battalion was engaged in holding the right flank and driving off a party of Boers who had crept round to the rear and attacked our baggage. In doing this Captain C. P. Higginson and three men were wounded.

At 1.25 P.M. a signal message from Krugersdorp conveyed an order from Lord Roberts cancelling the operations. This order was sent on account of the failure of the expected support from the Scots Greys, who, as stated, were engaged at Zilikats Nek.

Smith-Dorrien then gave the order to retire, but on receiving a message from Colonel Macbean that the Gordons could hold on all day, the order was cancelled and attempts were made to get the guns away.

Two teams were sent up, one taking cover under one of the small kopjes occupied by the Gordons, the other attempting to bring a gun away—but three horses being shot down the remainder of the team galloped off towards the Boer position and stopped grazing in the dead ground at the foot of it. Colonel Macbean then called for volunteers to drag the guns under cover—ten men came forward, but five were shot including Captain Younger, who was killed. The Gordons were now running short of ammunition and a fresh supply was taken to them by some men of 'A' company, who carried up the boxes under a heavy fire to the firing line.[1]

No further attempt was made to save the guns till nightfall, when they were pulled under cover by ropes and eventually got safely away, no doubt to the considerable annoyance of the Boers who, headed by Oosthuizen, had made a gallant charge to capture them, in which their brave commander was killed. He was described by a bookseller at Krugersdorp, who himself had been out on commando but was taken prisoner by us at Dundee, as a great friend of his and a nice, quiet, unassuming man, with red hair and a bull neck, and very brave. He was severely wounded at Dundee but was fighting three weeks afterwards. His weakness was his love for whisky!

We retired during the night to Krugersdorp in accordance with the orders received from Lord Roberts and arrived there at 5 A.M. In case of being waylaid we returned by a different route to that by which we had marched out the previous morning.

[1] The names of these men as far as can be ascertained at present are as follows:

Corporal Jarvis,	Pte. Roberts,
Lance-Corporal Clark,	Pte. Stevenson,
Pte. Boden,	Pte. Smith,
Pte. Farrington,	Pte. Watkins,
Pte. Palmer,	Pte. Crowe.

It is well that they should be recorded.

The following order was issued :

'The Major-General commanding the Brigade wishes to place on record how highly he appreciated

'(1) The gallantry of Captⁿ Gordon, Captⁿ Younger and the men of the Gordon Highlanders in their attempts to drag away the guns on the 11th inst. and how deeply he deplores the death of Captⁿ Younger while performing this brave deed.

'(2) The devoted bravery with which 2 Lt. Turner, R.A., and the men of the 78th Battery fought their guns until 15 out of 17 were wounded, Lieut. Turner himself, although three times wounded, being the last to take cover.

'(3) The cool courage shewn by the men of the Shropshire Light Infantry who for nearly a mile under a heavy fire carried ammunition to the Gordon Highlanders and returned.

'The Major-General feels that he owes all troops engaged a debt of gratitude for their steadiness and promptness in repelling attack throughout the day, and it gives him much pleasure to be able to report to the Field-Marshal C.-in-C. that, thanks to the behaviour of the splendid troops he had under him, he was able to inflict more loss on the enemy than they caused to his force.

'In conclusion he would like all ranks to understand that he had no intention of withdrawing that night, as never before have the 19th Brigade retired, but he received a peremptory order from Lord Roberts himself, who was unaware that the force was fighting, to return at once, as he had some fresh plan of operations in view.'

We remained at Krugersdorp till July 18, when we were joined by the 5th Fusiliers, the 58th and Loyal North Lancashire Regiments, General Douglas's brigade and the 3rd battalion of Imperial Yeomanry under Lord Chesham. The command of the whole force was taken over by Lord Methuen, who had been brought up from the neighbourhood of Lindley for another attack on Hekpoort and Olifants Nek, while Baden-Powell was ordered to co-operate with him from the north.

On July 18 we left Krugersdorp at 2.30 P.M. in two columns: (*a*) under Lord Methuen and General Smith-Dorrien—the Shropshires, Gordons, 4th Field Battery and two pom-poms with the 3rd, 5th and 10th battalions of Imperial Yeomanry under Lord Chesham;

(*b*) under General Douglas, Loyal North Lancashires, Northumberland Fusiliers, half of 3rd South Wales Borderers, four guns 20th Field Battery, two Howitzers and half a squadron of Imperial Yeomanry.

July 19. Marched to Hekpoort, a narrow pass through the Witwatersrand; but finding it occupied we went further north to Zeekoehoek. Arriving there at nightfall our waggons were unable to cross the drift; so we spent a cold and hungry night.

July 20. We went through the pass with but little opposition and marched next day against Olifants Nek, which was held by a force of 900 Boers and two pom-poms under Du Plessis.

The exit from this pass to the north was presumed to be held by Baden-Powell.

The Boers opened fire on our left flank about 8.30 A.M. and were attacked by the 5th Fusiliers. 'A' and 'H' companies of the battalion attacked a kopje on our right flank. This being evacuated by the enemy, the regiment attacked the pass—but the Boers had fled and their retreat to the north was not, as we had hoped, cut off.

The land in this part of South Africa is some of the richest in the country, a wooded and well-watered district, with groves of orange trees and fields of tobacco, and many pigs, geese and fowls. We dined well that night, encamped in the woods close to the river—hare-soup, sucking-pig, roast goose, plum-duff, oranges and rum-punch. The following day was Sunday, which was spent wandering round the orange groves which were heavy with fruit, sackfuls of which we carried away. We also espied two nice pigs and half a dozen suckers, which we marked down as our own. We returned to lunch and a party of yeomanry went out to burn the farm, which belonged to Du Plessis. After lunch we sent a cart for the pigs; but, alas! the yeomanry were before us and got the suckers! But we managed to get the pigs, and filled up the rest of the cart with oranges. But we were not fated to remain long in such luxurious surroundings, as Lord Methuen was ordered south again to stop Liebenberg's train-wrecking attacks on the Potchefstroom line.

Leaving the Loyal North Lancashires to hold Olifants Nek we started at 8.30 A.M. on July 23, and marched south forty-six miles to Bank station, which was reached on July 26. Here the line had been torn up and a train wrecked. Lord Methuen's force left us and proceeded to Potchefstroom and the Gordon Highlanders went to Krugersdorp. They had no boots, so we were left to bring a convoy on to Potchefstroom. July 28 we were joined by the City Imperial Volunteers and thirty-three sick and wounded, who were captured in the wrecked train and sent in by the Boers. The convoy arrived on the 29th and we proceeded towards Potchefstroom the same day.

An unfortunate accident, attended with considerable loss of life, happened on the way. A heavy train of supplies passed us just as we were reaching camp at Frederikstad on July 30; and seated on the trucks was a draft of ours which had lately arrived in the country, together with some sick men rejoining. Our yeomanry scouts had been ordered to inspect the line as they advanced, but they failed to perceive that on a sharp curve, where the line ran down a decline, some of the fishplates had been unscrewed, and the rails loosened but carefully replaced. The engine on reaching the spot was derailed and the heavy supply-trucks ran up one on the top of the other. Thirteen of our men were killed, together with the engine-driver, and forty-one injured.

Captain W. S. R. Radcliffe, Lieutenant E. P. Dorrien-Smith and Lieutenant H. P. Harris-Edge (the latter belonging to the Volunteer Company, who were on the train) had a narrow escape, having shortly before the accident moved from the front carriage, which was completely telescoped, to a rear carriage which was unharmed.

The following men of the battalion were killed in the accident and buried the next day, Colonel J. Spens reading the service:

1503	Private	A. Goodwin	3346	Private	T. Cadwallader
5928	,,	J. Livesay	1545	,,	J. Hobby [1]
3345	,,	E. Jones	1216	,,	E. Scott
5892	,,	T. Parry	3829	,,	J. Allen
7370	,,	T. Gough	7363	,,	H. Jones
7358	,,	J. Wright	5949	,,	H. Lippett
2779	,,	J. White			

On the morning after the accident, at 7 A.M., we received orders to fall in at once and were informed that a message had come in from the

[1] This man had been wounded at Hout Nek and was just rejoining.

Boers ordering us to surrender in half an hour, the conditions being that we were to hand in our arms and could then proceed to Potchefstroom.

The Boers had apparently crept up close to the City Imperial Volunteer camp, which was about a mile away, and presumably thinking they were the survivors of the wrecked train had sent in a summons for the immediate surrender of the whole force. But on discovering our strength they quickly departed.

It must be remembered that, on July 15, De Wet accompanied by President Steyn had broken away from the big 'round up' of the Free State forces in the Brandwater basin and had escaped through Slabberts Nek with a force of 2600 men, four guns and a Maxim, and a convoy nearly three miles long.

De Wet's plan was to make a safe passage for President Steyn, who wished to cross into the Transvaal and confer with the Boer Government at Machadodorp.

Lord Roberts' plan was to complete a semicircle round De Wet on the south and force him across the Vaal into the hands of Lord Methuen. The driving columns, who were under the command of Lord Kitchener, consisted of 11,000 troops.

De Wet's laager was near Reitzburg, twenty-five miles south of Potchefstroom.

North of the Vaal was Methuen's column at Potchefstroom, Smith-Dorrien's consisting of the City Imperial Volunteers and ourselves and 120 Imperial Yeomanry on the railway line between Potchefstroom and Krugersdorp, and Barton's force of 3000 men at Krugersdorp.

North of this line again, holding the passes of the Magaliesberg, was a force under Ian Hamilton. It will be recollected that one of these passes, Olifants Nek, was held by the Loyal North Lancashires left there by Lord Methuen on July 23—but this regiment had been withdrawn about August 9 by the order of Lord Roberts, who two days later ordered Ian Hamilton to re-occupy it.

We left the regiment on July 31 encamped at Frederikstad. The day was occupied in clearing the line of the wrecked train. Eight teams of 128 oxen were employed in trying to hoist the engine upright but without effect, the trek chains not being able to stand the strain.

On August 3 one company proceeded to Potchefstroom as escort for a convoy of supplies. 'A' and 'B' companies were on permanent outpost duty on a kopje two miles from the station from August 1 to August 8, a position which we strongly fortified. 'E,' 'F,' 'G,' and 'H' were with the Suffolk Yeomanry and two guns marched to Welverdiend station, the remainder of the battalion, with the City Imperial Volunteers, remaining at Frederikstad.

August 4 was spent in entrenching ourselves at Welverdiend, and on the 5th we marched to Bank station. The stationmaster's daughter sang and played hymns to us during dinner, the 'Old Hundredth' and 'Ora pro nobis' being amongst her repertoire. At breakfast the following morning she continued her programme—but, being Monday, she blossomed out into valses and lighter music. Two supply trains arrived at about 1 P.M., and two companies being placed in each train and with the yeomanry galloping alongside and the guns and baggage following we escorted them back to Welverdiend. One train proceeded on to Frederikstad that night, the second starting the following morning (August 7), but the latter returned after proceeding about four miles as the Boers were about. Later it started again and with a larger escort got safely through.

The engine being short of fuel we stacked it up with the palings round Welverdiend station.

On August 8 we heard that De Wet had crossed the Vaal, and that evening we received orders to return to Bank station.

On August 9 the City Imperial Volunteers and our four companies left at Frederikstad rejoined us, two of our companies ('A' and 'B') marching forty-three miles in thirty-two hours to do so. This was, however, as it turned out, an unfortunate move.

On August 10 about 5 P.M. we were ordered back to Welverdiend, the City Imperial Volunteers to go half-way. De Wet was reported to be at Losberg and Broadwood coming round to the east of him with Methuen to the west.

We started by train at the rate of three miles an hour, six companies on the train, two marching alongside and the yeomanry in front. The latter had on the way a hot skirmish with a party of Boers to the south of the line.

After going half-way to Welverdiend the train shook with the concussion of several heavy explosions, followed shortly afterwards by eight or nine smaller ones. Suspecting the line to be blown up ahead we had to go with extreme caution, and reaching Welverdiend at 1 A.M. on August 11 heard that a culvert and the line had been blown up half a mile further on.

The next day we heard that De Wet with 300 waggons had crossed the line between 9 P.M. and 1 A.M. He had again safely eluded the various columns sent to surround him.

That evening several large columns turned up under Lord Kitchener and we all camped at Welverdiend and cursed our luck.

In the meanwhile Methuen had pushed on and harassed De Wet's column in every way. We followed in hot pursuit the next morning.

A difficult drift over the Mooi River delayed us the first day, but on August 13, breakfasting at 1.45 A.M., we marched twenty-five miles through some of the worst country we had experienced, the whole countryside black and burnt, a strong head wind and clouds of dust and cinders. But all was excitement! De Wet was in front and, hustled by Methuen, was dropping waggons, abandoning guns, and releasing prisoners. In front of De Wet were the precipitous heights of the Magaliesberg with, as we presumed, all the passes held. Our only fear was that De Wet would double back into the Free State. But no! he was heading for Olifants Nek, with Methuen to the north of him heading him away from Magato Nek and Broadwood harassing his rear, we hastening behind as fast as we could.

But we were again doomed to disappointment; Olifants Nek was not held and De Wet with Steyn escaped safely through on the morning of August 14—whose mistake it was does not enter into this history. There was more work to be done and we were not left idle.

On August 15 news was received that a force of colonials under Colonel Hore, who had been holding a post at Brakfontein on the Elands River, was still holding out. It had been surrounded by De la Rey since August 4. Several attempts had been made to relieve it, but without success, and it was supposed to have been compelled to surrender.

The following order was published by Lord Kitchener:

'Colonel Hore and some 300 Bushmen and Rhodesian Horse have held out against De la Rey's Commandos for over a week, and to relieve them the G.O.C. is forced to call on the troops for one more arduous march commencing at 4 A.M.

'No words can describe the G.O.C.'s pride at the splendid way the infantry have marched. It has drawn forth the admiration of all the mounted troops.'

Lord Kitchener started off on August 15 with Broadwood's, Little's and Smith-Dorrien's brigades; but on hearing of our approach De la Rey decamped and the brave garrison was relieved.

We reached their camp at Brakfontein about 1 P.M. on the 16th. Their laager was a most interesting sight, surrounded by a trench cut practically out of the solid rock and strengthened in places with the supplies with which the camp was well furnished. A tattered Union Jack was flying over all.

The defenders had had about seventy casualties, mostly on the first day and before they had hewn their trench out of the rock.

The stench from the bodies of about 170 dead horses and oxen who were shot down by the Boers at the beginning of the siege was sickening and must have added to the difficulties of the defenders.

We remained here, feasting on full rations for a change, till August 18, when we started east towards Rustenburg, marching nineteen miles that day. On the way we passed Lord Methuen's column. He had forced his way through Magato Nek in the hopes of still catching up with De Wet, but had been ordered by Lord Roberts to take over charge of the Mafeking district, to which he was then making his way.

On the 19th we passed through Magato Nek and marched seventeen miles to Rustenburg.

On the 21st we halted at Wolhuters Kop, where Lord Kitchener, with half of the City Imperial Volunteers, left us and proceeded to Pretoria.

About this time De Wet accompanied by about 300 men broke back south and crossed over to the south side of the Magaliesberg by a precipitous pathway west of Commando Nek, the remainder of his force being for the time dispersed.

We marched north to Wolvekraal on 22nd and back to Wolhuters Kop on 23rd, and then to Pretoria via Commando Nek, where we encamped on the race-course on August 25.

Here we were fitted out with new boots and clothing and a few weakly men were weeded out. The men were, however, in fine condition and fit for anything; we had marched 1200 miles since leaving Graspan and were as hard as nails. On the 28th and 29th we started by train to rejoin the main army under Lord Roberts, who was then at Belfast. After considerable delay, owing to the block of traffic on the line, we reached that place on September 1.

Here we met again our old friends the Gordon Highlanders (75th); but we were not destined to see much of them, as on September 4 they proceeded north towards Lydenburg under General Ian Hamilton and we were to proceed south on the next day and join French, who, with two cavalry brigades, had reached Carolina, our eventual destination being Barberton.

The general situation at this period was that the main Boer army under Botha had been defeated and had retired from Bergendal on August 27. From here he retreated north to Lydenburg, out of which he had been driven by Buller and Ian Hamilton. East of Belfast the ground drops suddenly 3000 to 4000 feet, and the country from here to the coast is hot and unhealthy and cut up by precipitous mountains and deep gorges. Down one of these runs the railway to Komati Poort, at which point it enters Portuguese territory, and thence proceeds to Delagoa Bay.

President Kruger and his government had taken up their quarters on this line. Enormous accumulations of supplies were stored near Komati Poort and at Barberton, where also had been collected most of the rolling stock of the country. Two thousand of our prisoners were at Nooitgedacht.

Botha was east of Lydenburg, Viljoen on the railway line and Smuts and Fourie with the Ermelo, Bethel and Standerton commandos, holding the different passes between Carolina and Barberton.

Lord Roberts, hoping to finally defeat the Boer army by driving them up against the Portuguese frontier, had sent Buller and Ian Hamilton to Lydenburg, while Pole-Carew advanced in the centre along the railway line and General French proceeded to Carolina on the extreme right.

With General French were Dickson's and Gordon's cavalry brigades,

Mahon's and Hutton's columns, the Suffolk Regiment and ourselves, the two latter being under the command of Colonel Spens.

We left the regiment camped at Belfast, where we arrived on September 1. On the 2nd we attended divine service, where the parson and the band rivalled one another as to who should lead the hymns; the band usually started in one key, the parson with a very powerful voice coming in a bar late in a different one.

An amusing incident occurred here one day. An officer of the regiment, when visiting one of his picquets soon after the issue of rations, was surprised to see a newly-made grave. The earth, only lately filled in, was surmounted by a wreath of tin flowers and a wooden cross marked the spot.

On the cross was the following inscription:

'Sacred to the memory of an old trek ox
That faithfully did duty with the 19th Brigade from Graspan to Belfast, afterwards becoming the discarded dinner of four Soldiers.
As such he was buried near this spot.
"God's humbler servant of a meaner clay
"Must share the honours of that glorious day."
R.I.P.'

On September 5 we started with a large ox convoy carrying eight days' supplies for French's column. The oxen were in a very poor condition and the intense cold at night did not improve them. The country round Belfast is some of the highest in the Transvaal, being 5000 feet above the sea level, and the weather at this period of the year is bitterly cold at night. We must have lost sixty or seventy oxen the first two days.

On September 7 we reached Carolina, where we spent the morning digging up potatoes. One officer seeing some bean-stalks shouted 'Hurrah, beans!' and began hastily to dig them up, thinking they grew like potatoes!

On September 9 we started at daybreak for Barberton, and soon found ourselves in a very rough and mountainous country.

Hutton's column was on our extreme left, then came Mahon, then Spens, and on the right Gordon's cavalry brigade.

The enemy were encountered at Buffels Spruit holding the high ground west of the drift. Their left was attacked and driven back by Gordon, but only to retire into a stronger position further east, which was well protected by nature and had a good field of fire. Out of this they were driven by the Suffolk Regiment assisted by two of our companies ('B' and 'E') working round and so hastening their departure and leaving clear our descent to the drift.

The road from here onwards was very bad. On the 11th we marched to Silverkop, crossing five very bad drifts, which with our enormous train of ox waggons was a tedious business, and on the 12th we crossed the Komati River at Hlomohlom.

On the 13th Nelhoogte Nek had to be captured—it was held by the Boers under Smuts.

Gordon's cavalry were sent to the right and Mahon's to the left, while the precipitous heights immediately flanking the pass were attacked and captured by the regiment, which allowed two of our companies to advance up the road.

The pursuit was taken up by these two companies, and seven waggons full of supplies were captured by a small party of 'H' company, who returned to camp that evening with eggs and chickens, bags of sugar, tea, flour, coffee, Boer hats, shirts and boots and a gramophone with five tunes—'Watchman, what of the Night?' 'The Lost Chord,' 'The Anvil Chorus,' etc.—and other luxuries besides.

The capture of the pass placed Barberton at French's mercy, and on the same day he, with the 1st Cavalry Brigade, taking a goat-track down the hills, pushed boldly on and captured the town, forty-four engines, two trains and an immense amount of stores, together with eighty British prisoners and 2500 Boer refugees and the Mayor of Vryheid, with £10,000 in gold and notes, and later on fifty more engines.

We had been left behind, together with Dickson's and Mahon's brigades, and were engaged in hauling the waggons over the pass till September 19.

On the 20th we marched towards Barberton, reaching there September 21—Colonel Spens was appointed military governor of the town and the battalion was detailed to garrison it.

The following order was published by General French on the break up of the force:

'I cannot refrain from commenting particularly on the magnificent marching powers and gallant endurance of the 2nd Battalion King's Shropshire Light Infantry, who in every kind of weather have kept up with and supported the mounted troops.

'I feel that the hardest part of the work has inevitably fallen upon them and that they have splendidly upheld the traditions of the magnificent regiment to which they belong.'

On September 27 fifty men under Lieutenants Head and Marshall escorted prisoners to Pretoria.

On October 9 a draft of 100 men with Second-Lieutenant F. E. N. Wrench and J. H. A. Payne joined the battalion.

We remained at Barberton till relieved by the Welch Regiment on October 13, the time being spent, when not on outpost duty, in playing cricket, shooting buck and hunting with the Barberton Hounds.

Owing to the number of engines captured at Barberton engine-drivers were badly needed, and men with knowledge of machinery were in great request. Even a man in the regiment who had been a watch-maker applied for the billet of an engine driver, 'thinking his knowledge of machinery would be useful under the circumstances.' From this period onwards the campaign took the form of guerilla warfare. President Kruger had fled the country, and a great number of faint-hearted Boers together with most of the foreigners fighting on their side had surrendered to the Portuguese at Komati Poort. The bulk of the Boer guns had been sunk in the Komati River, huge masses of stores had been burnt and the woodwork of the rolling stock destroyed by fire, leaving little but the framework of the trucks and carriages which had been accumulated at this, the far end of the line. President Steyn who, it will be remembered, escaped with De Wet through Olifants Nek had after an adventurous journey joined the Boer forces at Machadodorp just previous to the battle of Bergendal. He, with his experiences of wriggling out of a tight place, put fresh heart into the disorganised Boer forces and together with Botha, Viljoen and all the Boer stalwarts retreated northwards from Komati Poort and regained the high veldt round Buller's left flank. The President continued his journey back to the Free State and joined De Wet on October 31; and they concocted a plan to invade Cape Colony so as to give the Transvaalers time to reorganise their scattered and disheartened forces and divert the attention of the British from them for the time being.

On October 13 and 14 we left Barberton by train in half-battalions for Belfast. After many stoppages and several break-downs, due probably to the inexperience of the engine-drivers, we arrived at Belfast on October 16, where we found our old Brigadier-General, Smith-Dorrien, in command, and the Gordon Highlanders, Suffolk Regiment, 1st Royal Irish, two squadrons 5th Lancers and the Canadian Dragoons forming the garrison. On October 18 the Volunteer Service Company under Lieutenant B. Head left us for England, having escorted 135 Boer prisoners as far as Barberton. We were sorry to lose them. They had joined us on May 2, and during the months they were with us we had marched many miles together.

It has been said that the campaign was now entering a new phase. It was assumed, the main Boer army being dispersed and the whole of the railway communications being in our hands, that it only remained to deal with scattered bodies of so-called 'rebels' and that our duties would be more those of police than soldiers until the country was pacified.

Lord Roberts was to give over the command to Lord Kitchener and a number of troops were shortly to be sent home—among them were the yeomanry and colonials, most of whom had only been enlisted for a year or till the end of the war. Many of these men considered that they had a right to go at the end of their year's engagement, and those who wished to do so were allowed to go home and not held to their liability of serving till the end of the war. The result was a great depletion of mounted troops just when they were particularly needed. Our infantry columns could march about the country with their ox convoys wherever they liked, but they were unable to catch the mounted Boer, unhampered as he now was with guns or waggons.

No regular system had yet been evolved to meet this new state of affairs and spasmodic marches were made with the idea of capturing and dispersing any Boer laager whose presence was heard of in the neighbourhood. But what could infantry do against such evasive troops?

On November 1, 1900, a combined march was made from Belfast

General Smith-Dorrien taking command of one column and Colonel Spens of the other.

Information had been received of a Boer laager on the Komati River half way between Belfast and Carolina.

Not long after starting the rain came down in torrents, and it rained the whole night, a perfect blizzard and intensely cold. The two columns combined at Van Wyk's Vlei, where it was decided to abandon the expedition as the weather was too thick to see anything distinctly. Now was the Boers' chance; galloping at the rear-guard, they drove in the cavalry and attacked the Gordon Highlanders who, however, managed to drive them off, with the help of the Canadian Dragoons. We then retired to Belfast.

On November 6, at 3.30 A.M., we again started out in the same direction under General Smith-Dorrien, and after marching about ten miles a number of Boers appeared on our left flank; but on being shelled by our guns they retired and hung about our flanks, eventually taking up a strong position near the Komati River. The regiment attacked them in front, being extended over a front of nearly three miles, while two companies of the Suffolk Regiment and the Canadian Mounted Rifles made a wide flanking movement against the enemy's left. The Boers on their flank being turned were forced to retire. Our casualties on this day were Colour-Sergeant Scouse and five men killed [1] and Major A. R. Austen and fifteen men wounded.

Nothing was to be gained by following up the enemy, and the following day orders were given to retire on Belfast. Again the Boers, now considerably reinforced, attacked our rear-guard and endeavoured to recapture the position they had been driven out of the previous day. In this they were forestalled by the Canadian mounted troops and two Canadian guns and two guns of the 84th Battery.

The Boers then galloped round our right flank, and the Canadian guns escaping them several hundred Boers charged impetuously, firing from the saddle. The Canadians sustained considerable losses in covering the retirement, but the guns were got safely away.

This mode of attack by the Boers was frequently used later in the war, but previous to this time was unknown.

[1] 1268 Pte E. Moran. 4725 Pte F. Elocks. 6185 Pte C. Spencer.
1539 „ J. Broderick. 5626 „ A. Watkins.

On November 9 we moved out of camp to the town of Belfast. Similar indecisive actions were taking place in other parts of the country with no effect except to raise the spirits of the Boers.

The bulk of our troops were chained to the railway while the Boers practically held possession of the rest of the country, their peace being occasionally disturbed by wandering columns whose operations were, owing to the difficulty of supplies, confined to a few days' march from the railway. The Boers on their part made frequent raids on the railway and lost no opportunity of injuring it. Meanwhile they were quickly reorganising their forces and beating up many a farmer who had returned to his home.

On November 29, 1900, the following special army order was issued by Lord Roberts, who handed over the command of the army to Lord Kitchener and left for England:

'SPECIAL ARMY ORDER.

HEADQUARTERS OF THE ARMY IN SOUTH AFRICA.

'JOHANNESBURG,
'*29th November* 1900.

'Being about to give up the command of the Army in South Africa into the able hands of General Lord Kitchener of Khartum, I feel that I cannot part with the comrades with whom I have been associated for nearly a year—often under very trying circumstances—without giving expression to my profound appreciation of the noble work they have performed for their Queen and Country and for me personally, and to my pride in the results they have achieved by their pluck and endurance, their discipline, and devotion to duty.

'I greatly regret that the ties which have bound us together are so soon to be severed, for I should like to remain with the Army until it is completely broken up, but I have come to the conclusion that, as Lord Kitchener has consented to take over the command, my presence is no longer required in South Africa, and that my duty calls me in another direction. But I shall never forget the officers and men of this force, be they Royal Navy, Colonials, Regulars, Militia, Yeomanry, or Volunteers; their interests will always be very dear to me, and I shall continue to work for the Army as long as I can work at all.

'The service which the South African Force has performed is, I venture to think, unique in the annals of war, inasmuch as it has been absolutely almost incessant for a whole year, in some cases more than a year. There has been no rest—no days off to recruit—no going into winter quarters as in other campaigns which have extended over a long period. For months together in fierce heat, in biting cold, and in pouring rain, you—my comrades—have marched and fought without a halt and bivouacked without shelter from the elements, and you frequently have had to continue marching with your clothes in rags and your boots without soles—time being of such great consequence that it was impossible for you to remain long enough in any one place to refit.

'When not engaged in actual battle, you have been continually shot at from behind kopjes by an invisible enemy, to whom every inch of the ground was familiar, and who, from the peculiar nature of the country, were able to inflict severe punishment while perfectly safe themselves.

'You have forced your way through dense jungles and over precipitous mountains, through and over which, with infinite manual labour, you have had to drag and haul guns and ox-waggons. You have covered, with almost incredible speed, enormous distances, and that often on a very short supply of food, and you have endured the sufferings, inevitable in war to sick and wounded men far from the base, without a murmur—even with cheerfulness. You have, in fact, acted up to the highest standard of patriotism, and by your conspicuous kindness and humanity towards your enemies, and your forbearance and good behaviour in the towns we have occupied, you have caused the Army of Great Britain to be as highly respected, as it must henceforth be greatly feared, in South Africa.

'Is it any wonder that I am intensely proud of the Army I have commanded, or that I regard you—my gallant and devoted comrades—with affection as well as admiration, and that I feel deeply the parting from you? Many of you—Colonials as well as Britishers—I hope to meet again, but those I may never see more will live in my memory and be held in high regard to my life's end.

'I have learnt much during the war, and the experience I have gained will greatly help me in the work that lies before me, which is, I conceive, to make the Army of the United Kingdom as perfect as it is possible for an army to be. This I shall strive to do with all my might.

'And now farewell! May God bless every member of the South African Army, and that you may be all spared to return to your homes and to find those dear to you well and happy is the earnest hope of your Commander,

'ROBERTS,
'*Field-Marshal.*'

With the departure of Lord Roberts a new era was to commence. Up till now battles had been fought with the idea of occupying positions. This method was to cease. We were in possession of the whole railway system and most of the large towns and nothing was to be gained by driving the Boers from positions which we did not afterwards wish to hold ourselves.

Successes were in future to be judged by the numbers of Boers killed or taken prisoner. Our fight for the rest of the war was to be against the inhabitants, bearing in mind the fact that a stubborn race of hardy fighters of European descent was not only to be conquered but absorbed within the empire. The bulk of the weaklings had already dropped out; but in spite of this the Boer forces at this period turned out eventually to number some 60,000 men, most of whom were mounted.

Now a man on foot is incapable of catching a man on horseback; so it followed that from this period till the end of the war the regiment was chiefly confined to garrison work, at first guarding a portion of the railway line and later building and occupying a section of the blockhouse line of which more anon. Companies were from time to time detached as escort to trains or convoys or to guard the baggage of a mounted column, but our work for the most part was of a defensive nature while that of the mounted troops was eventually to develop into a series of drives with the object of systematically clearing large tracts of country encircled by blockhouses which were joined together by many hundreds of miles of barbed wire fences.

Our headquarters remained at Belfast and our time was mostly spent in holding the outpost line to the south of that town—Monument Hill and Colliery Hill being the chief posts.

Viljoen was in command of the Boer forces to the north of the line (where the Boer Government also had its headquarters) while Botha was to the south of us, and both were actively engaged in rousing the enemy to a general offensive.

On Christmas Day, knowing that the staff had been fattening a turkey for some weeks, we asked them to come and dine with us and they brought the bird with them; and the following day, to shake down the Christmas dinner, we paraded at 3 A.M. with four of our companies, some of the 5th Lancers and mounted infantry, and proceeded to raid a farm seven miles away, during which process we lost one man (2842 Private H. Baylis).

On the night of December 28 Viljoen attacked and captured a post at Helvetia on the road from Machadodorp to Lydenburg and carried off a 4·7 gun; and on the following day 'E,' 'F,' 'G,' and 'H' companies under Major P. Bulman were sent to reoccupy the position, and there they remained till February 11, 1901, holding the road for the safe passage of the Lydenburg convoys.

Encouraged by this success Botha formed a grand scheme of attack on the garrisons of the Delagoa Railway from Pan to Machadodorp. A simultaneous attack was to be delivered by Botha from the south and Viljoen from the north on Pan, Wonderfontein, Belfast, Dalmanutha, Nooigedacht and Machadodorp. The attack took place on the night of January 7, a thick driving mist giving the enemy a good chance of surprising the garrisons.

The main attack was delivered against Belfast, where Smith-Dorrien held a perimeter of fifteen miles.

Since our departure for Machadodorp, Monument Hill had been held by a company of the Royal Irish Regiment while Colliery Hill was held by a party of the regiment under Lieutenant C. Marshall; between these posts lay the town defended by smaller picquets.

To the south of the town and on the other side of the railway were posts held by the Gordon Highlanders. Punctually at midnight the attack on the northern posts took place, Müller with the Johannesburg and Boksburg commandoes attacking Monument Hill while Wolmarans and the Staats Artillerie, now organised as a mounted corps, tackled Colliery Hill. Between these two points Viljoen, with a reserve, held himself ready to make a direct attack on the town as soon as the posts fell. Botha's attack on the southern posts took place somewhat later, it being undertaken by Christian Botha with the Ermelo and Carolina commandos. The Monument Hill position was captured after a sharp hand-to-hand fight in which Captain Fosbery was killed and thirty-nine of his men killed or wounded. The small post at Colliery Hill held out bravely for an hour, but was eventually overpowered after a strenuous résistance, Lieutenant Marshall and nine of his men being killed or wounded; amongst the former being Sergeant Connor, who was shot through the throat. The Boer attack to the south of the line was less successful, only one small picquet being captured.

It remained with Viljoen to press the attack home; but now the forces which had been in the Boer favour reacted against them. They lost their way in the fog and commenced firing on one another, and the size of the perimeter prevented Viljoen from finding out if Botha's attack had been successful. Concluding finally that the co-operation had failed, and dreading artillery fire when daylight arrived, he called off the whole of his men before dawn and retired.

The attacks on the six other garrisons were carried out with extraordinary punctuality, but were not all pressed with equal vigour and met with an even smaller measure of success. Thus Botha's big scheme had signally failed. With all the advantages of the initiative, with no mobile columns in the field to interfere with his movements, with a great preponderance of force over any single garrison, the Commandant-General threw all these advantages to the winds, and instead of concentrating against Belfast or any other garrison that he might desire to reduce, disseminated his large force in a series of isolated attacks.

The following order was published on the 12th:

'The G.O.C. wishes to express his appreciation of the steadiness of the troops on the morning of the 8th inst. He would especially mention the fine defence of the R. Irish Regt. picquet on Monument Hill under that gallant officer Capn. Fosbery until overwhelmed by vastly superior numbers, also that of the King's Shropshire Lt. Infy. picquet under Lieut. Marshall. He regrets the losses but does not consider them heavy considering the determined nature of the attack.'

On January 23 the death of Her Majesty Queen Victoria was

announced and all flags lowered to half mast, but were raised the following day and a royal salute of twenty-one guns fired in honour of His Majesty King Edward VII's accession to the throne.

On the 25th a salute of eighty-one guns was fired during the funeral of Her late Majesty and the following telegram from King Edward was received by Lord Kitchener and promulgated to the army in South Africa :

'Am much touched by your kind telegram of sympathy, and beg you to convey my warmest thanks to my gallant army in South Africa.'

To which Lord Kitchener replied :

'Your Majesty's gracious telegram has been communicated to the troops. On behalf of the Army in South Africa I humbly beg to express our feelings of the utmost loyalty and devotion to your Majesty.'

On February 11, 1901, the four companies at Helvetia left that place, having been joined by three companies of the Devons and some mounted troops under General Walter Kitchener, and proceeded to Dullstroom. Two other columns co-operated, one from Machadodorp and the other from Belfast, composed of 180 Mounted Infantry, two guns, 400 men of the Royal Irish, Gordons, and Shropshires, under Major Austen, with the idea of converging on Windhoek, Viljoen's head-quarters, but the operations came to nothing. The Boers as usual retired before us, and on the force retiring on Belfast followed us up and attacked the rear, during which the force had fifteen casualties.

About this period the first blockhouses began to appear and a description of them may not be amiss. They consisted of two skins of corrugated iron four and a half inches apart, the interior being filled with shingle. Loopholes were provided and the whole roofed in. They were cheap and required little transport. The usual garrison was a non-commissioned officer and six men.

As defensive works they were very successful. The Boers had now but few guns and the four and a half inches of shingle was sufficient to stop a bullet.

These blockhouses were at first built at important points on the railway but eventually were established at regular intervals down the whole extent of the line, thus doing away with the necessity of mounted patrols.

The system was gradually increased, a continuous fencing of barbed wire running along the line and elaborate entanglements surrounding each blockhouse, the telephone linking up the whole system. They were originally used to guard the railway from molestation but eventually served to convert the railway into a barrier against the free passage of the enemy.

The perfection of the system was a long process. After the railway line was completed, lines of blockhouses were built along the main roads and linking up together formed huge kraals.

In this work our men now engaged, erecting and occupying blockhouses towards Dalmanutha in one direction and towards Wonderfontein in the other.

In the *London Gazette* of April 19, 1901, and dating from November 29, 1900, the following rewards appeared:

'To be Companion of the Most Honourable Order of the Bath :—Lieut.-Colonel James Spens. To be Companion of the Distinguished Service Order :—Captain Cecil Higginson. Medal for Distinguished Service in the Field :—4956 Corporal W. W. Marsden.'

On September 15 an attack was delivered on the defences of the town of Belfast, at about 9 P.M. Firing lasted until about 3 A.M., when the Boers withdrew. During the attack a small number of the enemy penetrated into the concentration camp and fired from the tents on our men.

From the *London Gazette*, September 27, 1901:

'The King has been graciously pleased to give orders for the following appointments to the Distinguished Service Order, for the following Promotions in the Army, and for the grant of the Medal for Distinguished Conduct in the Field, to the undermentioned Officers and Soldiers, in recognition of their services during the operations in South Africa. The whole to bear date 29 November 1900, except where otherwise stated.

To be Companions of the Distinguished Service Order

Lieutenant E. P. Dorrien-Smith, The King's (Shropshire Light Infantry).

Major P. Bulman, The King's (Shropshire Light Infantry).

Captain R. R. Gubbins, The King's (Shropshire Light Infantry).

Lieutenant H. G. Bryant, The King's (Shropshire Light Infantry).

To be Brevet-Majors

Captain O. H. E. Marescaux, The King's (Shropshire Light Infantry).

,, R. A. Smith, The King's (Shropshire Light Infantry).

,, W. S. W. Radcliffe, The King's (Shropshire Light Infantry).

To be Brevet Lieutenant-Colonel

Major A. R. Austen.

To have the Distinguished Conduct Medal

Sergeant W. Harrison.

,, G. Powell (now Colour-Sergeant).

,, R. H. Talbot (now discharged).

Corporal A. Jarvis (now Sergeant).

,, W. Shaw (now Lance-Sergeant).

Lance-Corporal T. Avery.

To be Companions of the Distinguished Service Order

1st Battalion Mounted Infantry:

Lieutenant C. W. Battye (King's Shropshire Light Infantry).

4th Battalion Mounted Infantry:

Lieutenant H. M. Smith (King's Shropshire Light Infantry).

To have the Distinguished Conduct Medal

Coldstream Guards:

Colour-Sergeant J. Gardham (now Sergeant-Major, 2nd King's Shropshire Light Infantry).'

On November 2, 1901, orders were received that a trench was to be dug three feet deep by three feet wide, so as to make a continuous trench from station to station; also a barbed wire fence. Each man in the blockhouse had to work two-and-a-half hours daily in addition to his other duties.

About this time a local company of mounted infantry was formed by Major Dawkins, C.M.G., purely of volunteers from the battalion. Its object was to make night raids on the town and its surroundings, and capture, if possible, any Boers that might come in after dark for food, etc.

This small force was very successful both at Belfast and later on at Carolina, making a number of captures and causing many a Boer to sleep out instead of comfortably in a farm house.

On November 24 Lieutenant Leach, one sergeant and twenty-five rank and file left Belfast as a crew to No. 7 armoured train.

On November 28 headquarters, with 'B,' 'D,' 'F,' and 'H' companies, entrained for Wonderfontein. Headquarters and 'D' and 'H' companies (under Major Dawkins) joined Colonel Fortescue's column which formed an escort to convoys for columns operating south-east of Carolina.

'B' and 'F' companies, under Captain Luard, joined Major Williams' column and operated between Wonderfontein and Ermelo. The remainder of the battalion proceeded to Machadodorp, being relieved at Belfast by the Gordon Highlanders.

During the month of December the usual routine was carried out. The two companies with Fortescue's column, and those with Williams', continued trekking between Wonderfontein and Carolina and Ermelo respectively.

The remainder of the battalion were recalled from Machadodorp to Belfast for duty on the blockhouse line.

April 3, 1902. The battalion entrained for Wonderfontein, *en route* to Carolina, arriving there on April 6.

April 24, 1902. The Volunteer Service Company having been delayed on their way to the coast left the battalion for England after a year's service in South Africa.

The following letter was published by Major C. T. Dawkins, C.M.G.:

'On the occasion of the 2nd Volunteer Company embarking for England after a year's service at the front, the C.O. wishes to place on record his high appreciation of the patriotic spirit shewn by the members of the Company in volunteering for active service at a time when the original excitement raised by the war had to a great extent subsided, and also of their excellent behaviour while attached to the 2nd Battalion. The work which they have been called upon to perform has been of a hard and uninteresting nature, but they have carried it out with credit to themselves and to the Regiment, and the C.O. desires to thank Captain Treasure and the Officers, Non-Commissioned Officers and Men of the Company for the constant support they have afforded him.

'In bidding the Company good-bye, the C.O. hopes the experience they have gained in the field will not be lost, and that the recollection of the days spent in South Africa will tend to strengthen the ties which bind together the various Line, Militia and Volunteer Battalions of the Counties of Shropshire and Herefordshire.'

On May 6, 1902, Lieut.-Colonel A. H. J. Doyle joined from India to take over the command of the battalion.

Rumours of peace had been in the air for some time previously, and leading Boers had been granted passes through the blockhouse lines to consult with their leaders, who were in conference at Pretoria. Towards the end, a practical armistice was arranged, so there was no surprise but considerable relief when peace was at length formally declared on June 1, 1902. On the following day, Lieutenant Hanbury, Sergeant-Major Gardham, two non-commissioned officers and seven men were sent off to England to represent the regiment at the coronation of King Edward VII.

All the world knows that this did not take place as arranged.

On June 3 a message from the Boer General Grobelaar was received for delivery to Commandant Steyn of the Carolina commando, directing him to be at Twyfelaar Farm at 10 A.M., June 11, with all his burghers fully armed—none to be absent.

OFFICERS' KITCHEN, CAROLINA—
February 2, 1902.

GOODEHOOP—February 19, 1902.

CAMP, GOODEHOOP PAN, WHERE MANY DUCKS AND COOTS WERE SLAIN—
February 19, 1902.

THE MESS CAPE-CART AT NO. 17 BLOCKHOUSE, WITH PREVIOUS DAY'S BAG—March 8, 1902.

HOSPITAL AT CAROLINA—March 10, 1902.

On June 3, 1902, the following telegram was received by Lord Kitchener, Commander-in-Chief, from His Majesty the King:

'Hearty congratulations on the termination of hostilities, and also congratulate my brave troops under your command for having brought this long and difficult campaign to so glorious and successful conclusion.'

June 6, 1902. The following telegram, received by Lord Kitchener from the Secretary of State for War, was published:

'3rd June. His Majesty's Government offer to you the most sincere congratulations on the energy, skill, and patience with which you have conducted this prolonged campaign, and would wish you to communicate to the troops under your orders their profound sense of the spirit and endurance with which they met every call made upon them, of their bravery in action, of their excellent discipline, perseverance, and of the humanity shewn by them throughout this trying period.'

On June 9 General Lucas Meyer came and breakfasted with us —he complained of a pain in his chest and our doctor gave him a large black bottle of cough mixture. He died, it will be remembered, not long afterwards, of heart trouble.

On June 9 General Grobelaar and his secretary, De Wet, arrived and stopped the night, having handed over the Ermelo commando.

They left early the following morning for Twyfellaar Farm to meet the Carolina commando. We followed later and 320 burghers with all arms and ammunition surrendered to us.

The names and addresses of all were taken and passes issued. A plenteous meal, which we had brought out with us and which was no doubt heartily appreciated, was partaken of. The waggons of food returned piled up with rifles.

The General, his secretary, the field cornet, and a German officer who had been fighting for the Boers dined with us that night and a convivial night followed, the field cornet having to be carried off to bed and the German singing his native patriotic songs far into the night. But all joined in singing 'God save the King' before we broke up.

The following days were fully occupied in issuing rations to surrendered Boers and their families.

THE GAOL, CAROLINA, USED AS A FORT—March 11, 1902.

CAMP, WONDERFONTEIN—March, 1902.

OUR WATER SUPPLY, NEAR WONDERFONTEIN

SOLDIERS' GRAVES AT WONDERFONTEIN. On unmarked graves a bottle is placed with name, regiment, rank, etc., inside. The officer's grave has a helmet.

BOER BURIAL GROUND NEAR WONDERFONTEIN. THE GRAVES OF THE BREITENBACK FAMILY.—OUR CAMP—March 13, 1902.

On June 19 this letter was received from Colonel Spens:

'Colonel Spens, on leaving the 85th King's Shropshire Light Infantry after serving in it for nearly 30 years, wishes to thank the Officers, N.C.O.'s, and men for the cordial and ready support he has at all times received from them during his command of the Regiment.

'The good spirit which exists among all ranks has been instrumental in keeping up the very high reputation gained by the Regiment, whether in quarters or during the recent severe campaign in South Africa. In wishing the Regiment good-bye, Colonel Spens wishes them the best of luck in the future and God-speed.'

The battalion was now about to quit South Africa.

Its ranks were greatly thinned. On July 5, 1902, a draft of two non-commissioned officers and 185 men joined from the details at Mullingar. Shortly before this the headquarters were moved from Carolina to Wonderfontein. The Shropshire company, 4th Mounted Infantry, rejoined after an absence of two and a half years. Their adventures are detailed in the following chapter.

The first batch of reservists left Middelburg for home under the command of Captain Hudson, 3rd King's Shropshire Light Infantry, and other reservists were forwarded as soon as passages could be secured for them.

On August 7, 1902, the battalion entrained for Newcastle, Natal, arriving there two days later.

On October 29, 1902, a notification was received that the battalion would probably embark at Durban on January 10, 1903, and that their destination would be India.

The battalion in consequence entrained for Ladysmith on November 22, 1902, and there remained in the Tin Town barracks until the time arrived for embarkation to India.

CAMP, March 9, 1902.

CHAPTER XVIII

A Short Diary of the 85th Company of the 4th Mounted Infantry in South Africa, January, 1900—June, 1902

Towards the end of January, 1900, when the regiment was still at Orange River, orders were received to at once form a mounted company of 120 non-commissioned officers and men.

It had at length been realised that to meet the Boers on equal terms more mounted men must be raised and accordingly, amongst many others, this rather rough-and-ready plan was evolved. Every regiment then in South Africa furnished one company. Four such companies, with its own staff, composed a Mounted Infantry regiment It was intended to brigade four regiments together to form a Mounted Infantry corps, but owing to lack of time this did not become an accomplished fact till later on. The command of the 85th company was given to Captain J. J. White, and he took with him, as section commanders, Lieutenant H. M. Smith, Lieutenant E. A. Underwood, Second-Lieutenant P. F. FitzGerald, and Second-Lieutenant J. C. Hooper.

The company left Orange River for De Aar, where horses and equipment were to be drawn, on January 23. Here we were joined by companies from the Royal Warwickshire Regiment, the Yorkshire Regiment, and the Duke of Cornwall's Light Infantry, and thus constituted we went through the war under the title of the 4th Regiment of Mounted Infantry. Lieutenant-Colonel St. G. C. Henry, 5th Fusiliers, was appointed Commanding Officer and Captain H. J. Everitt, Somerset Light Infantry, Adjutant. Eight days after our arrival at De Aar, that is to say on February 1, we were actually on march route to Orange River as fully fledged Mounted Infantry. In a short account (as this must be) it is impossible to convey any sort of idea of the amount of work, hurry, and bustle, that was crowded into these only too short eight days. The weather was extremely hot and no rain had fallen for a long time, conditions under which dust and flies reigned supreme. From January 24th to the 27th our time was fully occupied in taking over equipment, saddlery and horses; from the 28th to the 31st every minute of the day was devoted to drill, telling men off to their sections, selecting suitable men to the various horses, fitting saddlery, and generally to getting things into working order. It was with a light heart that we left De Aar behind us. The greater part of the cobs issued to us were Argentines, an excellent stamp of Mounted Infantry pony when broken, but possessed of a will of his own, and some proved a holy terror to the poor foot-soldier. The remainder of our cobs were the ordinary country-bred.

On February 1, as I have mentioned before, our march to Orange River began. I cannot pass on, however, without a word concerning the material of which this new mounted force was composed. The first men selected were naturally those who had already been trained in Mounted Infantry duties; but I think I am safe in saying that of these there were barely 10 per cent. per company. It must be remembered that the trained section was already in the country and that that section had taken nearly all the trained men. The remainder of the men were picked out as being of the right size and build, or because they had been used to horses when in civil employment. The great majority had never put leg across a horse in their lives, and knew nothing as regards feeding, grooming, or saddling a horse. Eight days, it must be allowed, was a short time indeed to mould into shape a mounted force of such material, and success could only possibly be achieved by every officer and man putting his whole soul into the work, and on Feb. 11, or only seventeen days after first having taken over our horses, we were fighting the enemy in the then Orange Free State, having previously marched day and night for four days. I lay stress on these points, as hard things have been said and written by the critics of the early days of the Mounted Infantry which undoubtedly are true, and it is comforting to many who served with that force to point out how great were the difficulties with which they contended.

The distance between De Aar and Orange River is seventy miles, and we arrived there on February 7; the march being done in easy stages, which gave both officers and men an opportunity of settling down to their new work, and the horses, a great part of which had not long been landed from a lengthy sea voyage, a chance to get into some sort of fettle for the hard work before them. We rested at Orange River for forty-eight hours, being camped near the bridge over the river. On the afternoon of February 9 camp was struck, and in conjunction with the remainder of the troops under Lord Roberts' immediate command, the great advance into the enemy's country began. From that date until the 14th we marched practically day and night, a halt of three or four hours being the most allowed us. On the 11th we had our first skirmish with the enemy. On the 14th we reached Wegdraai at 10 A.M., having marched through the night. From thence we proceeded to Jacobsdal, about five miles distant, and occupied that small town without resistance. The telegraph office, however, apprised General Cronje of our entry and a small party of Boers was despatched from Brown's Drift. This party attacked us as we retired from the town, and the skirmish which followed continued well into the evening. A few officers and men were within an ace of meeting these Boers face to face. So peaceful was the appearance of Jacobsdal that leave had been given to a certain proportion to remain behind, after the main body of the regiment had retired, to buy what food could be found, and it was only a timely warning of danger ahead from the commanding officer which cut short the interesting shopping. Scarcely had we passed the last house of the town before desultory shooting began and bullets were dropping uncomfortably close. So loaded were all, with loaves of bread, eggs, cigarettes, etc., that hurrying was a matter of extreme difficulty; but, on the other hand, so hungry were the friends awaiting us that we were extremely loath to part with our precious, but most inconvenient, burdens. I regret to say that the greater part of our purchases had finally to be deposited on the ground. Few of us had ever troubled to solve the problem of mounting a restive pony encumbered by four good loaves of bread and, as far as I remember, only one soldier-groom was clever enough to surmount the difficulty. Needless to say his welcome home was deservedly more warm than ours, and he and his friends enjoyed a better supper that evening. Unfortunately Colonel Henry was wounded in this skirmish and taken prisoner, but rejoined us a few days later at Paardeberg. On the 16th we arrived at Klip Drift, and here, for the first time, got into close touch with the rear-guard of General Cronje's force. A long and tedious day's fighting took place in the vicinity of this drift, and it is with regret that one has to confess that the Mounted Infantry did not cover themselves with glory. The Riet River, at this point, flows between steep wooded banks and an endeavour to utilise this to protect the horses was our downfall. The weary and overladen ponies were absolutely incapable of getting a footing on the steep banks and many slipped straight into the river. Others, terrified by the noise of the firing, broke loose only to share the same fate at some other point in the river. At dusk, part of the regiment received an order to retire back to the main body, but this did not reach all the companies. By those who remained the night and early morning were spent in endeavouring to get the horses out of the river-bed; and, after the most strenuous efforts, all but six were got safely to the bank. Daybreak found us again on the move, and, with only slight opposition, we bivouacked that night within two miles of General Cronje's main laager. Lord Kitchener with an aide-de-camp shared our bivouac on that night.

The following day found Cronje at bay near Paardeberg Drift, and the regiment was present at the action on that and the following days. It played no great part in the attack on the laager, but it is interesting to note that Kitchener's Kop was seized and held by the regiment from a very early hour, on the morning of the 18th. A section of the Warwick Company under Lieutenant Hankey took part in Colonel Hannay's charge on the laager in the afternoon, and was nearly wiped out, Lieutenant Hankey being killed. From February 19th to 27th, the day on which Cronje surrendered, we were constantly on the move, but on that date we marched to Kameelfontein, a small farm about four miles north of the drift at Paardeberg, and there we remained until all was ready for the move forward which took place on March 6. The long halt caused by the operations round Paardeberg, although undoubtedly detrimental to our cause, as giving the enemy time to collect both his scattered brains and men, was most welcome to the troops, and to the newly raised Mounted Infantry more than any. We now, for the first time, were able to get things into working order. The constant marching and fighting of the preceding fortnight had been a severe trial to all the Mounted Infantry regiments and a halt of a few days was almost imperative to their future success.

From this time forward all the regiments began to do good work, and were now a force to be reckoned with.

Lieutenant-Colonel Henry, having been left in the hospital by the Boers when they finally evacuated Jacobsdal, rejoined us about this time, although not quite recovered from his wound, and it was under his command that the regiment left Kameelfontein on March 6. On the 7th the action at Poplar Grove was fought. The 4th Mounted Infantry covered the left flank, and although some distance from the main action, got through a good deal of desultory fighting.

Lieutenant H. M. Smith was wounded in this action, and his valuable services lost to the company for nearly a year. He was sent with his section to seize an isolated kopje on our left flank and succeeded in doing so with little opposition. He had only been in possession a short time when a party of about thirty Boers appeared on the kopjes in front of him, and making very skilful use of dry nullahs and dead ground, managed to get to within close range without suffering any severe casualties.

A report from one of his men to the effect that a small party were gaining their flank caused him to move to that flank to see what was really happening. He had only been there a short time when a bullet struck a rock just in front of his head, causing the splinters to fly up in his face and temporarily blind him. Having already realised that the position had become untenable, he gave the order for the men to retire, which order was well carried out. In consequence of his wound he was unable to move, and on the arrival of the enemy was taken prisoner. That evening he was taken to the little town of Boshof, some twenty miles distant, and well treated by the Boers. A week later the town surrendered to Lord Methuen's force and Smith was at once sent to Kimberley, where it was found necessary to remove one eye. He was afterwards invalided to England, and rejoined us again at Komati Poort. For his meritorious services on this and other occasions he was awarded the Distinguished Service Order.

The regiment remained at Poplar Grove until the 11th, being under orders to escort a convoy coming from Kimberley. After three long and wearisome marches, during which time many rumours of threatened attacks were received, we finally got in touch with General Tucker's Division on the 14th at Driekop. Here we were informed that Lord Roberts had entered Bloemfontein without opposition on the morning of the 13th. On the 17th the regiment received orders to proceed to a small farm called Villambrosia, about ten miles east of Bloemfontein, and take up an outpost line, and we remained on this line during the greater part of the halt of the army at Bloemfontein. Our short period of comparative peace and rest was rudely disturbed on March 30, for on that evening we received orders to be ready to march at 3 A.M. the following morning to the waterworks near Bloemfontein. De Wet, it was found, was responsible for disturbing our peace, having laid a trap at Sannah's Post for General Broadwood's force which was returning to Bloemfontein from Thaba' Nchu. We came into action at about 9 A.M. on the 31st, and by 11 A.M. were in position holding Waterval Drift on the Modder River. Fighting was severe until 4 P.M., when General Colvile's Division put in an appearance, having marched from Bloemfontein. Owing to Colonel Martyr's failing to thoroughly appreciate the situation, the regiment never had any real chance of affecting the situation at the waterworks. Lieutenant Cantan of the Cornwall company and a few men were unfortunately taken prisoners during the course of this action. We remained in the vicinity of the waterworks the two following days, and on April 3 returned to our old camp at Villambrosia. On the 8th the whole regiment marched back to Bloemfontein, and bivouacked at Rustfontein. Stores and horses had by this time arrived in Bloemfontein, and we now started re-equipping in earnest. The boots and clothing of all the men were worn out and heavy casualties in horse-flesh had to be replaced. Over forty horses were required to complete the 85th company alone. Advantage was also taken of this halt to re-organise the whole force of Mounted Infantry. It was with great regret we learnt that in this reorganisation we were to lose our commanding officer, Colonel St. G. C. Henry, and his adjutant, Captain H. J. Everitt—the former having been appointed to command one of the Mounted Infantry corps, and the latter going with him as his staff officer. Major Handcock of the Yorkshire Regiment succeeded Colonel Henry in command of the regiment. The new corps to which we henceforward belonged was made up of the following regiments: 1st Victoria

Mounted Rifles, 1st Tasmanian Mounted Rifles, South Australian Mounted Rifles, 'J' Battery, Royal Horse Artillery, and ourselves. The corps was commanded by Colonel St. G. C. Henry, our late commanding officer. On April 21 we left our camp at Rustfontein, having been thoroughly flooded out during the last week of our stay there, and marched to Karree Kloof, arriving there on the 23rd. From this place we carried out two reconnaissances, and on May 2 the general advance on Pretoria began. Throughout this march the 4th corps of Mounted Infantry formed the advanced guard to Lord Roberts' centre column. Long marches were now the order of the day, but the fighting was never of a severe order. The Mounted Infantry, however, were engaged daily in clearing away parties of the enemy. Brandford was occupied with little trouble on May 3. The Vet River was crossed on the 5th after some little resistance had been offered, and on the 10th the mounted troops were engaged with the enemy in effecting the crossing of the Zand River. On the 12th Kroonstad was occupied after an engagement at Boshplat. We advanced from Kroonstad on the 22nd, and during the rest of our march through the Free State were practically unopposed. At Vereeniging we again met the enemy, but they fell back after offering but slight resistance. On the 28th we arrived within striking distance of Johannesburg; and during that and the following day the regiment was engaged in running fights in the neighbourhood of Natal Spruit station and Boxburgh, the work here being very well carried out and especially noted in Lord Roberts' general orders.

On the 30th we were camped at Orange Grove, just north of Johannesburg, and on the 31st Lord Roberts made his entry into the town. The march to Pretoria was at once begun, and with only slight opposition at Six Mile Spruit the capital was entered on June 5.

The enemy, under command of General Botha, now took up a position on the Donkerhoek ridge about sixteen miles east of Pretoria, and on June 10 Lord Roberts moved out to attack him. Severe fighting took place during the whole of the 11th and 12th. The enemy retired during the night of the 12th, having been compelled on the previous evening to evacuate Diamond Hill, which was the key of their position. In this action the 4th corps of Mounted Infantry formed part of the central column under General Pole-

Carew. On the enemy evacuating this position all the troops were withdrawn to Pretoria with the exception of the 4th corps of Mounted Infantry, who remained along the Donkerhoek and Diamond Hill ridges. The outpost work on this line was trying, the enemy being in close proximity and the line to be protected nearly twenty miles long.

We remained in this position until July 8, when we moved to Edendale, but the general advance did not commence until July 23. Still forming the advanced guard to the central column, the move on Middelburg commenced on this date. We arrived at the latter place on August 5, having met with little or no opposition. Immediately on our arrival the corps was despatched to Doornkop, a kopje about three miles north of Middelburg, and there took up an outpost line, holding it until August 19, when the general advance was resumed. By August 24 we had arrived at Belfast, and here the last pitched battle of the war took place. The action in and around Belfast continued from the 24th to the 27th, and the 4th Mounted Infantry were heavily engaged, more especially on the first day in the vicinity of Monument Hill. Lieutenant Tarbet of the Yorkshire Company and adjutant of the regiment was slightly wounded for the second time during the war.

No further opposition was offered to the central column during the march to Waterval Boven, where it arrived on August 31 and relieved General French's cavalry division, which had arrived a day or two earlier.

We rested here a few days, while the Commander-in-Chief was arranging a general shuffle of the cards with a view to endeavouring to corner the enemy before they could manage to escape over the border, or break back behind us.

On September 6 the 4th corps of Mounted Infantry left General Pole-Carew's division, for the time being, and joined General Hutton's force at Machadodorp. From here we plunged into the Kaap mountains, arriving seven days later at Kaapsche Hoop. The object of this march was to protect General French's left flank (who was now on his way to Barberton) and also to endeavour to turn the flank of the enemy under General Viljoen, who was barring the advance of General Pole-Carew on the railway. Both these objects were accomplished. A sharp action was fought at Weltevreden on the 8th, but after that we only encountered slight opposition. By far the greater difficulties were those offered by the nature of the country. The roads were merely mountain tracks, and it was only by the most strenuous efforts that the guns were got along. All other wheeled transport had to be left behind. September 13 was a particularly strenuous day, and although we were moving from early morn till dewy eve we barely covered six miles as the crow flies. On the 14th we arrived at Kaapsche Hoop, only twelve miles from our last bivouac, but it took us from 5 A.M. to 8 P.M. to accomplish this journey. On the 18th the 4th corps rejoined General Pole-Carew at Godwan River station and resumed its former duties of advance guard to his division. The remainder of the march to Komati Poort was uneventful. At Hector Spruit a great quantity of gun ammunition was found in the river and some half-dozen guns, but the latter had been destroyed. The railway line was absolutely blocked with engines and railway stock, a large proportion of which had been burnt or destroyed in other ways. Since leaving Godwan River station all wheeled transport had been left behind, which necessitated a cutting down of rations both for men and horses. The horses suffered severely, as at this time of year there is little or no grazing, and the best we could do for the poor hungry animals was to mix flour, which was found in large quantities in some of the railway trucks, with their scanty ration of corn. Komati Poort was reached on September 24, and on our arrival there we learnt that all the enemy who had been in front of us had been crossing the border for some days and were now safely in Lourenço Marques. Their camp was left standing, but contained little worth the trouble of carrying away.

We now fondly imagined that the war was to all intents and purposes over, and were buoyed up in this hope by the departure, shortly after our arrival at Komati Poort, of the colonial regiments of Mounted Infantry who had formed part of the 4th corps since our departure from Bloemfontein. The men of these corps had enlisted for a year, and their time being now very nearly up the great majority took their discharge, and very shortly sailed for their homes. From this time, therefore, the 4th corps of Mounted Infantry ceased to exist and until the end of the war we worked as a single regiment. Before continuing my story, a word in praise of these stout colonial corps, with whom it was our privilege to serve, seems in season. The officers and men were the pick of the colonies of Victoria, Tasmania, and South

Australia. They were chosen from amongst hundreds, who were at that time clamouring for the honour of helping the Mother Country in the hour of need. Knowing the material, it seems almost unnecessary to say they were the best of comrades both in the field and in bivouac. It was with many regrets that we bid them good-bye. The Brigade of Guards also left Komati Poort, as they then believed, *en route* for England. As they marched out, many a Guardsman was heard to remark with glee to his less fortunate brother-in-arms, 'First stop, Hyde Park Corner.' But on their arrival at Middelburg previous orders were cancelled, and they were destined to do many more months of good work in South Africa.

In about ten days' time the great army which had invaded Komati Poort on September 24 had dissolved away. The bulk of the troops returned to Pretoria with Lord Roberts and were soon more busy than ever in every part of the country. An expedition, under the command of Major Walker, was despatched a day after our arrival at Komati Poort along the Selati railway line to try to capture a party of 200 Boers, who, more bold than their brothers, had disdained the shelter of Portuguese territory. About 100 of the 4th Mounted Infantry went with this expedition; but, owing to the condition of the horses, it was not successful.

Headquarters and three companies of the 4th Mounted Infantry now left Komati Poort. One company had orders to proceed to Balmoral, the other two to Pretoria, where they were to be re-equipped. The 85th company, however, remained and took over various outposts. Captain White, who had been in command of the 85th company since the commencement, was, about this time, taken ill and invalided to Pretoria, eventually being sent to England; but Captain Smith, who had been promoted since he left us, returned and took over command. A large draft was sent to us from the regiment to replace casualties, and as we had taken over the fittest of the horses from the other companies of the regiment before their departure, we were at this time a strong and well-mounted company.

We now settled down to a more or less peaceful life. Two posts, at a considerable distance from Komati Poort, were furnished by the company; but the remainder were in comfortable quarters.

On November 30 we received orders to proceed on the following day to Kaapsche Hoop. The company had suffered severely during the last month from fever and horse sickness, and a move was welcomed by all. On arrival at Kaapsche Hoop we joined the headquarters of the regiment. The Yorkshire company was already there. The other two companies were on detachment, one at Barberton, the other at Waterval Boven. Major Handcock, who had been in command since Bloemfontein, was now given command of the Mounted Infantry depot in Pretoria, and Major Walker of the Cornwall Light Infantry took over command.

During the next few months no events of any importance took place. The regiment was kept busy endeavouring to prevent the enemy making raids on and blowing up the railway line. In January Captain Smith, Lieutenant FitzGerald, and Lieutenant Liddon, with fifty men of the 4th Mounted Infantry, were attached to General Smith-Dorrien's column, which moved south from Wonderfontein and worked through the Carolina and Ermelo districts in conjunction with other columns. They were present when General Botha made his night attack at Lake Chrissie on February 6. The operations of these columns were very much hampered by wet weather, the rivers becoming so flooded that it was impossible for the supply columns to keep in touch. It was during these operations that Lieutenant FitzGerald was severely wounded in the leg, and had eventually to be sent home, the company losing his services for the remainder of the war. At the end of March, General Smith-Dorrien's column marched into Wonderfontein and the detachment rejoined the 4th Mounted Infantry, which was now concentrated at Machadodorp.

On April 8 we left Machadodorp *en route* for Lydenburg. Here we were to join the column under command of General W. Kitchener and take part in the extensive operations which had for their object the capture of the force under General Viljoen, which had been left for some time in comparative peace in the Pietersburg district.

As in most cases in the guerilla warfare which may be said to have commenced after the occupation of Komati Poort, there was no fighting of a severe nature. Scarcely a day passed without an exchange of shots, but seldom were we able to pin the enemy to a position and compel them to fight it out. The marching, on the other hand, was of the most strenuous order, and in the operations at present under discussion we averaged fifteen miles a day. It was not until May 8 that we again touched the railway line at Middelburg, so that, roughly, we covered 450 miles during that month. From this date until July 2 the regiment was on the lines of communication. With headquarters at Middelburg, the regiment sent small detachments along the railway line from Belfast to Balmoral, their duties being to patrol on either side of the line and to keep communication between the infantry posts. During this time the headquarters of the 85th company were at Belfast with detachments at Pan and Wonderfontein. Early in May Second-Lieutenant Surtees joined the company, replacing Lieutenant FitzGerald, who had been sent to England. Second-Lieutenant Williams, an officer belonging to the 3rd battalion, who had come to South Africa as a Special Service officer, was also sent to us, filling the place of Captain H. M. Smith who had taken over command of the company. It was also at about this time that Sergeant Crooke joined, who was shortly afterwards appointed colour-sergeant of the company, in place of Colour-Sergeant Cadmore, who had been invalided to England, and he held this appointment to the end of the war.

On July 1 orders were received for the whole regiment to concentrate at Wonderfontein by July 6. For almost the first time during the war, the regiment was now at full strength, both as regards men and horses. Major Walker had made every use of our long halt to get everything into thorough working order, and we now looked forward eagerly to our 'trek.'

To the great delight of the 85th company we learnt at Wonderfontein that we were to form part of General Spens' column, and on the 9th we moved off. Nothing of particular interest happened during this 'trek.' We explored the country in the Blinkwater district, rounded up a number of cattle, but of the enemy we saw very little. On July 20 we were once more camped at Middelburg. This time, however, we were not destined to make a long stay, and on the 23rd we were once more on the move. To our regret we found we were to leave General Spens' column and join one under Colonel Park, who was now operating in the vicinity of Zwarts Kopje.

Escorting a large convoy we marched via Belfast and Zwarts Kopje and finally joined him at Dullstroom on July 28. From this date until September 11, when we returned to Middelburg, the column, in conjunction with others, all under the command of General F. W. Kitchener, operated in the neighbourhood of Roos Senekal and Dullstroom. No marked success crowned our efforts, but large herds of sheep and cattle were driven in and shots were exchanged daily with roving parties of the enemy. The marching, as usual, was very hard and the Mounted Infantry came in for the lion's share.

On September 16 we were once more on the move, still forming part of Colonel Park's column. Our orders were now to march to Lydenburg, where we arrived, after an uneventful march, on the 29th. General Viljoen for some months past had made his headquarters at Pilgrim's Rest, and owing to the difficult nature of the country had been allowed to escape the attentions of the numerous columns which were causing his brothers-in-arms so much annoyance. This immunity from attack seems to have emboldened his commando; for of late they had constantly been annoying the various posts round Lydenburg and had carried off a number of cattle when grazing round the outskirts of the town. The arrival of Colonel Park's column on the 29th was to be taken advantage of to put an end to this annoyance. The intelligence department had located a small commando of Boers at a farm half-way between Lydenburg and Pilgrim's Rest and, with a view to taking this party by surprise and also of advancing over a difficult part of the country under cover of darkness, a night march was decided on. Accordingly at 1 A.M. on the morning of October 1 the column moved off. The 85th company were leading, and all went well for the first two hours; but on this occasion the Boers had been wise enough to leave a very wide-awake picquet on the road at a considerable distance from their headquarters. With no word of warning this picquet emptied their rifles into us, then jumped on their ponies and vanished into the darkness, leaving behind them only a very old blanket or two and the remains of an uninteresting supper. A minute's confusion ensued, but order was rapidly regained. The minute was unfortunately a minute too much; for it allowed the picquet time to make good its escape and warn their friends of our near approach. We now pressed on more rapidly in the hope of making the farm before the enemy had time to be thoroughly awakened. It was not to be; the picquet had done its work right well, and on arrival at the farm just before daylight we found it deserted. Colonel Park, anxious to be in possession of the high ground before daylight, at once ordered the 85th company to gallop on and seize the ridge overlooking Kruger's Post. The first faint glimmer of day was now showing, and hardly had we started when we saw the flicker of a match being lighted high up on the ridge for which we were to make. Those of us who were leading knew at once that the ridge was held. Owing to the rugged nature of the country the road was our only possible line of advance, and Smith, who was leading, at once saw that the quicker we did so the better. The order was passed down to trot, then gallop, and sooner than it takes to tell we were within measurable distance of the ridge. At that moment, we being about 400 yards off, the Boers opened a heavy fire. A number of horses at once fell wounded and several men, two, I regret to say, mortally. For a short time the road was blocked by the horses that had fallen, and the remainder of the company coming on behind were unable to avoid them and were thrown from their horses. In this way a certain amount of confusion was caused, but the company was soon pulled together and by daylight the ridge was in our possession. The enemy, however, deeming that discretion was the better part of valour, retired and, for the time being, escaped unscathed. For the remainder of that day we bivouacked under cover of the ridge. The following day the regiment advanced into the Orighstad Valley, the bivouac remaining where it was. A number of farms were destroyed and there was fighting all day; Lieutenant Liddon of the Yorkshire company and two or three men were wounded.

The following day the valley was again visited and more farms burnt, but the enemy posted on all the surrounding hills made any progress a matter of extreme difficulty. Captain Stokoe of the Cornwalls and Lieutenant Crooke of the Warwicks and several men were wounded. Colonel Park, deeming that he had made the valley a less comfortable resting-place, now decided to return to Lydenburg, and on the 7th the return journey began.

From October 9 until December 18 we camped at Lydenburg, our chief occupation being to escort convoys coming from

Machadodorp; but several night raids were made, in the endeavour to capture small parties of the enemy, which met with some success.

Marching from Lydenburg in the early morning of the 18th, we arrived at Eland's Spruit on the 20th without adventure. The enemy had been seen in fairly large numbers throughout the march on the 20th but never came within striking distance. At about 10 P.M. that night, about 500 Boers under the command of Commandants Müller and Trichardt made a determined attack on our camp. A picquet of the Manchester Regiment was severely handled, but held on to its post, and after about three hours' fighting the enemy was driven off. The casualties in Colonel Park's column were three officers wounded, seven men killed and fourteen wounded. We buried ten of the enemy and four wounded were left in our hands. A great number of our horses and trek oxen were killed or wounded. On the following day we arrived at Dullstroom, our camp remaining there until 23rd. On the 22nd the mounted troops were out all day and had a running fight with the enemy.

On the evening of the 25th we arrived at Machadodorp, and marching along the railway line on the two following days arrived at Belfast on December 27. We camped here in comparative peace until January 14.

On the night of the 15th we marched into the Steelport Valley with the object of rounding up Boers who were reported in the farms in that neighbourhood, but the quarry had gone before our arrival. The following days found us in our old hunting ground in the vicinity of Dullstroom. There were still many of the enemy in small bands in this part of the country and sniping was of daily occurrence. The hilly nature of the ground made it extremely difficult either to capture or to inflict punishment on them. Captain Menzies of the Manchesters was mortally wounded on the 21st and several men were killed during the trek. On the 29th we once more reached Lydenburg. On arrival there we heard that General Viljoen had been captured, close to Lydenburg, on the 25th; and on the 31st Colonel Park's column escorted him to Machadodorp, arriving there on February 3. On the 5th we started our return journey to Lydenburg, escorting a large convoy of supplies, and arrived there on the 8th.

On the 10th we were on the move again, and once more traversed the now well-known road to Machadodorp; but this time there was to be no halt there, and we marched straight down the line to Pan station, just east of Middelburg, where we arrived on the 18th.

A long night march on the 20th brought us by daybreak in the vicinity of the Bothasberg and there, in conjunction with columns under the command of Colonels Ingouville-Williams and Urmston, we completely surprised a large part of Müller's commando. One hundred and sixty prisoners, including three field-cornets, were taken. This commando was holding the line of the Bothasberg in order to protect the wandering Boer Government officials. They, however, had already taken alarm and were now hurrying westward.

Colonel Park's column was turned towards the railway as escort to the prisoners. On the 21st, fifteen Boers surrendered themselves to Colonel Park, and Colonel Williams captured twenty-eight. The total number, therefore, which were handed over on our arrival at Pan on the 22nd was over 200.

On the 23rd we again trekked east along the railway line and arrived at Belfast on the 24th. On the 26th orders were received by Colonel Park to prepare a flying column of about 600 men. This force was to endeavour to surprise and to capture the officials of the Boer Government who were known to be camped somewhere north of Rhenoster Kop. The column was made up of 200 men of the 4th Mounted Infantry, 200 of the 18th Mounted Infantry, and 200 National Scouts. One hundred of the Manchester Regiment accompanied us as escort to the few carts which accompanied the column. We were to travel by train as far as Bronkhurst Spruit, and at 3.30 A.M. on the 27th the first train left Belfast; by 7 P.M. the whole force had assembled there. At 12.30 A.M. we moved off from Bronkhurst Spruit and marched through the night without a halt. As day broke our pace increased and until about 10 A.M. we pressed forward, alternately trotting and walking. To the disappointment of all, our efforts had been in vain. The Boer Government, although unconscious of our approach, had moved on a day or two previously, partly because the grazing for their animals was not good and partly because, in these days of hustle, no camp was entirely safe for twenty-four hours together and they made it a rule to be always on the move.

Our halt at 10 A.M. was of short duration, and although the enemy had not been found where they were expected we made a long cast round in the hope of picking up stragglers; but as we had to return to Rhenoster Kop, where the infantry and waggons were camping, a further advance was impracticable.

We finally reached camp at 3.30 P.M. on the 28th, having been fifteen hours in the saddle and marched sixty-five miles. Since 3.30 A.M. on the 27th, when we left Belfast, there had been no rest for any one. Our bag of seventeen prisoners was a poor return for such hard work.

On March 1 wereturned to Bronkhurst Spruit and heard that the remainder of Colonel Park's column were being railed from Belfast to meet us there. By the 4th we were once more together again. From March 5 to 22 we were incessantly on the move in the district to the north of the line between Wilge River station and Brugspruit station. The greater part of our work was now carried out by night and the marching was exceedingly hard. On the 12th we surprised a small laager at daybreak near Rietfontein and took twenty-two prisoners and some 600 head of cattle. On the 14th eleven more prisoners were taken.

On the 22nd we and a column under Colonel Williams were at Maggi's farm, our column having crossed to the south of the line at Brugspruit on the 18th. The general impression was that we were now going into Middelburg, but at 4.30 P.M. that day orders were received for both columns to turn about and combine with Colonel Wing's column in a drive due south to the Standerton line. By 6.30 P.M. both columns were on the move, getting into position. A long night march, followed by a march of fourteen miles on the next day, brought us into our position on the right of the line, and on the morning of the 24th the drive commenced. At 9.30 P.M. we reached Vlaklaagte station, having drawn blank. The whole of the Boers who were to have been in the net had apparently broken through on our left before Colonel Wing's column could get into position.

The columns remained at Vlaklaagte until the 29th. On the 30th another drive commenced, this time taking us through Bethel Ermelo and Carolina, and back again to Val station, which is the station just west of Vlaklaagte, on the Standerton line. Here we arrived on April 14, having been on the march every day since March 30. The result was not encouraging, very few prisoners were taken and there had been little or no fighting. The truth was that the columns were not really strong enough to cover the ground they were trying to. Extensive gaps in the line occurred and the Boers broke through with ease.

On the 16th we left Val station and again worked up in line to the Delagoa Bay railway, which we touched at Witbank on the 20th, but with no result.

On April 9 Second-Lieutenant Mellor joined the 85th company, taking the place of Lieutenant Underwood, who had been invalided to England.

On the 21st we moved east along the railway line to Groat Oliphants. On the 23rd we were at Steenkool Spruit, where we remained until the 25th; and from this point another drive was commenced in the direction of Middelburg, where we arrived on April 30.

From this time until the end of the war Colonel Park's column did little or nothing. We remained for some time in Middelburg and were re-equipped and got remounts. We then moved by easy stages along the railway line to Belfast. On May 22 the column moved to Wonderfontein, proceeding to Pan on the following day.

Before leaving Belfast, the Manchester Regiment (which had formed part of Colonel Park's column for over a year and done a great deal of very heavy marching) was relieved by the King's Liverpool Regiment from Waterval Boven. It was with great regret that we bid farewell to this fine regiment with which we had been so closely associated.

We remained at Pan, and heard on June 1 of the successful issue of the Conference then being held at Vereeniging.

On June 7 the column marched into Middelburg and was finally broken up on the 20th. The regiment, however, was left intact. The reservists only at once returned to their line battalions to be sent home as early as possible, but the nucleus of the regiment remained. On July 3 we moved our camp some two or three miles out of Middelburg as the grazing for the horses was better there. It was about this date that Lieutenant FitzGerald rejoined us, having recovered from his wound.

A long period of peace and quiet now followed; but on August 7 orders were received for the 85th company to rejoin the regiment at Middelburg *en route* for Newcastle, and after a period of two years and eight months with the 4th Mounted Infantry the company left on August 9.

The 4th Mounted Infantry were present at and received clasps for the following engagements:

(1) Paardeberg	Queen's Medal
(2) Relief of Kimberley	
(3) Driefontein	
(4) Johannesburg	
(5) Diamond Hill	
(6) Belfast	
(1) South Africa, 1901	King's Medal
(2) South Africa, 1902	

The following officers of the 85th Regiment served in the 85th company of the 4th Mounted Infantry:

Captain J. J. White, in command from January 1900 to September 1900. Invalided to England; mentioned in Lord Roberts' Despatches, November 1900.

Captain H. M. Smith, as senior subaltern, from January 1900 to March 7, 1900, when he was severely wounded and invalided to England. Rejoined in October 1900, and took command *vice* Captain J. J. White, and commanded until the end of the war; mentioned in Lord Roberts' Despatches, November 1900, and received the D.S.O.

Lieutenant E. A. Underwood, from January 1900 to December 1901, when he was invalided to England.

Lieutenant P. F. FitzGerald, from January 1900 to February

1901, when he was severely wounded and invalided to England. Re-joined in July 1902; mentioned in Lord Roberts' Despatches, November 1900.

Lieutenant J. C. Hooper, from January 1900 to the end of the war; mentioned in Lord Roberts' Despatches, November 1900.

Second-Lieutenant R. L. Surtees, joined in May 1901 and remained till the end of the war.

Second-Lieutenant Williams (3rd Battalion Shropshire Light Infantry), joined in May 1901 and remained until the end of the war.

Second-Lieutenant Mellor, joined in April 1902. Invalided with enteric fever in June 1902.

NEAR WONDERFONTEIN STATION.

THE SOUTH AFRICAN WAR MEMORIAL, SHREWSBURY.

CHAPTER XIX

INDIA, 1903

THIS, the concluding chapter of the Military History of the 85th, will be but brief.

We left the battalion under orders to proceed from South Africa to India.

The embarkation accordingly took place at Durban, January 8, 1903, on board the steamship *Syria*, the strength of the battalion being as follows : 11 officers, 36 sergeants, 32 corporals, 14 buglers and 341 privates.

Owing to lack of accommodation on board the *Syria*, two officers, four sergeants, one bugler, three corporals and forty men of ' B ' company were left behind at Pietermaritzburg. Two buglers and forty-six men were also left behind in hospital.

On board the *Syria* there was already a draft from the details of the King's Shropshire Light Infantry (13th provisional battalion) of three officers, two warrant officers, three sergeants, one corporal, eighty-nine privates and boys, as well as the women and children of the battalion. These had sailed from Southampton on December 9, 1902.

The *Syria* arrived at Bombay on January 23, and disembarked the following morning. Thence they proceeded by half-battalions to Deolali, where they arrived early on the 25th. On the 29th, 13 sergeants, 20 corporals, 501 privates, 9 women and 5 children joined from the 1st battalion, upon which date the battalion (with the exception of those who had joined from the 1st battalion) left Deolali for Bareilly under the command of Lieut.-Colonel A. H. J. Doyle. They arrived at Bareilly on February 3. The non-commissioned officers and men from the 1st battalion, under the command of Major S. G. Moore, joined on the following day.

From the *London Gazette*, February 9, 1903, we read:

'To be Brevet Lieut.-Colonel: Major C. T. Dawkins, C.M.G.

To be Brevet Major: Captain C. A. Wilkinson.

To have the Distinguished Conduct Medal: Private R. Meredith.'

The 'B' company (consisting of two officers and fifty-five non-commissioned officers and men) joined on February 9, having sailed from South Africa in the s.s. *Dunera*.

On the following day 'C' company (strength: 3 officers and 100 non-commissioned officers and men) left Bareilly to proceed to Ranikhet, there to take over barracks from the 1st Devonshire Regiment.

On February 25 the battalion was inspected at Bareilly by Major-General G. H. More-Molyneux, C.B., D.S.O., commanding the Rohilkand district. The regiment was found to be in a highly efficient state and fit for service. The inspecting officer added that:

'The way in which the Regiment, consisting as it did of men from South Africa, drafts from England and men transferred from the 1st Battn., has been pulled together, and the manner in which the men drilled, reflected great credit on all concerned.'

The total strength of the battalion at headquarters on the occasion was 914 of all ranks (efficient). The total strength of the battalion (including those on detachment, on leave, sick in India, and sick in South Africa) was 1147 officers, non-commissioned officers and men.

The battalion left Bareilly for Ranikhet on March 2, 1903, in two parties—the right half-battalion proceeding thither on that day, to be followed on March 3 by the left half-battalion. The journey was made by train to Kathgodam and thence by route-march to Ranikhet. The half-battalions arrived at their destination on March 7 and 8 respectively. A draft consisting of one sergeant, one corporal, and seventy-two privates joined the battalion from England on April 24.

The total effective strength of the battalion at the end of the month was as follows: 33 officers, 2 warrant officers, 51 sergeants, 16 buglers, 48 corporals and 1043 privates. On July 23 an outbreak of cholera occurred at Ranikhet. In twenty days the battalion unhappily lost one officer (Lieutenant J. M. Carter), sixteen men and one boy. To check the epidemic the battalion was broken up and went into segregation camps on the hills. These camps were situated at Bessua, Tarakhet, Seoni, Majkali, and Upat.

The season for winter quarters having arrived, the battalion was concentrated again at Ranikhet and left for Bareilly on November 20. *En route* they took part in hill manœuvres with the 1st Devonshire Regiment and the Gurkhas from Almora. The battalion having arrived at Kathgodam entrained on November 20 and reached Bareilly the same day.

On January 1, 1904, a draft of two corporals and seventy-one privates joined at Bareilly from England.

On January 12, 1904, the battalion was inspected by Brigadier-General A. R. Martin, C.B., commanding the Rohilkand District, at Bareilly; the strength of the battalion on the occasion being 24 officers, 2 warrant officers, 35 sergeants, 16 buglers, 38 corporals and 866 privates.

The total strength of the battalion (including those on detachment duty, on leave and absent, sick) amounted to 1225 officers, warrant officers, non-commissioned officers and men. The report of the Brigadier-General was as follows:

'A fine Battalion—well drilled and in good order—an excellent spirit prevails—In skirmishing and hill manœuvres the Regiment is especially good—Musketry efficiency is good and the N.C.O.'s have been carefully worked up—The Battalion is in all respects fit for active service.'

The following extract from the remarks of the Lieutenant-General Commanding the Forces, Bengal, is here given:

'Report satisfactory: there is a fine healthy tone in the Regiment, their training is good and the progress made during the year most marked; full advantage has been taken of their being stationed in the hills to carry out hill work. I consider that the Regiment should do well on service.'

The battalion left Bareilly for Ranikhet in two parties of half-battalions on March 24 and March 25, arriving at their destination on March 29 and March 30 respectively.

From October 29 to November 5 the battalion took part in hill manœuvres between Almora and Ranikhet under Brigadier-General A. R. Martin, C.B. The other troops engaged in the manœuvres were the 1st Royal Sussex Regiment, the 1st Devons, and the 1st and 3rd Gurkhas.

On November 17, 1904, the battalion left Ranikhet for Fyzabad. They arrived there on December 22, and went into camp. On December 26 they left again by route-march for Rae-Bareli manœuvres and arrived at Salarpur Camp on December 3. Here they took part in the manœuvres around that district under Lieut.-General Sir E. Locke Elliot, K.C.B., D.S.O., commanding the 8th (Lucknow) Division. The manœuvres lasted until December 21. The battalion returned to Fyzabad by rail on the following day.

The force was composed as follows: 'O' Battery, Royal Horse Artillery; 74th and 77th Batteries, Royal Field Artillery; 1st Royal Dragoons; 2nd Battalion East Surrey Regiment; 1st Gloucestershire Regiment, 2nd South Staffordshire Regiment, the Mounted Infantry class from the Mounted Infantry School at Fatehgarh.

Of native troops there were the 2nd and 4th Lancers, 6th Cavalry, 7th Rajputs, 10th Jats and 62nd Punjabis.

The report on the battalion was as follows:

'The Battalion is well commanded and has excellent Officers; the N.C.O.'s and Rank and File are a fine, well-set-up body of men. The Battalion has done very good work throughout the summer, carrying out its field-training in the hills. The men are in hard condition and excellent marching order. A fine spirit pervades all ranks and willingness and good work on fatigue duty are very marked. Musketry has been satisfactorily performed in spite of difficulties connected with the range. Field-firing has received great attention with good results. The general musketry efficiency is good.'

A draft consisting of 2 corporals and 121 men joined the battalion from England on December 23, 1904. On February 18, 1905, another arrived under the command of Captain and Brevet-Major W. S. W. Radcliffe consisting of sixty-seven men and one boy.

April 1905. The following is a copy of the remarks by the

Lieutenant-General Commanding the Eastern Command on the annual inspection of the battalion:

'I have seen this Battalion at work, both on hills and plains. The Officers are very keen and I should say an exceptionally experienced body; most of them have seen considerable field service.

'I consider the Battalion well trained and efficient, with a very good tone prevailing throughout.'

Towards the end of the year two drafts arrived from England.

December 9, 1905, the battalion left Fyzabad by road for Bahramghat and took part in the 8th (Lucknow) divisional manœuvres around that district, and returned by rail.

The other troops engaged in the manœuvres were the 74th, 77th and 79th Batteries, Royal Field Artillery, 1st Royal Dragoons, Durham Light Infantry and Oxford Light Infantry; native troops: 2nd Lancers, 6th Cavalry, 10th Lancers, 24th Punjabis, 62nd Punjabis and 10th Jats.

The battalion was inspected on January 8 and 12, 1906, by Colonel H. S. Wheatley, C.B., commanding Fyzabad Brigade.

An extract from his inspection report is as follows:

'The Battalion is quite fit for War. A very good tone prevails throughout. A large proportion of the Officers have seen War Service and are capable instructors. Special attention has been paid to signalling, and the Cyclists do very good work at Manœuvres, and on Minor Tactical Field Days. The comfort and efficiency of the soldier are well looked after in this Battalion.'

A draft consisting of 29 men, under command of 2nd Lieutenant E. F. Robinson, joined the battalion from England on February 2, 1906, and another of one corporal and 39 men joined on March 25.

The following is the remark by His Excellency the Commander-in-Chief on the report on the battalion for 1905–6: 'Very satisfactory.'

During 1906 and 1907 several drafts joined the 2nd battalion from England, from the 1st battalion. Total: 4 non-commissioned officers, 1 bugler, 139 privates and 5 boys.

On December 30, 1908, the battalion left Fyzabad (change of station) for Dinapore and Dum Dum; headquarters with 'A,' 'B,'

'E' and 'F' companies, under Lieut.-Colonel I. L. Pearse (strength, 13 officers and 411 rank and file), proceeding to the former station, and 'C,' 'D,' 'G' and 'H' companies, under Major S. G. Moore (strength: 7 officers, 323 rank and file), proceeding to the latter.

There are no events of importance to record for the year 1909.

On March 4, 1910:

'His Majesty the King has been pleased to approve of the grant of a silver medal, without an Annuity, to Sergeant-Major William Henry Austin, a Yeoman of the Guard (formerly of the 85th Foot), as a reward for his long and highly meritorious service—including Crimean Campaign.'

November 15, 1910. The wing stationed at Dum Dum, consisting of 'C,' 'D,' 'G' and 'H' companies, attended a review held by His Excellency the Earl of Minto, the Viceroy of India, at Calcutta. They were commanded by Captain H. M. Smith, D.S.O., and formed part of the 1st Infantry Brigade, under Lieut.-Colonel W. T. Grice, 1st Battalion Calcutta Volunteer Rifles.

The following officers attended: Captains H. M. Smith, D.S.O., R. Masefield, A. T. C. Rundles and P. L. Hanbury; Lieutenants R. H. Poyntz, C. D. Harris and L. H. Torin.

The strength on parade was as follows: 7 officers, 11 sergeants, 11 corporals, 2 buglers, and 228 privates—total 259. The wing going to Calcutta and back marched some nineteen miles.

On February 6, 1911, the headquarters wing of the battalion left Dinapore, proceeding to Calcutta, *en route* to Secunderabad. Strength: 12 officers, 3 warrant officers (including army schoolmaster), and 443 non-commissioned officers and men. The wing from Dum Dum left that station on February 7. Strength: 7 officers and 391 non-commissioned officers and men. On arrival at Calcutta on the 7th the battalion embarked at the Kidderpore Docks on board R.I.M.S. *Northbrook*, and sailed for Madras on the morning of the 8th. They arrived at Madras on the 11th and disembarked the same day. They entrained immediately for Secunderabad, and arrived there on February 13, 1911.

Strength: 19 officers, 3 warrant officers, and 831 non-commissioned officers and men. One man died on board the R.I.M.S. *Northbrook* and two men were left in hospital at Madras.

On being joined by a draft and details at Secunderabad the strength was 21 officers, 3 warrant officers, and 959 non-commissioned officers and men.

The battalion joined the 1st Infantry Brigade, commanded by Major-General W. R. Kenyon-Slaney, C.B., and formed a part of the 9th (Secunderabad) Division under Lieutenant-General Sir James Wolfe Murray, K.C.B. They were quartered in the entrenchments at Trimulgherry.

The 1st Infantry Brigade was then composed of the 2nd Battalion The King's (Shropshire Light Infantry), 1st Brahmans, 6th Jat Light Infantry, 95th Russell's Infantry.

The records which have been placed at the disposal of the Editor conclude here, and in consequence the military history of the 85th King's Light Infantry in this volume ends at this date.

CHAPTER XX

Dress and Equipment, 1793–1912

By P. W. Reynolds

When the 85th Foot was raised, in 1793, the British infantry was wearing a form of the typical 18th-century military costume—a cocked hat, a long coat worn over a waistcoat, breeches, and gaiters up to the knee. Successive changes in 1796, 1797, and 1800 radically altered this style. The coats were closed in front down to the waist, so that the waistcoat could not be seen, and need not necessarily be worn with the coat. Hence it became a fatigue dress for the men and the parent of the 'shell-jacket.' The men's hats were replaced by 'caps' of lacquered felt, with a peak in front (a sort of stove-pipe shako), while those of the officers were changed in shape and worn with the long peaks fore and aft instead of at the sides. Only the breeches and gaiters remained, and for the officers these were more and more replaced by 'pantaloons and half-boots' or Hessians.

From 1800 to 1855 may be termed the coatee period, though the word 'coatee' did not appear for some time. The coat, that is to say, ended at the waist in front and had skirts or tails (of very varying proportions) at the back only. Throughout this period, in the British infantry, the officers had double-breasted and the rank and file single-breasted coats. Down to 1829 the officers' coats had, in full dress, lappels of the facing colour, often with loops or bars of lace thereon; but when buttoned across they were usually quite plain; and after 1829, when the lappels were abolished, officers' coats were always plain in front. But during the whole period the men's single-breasted garments were ornamented with loops of worsted tape or 'lace' across the chest. The head-dresses were shakos of various shapes and dimensions, except that cocked hats were worn by line infantry officers until 1812 or so, and by certain regimental staff officers until quite modern times. The breeches and gaiters were officially abolished in 1823, trousers having long been in use in many regiments, and the trousers became the only leg-covering. At the same time the men's shoes (and ankle-gaiters if worn with trousers) were replaced by boots.

While clothing was undergoing these changes, equipment and armament were very little altered. Except that light muskets may have been used by flank company officers for a few years after 1793, all officers had only their swords. The rank and file had smooth-bore muskets, the only really important change in which was the substitution of percussion locks for flint (or 'fire') locks about 1842. The men were equipped, in 1793, with cross-belts, supporting ammunition-pouch and bayonet, and a knapsack with the usual straps, including one across the chest. Belts, pouches, and knapsacks were slightly varied in pattern; but no essential alteration was made in this equipment until almost the end of the period, when, at the time of the Crimean War, both armament and equipment underwent important modifications.

OFFICER'S 'BREASTPLATE,' 1815.

Leaving the tunic period, which began in 1855, for a few remarks later on, the regimental distinctions and peculiarities of the 85th Foot during the coatee period will now be considered.

Regiments differed from one another in the colours of the facings, in the pattern of the men's tape lace and of the officers' gold or silver lace or embroidery (with the additional variation that officers often did not wear lace at all), and in the number and spacing of the buttons and the shape of the 'loops' or bars of lace which were connected with the buttons or buttonholes. The officers' buttons were either plated or gilt; but those of the men were of white metal in all regiments.

The facings of the 85th were yellow from the formation till

1821. The officers' buttons were silver, their epaulettes or wings also silver, the buttons were no doubt set on in pairs, like those of the men. Their buttonholes were of yellow silk twist, no silver lace being worn.

In 1803 (Bossett's diagrams) and in 1812 (C. H. Smith's diagrams) the men's lace had a white ground with a red ornament running along the centre and a blue or black stripe on each side; and their jackets were ornamented with square-ended 'loops' of this lace, set on in pairs. This pattern of privates' lace was, however, afterwards changed for a more ordinary type—white lace with a narrow red stripe running along one edge, which was kept up until, in 1836, coloured regimental laces were abolished and a universal plain white tape substituted. Sergeants, it may be remarked, wore plain white lace to distinguish them more from the men. When white lace was given to the latter in 1836, therefore, the sergeants were given double - breasted coatees, made like those of the officers.

OFFICER'S SHAKO PLATE, 1820–8.
The same without the crown was placed upon a large gilt star, 1829–45.

In 1808 the 85th was made a Light Infantry regiment. This involved the disappearance of the grenadier and 'battalion' companies, all companies being henceforward dressed alike. The plumes became green throughout, and all officers wore wings instead of epaulettes, field officers, however, wearing epaulettes over their wings, which, it must be confessed, looks in portraits rather a clumsy arrangement, but which was maintained up to 1829. Sergeants' pikes also disappeared in 1808 in the 85th, Light Infantry sergeants being armed with firelocks. The men of all companies also would henceforward wear wings on the shoulders. It seems that it was a general custom with Light companies and regiments for the company officers to wear the stove-pipe shako like their men, instead of the cocked hat; and that this type of shako was used by Light Infantry *regiments* right up to 1815, the high-fronted cap, with feather at the side, never being adopted by them apparently. But there is no positive evidence as to the 85th.

In 1819 Marcuard's diagrams of the officers' uniforms of the Army give, for the 85th, a shako with silver lace at top and bottom, and an upright green plume, a scarlet coat with silver buttons in pairs, plain yellow facings, and silver wings. A date is now reached when fuller evidence is available as to dress, at any rate for the officers. After Waterloo a new large-topped shako had been introduced, imitated from the French and Austrians. It was seven and a half inches high and eleven inches in diameter at the top, made of black beaver with a leather top and peak, and had, normally, two-inch lace round the top, three-quarter-inch lace round the base, chin scales corresponding to the metal of the lace, a small plate in front, and a twelve-inch upright feather. The earliest recorded shako plate is found in one of Messrs. Jennens' invaluable old pattern-books, dating from 1817 to 1822, and is a silver eight-pointed star about three and a half inches in greatest length with a gilt garter and crown in the centre, having the regimental motto on the garter, and within the garter the numerals '85.' The illustration dated 1820 is taken from a drawing in the collection of the late Mr. S. M. Milne, one of a series executed about 1860, perhaps, by someone who certainly knew a great deal about English uniforms, though it is difficult to say how much confidence ought to be reposed in his figures in cases where it is impossible to check them by other evidence. The dress shown is practically in complete accordance with the tailors' books, except the peculiar

OFFICER'S' 'BREASTPLATE,' 1825–45.

shako decoration. This may be meant for black silk lace (which was worn by one or two other regiments), but has the appearance of being an ornament very similar to that on the pouch-belt of the 10th Hussars, only in black. It is possible at least that such a shako may have been worn at some time between 1816 and 1821. The wings and wing-straps were of a pattern called 'dotted scale,' all bright plated metal in the form of overlapping scales, corded round with silver, with a one and a quarter inch bullion fringe, and with 'small gilt bugles and laurel and scroll' on the top. Field officers had epaulettes with the same scaled straps and silver fringe three and a half inches deep, and wing-points to correspond showing underneath. These were on the short-tailed coatee technically known as the 'jacket.' At Court, officers had to wear a long-tailed coat, and with this Light Infantry officers seem to have always worn epaulettes, discarding their wings altogether. This is confirmed for the 85th about 1820, by an entry in one of the old tailors' books. The only other distinctive ornament on the coat or jacket was one on each tail at the junction of the white 'turnbacks.' This consisted of a scarlet garter inscribed 'BUCKS LIGHT INFANTRY,' and a silver bugle within it, on a ground of yellow cloth.

BREASTPLATE, 1846.

At the change of title and facings in 1821, silver was retained as the metal of the officers' buttons, but loops (of silver embroidery) were added to the collar and cuffs. The new blue lappels were, however, left plain apparently, like the old yellow ones. One of Del Vecchio's prints (about 1826) shows this costume, with embroidered collar and cuff and plain plastron, correctly enough, though the figure is almost a caricature. The 'dotted scale' wings were retained, and the design of the skirt ornament was the same as before, but the silver bugle was now on a blue ground, and the red garter was inscribed 'KING'S LIGHT INFANTRY.' The shako was now undoubtedly laced with silver, two and a quarter inches wide round the top and five-eighths inch round the base; the large green upright feather was adhered to. Some of the regimental 'etceteras' at this period are worth recording. The blue-grey trousers had one and three-eighths inch, or according to another version one and seven-eighths inch, silver 'fine check' lace with 'vellum edge.' Possibly field officers and company officers had different widths of lace. The white shoulder-belt was still made with slings, and a 'sabre' attached to them; the sabre-knot had 'sattin guard and ruff, pearl ball' (whatever that may mean exactly). There were two sashes: a 'field sash' of the usual Light Infantry pattern—a band round the waist and a pair of cords one and a quarter yards long ending in 'basket balls' which were brought round the right side and hooked up in front, all crimson—and also a 'dress sash,' like that worn by Hussars, of cords and 'barrels,' with long cords like the other, with six bullion slides and two oval balls. In 1828 a new pattern of wing was introduced, having three rows of curb chain and centre-plate with gilt bugle and laurel. As to the rank and file, the facing colour had been changed to blue, but the square-ended lace loops set on by pairs were retained.

OFFICER'S BUTTON, 1840–55.

The great variety in infantry officers' dress was, however, about to be curtailed by the authorities. In December 1828 a new pattern shako was approved, reduced to six inches in height, though with the same large top as before, without any lace, and with a universal pattern large star-plate, with regimental devices in the centre; and in February 1829 a new coatee was ordered, double-breasted without lappels, with a red flap added to the cuff, the cuff itself plain, but the cuff-flap and the collar laced or embroidered for all regiments; the pocket flaps on the tails were at the same time made vertical instead of horizontal, and laced or embroidered to correspond with the sleeve flaps. The collar was to have two loops on each side in all regiments, and this affected the 85th, which had had only a single loop. In September 1830 an order was issued that only gold was in future to be used for the buttons and lace of officers of the Regular Army, another step in the direction of uniformity, and the 85th had to change from silver to gold accordingly. This change took place in November 1831. Their embroidery was a design of laurel leaves and berries. For the new shako plate they superposed their old small silver star on the large universal plate, but, as a large crown surmounted the new plate, the small crown was removed from the silver star and a bugle substituted. The dark grey 'Oxford mixture' trousers (which were gradually supplied darker and darker until they finally became black) were substituted for the blue grey. Altogether the British infantry uniform of 1830 was a close imitation of the Prussian style. Ball-tufts were directed to be substituted for feathers by a circular of 1835, but Light Infantry had mostly had them already for several years. In 1838 we find the 85th wings had gilt treble chain straps and points, and the skirt ornaments are described as 'gold laurels on scarlet, bugle in centre, crown above.' Shell-jackets were used by the officers on parade about this time; they were probably of regulation pattern, and the shoulder straps consisted of three gold cords, with a small silver bugle. In 1841 the sergeants discontinued the chains and whistles hitherto worn on their pouch-belts. A portrait of Lieut. T. E. Knox, which, if taken while he was a lieutenant, might be about 1842, shows wings covered with embroidery, instead of the standard curb-chain pattern which seems to have been worn in 1838, and indicates that the regimental custom was to loop the 'bell rope' sash ends so that the tassels hung about level with the waist line.

OFFICER'S BUTTON, 1855–81.

In January 1842 black pack-straps were taken into use, instead of the ordinary white ones, and the men's shell-jackets were made with red collars and a white bugle thereon. These jackets, white up to 1830, were presumably red in 1842, and would ordinarily have plain blue collars and cuffs. It was apparently in this year that the modified pattern of shako, approved in 1839, was taken into wear. This was three quarters of an inch higher than that of 1828, with rather straighter sides; it thus appeared less 'belled.' For officers the shako had a curb-chain instead of the chin-scales; while for the men these were replaced by a black leather strap, an inverted peak or 'fall' was added to the back, and the large star-plate was changed for a much smaller and plainer circular plate surmounted by a crown.

In 1843 another pattern of shako was approved, a little smaller in diameter at top than at bottom, thus changing the appearance altogether, with peaks both in front and at back. This time it was the officers' turn to have a new plate in front (still a star, but smaller than the previous one, and all gilt), while the men's remained unaltered.

The appearance of the band and buglers between 1846 and 1854 may be seen in the groups taken from the late Mr. Ebsworth's sketches. The summer trousers of all ranks, in temperate climates, were now of a light blue-grey woollen material instead of the old white linen. For undress, officers had worn a single-breasted blue frock coat since the second decade of the century, only slightly varied from time to time. This was abolished in 1848 in favour of the shell-jacket.

OFFICER'S SHAKO PLATE, 1855–1862.

Nothing has yet been said of a rather interesting regimental ornament, the 'breastplate' or belt-plate worn by both officers and men on the white belt which passed over the right shoulder. The earliest pattern traceable was an oval plate bearing an eight-pointed star, with '85' in the centre within a circular band, inscribed 'Bucks Regiment.' On the top portion of the star, above the circle, is a small bugle which has the appearance of having been an addition to the original pattern drawing. This may be the 1793 plate itself; at all events it is before 1808, and the bugle may have been the only modification immediately introduced. This plate was entirely silver. The men's pattern is not known; but it would be of brass and almost certainly oval, like the officers'. The next pattern belongs to the period after 1821; and, considering the changes of title and also of fashion in the interval, there must probably have been another plate used (of which the design is missing) between the oval one and this. It is rectangular, bearing a garter surmounted by a crown, and a straight label, slightly ornamented, at the lower part under the garter; the garter is inscribed 'THE KING'S LIGHT INFANTRY,' the label 'PENINSULA,' and within the garter is '85' with a small bugle above. It is entirely silver. In 1832 the plate itself became a gilt plate, the same ornaments being continued in silver, and this pattern was still in wear in 1837. In 1846 the last pattern of breastplate was brought into use. In this a garter bearing the regimental motto is surrounded by a wreath and the straight label is replaced by a curved scroll with the regimental title. The so-called 'bugle,' really a conventionalised powder-horn, is reversed, a singular fact. This, again, was a gilt plate with silver mounts. A specimen entirely in brass is now in the United Service Institution Museum, and probably belonged to a sergeant or private. A letter from an officer, dated February 1847, mentions that he 'must be provided at once with the new pattern breastplate, which is exactly similar in device to that worn by the men.'

SHAKO PLATE.

The British infantry had retained the coatee longer than that of any other large European power, and the disappearance of this article of dress practically coincided with that of the smoothbore musket. When the change was made, in 1855, it was made as thorough as possible. With the old garment went all its ornaments. All epaulettes, wings, and other shoulder adornments were abolished. The men's white lace loops and the regimental ways of wearing them, and the corresponding regimental variations in the officers' clothing were all discarded. Although gold lace was retained for the officers' collars and cuff-flaps, a new way of putting it on was invented; the cuff-flap was made of the facing colour, apparently because the old one had been red, and was to have three buttons because the old one had had four. A change was already in progress in the soldier's equipment, substituting a waistbelt for the bayonet shoulder-belt. A waistbelt now also replaced the officer's sword-belt, thus entirely getting rid of the old breastplate. Sashes, which had been worn round the waist for nearly a century, were now to be worn over the shoulder, and a new pattern of shako was brought in. In fact the only regimental variations which remained were the facing colour (most of the facing colours being, of course, common to many regiments) and certain small differences in buttons and shako-plates and waist-plates, generally indistinguishable at a few yards' distance.

OFFICER'S HELMET PLATE, 1880.

With regard to the 'tunic' period, therefore, which may be said to extend to the present day, there is, from the regimental point of view, very little to be said The old style and the new, in the 85th, may be seen contrasted in the sketch of the landing in South Africa in 1856, and some of the principal variations in the home service kits during the last fifty years are shown in the last coloured 'uniform' illustration. The distinctive marks of a Light Infantry Corps were the green plume or pompon (candidly called black from 1874) in the shako; the green colour of this head-dress in its later years, and of the helmet which replaced it; and of the undress cap; and a small bugle joined to the regimental number on the dress and undress caps and on the waistplate. For the 1855 shako the 85th had, in the centre of the star plate, '85' within a gilt bugle on a black ground. This was not affected by the change of plate in 1862, the solid gilt centre with pierced numerals not applying to Light Infantry and other regiments with special badges. For the helmet, after 1881 the centre of the plate showed a silver bugle with a gilt 'K.L.I.' (in place of the numerals) on a dark green velvet ground which has since been altered to dark green enamel. The rank and file received a star plate for their shako in 1855, and have ever since continued to wear plates on the head-dress similar in general appearance to those of the officers. The bugle badge was placed on the men's collars in accordance with a circular of 1875, the regimental buttons of the army having been abolished The officers' collars had the rank-badges on them ; when these were removed to the shoulder cords, in 1880, regimental badges appeared on their tunic-collars also. The numerous changes in the rifles and in the patterns of knapsack and valise equipments are not here particularised, being common to the whole army.

As to bandsmen and drummers (or buglers), the drummers up to 1821 would wear yellow jackets faced with red and ornamented with additional lace of a special pattern, and from 1821 red faced with blue, and special lace as before. The 85th pattern of drummers' lace was white with a fairly broad red stripe along each side, but leaving two or three threads of white on the actual edge. A universal pattern was substituted early in the 'seventies.' The band, when one was formed, may have also worn yellow jackets faced red, but great licence was permitted as to the costume of bands. From 1830 they were ordered to be dressed in white coatees with regimental facings, and in 1867 white band tunics were abolished in favour of red ones.

APPENDICES

APPENDIX I

OFFICERS OF THE ROYAL VOLONTIERS. 1759–1763.

THE LIST HERE GIVEN CONTAINS THE NAMES OF ALL THE OFFICERS OF THE 'ROYAL VOLONTIERS' AS FAR AS IT HAS BEEN POSSIBLE TO OBTAIN THEM. THEIR CAREERS, HOWEVER, ARE ONLY CONSIDERED AS FAR AS THEIR CONNECTION WITH THE REGIMENT IS CONCERNED, EXCEPT IN A FEW INSTANCES IN WHICH BRIEF NOTES HAVE BEEN ADDED. FOR CONVENIENCE THESE NAMES ARE ARRANGED ALPHABETICALLY. IT IS CURIOUS TO NOTE THE NUMBER OF OFFICERS (MORE THAN TWELVE) WHO LEFT THE REGIMENT FOR SOME UNEXPLAINED CAUSE IN 1761; AND ALMOST ALL SUBALTERNS. IT HAS NOT BEEN FOUND POSSIBLE TO OBTAIN ANY CLUE TO THE REASON. HOWEVER, SHOULD LATER RESEARCH AFFORD A SOLUTION, A FOOTNOTE WILL BE ADDED.

ACTON, WILLIAM.—First Lieutenant August 31, 1759; resigned.

ADAMS, JOSEPH.—First Lieutenant August 8, 1759; Captain-Lieutenant June 2, 1762, vice St. Clair exchanged; half-pay 1763.

ADAMSON, MARTIN.—Second Lieutenant December 20, 1759; half-pay 1763.

ALLEN, MILES.—First Lieutenant August 28, 1759; dismissed the service 1761.

BARRY, ROBERT.—First Lieutenant August 19, 1759; half-pay 1763.

BARRY, ROBERT.—Second Lieutenant September 8, 1761; half-pay 1763; here is an ambiguity. Possibly this is an error in the Commission Register.

BATHURST, —— .—Second Lieutenant August 7, 1759. There is some doubt as to the identity of this officer, whose name does not occur again. He is not, however, to be confounded with Peter Bathurst.

BATHURST, PETER.—Captain August 7, 1759; left the regiment 1761.

BENDINELL, ROBERT.—Second Lieutenant December 22, 1761; half-pay 1763.

BOYCOTT, RICHARD.—Second Lieutenant August 5, 1759; first Lieutenant October 24, 1761; half-pay 1763.

BRUCE, ANDREW.—First Lieutenant August 9, 1759; Captain December 22, 1761; exchanged to 52nd Foot.

BRYDON, PATRICK.—Second Lieutenant October 14, 1759; half-pay 1763.

BURGOYNE, JOHN.—Captain March 4, 1761; half-pay 1763.

BUXTON, JOHN.—Second Lieutenant October 29, 1761; half-pay 1763.

CALCRAFT, MR.—Agent (Channel Row, Westminster) 1759.

CAWTHORN, WILLIAM.—First Lieutenant August 4, 1759; Captain-Lieutenant October 29, 1761, vice Maxwell; had been Quartermaster in the 7th Dragoons 174$\frac{3}{4}$; half-pay 1763.

COOPER, CHARLES.—Captain August 3, 1759; left the regiment 1761.

COOPER, JOHN.—First Lieutenant August 22, 1759; half-pay 1763.

COPE, HENRY.—Captain (Army) August 27, 1760; Captain (Royal Volontiers) March 18, 1761; half-pay 1763.

CORNWALLIS, CHARLES.—Commonly called Lord Charles Brome; Captain August 7, 1759; left the regiment 1761. He was afterwards Marquis Cornwallis, K.G.—Ensign 1st Foot Guards 1756; Captain 85th 1759; Lieut.-Colonel 12th Foot 1761; Colonel 33rd Foot 1766. Served with distinction in Germany and North America. Commander-in-Chief in India 1786–93; do. Ireland 1798–1801; as Ambassador to France negociated the Peace of Amiens; Governor-General of India 1805; died at Ghazeepore 1805.

CRAUFORD, JOHN.—First and only Colonel of the regiment July 21, 1759; wounded and taken prisoner at Belleisle. After long service in the junior ranks this officer was appointed Lieut.-Colonel of 13th Foot 1749; Colonel 85th 1759; Governor of Belleisle 1762; Colonel 3rd Buffs and Governor of Minorca 1763; where he died of a fever fifteen months later. A monument to this officer's memory is believed to have been erected in Minorca, but its present condition is unknown.

DAWSON, ——.—Second Lieutenant May 28, 1762; no further information.

DAWSON, HATTON.—First Lieutenant August 10, 1759; Senior First Lieutenant in 1762; half-pay 1763.

DAWSON, JAMES.—First Lieutenant August 5, 1759; Captain-Lieutenant December 16, 1761; Captain February 12, 1762; half-pay 1763.

DONALDSON, PRYSE.—First Lieutenant August 11, 1759; half-pay 1763.

DOUGLAS, CHRISTOPHER.—Surgeon May 10, 1760; no further information.

DOUGLAS, J. STUART.—From the 4th Foot; Major July 16, 1762, vice Heywood; half-pay 1763.

DUME, GEORGE.—Second Lieutenant December 25, 1761; half-pay 1763.

EGERTON, CHARLES.—Captain August 13, 1759; half-pay 1763.

FLEETWOOD, JOHN GERRARD.—Second Lieutenant May 15, 1761; half-pay 1763.

FORRESTER, WILLIAM.—Captain August 14, 1759; half-pay 1763.

FREEMAN, THE REV. ANTHONY.—Chaplain January 7, 1761; no further information.

GODFREY, ROWLAND.—Second Lieutenant December 16, 1761; half-pay 1763.

GORDON, LOCKHART.—Captain August 10, 1759; half-pay 1763.

GOUGH, RICHARD.—Adjutant August 2, 1759; promoted Second Lieutenant October 23, 1760; half-pay 1763.

GRAYDON, GEORGE.—Second Lieutenant April 13, 1761; he was the senior Second Lieutenant in 1762; half-pay 1763.

GREEN, WILLIAM.—First Lieutenant August 7, 1759. It is believed that this officer afterwards became General Sir William Green, Bart. (Colonel, Royal Engineers).

GWYNNE, JOHN.—First Lieutenant August 7, 1759; left regiment 1761. (This officer's name also appears as 'Glynne' and 'Glynn.')

HARIS, JOHN (? HARRIS).—Second Lieutenant March 22, 1762; half-pay 1763.

HATTON, HENRY.—First Lieutenant August 16, 1759; half-pay 1763.

HEYWOOD, NATHANIEL.—Captain July 30, 1759; Major January 6, 1761; this officer retired as a Colonel from the old 21st or Royal Windsor Forester Light Dragoons; was believed to be the last surviving officer of the Royal Volontiers.

HUGHES, RICHARD.—Junior First Lieutenant in 1762; half-pay 1763.

HUMPHREYS, FRANCIS RICHARD.—Second Lieutenant August 10, 1759; First Lieutenant April 25, 1761; exchanged to 111th Foot.

HUSSEY, RICHARD.—Captain (Army) May 2, 1758; Captain (Royal Volontiers) March 3, 1761; half-pay 1763.

JEFFREYS, ST. JOHN.—Captain August 11, 1759; left the regiment 1761.

JOHNES, JAMES.—Second Lieutenant August 2, 1759; First Lieutenant May 1, 1760; half-pay 1763.

JONES, CHARLES.—First Lieutenant August 27, 1759; half-pay 1763.

LAMBERT, OLIVER.—Second Lieutenant August 8, 1759; half-pay 1763.

LANGHAM, JAMES.—Captain August 4, 1759; was the senior Captain in 1762; half-pay 1763.

LLOYD, JAMES.—First Lieutenant August 2, 1759; Captain March 24, 1761; resigned December 16, 1761.

LOFTUS, THOMAS.—Second Lieutenant December 17, 1761; half-pay 1763.

LUCAS, RICHARD.—Second Lieutenant August 11, 1759; First Lieutenant December 16, 1761; half-pay 1763.

MACKENZIE, ——.—Second Lieutenant June 2, 1762; half-pay 1763.

MATTHEWS, THOMAS.—Captain August 15, 1759; resigned 1761; succeeded by George Maxwell.

MAXWELL, GEORGE.—Adjutant August 4, 1759; First Lieutenant August 2, 1759; Captain-Lieutenant vice Thorpe March 24, 1761; Captain vice Matthews October 29, 1761; half-pay 1763.

MERSON, NATHANIEL.—Second Lieutenant August 4, 1759; left regiment 1761.

MESTER, ——.—Captain of one of the four additional Companies August 7, 1759. The name of this officer is open to doubt.

MEYRICK, OWEN.—Second Lieutenant August 7, 1759; First Lieutenant May 15, 1761; half-pay 1763.

MOORE, J——.—Captain August 9, 1759; half-pay 1763.

NEWSHAM, CHRISTOPHER.—First Lieutenant August 25, 1759; had left the regiment by 1761.

NUGENT, EDMUND.—Captain August 9, 1759; half-pay 1763; subsequently on full pay and rose to the rank of General. Possibly one of the O'Reilly Nugents of Ballinlough Castle, Co. Westmeath.

NUGENT, OLIVER.—First Lieutenant August 21, 1759; half-pay 1763.

PEMBERTON, THOMAS.—First Lieutenant August 13, 1759; no further information.

PERCY, HUGH (commonly called Hugh, Lord Warkworth).—Son of Sir Hugh Smithson, first Duke of Northumberland, Captain August 7, 1759; afterwards second Duke of Northumberland, K.G.; when Earl Percy, greatly distinguished himself as Major-General in the early campaigns in North America. General and Colonel Royal Horse Guards; died 1817.

PRICE, ROGER.—First Lieutenant December 20, 1759; half-pay 1763.

PULTENEY, WILLIAM, VISCOUNT.—First Lieutenant-Colonel of the regiment July 21, 1759; served both at Belleisle and in Portugal. This promising young officer died in 1763 in the twenty-fifth year of his age. He was the son of the 'Great Earl of Bath' whose name is so familiar to the reader of Horace Walpole's letters and other chronicles of the time. He appears to have held the Lieut.-Colonelcy of the newly raised regiment of Shropshire Militia as well as that of the 85th Light Infantry. Much dissatisfaction was caused at this period by the enforcement of the Militia Act. The subject gave rise to endless squibs and broadsides. An amusing effusion of this class preserved in Owen and Blakeway's 'History of Shropshire' records the formation of the Shropshire Militia and Lord Pulteney's connection with it in the following strain :

'As the Militia of Salop is now to be raised
The Militia of Salop by me must be praised,
While others but trot, my muse goes full gallop,
To sing to some tune the Militia of Salop.'

We need not quote the whole ballad, but the last verse may be inserted

'And now of my ballad, pray don't make a jest,
To honour the County I've done my best.
But fill up a glass of Joe Laurence's beer,
And drink the Militia of merry Shropshire.'

Joe Laurence, it would seem, was host of 'the Raven in this good town of Shrewsbury,' erst the billet of the famous Sergeant Kite.

RICE, WOODFORD.—First Lieutenant August 12, 1759 ; half-pay 1763.

ROBERTS, JOHN.—Second Lieutenant April 19, 1762 ; half-pay 1763 ; possibly the John Roberts who was Quartermaster August 4, 1759.

ROOKE, JAMES.—First Lieutenant August 15, 1759 ; left the regiment 1761 ; afterwards rose to the rank of General and was Colonel of the 38th Foot.

ST. CLAIR, GEORGE.—First Lieutenant August 6, 1759 ; promoted Captain February 12, 1762, vice Dawson ; exchanged to 111th Foot June 2, 1762.

SANDYS, GEORGE.—First Lieutenant August 14, 1759 ; half-pay 1763.

SHIPLEY, RICHARD.—First Lieutenant August 18, 1759 ; exchanged to 109th Foot.

SKINNER, WILLIAM.—Captain August 2, 1759 ; Major February 11, 1761 ; half-pay 1763.

SLEIGH, JOHN.—Surgeon August 20, 1759 ; no further information.

STEWART, JAMES.—First Lieutenant August 17, 1759 ; exchanged to 114th Foot.

STUART, ARCHIBALD.—Second Lieutenant April 23, 1761 ; resigned December 25, 1761.

TEMPLE, ROBERT.—Second Lieutenant January 12, 1761 ; First Lieutenant December 22, 1761 ; half-pay 1763.

THACKERAY, FREDERICK.—Second Lieutenant August 9, 1759 ; First Lieutenant October 29, 1761 ; half-pay 1763.

APPENDIX II. 1779–1783

THE LIST HERE GIVEN CONTAINS THE NAMES OF ALL THE OFFICERS OF THE 'WESTMINSTER VOLONTIERS' AS FAR AS IT HAS BEEN POSSIBLE TO OBTAIN THEM.

ANDERSON, CHARLES F.—Ensign half-pay on disbandment 1783.

ANDERSON, EVELYN.—Major March 20, 1783; half-pay on disbandment 1783.

BARKER, JOSEPH.—To be Captain-Lieutenant vice Hill April 12, 1783; half-pay on disbandment 1783.

BAYLEY, JAMES.—An officer of this name is on the Half-Pay List for 1785 as an Ensign of 85th Regiment.

BLOXHAM, CHARLES.—Ensign March 14, 1782; half-pay on disbandment 1783.

BOOTH, JOHN.—Quartermaster November 26, 1781; drowned on voyage home.

BRICE, A. H.—Ensign February 26, 1783; half-pay on disbandment 1783.

BRICK, DANIEL.—From 22nd Regiment to be Captain, vice Alexander Salans deceased, March 29, 1783; half-pay on disbandment 1783.

BROOKE, RICHARD.—From 34th Foot; Lieutenant September 2, 1779; drowned on voyage home.

BROWN, GEORGE.—Major December 13, 1780.

BROWN, WILLIAM.—From 14th Foot; Captain September 1, 1779.

BULKELEY, FREKE.—Ensign September 24, 1780; Lieutenant March 6, 1782.

COOK, GEORGE.—Ensign February 7, 1780.

CRANE, WILLIAM.—Lieutenant half-pay on disbandment 1783.

CREWE, RICHARD.—Major 85th Regiment August 31, 1779; had left the regiment by 1781; later he raised and commanded the 125th Foot or Waterford Regiment, afterwards disbanded.

DALTON, EDWARD.—Ensign March 6, 1782; Lieutenant May 14, 1782; half-pay on disbandment 1783.

DE SALANS, BARON ALEXANDER.—Ensign 9th Foot; Lieutenant 85th Westminster Volontiers September 3, 1779; Captain-Lieutenant and Captain May 19, 1780; drowned on voyage home.

DAVIS, EDWARD.—Ensign December 13, 1780.

DICKSON, RICHARD.—From 48th Foot; Lieutenant August 6, 1779; Captain March 6, 1782.

DOBBLYN, MICHAEL.—From 33rd Foot; Lieutenant September 1, 1779; drowned on voyage home.

DONALD, ALEXANDER.—A Lieutenant half-pay late 89th Regiment to be Lieutenant 85th Regiment vice John Duke; half-pay on disbandment 1783.

DUKE, JOHN.—Ensign August 3, 1779; Lieutenant September 24, 1780; drowned on voyage home.

EATON, THE REV. S.—Chaplain vice Tickell May 3, 1783; half-pay on disbandment 1783; evidently an exchange.

FINNEY, JOHN.—Ensign September 4, 1779; Lieutenant December 12, 1780; drowned on voyage home.

FITZGERALD, LORD HENRY.—From 66th Foot; Captain September 5, 1779; (junior) Major September 4, 1781; Lieut.-Colonel Commanding March 12, 1783; half-pay 1783; retired from the army shortly afterwards; apparently this officer was the fourth son of the first Duke of Leinster and brother of the ill-fated rebel Lord Edward Fitzgerald; Lord Henry Fitzgerald died in 1830.

GIBBON, EDWARD.—Ensign August 13, 1780; no further information.

GIBSON, GODFREY.—Ensign vice McDowall April 5, 1783; half-pay on disbandment 1783.

GOULD, JOHN.—Ensign August 6, 1779; Lieutenant December 12, 1780; drowned on voyage home.

GREY, THOMAS.—Quartermaster December 1780; fate unknown.

GROSE, FRANCIS.—From the 52nd Foot to be Captain 85th Foot September 2, 1779; he was senior Captain in the regiment in 1783; half-pay on disbandment in 1783; later restored to full pay, he rose to be Colonel of the 102nd (late) or New South Wales Regiment and Deputy Governor of Botany Bay; he ultimately attained the rank of General. He was the eldest son of the celebrated Francis Grose the antiquary, among whose many illustrated topographical and antiquarian works may be numbered the 'Military Antiquities respecting a History of the English Army from the Conquest to the Present Time' (1786). The antiquary was at one time Adjutant and Paymaster, and subsequently Captain and Adjutant of the Surrey Militia; hence his military title. It was of Grose, while engaged on his work on Scottish Antiquities, that Burns wrote the well-known and oft-quoted lines:

'A chiel's amang ye takin' notes.
And, faith, he'll prent it.'

Much to the annoyance, too, of Grose the piece which begins:

'Ken ye ought o' Captain Grose?
Igo and ago.
If he's among his friends or foes?
Iram, coram, dago.'

Some of the following stanzas are not too complimentary. They were handed round and widely read. Burns also wrote a somewhat scurrilous epigram upon Grose, why is not known.

HELY, JOHN.—From 50th Foot; Captain August 31, 1779; fate unknown.

HILL, JOHN FORSTER.—Ensign and Adjutant September 14, 1779; Lieutenant September 27, 1779; Captain-Lieutenant and Captain April 8, 1782; Captain April 12, 1783, vice Grose; half-pay on disbandment 1783.

KEMPTHORNE, SAMUEL.—Ensign 50th Regiment; to be Lieutenant 85th Regiment vice Dobblyn March 29, 1783; half-pay on disbandment 1783.

KINKEAD, MOSES.—Ensign August 5, 1779.

LASCELLES, EDWARD.—Ensign September 14, 1779; Lieutenant February 28, 1780.

MACKAY, HECTOR.—Ensign February 26, 1783; half-pay on disbandment 1783.

MADGETT, ROBERT.—Ensign May 1, 1782; half-pay on disbandment 1783.

MAUNSELL, SEWELL.—Captain (Army) May 23, 1781; Captain October 24, 1781.

MAWHOOD, THOMAS.—From 19th Foot; Captain September 4, 1779.

MAXWELL, JAMES.—Lieutenant September 6, 1781; half-pay on disbandment 1783.

McDOWALL, JOHN ALEXANDER.—Quartermaster August 30, 1779; Ensign March 13, 1782; Lieutenant March 29, 1783, vice Brooke. Half-pay on disbandment 1783.

MALLOR, DANIEL.—Lieutenant; half-pay on disbandment 1783.

MANLEY, JOHN.—Half-pay on disbandment 1783.

MAUNSELL, SEWELL.—Ensign 29th Foot July 17, 1771; Lieutenant do. June 30, 1774; Captain 85th Regiment October 24, 1781.

MELLER, WILLIAM.—Ensign October 27, 1779.

MILLAR (? MALLOR), THOMAS.—Lieutenant December 13, 1780; Captain September 5, 1781; half-pay on disbandment 1783.

MOLESWORTH, ROBERT.—Ensign September 14, 1779; Lieutenant May 19, 1780; half-pay on disbandment 1783.

MULLER, DANIEL.—Lieutenant December 23, 1780; Adjutant February 15, 1782; resigned.

MULLER, WILLIAM.—Adjutant vice Daniel Muller April 8, 1783; half-pay on disbandment, 1783.

MULLHALLEN, JOHN.—Ensign May 19, 1780; Lieutenant September 7, 1781; retired.

MUNRO, GEORGE.—Captain (Army) October 18, 1779; Captain (85th) September 29, 1780; half-pay on disbandment, 1783.

MURRAY, JOHN.—Captain 83rd Regiment to be Major 85th Regiment vice Samuel Pole; half-pay on disbandment 1783.

PARSONS, WILLIAM.—Lieutenant September 8, 1779; drowned on voyage home.

PESHALL, SIR JOHN, BART.—Ensign December 12, 1780.

PHILLIPS, ——.—Surgeon July 9, 1781; drowned on voyage home.

PHILLIPS, RALPH.—Major October 4, 1780.

PHIPPS, THE HON. EDMUND.—Ensign March 17, 1780; remained but a short time in the regiment; born April 7, 1760; 4th son of 1st Baron Mulgrave; ultimately rose to rank of General and was Colonel Commandant of 60th regiment; died unmarried September 14, 1837.

PHIPPS, THE HON. HENRY.—Afterwards 1st Earl (4th Baron) Mulgrave; Ensign 1st Foot Guards June 8, 1775; Lieutenant and Captain 1778; exchanged to 85th Foot as Major August 30, 1779; October 4, 1780 Lieut.-Colonel 88th (Connaught Rangers); exchanged to 45th Foot 1782; Colonel 31st Foot February 8, 1793; from Gibraltar he was sent to Toulon with three regiments; assisted in fortifying that place; the command being assumed by O'Hara, Phipps declined to serve under him and returned home; served next in Flanders; succeeded to Irish Barony of Mulgrave 1792; raised to the peerage August 13, 1794 as Baron Mulgrave of Yorkshire; Major-General same year (October 3); Lieut.-General January 1, 1801; General October 25, 1809; his military career ended in 1801 and political matters occupied his attention; he declined the offer of the Irish Command on April 7, 1801; G.C.B. 1820; died April 7, 1831; Lord Mulgrave was a liberal patron of the fine arts; he was an Elder Brother of the Trinity House and

Vice-Admiral of Yorkshire; it is stated that he was one of the first military men at home to discern the talents of Wellington in the earlier portion of the Peninsular Campaign. Among other offices held by Lord Mulgrave were those of First Lord of the Admiralty 1807–12, Master-General of Ordnance 1812–13 and Governor of Scarborough Castle; created Earl of Mulgrave and Viscount Normanby September 7, 1812.

POLE, SAMUEL.—From 56th Foot, Captain September 3, 1779; (Senior) Major September 3, 1781; drowned on voyage home.

POULETT, THE HON. VERE.—From 62nd Foot; Lieutenant September 7, 1779; Captain December 13, 1780; 2nd son of 3rd Earl Poulett; born 1761; ultimately a Lieut.-General in the Army and some time M.P. for Bridgewater.

ROW, JOHN.—From 9th Foot; Captain-Lieutenant August 30, 1779.

RUTHERFORD, JOHN.—Surgeon August 30, 1779.

ST. JOHN, THE HON. FREDERICK.—Ensign September 14, 1779; Lieutenant February 7, 1780; did not remain long in the regiment; was the 2nd son of the 3rd Viscount St. John and 2nd Viscount Bolinbroke; born December 20, 1765; became a General in the Army and died November 19, 1844.

SHERBROKE, JOHN COAPE.—Captain March 6, 1783; vice Richard Dickson.

STANHOPE, CHARLES, EARL OF HARRINGTON.—Born 1753; Ensign Coldstream Guards November 1769; Captain (Light Company) 29th Foot August 1773; 1774–1776 M.P. for Thetford, and later for Westmeath till 1779. By favour he exchanged his Light Company for a Grenadier Company in the 29th. In 1776 he served at Quebec and was present at the battle on the Heights of Abraham; he also served on the St. Lawrence under Carleton, afterwards Lord Dorchester; he accompanied General Burgoyne as A.D.C. on the campaign which ended in disaster at Saratoga; thence he brought home the official despatches, but arrived in London December 24, 1777, after the news had been received; 1778 Captain and Lieut.-Colonel in the 3rd Foot Guards; succeeded to the Peerage in April 1779; raised and commanded as Lieut.-Colonel the 85th Westminster Volontiers; accompanied the regiment to Jamaica with the rank of Brigadier-General; here he assisted John Dalling the Governor in fortifying the island; ill-health compelled his return to England within a year and he went home with his wife and child; promoted Colonel and A.D.C. to the King November 26, 1782; March 1783 Colonel of the 65th Foot, in which regiment the new system of tactics invented by Dundas was first tried; January 29, 1788, Colonel of his old regiment the 29th; December 5, 1792, Colonel 1st Life Guards and Gold Stick; Major-General October 1793;

Lieut.-General 1798; General September 25, 1802; Privy Councillor October 24, 1798; Constable and Governor of Windsor Castle March 14, 1812; G.C.H. 1816; he was also from July 1803 to October 1805 Second in Command of the Staff of the London District and from October 31, 1805, to 1812 Commander-in Chief in Ireland. He was instrumental also in introducing into the Army the new sword of 1792; he died September 15, 1829.

STANHOPE, PHILIP, 5th EARL OF CHESTERFIELD.—Captain September 6, 1779; but it is very doubtful whether this officer ever joined the regiment. He had succeeded to the earldom in 1773; he was born in 1755 and died in 1815. Lord Chesterfield had a curious career. His education was a strange one; among his private tutors was the notorious Dr. William Dodd, who later in 1777 forged the name of his former pupil to a bond for £4,200. Dodd was detected, prosecuted, convicted and hanged.

STEELE, THOMAS.—From 63rd Foot; Lieutenant September 4, 1779; Captain September 4, 1781; half-pay on disbandment 1783.

STONEY, VIGOE ARMSTRONG.—Ensign May 14, 1782; half-pay on disbandment 1783.

STUART, THOMAS.—Junior Lieutenant; half-pay on disbandment 1783.

TEARE, JOHN.—Surgeon vice Phillips April 19, 1783; half-pay on disbandment 1783.

TICKLE, THE REV. JOHN.—Chaplain August 30, 1779.

WALLIS, ROBERT.—From 40th Regiment; to be Captain vice Sewell Maunsell March 29, 1783; half-pay on disbandment 1783.

WEBB, HENRY.—From half-pay late 47th Foot; Lieutenant August 31, 1779; Captain-Lieutenant September 24, 1780.

WHITE, ROBERT.—Ensign April 8, 1782.

WILSON, JOSEPH.—To be Quartermaster vice John Booth March 29, 1783; half-pay on disbandment 1783.

APPENDIX III

REGISTER, WITH COMMISSION DATES, SERVICES, HONOURS, AND DISTINCTIONS, OF THE OFFICERS OF THE REGIMENT, 1793–1911.

À COURT, THE HON. EDWARD ALEXANDER HOLMES.—Ensign February 26, 1864; Lieutenant December 4, 1866; Captain February 13, 1874; Major September 30, 1882; retired April 7, 1886. *Afghan War*, 1879–80.—With the Kuram Division, Zaimusht expedition and assault of Zawa, operations in Khoormana Gorge. Mentioned in despatches; medal.

À COURT, FREDERICK HOLMES.—Ensign June 19, 1857; Lieutenant September 23, 1862; retired August 16, 1864.

ADDISON, CHARLES.—Lieutenant June 4, 1801; half-pay 1802; 26th Foot July 9, 1803; Captain 26th Foot April 3, 1806; half-pay 62nd Foot July 23, 1818; Captain 80th Regiment March 16, 1820; half-pay unattached August 4, 1825; Brevet-Major August 12, 1819; died in Edinburgh 1838.

AIDÈ, CHARLES HAMILTON.—Ensign May 17, 1844; Lieutenant April 1, 1847; Captain December 3, 1852; retired 1853.

AINSLIE, GEORGE ROBERT.—Ensign June 14, 1793, 19th Foot; Lieutenant 85th Foot 1793; Captain do. April 15, 1794; Major do. July 25, 1799; Lieut.-Colonel do. January 11, 1800 (late Birmingham Fencibles); Lieut.-Colonel 25th Regiment May 21, 1807; Colonel July 25, 1810; Major-General June 4, 1813; Lieutenant-General May 27, 1825; died in Edinburgh April 16, 1839; served in Flanders in 1793; at Walcheren 1794; present at St. André, under Sir Ralph Abercromby, and at Thuyl on the Waal; Retreat from the Rhine; Expedition to Holland 1799.

ALEXANDER, JOHN.—Ensign March 9, 1800; Lieutenant June 25, 1803; no further information.

AMIEL, JOHN ARCHIBALD.—Ensign August 9, 1799; Lieutenant April 24 1800; half-pay 1802; Lieutenant 27th Regiment October 12, 1804; Captain 27th Regiment July 11, 1811; half-pay 94th Regiment June 8, 1820; died at Guernsey November 16, 1829.

ANSTRUTHER, WINDHAM GEORGE CONWAY.—Ensign July 25, 1865; retired August 9, 1867.

ANTROBUS, JAMES CUTHBERT.—Ensign December 25, 1805; Lieutenant 60th Regiment December 7, 1807; died 1810.

APPELIUS, LEWIS.—Ensign 60th Regiment June 1, 1804; Lieutenant 85th Regiment August 29, 1805; dismissed the service 1812.

APPLEYARD, FREDERICK ERNEST.—Ensign 80th Regiment June 14, 1850; Lieutenant do. October 12, 1852; Captain 7th Foot December 29, 1854; Major December 26, 1856; appointed Major 85th Regiment February 5, 1861; Lieut.-Colonel do. March 6, 1867; Colonel (Honorary Major-General) retired pay. *Burmese War*, 1852.—Storming and capture of Martaban, Rangoon, and Prome; medal with clasp. *Crimean Campaign*, 1854–5.—Battles of Alma (wounded) and Inkerman, and siege and fall of Sevastopol, including repulse of sorties of April 5 and May 9, defence of Quarries on June 7, and assault on Redan June 18, 1855 (wounded). Mentioned in despatches; medal with three clasps; Turkish medal; Brevet-Major; Chevalier of Legion of Honour; 5th class of Medjidie. *Afghan War*, 1878–9.—Commanded 3rd Brigade, 1st Division, Peshawar Valley Field Force, assault and capture of Ali Musjid (mentioned in despatches) and Bazar Valley Expedition. Despatches, *London Gazette*, November 7, 1897; medal with clasp.

ARMITAGE, JOHN.—Ensign November 10, 1848; Lieutenant March 19, 1852; no further information.

ARMSTRONG (or WALLER), ARTHUR SAVORY.—Ensign August 21, 1849; Ensign 49th Regiment September 4, 1849; no further information.

ARMYTAGE, EDWARD JOHN.—Ensign Rifle Brigade February 20, 1858; Ensign 39th Foot April 15, 1859; exchanged to 85th Regiment and retired September 23, 1862.

ASHE, WALKER.—Ensign January 21, 1853; Captain December 7, 1858; exchanged to 1st Dragoon Guards November 1860.

ASHTON, GEORGE.—Ensign December 25, 1813; Lieutenant March 21, 1815; half-pay December 25, 1818; died at Carnarvon June 9, 1826; served in America.

ATCHISON, CHARLES ERNEST.—Second Lieutenant February 20, 1895; Lieutenant September 7, 1898; Captain February 11, 1902. *South African War*, 1899–1902.—Slightly wounded. Operations in the Orange Free State, February to May 1900, including operations at Paardeberg (February 17 to 26). Operations in Cape Colony, south of Orange River, 1899–1900. Operations in the Transvaal September 1901 to May 31, 1902. Queen's medal with three clasps. King's medal with two clasps.

ATHROPE (or ATHORPE), JOHN.—Ensign September 16, 1851; Lieutenant December 17, 1852; Captain October 8, 1855; died of apoplexy in Natal March 3, 1861, aged twenty-seven.

ATKINSON, GUY NEWCOMEN.—Ensign February 21, 1865; Lieutenant June 12, 1867; Captain March 10, 1875; Major September 30, 1882; Lieut.-Colonel October 27, 1886; to command battalion July 1, 1887; died while in command of the battalion at Kilkenny February 10, 1890.

ATKINSON, SAMUEL H.—Ensign 27th Regiment August 27, 1807; Lieutenant 85th Regiment October 26, 1808; Lieutenant 7th West India Regiment January 18, 1810; retired October 17, 1811.

AUSTEN, ARTHUR ROBERT.—From 1st battalion, Captain December 6, 1889; Major November 16, 1898; Brevet-Colonel November 29, 1906; Lieut.-Colonel June 17, 1908; retired June 17, 1908. *Egyptian Expedition*, 1882.—Defence of Alexandria, occupation of Kafr Dowar, surrender of Damietta; medal; bronze star. *Soudan Expedition*, 1885.—Suakin; served in the Camel Corps (wounded); clasp. *South African War*, 1899–1901.—(Slightly wounded). In command 2nd Battalion Shropshire Light Infantry from February 15 to May 15, 1901. Operations in the Orange Free State, February to May 1900, including operations at Paardeberg (February 17 to 26); actions at Poplar Grove, Dreifontein, Houtnek (Thoba Mountain), Vet River (May 5 and 6), and Zand River. Operations in the Transvaal in May and June 1900, including actions near Johannesburg and Pretoria. Operations in the Transvaal, east of Pretoria, July to November 29, 1900. Operations in the Transvaal, west of Pretoria, July to November 29, 1900, including actions at Elands River (August 4 to 16). Operations in Orange River Colony May to November 29, 1900, including action at Rhenoster River. Operations in Cape Colony, south of Orange River, 1899–1900. Operations in the Transvaal November 30, 1900, to May 1901. Performed duties of Commandant, Belfast. Despatches, *London Gazette*, September 10, 1901. Queen's medal with five clasps. Brevet of Lieut.-Colonel.

AUSTIN, JOHN.—Lieutenant 85th Regiment July 22, 1800; Captain 58th Regiment November 28, 1805; Paymaster 85th Regiment August 20, 1803; Brevet-Major September 4, 1807; Brevet Lieut.-Colonel February 25, 1813; Major October 25, 1814; half-pay, Spanish Staff, December 25, 1816; 97th Regiment March 25, 1824; Lieut.-Colonel June 8, 1826, unattached; exchanged to 17th Foot June 30, 1829; retired August 13, 1829; died at Bath March 24, 1860.

AYLMER, MATTHEW, LORD, G.C.B.—Ensign 49th Regiment October 19, 1787; Lieutenant do. October 26, 1791; Captain do. August 8, 1794; Major 85th Regiment October 9, 1800; Lieut.-Colonel

March 25, 1802; half-pay 1802; exchanged to Coldstream Guards June 9, 1803; Colonel July 25, 1810; Major-General June 4, 1813; Adjutant-General in Ireland, December 22, 1814; Lieutenant-General May 27, 1825; Colonel 56th Regiment October 27, 1827; Commander of the Forces in North America, August 11, 1830; Colonel 18th Regiment July 23, 1832; General November 23, 1841; died in Eaton Square, London, February 23, 1850. Served upwards of three years in the West Indies, eleven months of which he was at St. Domingo; present at the first and second attacks upon Tiburon; at the storming of Fort de l'Acul Leogone; at the affair of Bombard, near Cape Nicola Mole; at the reduction of Port-au-Prince in 1798; present at the descent near Ostend, taken prisoner and detained six months; in 1799, present at the Helder; attack on the British lines September 10; battles of September 19 and October 2; Expedition to Hanover 1805; Expedition to Copenhagen; Peninsula; gold medal for Talavera, Busaco, Fuentes d'Oñoro, Vittoria and Nive.

AYLMER, MICHAEL.—Lieutenant 85th Regiment November 23, 1793; Captain 41st Regiment January 26, 1796; Captain 11th Battalion of Reserve July 9, 1803; half-pay 1805; Major (Army) April 25, 1808; Major 24th Regiment October 23, 1806.

AYTOUN, JAMES.—Ensign 1st Foot October 22, 1847; Cornet 7th Hussars April 20, 1849; Lieutenant do. October 31, 1851; half-pay November 10, 1856; Captain (Army) January 24, 1856; full pay Captain 7th Hussars August 14, 1857; exchanged to 85th Foot May 12, 1863; retired as Major on half-pay April 25, 1865.

BACKHOUSE, THOMAS.—Lieutenant November 21, 1793; Captain September 3, 1795; removed from the service 1801.

BAGOT, WILLIAM HENRY.—Ensign April 2, 1807; no further information.

BAILEY, JOHN HENRY.—Second Lieutenant March 15, 1893; Lieutenant April 6, 1898; Captain June 11, 1901.

BAKER, WILLIAM THOMAS.—From 44th Foot; Lieutenant February 18, 1853; Captain August 29, 1856; exchanged March 22, 1864.

BALL, WILLIAM CHICHESTER—Ensign 62nd Regiment September 3, 1803; Lieutenant do. November 30, 1804; Captain do. March 13, 1806; Captain 37th Regiment September 21, 1809; Captain 85th Regiment January 25, 1813; Brevet-Major September 25, 1814; died at Genoa on board H.M.S. *Redpole*, on passage from Malta, November 5, 1823.

BANON, FREDERICK LIONEL.—Lieutenant 1st Battalion September 9, 1882; attached to Army Service Corps April 2, 1889, to April 1, 1894; Captain January 1, 1890; Major August 11, 1900; Imperial Yeomanry May 15, 1901, to August 28, 1902; Lieut.-Colonel, half-pay, December 23, 1905; full pay same date; Brevet-Colonel

December 23, 1908; Substantive Colonel November 9, 1909; half-pay December 23, 1909, to July 14, 1910; Deputy Assistant Quarter-master-General, Dublin District, 1902–3; do. 3rd Army Corps 1903–5; do. Irish Command; Deputy Assistant Adjutant-General, Staff College; Deputy Assistant Adjutant-General; General Staff Officer, 2nd Grade; do. 1st Grade, Staff College, 1905–8; Assistant Adjutant-General War Office July 14, 1910. *Soudan Expedition*, 1885.—Suakin; medal with clasp; bronze star. *South African War*, 1899–1902.—Assistant to Military Governor, Johannesburg, from June 6 1900; temporarily Commanding Columns from June 3 to 23, 1901, and from December 24, 1901, to January 2, 1902; Commanding 17th Battalion, Imperial Yeomanry, May 16, 1901, to May 31, 1902. Operations in the Orange Free State, February to May 1900, including operations at Paardeberg (February 17 to 26); actions at Poplar Grove, Driefontein, Houtnek (Thoba Mountain), Vet River (May 5 and 6) and Zand River. Operations in the Transvaal in May and June 1900, including actions near Johannesburg. Operations in the Transvaal July to November 29, 1900. Operations in Cape Colony, south of Orange River, 1899–1900. Operations in the Transvaal November 1900 to May 1901. Operations in Orange River Colony and Cape Colony May 1901 to May 31, 1902. Despatches, *London Gazette*, July 29, 1902. Queen's medal with four clasps; King's medal with two clasps.

BARLOW, MAURICE.—Ensign July 21, 1814; Lieutenant March 23, 1815; half-pay December 25, 1818; Lieutenant 3rd Foot July 29, 1819; Captain do. December 20, 1821; Major do. June 12, 1828; Major 14th Foot June 25, 1830; Brevet Lieut.-Colonel December 25, 1847.

BARRINGTON, CHARLES.—Ensign December 9, 1795; Lieutenant September 20, 1798; Lieutenant 11th Battalion of Reserve November 25, 1803; retired September 24, 1802.

BARRY, GAYNOR.—Ensign 92nd Regiment June 14, 1794; Lieutenant 85th Regiment December 11, 1795; severely wounded in Holland August 27, 1799.

BARTON, MONTAGUE.—Ensign April 16, 1852; Lieutenant July 15, 1853; Captain January 8, 1856; Major April 9, 1870; retired October 7, 1871.

BATEMAN, ALLEYNE SACHEVERELL.—Ensign September 25, 1823; half-pay 44th Regiment April 4, 1825; retired November 1833.

BATTYE, CLINTON WYNYARD, D.S.O.—Second Lieutenant March 7, 1894; Lieutenant June 18, 1898; Captain January 22, 1902. *South African War*, 1899–1902.—Employed with Mounted Infantry. Slightly wounded. Despatches, *London Gazette*, September 10, 1901. Queen's medal with four clasps; King's medal with two clasps; 'D.S.O.'

Bayley, Alexander Ross.—Ensign 85th Regiment April 2, 1847; Ensign 8th Foot August 15, 1848; Lieutenant do. May 6, 1858; Captain do. May 11, 1858.

Bayntun, Edward.—Cornet 6th Dragoon Guards March 31, 1804; Lieutenant do. March 28, 1805; 2nd Life Guards July 1, 1811; 85th Regiment May 18, 1812; 18th Foot January 25, 1813; 4th Garrison Battalion (afterwards 2nd Garrison Battalion) June 10, 1813; half-pay October 24, 1816; died at Bromham, Wilts, September 18, 1854.

Beadon, Edward Musgrave.—Ensign March 12, 1853; Lieutenant March 2, 1855; Captain October 9, 1863; Major June 10, 1874; Lieut.-Colonel June 21, 1880; Colonel June 21, 1884; Hon. Major-General June 21, 1885; retired. *Afghan War*, 1879–1880.—With the Kuram Division; medal.

Belcher, George Berkeley.—Ensign January 28, 1826; Lieutenant July 6, 1829; retired February 17, 1832.

Belhaven, Robert M., Lord.—Ensign 26th Regiment January 24, 1811; Lieutenant 101st Regiment June 4, 1812; Lieutenant 85th Regiment January 25, 1813; Captain 40th Regiment, February 23, 1815; half-pay February 25, 1816; Captain 11th Foot July 8, 1819; half-pay October 25, 1821; Captain 2nd Life Guards January 2, 1824; served in the Peninsula; silver medal for St. Sebastian (assault and capture), Nivelle and Nive.

Bell, Henry Corbould Forbes.—Second Lieutenant January 18, 1902; Lieutenant August 29, 1906; to Indian Army January 18, 1911. *South African War*, 1902.—Operations in the Transvaal April to May 31, 1902; Queen's medal with four clasps.

Bell, William Robert.—Ensign September 15, 1854; Lieutenant 18th Foot 1855.

Belstead, Henry.—Ensign 6th Foot August 29, 1811; Lieutenant 85th Foot June 9, 1813; half-pay 89th Regiment July 7, 1829; Lieutenant 96th Regiment January 29, 1841; retired same date; served in Peninsula, slightly wounded at the Nive, December 10, 1813.

Bennet, The Hon. H. Astley.—Ensign Coldstream Guards October 17, 1774; Captain do. May 15, 1778; Captain 16th Light Dragoons January 24, 1780; Captain and Lieut.-Colonel 1st Foot Guards January 26, 1791; Colonel August 21, 1795; Major-General June 18, 1798; Lieut.-Colonel 85th Regiment August 15, 1798; Lieutenant-General October 30, 1805; retired April 8, 1813; died 1815.

Bennett, Valentine.—Ensign 6th Foot July 4, 1811; Lieutenant Sicilian Regiment June 25, 1812; Lieutenant 6th Regiment July 16, 1812; Lieutenant 85th Regiment January 25, 1813; retired

September 23, 1813; appointed Adjutant King's County Militia December 14, 1813.

BENSON, RALPH.—The identity of this officer cannot be traced.

BERKELEY, THE HON. CRAVEN F.—Ensign February 13, 1823; Lieutenant (unattached) June 30, 1825; Cornet 2nd Life Guards October 29, 1825; Lieutenant 2nd Life Guards January 29, 1827; Captain do. March 22, 1831; half-pay unattached August 25, 1837; exchanged to 32nd Regiment March 6, 1840; retired same date.

BEWES, CECIL EDWARD.—Ensign 8th Foot August 7, 1835; Ensign 85th Regiment November 24, 1835; Lieutenant do. October 4, 1839; Captain do. October 27, 1843; Captain 69th Regiment May 3, 1844; retired June 29, 1849.

BICKFORD, MAURICE HERBERT.—Second Lieutenant November 21, 1906; Indian Army February 21, 1909.

BIRCH, ROBERT JONES.—Ensign December 28, 1855; exchanged.

BLACKALL, SAMUEL WANSLEY.—Ensign June 26, 1827; Lieutenant February 17, 1832; retired February 1, 1833; Major Longford Militia January 1, 1833; Lieutenant-Governor of Dominica 1851; Lieutenant-Governor of Queensland; died at Brisbane January 2, 1871.

BLACKALL, NEWCOMBE EDWARD.—Ensign February 1, 1833; Cornet 10th Hussars June 20, 1834; Lieutenant do. August 21, 1835; exchanged to 90th Regiment April 10, 1840; retired June 25, 1841.

BLACKBURN, JOHN.—Ensign 56th Regiment June 15, 1832; Ensign 85th Regiment October 22, 1833; Lieutenant do. September 1, 1837; Captain do. November 16, 1841; Major do. December 29, 1846; retired same date.

BLAKE, FREDERICK RODOLPH.—Ensign June 30, 1825; Lieutenant August 14, 1827; Lieutenant 94th Regiment January 29, 1829; Captain do. August 23, 1831; Captain 33rd Regiment January 18, 1833; Major do. April 14, 1843; Lieut.-Colonel do. October 3, 1848; Colonel November 28, 1854.

BLAKELEY, WILLIAM AUGUSTUS.—Ensign 29th Regiment; Lieutenant 85th Regiment June 12, 1800; no further information.

BLIGH, THE HON. EDWARD.—

BLOSSE, FRANCIS LYNCH.—Ensign October 19, 1849; Lieutenant October 29, 1852; Retired 1853.

BODEN, EDWARD HENRY.—Ensign August 10, 1867; retired December 14, 1870.

BOND, ALEXANDER VALENTINE.—Ensign 36th Regiment July 24, 1846; Ensign 85th Regiment August 14, 1846; Lieutenant do. March 14,

1851; Captain in West Kent Militia November 27, 1854; Adjutant Kent Rifles August 1860; retired August 26, 1853.

BOOTT, KIRK.—Ensign 49th Regiment September 26, 1811; Lieutenant 85th Regiment May 26, 1813; retired March 13, 1817, and died in the United States.

BOSANQUET, GEORGE WILLIAM.—Ensign April 19, 1864; retired June 29, 1866.

BOURNE, GEORGE.—Ensign 7th Garrison Battalion March 5, 1807; Ensign 85th Regiment August 20, 1807; Lieutenant do. September 21, 1808; Lieutenant 2nd Life Guards May 18, 1812; retired September 21, 1812; died at Hatton Holgate, near Spilsby, April 4, 1865, aged eighty-four.

BOWEN, EDWARD COLE.—Ensign March 1800; Lieutenant July 21, 1800; half-pay 1802; Lieutenant 40th Regiment September 5, 1805; Captain do. November 7, 1811; retired March 6, 1823; fought at Waterloo.

BOWEN, THOMAS.—Ensign (wanting); Lieutenant March 6, 1794; Captain September 19, 1795; wounded near Bergen, October 2, 1799.

BOWLES, GEORGE HERBERT.—Ensign December 27, 1855; died 1857.

BOYES, JAMES WATSON.—Lieutenant 88th Regiment September 1, 1808; half-pay 96th Regiment 1809; 4th Garrison Battalion May 9, 1811; Lieutenant 85th Regiment April 29, 1813; exchanged to 38th Regiment September 4, 1823; half-pay 21st Regiment September 23, 1824; exchanged to 8th Regiment December 31, 1829; Captain unattached October 26, 1830; Captain 55th Regiment August 9, 1833; retired August 7, 1835; died at Dundrum, Ireland, September 1836. Served in North America; slightly wounded at New Orleans.

BOYES, ROBERT NAIRNE.—Ensign October 13, 1814; Lieutenant June 8, 1815; half-pay December 25, 1818; half-pay 55th Regiment October 7, 1819; Captain do. August 9, 1831; half-pay 3rd West India Regiment August 2, 1833; Barrack Master at Grahamstown, 1836; Military Knight of Windsor; died November 1872.

BOYLE, THE HON. EDMUND JOHN.—Ensign 69th Regiment September 12, 1848; Ensign 85th Regiment September 13, 1848; Lieutenant do. December 12, 1851; Captain do. December 16, 1853; Brevet-Major March 6, 1867.

BOYLE, THE HON. ROBERT EDWARD.—Ensign 68th Regiment November 14, 1826; Lieutenant 34th Regiment July 16, 1829; half-pay unattached August 1829; exchanged to 79th Regiment September 3, 1829; Captain unattached August 23, 1833; exchanged to 14th Foot March 7, 1834; Captain 85th Regiment September 5, 1834; exchanged to Coldstream Guards June 17,

1836; Captain and Lieut.-Colonel Coldstream Guards December 10, 1847; died 1854.

BRACKENBURY, FREDERICK.—Ensign March 20, 1806; Lieutenant 57th Regiment February 7, 1808; died at Spilsby, Lincolnshire, July 4, 1811.

BREBNER, JOHN.—Ensign January 18, 1859; exchanged to 79th Regiment November 1859.

BROCK, JAMES LOFTHOUSE.—Ensign May 28, 1807; Lieutenant September 20, 1808; Lieutenant 6th Foot January 25, 1813; half-pay October 2, 1817; died October 18, 1854; dangerously wounded at Fuentes d'Oñoro May 5, 1811; silver medal.

BROCK, JOHN.—Lieutenant of an Independent Company June 14, 1793; exchanged to 3rd Foot October 23, 1793; Captain 92nd Regiment July 1, 1795; Captain 5th West India Regiment February 9, 1797; Captain 85th Regiment May 19, 1800; retired September 10, 1803; wounded at Flushing.

BROCKMAN, GEORGE.—Ensign 60th Regiment October 21, 1824; Ensign 85th Regiment January 6, 1825; Lieutenant do. January 28, 1826; Captain do. April 5, 1831; retired October 4, 1839; Colonel East Kent Militia September 15, 1852; died at Folkestone April 17, 1864.

BROMLEY, GEORGE.—Lieutenant 49th Regiment October 25, 1798; Captain 85th Regiment May 27, 1801; Captain 5th West India Regiment November 18, 1802.

BROOK, HENRY GEORGE. — Ensign November 17, 1814; Lieutenant August 24, 1815; half-pay December 25, 1818; exchanged to 75th Regiment January 13, 1820; died at Clifton July 13, 1823.

BROWN, SIR GEORGE, K.C.B., K.H.—Ensign 43rd Regiment January 23, 1806; Lieutenant do. September 18, 1806; Captain 3rd Garrison Battalion June 20, 1811; Captain 85th Regiment July 2, 1812; Major do. May 26, 1814; Brevet Lieut.-Colonel September 29, 1814; Lieut.-Colonel July 17, 1823, unattached; exchanged to Rifles February 5, 1824; Colonel May 6, 1831; Major-General November 23, 1841; Deputy Adjutant-General same date; Adjutant-General April 8, 1850; Colonel of 77th Regiment April 11, 1851; Lieutenant-General November 11, 1851; resigned office of Adjutant-General December 12, 1853. Served at Copenhagen 1807; served in Peninsula; severely wounded at Talavera; served in North America; severely wounded at Bladensburg; silver medal for Vimiera, Talavera, Busaco, Fuentes d'Oñoro, San Sebastian (assault and capture), Nivelle and Nive.

BROWNE, THE HON. CAVENDISH.—Ensign November 5, 1847; Lieutenant 7th Fusiliers September 25, 1849; Captain August 4, 1854; killed 1855.

LIEUT.-GENERAL SIR GEORGE BROWN, K.C.B., K.H.

BROWNE, HERBERT RICHARD.—Second Lieutenant from 70th Foot June 12, 1878; Lieutenant July 1, 1881; to Indian Staff Corps.

BROWNE, HENRY SABINE.—Ensign 60th Regiment June 23, 1825; Ensign 85th Regiment April 8, 1826; Lieutenant June 11, 1830; Captain September 30, 1836; retired September 2, 1842; died at Stonehouse February 24, 1843.

BROWNE, LYDE.—Cornet 3rd Dragoons, June 11, 1777; Lieutenant do. April 25, 1781; Captain 20th Dragoons; half-pay 1783; Captain 40th Regiment May 30, 1794; Brevet-Major March 1, 1794; Major January 18, 1797, 4th West India Regiment; Brevet Lieut.-Colonel January 1, 1798; appointed to 90th Regiment July 11, Lieutenant-Colonel May 30, 1800; transferred to 85th Regiment April 10, 1801; do. 35th Regiment June 25, 1802; murdered in Dublin by the rebel mob July 23, 1803. The Annual Register tells us that on July 23, 1803, about dusk, the rebellion broke out in Dublin. A Mr. Clarke was very severely wounded by a gunshot 'in the midst and most frequented part of the city.' The victim had become aware of the intentions of the mob and had thought it his duty to apprise the secretary of the Lord Lieutenant. As he rode home a blunderbuss was discharged at him by one of his own workmen. Later on, the signal being given by means of a cannon and a rocket, 'Emmet, at the head of a chosen band, sallied forth from the obscurity of his headquarters in Marshalsea Lane, and drawing his sword in the street with a flourish incited his ruffians to action; before they reached the end of the lane in which they were arranged, a confidential member of the party discharged his blunderbuss at a person arrayed as an officer hastily passing along; and thus, by a base and unprovoked act of assassination, perished Colonel Browne, a most respectable and meritorious officer. On the same evening the venerable Lord Kilwarden, the Lord Chief Justice, was dragged from his coach and massacred.'

BROWNE, THE HON. RICHARD HOWE.—Ensign 85th Regiment November 6, 1827; Lieutenant unattached August 10, 1832; Lieutenant 85th Regiment August 17, 1832; appointed to 8th Hussars December 6, 1833; Captain 8th Hussars November 15, 1839; retired May 21, 1843.

BRUCE, ALEXANDER.—Ensign 48th Regiment September 1, 1796; Lieutenant do. October 17, 1799; Lieutenant 37th Regiment February 20, 1800.

BRUCE, ROBERT CAIRNES.—From the 92nd Foot to be Captain 85th Regiment February 21, 1855; retired 1858.

BRYANT, HENRY GRENVILLE, D.S.O.—Second Lieutenant from Bedfordshire Regiment September 26, 1894; Lieutenant July 3, 1898; Captain

January 22, 1902. *South African War*, 1899–1902.—On staff. Slightly wounded. Operations in the Orange Free State, February to May 1900, including operations at Paardeberg (February 17 to 26); actions at Poplar Grove, Dreifontein, Houtnek (Thoba Mountain), Vet River (May 5 and 6), and Zand River. Operations in the Transvaal in May and June 1900, including actions near Johannesburg and Pretoria. Operations in the Transvaal, east of Pretoria, July to November 29, 1900. Operations in the Transvaal, west of Pretoria July to November 29, 1900, including actions at Elands River (August 4 to 16). Operations in Orange River Colony, May to November 29, 1900, including action at Rhenoster River. Operations in Cape Colony, south of Orange River, 1899–1900. Operations in the Transvaal November 30, 1900, to May 31, 1902. Despatches, *London Gazette*, September 10, 1901. Queen's medal with four clasps; King's medal with two clasps. 'D.S.O.'

BULMAN, PHILIP, D.S.O.—Sub-Lieutenant and Lieutenant September 11, 1876; Captain June 9, 1885; Major February 11, 1894; Lieutenant-Colonel (Brevet-Colonel) August 19, 1904; retired October 24, 1906. *Afghan War*, 1879–1880.—With the Kuram Division in the Yarmusht Expedition; medal. *South African War*, 1899–1901.—In command 2nd Battalion Shropshire Light Infantry from May 1 to June 5, 1900, also from January 22 to March 13 and from May 16 to September 4, 1901. Operations in the Orange Free State, February to May 1900, including operations at Paardeberg (February 17 to 26); actions at Poplar Grove, Driefontein, Houtnek (Thoba Mountain), Vet River (May 5 and 6) and Zand River. Operations in the Transvaal in May and June 1900, including actions near Johannesburg and Pretoria. Operations in the Transvaal, east of Pretoria, July to November 29, 1900. Operations in the Transvaal, west of Pretoria, July to November 29, 1900, including actions at Elands River (August 4 to 16). Operations in Orange River Colony, May to November 29, 1900, including action at Rhenoster River. Operations in Cape Colony, south of Orange River, 1899–1900. Operations in the Transvaal, November 30, 1900, to September 1901. Despatches, *London Gazette*, September 10, 1901. Queen's medal with five clasps. 'D.S.O.'

BURBRIDGE, ——.—Ensign 85th Foot 1803; Lieutenant 14th Foot same year.

BURRELL, EDWARD.—Ensign from 84th Regiment March 6, 1867; Lieutenant October 28, 1871; Captain March 1, 1882; Major September 13, 1886; retired August 15, 1888.

BURRELL, JOHN.—Ensign 6th Foot March 28, 1811; Lieutenant do. February 20, 1812; Lieutenant 85th Regiment January 25, 1813; Captain do. May 18, 1815; Captain 60th Regiment June 2, 1815;

half-pay do. September 25, 1818; Captain 88th Regiment April 8, 1825; exchanged to half-pay unattached June 12, 1828; Captain 37th Regiment September 7, 1832; retired same date; severely wounded at Bladensburg.

BUSTEED, CHARLES.—Lieutenant 85th Regiment September 22, 1808; exchanged to 3rd Garrison Battalion December 20, 1810; (afterwards numbered 1st Garrison Battalion); half-pay do. September 25, 1817; died October 28, 1845.

BUSTEED, CHRISTOPHER.—Ensign June 29, 1809; Lieutenant January 15, 1813; Lieutenant 69th Regiment January 25, 1813; died at Mullingar November 4, 1827; severely wounded at Quatre Bras June 16, 1815; 'W' against name for Waterloo Campaign.

BUTLER, HENRY WILLIAM PAGET.—Lieutenant 7th Foot July 9, 1852; Captain 85th Regiment February 26, 1856; exchanged same year.

BUTLER, LORD JAMES.—Ensign November 7, 1834; Lieutenant 7th Fusiliers February 16, 1838; Captain do. June 3, 1842; retired April 28, 1846.

BUTLER, RICHARD ALEXANDER.—Ensign July 29, 1819; Lieutenant September 25, 1823; Captain unattached September 10, 1825; Captain 12th Foot March 20, 1827; exchanged to half-pay unattached November 5, 1829.

CAMERON, JAMES.—Ensign April 13, 1809; Lieutenant July 11, 1811; Lieutenant 79th Regiment January 25, 1813; died 1818.

CAMPBELL, FREDERICK.—Ensign February 27, 1796; Lieutenant 94th Regiment February 23, 1800; Captain (Army) December 21, 1809; Captain 94th Regiment January 25, 1813; regiment disbanded, placed on half-pay of it, December 25, 1818; died at Campbeltown, Argyllshire, October 31, 1829.

CAMPBELL, JAMES.—Ensign (wanting); Lieutenant December 29, 1795; Captain August 30, 1804; died in London 1809.

CAMPBELL, JAMES RAMSEY.—Sub-Lieutenant November 12, 1873; Lieutenant November 12, 1873; Captain February 21, 1884; Major September 13, 1886; Lieut.-Colonel October 4, 1893; retired pay same date.

CAMPBELL, JOHN.—Ensign March 30, 1815; Lieutenant 46th Regiment January 11, 1821; Captain unattached May 26, 1825; Captain 92nd Regiment June 8, 1826; Major unattached June 15, 1832; exchanged to 38th Regiment May 23, 1845; Brevet Lieut.-Colonel November 9, 1846.

CAMPBELL, MICHAEL HORACE.—Lieutenant Royal African Corps August 15, 1804; Lieutenant 85th Regiment October 17, 1805; Captain 5th Garrison Battalion June 18, 1807; Captain 99th Regiment March 30,

1808; half-pay, New Brunswick Fencibles April 11, 1816; Captain 1st Veteran Battalion February 13, 1823; half-pay 21st Fusiliers April 1, 1824; Brevet-Major July 19, 1821; Brevet Lieut.-Colonel January 10, 1837; died at St. George's Hospital June 14, 1838, aged sixty-eight.

CAMPBELL, THOMAS MAITLAND.—Ensign 21st Foot September 17, 1861; Ensign 85th Regiment September 23, 1861; retired February 14, 1865.

CAPEL, ARTHUR A.—Ensign November 19, 1858; exchanged.

CAPPER, WILLIAM BAUME.—Sub-Lieutenant February 12, 1876; Lieutenant February 12, 1876; Captain May 15, 1884; Adjutant July 10, 1886; Major August 19, 1893; Lieut.-Colonel to command Northamptonshire Regiment August 6, 1898; Colonel December 1, 1901; Commandant (General Staff Officer, 1st Grade) Royal Military College, Sandhurst, January 15, 1907. *Afghan War*, 1879–1880.—Zaimusht Expedition, assault on Zawa; medal. *Egyptian Expedition*, 1882. —In the Transport Service; medal; bronze star. *Soudan Expedition*, 1884–5.—Nile. Action of Abu Klea; two clasps; Brevet-Major.

CAREY, WILLIAM.—Ensign 28th Regiment May 23, 1799; Lieutenant 62nd Regiment December 31, 1799; Captain April 3, 1801; half-pay of the regiment 1802.

CARNE, THOMAS PIKE.—Ensign August 26, 1813; retired September 8, 1814.

CARTER, FREDERICK MARK.—From 1st Battalion, Lieutenant September 1, 1883; to 1st Battalion June 1, 1884.

CARTER, JAMES.—Lieutenant December 6, 1798; no further information.

CARTER, JOHN MARK.—Lieutenant January 30, 1901; died of cholera at Ranikhet July 1903.

CASH, HENRY CHRISTMAS.—Ensign 96th Regiment 1808; Lieutenant 85th Regiment January 26, 1809; Lieutenant 12th Foot January 25, 1813; Captain 41st Foot February 25, 1814; half-pay 41st Foot 1814; exchanged to 2nd Foot November 9, 1815; Major do. January 26, 1825; Lieut.-Colonel unattached August 15, 1826; exchanged to 29th Regiment May 31, 1839; retired June 1, 1839; died at Belville, Rockestown Avenue, January 19, 1843.

CASSAN, MATTHEW.—Lieutenant July 17, 1800, from the Queen's County Militia in which he was an Ensign; shot in a duel at Jamaica 1804.

CAVANAGH, NATHANIEL.—

CAVENDISH, SPENCER FREDERICK GEORGE.—Lieutenant from 1st Battalion October 15, 1881; Captain September 1, 1886; returned to 1st Battalion. *Nile Expedition* 1884–5; attached to 1st Battalion Royal Sussex Regiment; medal with clasp and Khedive's star.

CHADWICK, RICHARD.—Lieutenant July 24, 1800; no more information.

CHARLETON, ANDREW ROBERT.—Ensign 8th Foot June 7, 1811; Lieutenant 89th Regiment August 13, 1812; Lieutenant 85th Regiment January 25, 1813; Captain do. May 22, 1817; Major 92nd Regiment September 25, 1823; died in Jamaica August 1828; severely wounded at New Orleans December 28, 1814.

CHARLTON, ROBERT.—Lieutenant 85th Regiment September 23, 1813; Captain do. July 17, 1823; half-pay unattached April 21, 1825; died in Derbyshire 1831.

CHESTER, JOHN.—Ensign 53rd Regiment October 2, 1840; Lieutenant do. July 1, 1842; Captain do. December 3, 1847; exchanged to 85th Regiment December 6, 1850; Adjutant Parkhurst Depot Battalion April 7, 1854; exchanged to half-pay York Chasseurs, May 4, 1855; appointed to 7th Lancashire Regiment; appointed Staff Officer of Pensioners May 1, 1858, at Carmarthen; Brevet-Major February 20, 1859; retired December 1864; served with the 53rd Regiment in the Sutlej Campaign (medal); present at Buddiwal, Aliwal and Sobraon (severely wounded).

CHICHESTER, HENRY MANNERS.—Ensign February 18, 1853; Lieutenant August 11, 1854; retired April 19, 1864.

CHURCHILL, LORD CHARLES S.—Ensign 68th Regiment December 26, 1811; Ensign 95th Rifles May 8, 1812; Lieutenant 52nd Regiment September 9, 1813; Captain 60th Regiment March 9, 1815; exchanged to 85th Regiment June 2, 1815; half-pay 85th Regiment April 24, 1823; Captain 75th Regiment October 28, 1824; Major June 16, 1825 (unattached); exchanged back to 75th Regiment December 8, 1825; Lieut.-Colonel unattached December 31, 1827; exchanged to Scots Fusilier Guards August 24, 1832; retired August 31, 1832; died in London April 29, 1840; served in the Peninsula.

CLEMENTS, THE HON. CHARLES SKEFFINGTON.—Ensign December 8, 1825; Lieutenant unattached September 9, 1828; exchanged to 37th Regiment November 21, 1828; Captain 37th Regiment November 13, 1832; retired on half-pay 35th Regiment January 17, 1834; exchanged to 69th Regiment August 31, 1838; retired September 1, 1838; Captain Donegal Militia April 17, 1845; Assistant Poor Law Commissioner for Ireland from 1838 to 1846; died September 29, 1877.

COAPE, ARTHUR.—Ensign 84th Regiment November 19, 1830; Lieutenant do. October 22, 1833; Captain do. August 28, 1840; Captain 85th Regiment April 8, 1842; died at St. Vincent's January 12, 1846.

CODD, GARLIKE PHILIP ROBERT.—Ensign 67th Regiment April 1, 1813; Lieutenant 7th Fusiliers May 27, 1813; Lieutenant 85th Regiment July 22, 1813; killed at Bladensburg August 24, 1814.

COLE, ROBERT.—Ensign 60th Regiment November 21, 1816; half-pay do. March 1817; exchanged to 85th Regiment May 21, 1818; Lieutenant do. December 26, 1822; Captain do. August 14, 1827; exchanged to half-pay 96th Regiment February 24, 1832; exchanged to 48th Regiment July 29, 1836; Brevet-Major November 23, 1841; Major November 19, 1845 (48th Regiment); Lieutenant-Colonel December 31, 1847 (unattached); died in London April 17, 1869.

COLLETTE, CECIL HENRY.—Ensign January 8, 1868; Lieutenant October 28, 1871; Adjutant June 10, 1874; Captain September 30, 1882; Major August 19, 1885; Colonel retired pay. *Afghan War*, 1879–1880.—With the Kurum Division Zaimusht expedition, and assault of Zawa; mentioned in despatches; medal.

COLLINS, HAMILTON STRATFORD.—Second-Lieutenant October 6, 1906; Lieutenant March 29, 1909.

COLVILLE, CHARLES JOHN, LORD.—Ensign May 27, 1836; Lieutenant February 21, 1840; exchanged to 40th Regiment October 29, 1841; appointed to 9th Lancers April 22, 1842; Captain Canadian Rifles May 31, 1844; appointed to 11th Hussars June 25, 1844; exchanged to Coldstream Guards April 14, 1846; exchanged to half-pay unattached July 13, 1847; appointed to 15th Foot July 14, 1854; retired August 4, 1854; appointed Lieutenant-Colonel of the Hon. Artillery Company January 6, 1860.

COMPTON, LORD SPENCER SCOTT.—Ensign November 1, 1839; Lieutenant October 20, 1843; Captain December 29, 1846; appointed to 15th Hussars October 27, 1848; died at Exeter May 21, 1855.

CONNOLLY, HUNT.—Ensign 18th Foot August 11, 1794; Lieutenant do. September 2, 1795; exchanged to half-pay Irish Brigade August 16, 1799; exchanged to 85th Regiment May 8, 1801; half-pay 85th Regiment 1802; appointed to 81st Regiment November 18, 1803.

CONNOR, JOHN.—Ensign 1st Foot August 21, 1806; Lieutenant do. January 7, 1808; appointed to 85th December 29, 1808; appointed to 103rd Regiment January 25, 1813; appointed to 5th Veteran Battalion March 7, 1816; reduced May 24, 1816, and placed on retired full pay of it; appointed to 70th Regiment March 1, 1832; retired January 11, 1833.

CONNOR, WILLIAM.—Lieutenant 85th Regiment July 20, 1800; placed on half-pay of the regiment 1802; died in Dublin August 19, 1823.

CONOLLY, WILLIAM.—Ensign 35th Foot January 31, 1865; Ensign 7th Foot June 30, 1865; Lieutenant do. February 7, 1871; Captain Royal Fusiliers October 20, 1878; Captain 85th Regiment July 12th, 1882; Hon. Major and retired February 4, 1885; served in the Afghan War 1879–80; present at the defence of Kandahar and sortie on Deh Khoja; slightly wounded; medal.

COOKE, WILIAM BRYAN.—Ensign April 10, 1825; Lieutenant June 10, 1826; retired March 2, 1832.

COOKSON, JOSEPH.—Captain 85th May 17, 1800; no further information.

COOPER, THE HON. A. H. ASHLEY.—Ensign April 8, 1825; Lieutenant May 20, 1826; Captain June 21, 1831; retired on half-pay of Staff Corps September 5, 1834; exchanged to 9th Lancers May 18, 1841; retired same date; died at Windsor November 20, 1858.

COOPER, LOFTUS LEWIS ASTLEY.—Ensign April 30, 1858; Lieutenant October 9, 1863; Captain August 7, 1867; exchanged to 98th Foot August 21, 1869.

COPLEY, JOHN ALEXANDER.—Ensign 24th Regiment 1809; Lieutenant 3rd Garrison Battalion August 3, 1809; exchanged to 85th December 20, 1810; appointed to 4th Foot January 25, 1813; died May 4, 1814.

CORYTON, AUGUSTUS; Ensign (unattached) May 12, 1825; Ensign 66th Regiment June 29, 1826; Ensign 85th Regiment, July 20, 1826; Lieutenant do. June 21, 1831; Captain do. May 12, 1838; retired November 9, 1838; appointed Lieut.-Colonel of the Cornwall Rangers Militia October 4, 1852.

COSENS, GEORGE WEIR.—Ensign 80th Regiment October 2, 1855; Lieutenant do. March 11, 1859; Lieutenant 45th Regiment April 29, 1859; Captain 85th Regiment October 11, 1864; retired June 15 1866.

COTTINGHAM, EDWARD.—Ensign 28th Regiment April 20, 1809; Lieutenant do. July 19, 1810; Captain 85th Regiment July 28, 1813; exchanged to York Chasseurs June 23, 1814; half-pay do. March 2, 1815; exchanged to 99th Regiment August 16, 1842; Brevet-Major January 10, 1837; retired August 16, 1842; served in Peninsula; silver medal for Busaco, Albuhera, San Sebastian (assault and capture) and Nive; slightly wounded at Albuhera.

COTTON, RICHARD GODMAN TEMPLE.—Ensign 53rd Foot August 30, 1864; Lieutenant do. September 14, 1866; Captain 1st Battalion Shropshire Light Infantry March 2, 1878; half-pay April 13, 1884; Major September 3, 1884 (2nd Battalion); Lieut.-Colonel half-pay April 13, 1892.

COUSSMAKER, ARTHUR LANNOY.—Captain from the 3rd Foot August 19, 1858; retired same year.

COWELL, F. LUKE GARDINER.—Ensign 62nd Regiment November 15, 1800; Lieutenant do. November 26, 1801; half-pay of regiment 1802; appointed to 2nd West India Regiment November 28, 1802; exchanged to 85th Regiment January 16, 1806; exchanged to 4th Garrison Battalion September 8, 1808; appointed to 23rd Fusiliers December 28, 1809; died November 24, 1815.

2 I 2

CREAGH, WILLIAM.—Ensign February 4, 1800; Lieutenant 54th Regiment December 18, 1800; no further information.

CROFTON, EDWARD, LORD.—Ensign December 5, 1822; Lieutenant April 6, 1825; Captain unattached May 20, 1826; exchanged to 7th Hussars July 12, 1827; retired September 13, 1831; died at Mote Park, Roscommon, December 27, 1869.

CROFTON, THE HON. WILLIAM.—Ensign April 27, 1832; Lieutenant September 30, 1836; died at Halifax, Nova Scotia, April 16, 1838.

CROUCHLEY, THOMAS.—From the 2nd Lancashire Militia; Lieutenant 85th Regiment December 25, 1813; placed on half-pay of regiment 1815; died December 29, 1819; severely wounded at Bladensburg.

CRUICE, PATRICK.—Lieutenant Irish Brigade October 1, 1794; half-pay do. 1798; appointed to 85th Regiment July 3, 1799; Captain 31st Regiment August 4, 1804; exchanged to half-pay 6th West India Regiment January 29, 1818; Brevet-Major June 4 1814; died January 29, 1829; wounded on the heights of Albaro April 13, 1814.

CULLEY, JAMES.—Ensign October 27, 1799; Lieutenant May 17, 1800; Captain 5th Foot January 24, 1805; Brevet-Major March 3, 1814; Major January 9, 1822 (5th Foot); retired January 24, 1828.

CURTIS, HENRY CHARLES.—Ensign 29th Regiment December 30, 1831; Ensign 85th Regiment February 17, 1832; Lieutenant do. March 4, 1836; retired July 2, 1841; died March 7, 1861.

CUYLER, HENRY.—Ensign 30th Regiment December 1, 1782; Lieutenant do. September 3, 1788; Captain do. December 18, 1793; Brevet-Major September 9, 1797; Major 27th Regiment November 3, 1797; Lieut.-Colonel do. May 16, 1800; half-pay do. 1803; exchanged to 3rd Foot February 6, 1805; appointed to 85th Regiment November 13, 1806; Colonel, July 25, 1810; appointed to 46th Regiment January 25, 1813; retired February 4, 1813; died at Boley, near Rochester, November, 1842; fought in the Peninsula.

D'AGUILAR, SIR GEORGE CHARLES, K.C.B.—Ensign 85th Regiment October 2, 1800; left the regiment 1802; died subsequently to 1851 a Lieutenant-General.

DALY, RICHARD.—Ensign 60th Regiment July 10, 1800; Lieutenant 85th Regiment June 26, 1801; half-pay do. 1802; full pay 60th Regiment February 10, 1803; Lieutenant 34th Regiment June 27, 1805; Captain do. October 17, 1811; half-pay 53rd Regiment August 25, 1818; died at Pondicherry August 13, 1821.

DANIELL, RALPH ALLEN CHARLES.—Ensign 66th Regiment April 11, 1834; Lieutenant do. February 2, 1838; Captain do. October 20, 1843; exchanged to 85th Regiment December 31, 1845; retired March 14, 1851.

D'ARCY, PETER (? RICHARD).—Ensign November 26, 1793; Lieutenant same date; Captain September 23, 1799; Major 7th Garrison Battalion November 26, 1806; half-pay do. 1810; Brevet Lieut.-Colonel June 4, 1813; Colonel July 22, 1830; died November 10, 1840.

DARELL, HENRY JOHN.—Ensign May 5, 1837; Lieutenant July 2, 1841; Captain January 12, 1844; exchanged to 60th Regiment May 2, 1854.

DAYRELL, RICHARD.—Stated to have belonged to the regiment, but no records exist to identify the officer.

DAVIDSON, JAMES.—Ensign April 20, 1809; Lieutenant September 26, 1811; appointed to 89th Regiment January 25, 1813; exchanged to half-pay of the regiment February 19, 1818; appointed to 98th Regiment March 25, 1824; retired October 29, 1829.

DAVISON, JOHN.—Ensign March 12, 1861; Lieutenant February 21, 1865; Captain January 25, 1871; exchanged to 3rd Dragoon Guards.

DAWKINS, CHARLES TYRWHITT, C.M.G.—Second-Lieutenant May 1, 1878; Lieutenant July 1, 1881; Adjutant February 22, 1882; Captain September 13, 1886; Major March 23, 1898; Brevet Lieut.-Colonel August 22, 1902; Lieut.-Colonel August 19, 1905; Brevet-Colonel February 9, 1907; Substantive Colonel August 19, 1909; Assistant Quartermaster-General, Eastern Command, April 27, 1910; half-pay August 19, 1909; full pay April 27, 1910; Aide-de-Camp to Governor and Commander-in-Chief, Cape of Good Hope, March 6, 1884, to September 12, 1886; Military Secretary do. September 13, 1886, to April 30, 1889; Adjutant of Militia January 1, 1892, to April 21, 1895; Military Secretary to Governor and Commander-in-Chief, Cape of Good Hope, May 11, 1895, to May 4, 1897. *Afghan War*, 1879–1880.—With the Kuram Division, Yarmusht Expedition and assault of Zawa; medal. *South African War*, 1899–1902.—Severely wounded. Operations in the Orange Free State, including operations at Paardeberg (February 17 to 26); actions at Poplar Grove, Dreifontein, and Houtnek (Thoba Mountain). Operations in Cape Colony, south of Orange River, 1899–1900. In command 2nd Battalion Shropshire Light Infantry from September 5, 1901, to May 5, 1902. Operations in the Transvaal August 1901 to May 31, 1902. Despatches, *London Gazette*, September 10, 1901, and July 29, 1902. Brevet Lieut.-Colonel; Queen's medal with four clasps; King's medal with two clasps.

DAWSON, RICHARD.—Ensign September 23, 1799; Lieutenant 3rd Foot February 28, 1800; Captain do. February 3, 1804; retired 1807; died at Lisbon September 28, 1814.

DAY, ROBERT LADBROKE.—Ensign 63rd Regiment October 18, 1833; Lieutenant do. August 7, 1835; exchanged to 85th Regiment May 13, 1842; Captain do. May 19, 1846; retired August 1, 1851.

DE BATHE, SIR WILLIAM P., BART.—Ensign 4th Garrison Battalion December 17, 1807; Ensign 27th Regiment March 3, 1808; Lieutenant do. September 21, 1809; Captain 3rd West India Regiment March 5, 1812; exchanged to 94th Regiment July 30, 1812; appointed to 85th Regiment January 25, 1813; Brevet-Major October 27, 1814; Major 85th Regiment June 24, 1819; Lieut.-Colonel April 9, 1825; exchanged to 53rd Regiment February 28, 1828; exchanged to half-pay unattached April 2, 1829; exchanged to 8th Foot September 25, 1835; retired October 2, 1835; served in Sicily (defence) 1808; Peninsula, silver medal for Nivelle and Nive; present at Washington, Baltimore; and New Orleans, in North America.

DELMÉ-MURRAY, GEORGE ARTHUR.—Second-Lieutenant November 15, 1899; Lieutenant March 9, 1901; Captain November 17, 1909. *South African War*, 1899–1902.—Assistant Provost-Marshal and Assistant Press Censor. Operations in the Orange Free State February to May 1900, including operations at Paardeberg (February 17 to 26); actions at Poplar Grove, Dreifontein, Houtnek (Thoba Mountain), Vet River (May 5 and 6), and Zand River. Operations in the Transvaal in May and June 1900, including actions near Johannesburg and Pretoria. Operations in the Transvaal, east of Pretoria, July to November 29, 1900. Operations in the Transvaal, west of Pretoria, July to November 29, 1900, including actions at Elands River (August 4 to 16). Operations in Orange River Colony May to November 29, 1900, including action at Rhenoster River. Operations in Cape Colony, south of Orange River, 1899–1900. Operations in the Transvaal March 1901 to May 31, 1902. Operations in Cape Colony November 30, 1900, to March 1901; Queen's medal with four clasps; King's medal with two clasps.

DERING, CHOLMELEY EDWARD.—Ensign May 9, 1834; Lieutenant April 17, 1838; Captain September 2, 1842; retired April 13, 1849; appointed Lieutenant in the East Kent Yeomanry May 8, 1854.

DERING, EDWARD CHOLMELEY.—Ensign April 18, 1851.

DESHON, PETER.—Lieutenant 43rd Regiment September 1, 1795; Captain do. June 25, 1803; Major do. August 16, 1810; exchanged to 85th Regiment January 25, 1813; retired May 22, 1817; served in the Peninsula, gold medal for Nive, silver medal for Busaco, San Sebastian (assault and capture) and Nivelle.

DEWGARD, RICHARD ALCOCK.—Lieutenant 85th July 16, 1800; Captain 2nd West India Regiment May 11, 1805; appointed to 6th Foot October 15, 1807; Brevet-Major August 12, 1819; retired June 22, 1820.

DEY, ELI.—Ensign 11th West India Regiment 1798; Lieutenant do. September 1, 1801; half-pay do. 1802; appointed to 85th December 17, 1803.

DISGROVE, EDWARD AMSLING.—Ensign 85th Regiment March 12, 1852; exchanged next year to Coldstream Guards.

DIXON, GEORGE COCHRANE.—Ensign April 5, 1831; Lieutenant November 24, 1835; Captain November 1, 1839; exchanged to 84th Regiment April 8, 1842; died at Moulmein, Madras, November 17, 1842.

DIXON, HENRY.—Ensign 29th Regiment August 20, 1812; Lieutenant do. December 21, 1815; half-pay do. 1817; exchanged to 85th Regiment November 27, 1817; half-pay do. December 25, 1818; exchanged to 81st Regiment February 24, 1820; Captain do. November 21, 1828; retired on full pay May 10, 1844; Brevet-Major November 23, 1841; appointed Major 3rd West York Militia April 1851; died October 26, 1874, aged seventy-eight.

DORRIEN-SMITH, EDWARD PENDARVES, D.S.O.—Second-Lieutenant May 15, 1899; Lieutenant July 10, 1900; Captain February 15, 1908; Adjutant same date. *South African War*, 1899–1902.—Acting (afterwards extra, May 2 to July 9) A.D.C. to General Officer Commanding Infantry Brigade. Operations in the Orange Free State, February to May 1900, including operations at Paardeberg (February 17 to 26) actions at Poplar Grove, Dreifontein, Houtnek (Thoba Mountain), Vet River (May 5 and 6), and Zand River. Operations in the Transvaal in May and June 1900, including actions near Johannesburg and Pretoria. Operations in the Transvaal, east of Pretoria, July to November 29, 1900, including actions at Lydenberg (September 5 to 8). Operations in the Transvaal, west of Pretoria, July to November 29, 1900, including actions at Elands River (August 4 to 16). Operations in Orange River Colony May to November 29, 1900. Operations in Cape Colony, south of Orange River, 1899–1900. Operations in the Transvaal November 30, 1900, to July 1901. Operations in Orange River Colony July to August 1901, and November 1901 to May 31, 1902. Operations on the Zululand Frontier of Natal in September and October 1901. Despatches, *London Gazette*, September 10, 1901; Queen's medal with four clasps; King's medal with two clasps. 'D.S.O.'

DOUESPE, HENRY DE LA.—Ensign 10th Foot March 26, 1794; Lieutenant do. May 27, 1795; exchanged to 77th Regiment January 20, 1800; Captain do. May 21, 1801; appointed to 69th Regiment December 25, 1802; Major 69th Regiment June 10, 1812; Lieut.-Colonel do. January 1, 1819; died at Bangalore April 19, 1820.

DOUGHTY, CHESTER.—Ensign November 30, 1855; Lieutenant August 7, 1866; transferred to 23rd Foot August 7, 1867.

DOUGLAS, NORMAN.—Lieutenant 92nd Regiment September 16, 1795; appointed to 85th December 15, 1795; half-pay do. 1802; Captain 61st Regiment July 9, 1803.

DOUGLAS, WILLIAM.—Ensign 21st Fusiliers February 23, 1768; Lieutenant do. March 10, 1775; Captain do. December 24, 1785; appointed to 77th Regiment August 27, 1788; Brevet-Major May 6, 1795; Major 74th Regiment December 4, 1795; Brevet Lieut.-Colonel January 1, 1800; appointed to 85th Regiment May 17, 1800; appointed Inspecting Field Officer of Yeomanry and Volunteers October 1803.

DOWD, JAMES.—Ensign December 8, 1804; Lieutenant 23rd Light Dragoons May 15, 1806; exchanged to 62nd Regiment July 19, 1810; half-pay of Staff Corps 1811; exchanged to 21st Light Dragoons November 9, 1815; half-pay of 1st Foot June 25, 1818; appointed to 1st Veteran Battalion July 6, 1810; battalion disbanded 1821; placed on retired full pay of it; died November 25, 1831.

DOWIE, ADAM.—Ensign February 14, 1805; Lieutenant December 24, 1805.

DOYLE, ARTHUR HAVELOCK J.—Second-Lieutenant January 30, 1878; Lieutenant July 19, 1880; Captain June 21, 1885; Major August 19, 1897; Lieut.-Colonel February 11, 1902; Brevet-Colonel February 11, 1905; half-pay February 11, 1906. *Afghan War*, 1879–80.—With the Kuram Division, Yarmusht Expedition and assault of Zawa; medal. *West Coast of Africa*, 1892.—Attack on Tambi (March 14). *South African War*, 1899–1900, 1902.—Operations in the Free State February to May 1900, including actions at Houtnek (Thoba Mountain), Vet River (May 5 and 6), and Zand River. Operations in the Transvaal in May and June 1900, including actions near Johannesburg and Pretoria. Operations in the Transvaal, east and west of Pretoria July to November 29, 1900, including actions at Elands River (August 4 to 16). Operations in Orange River Colony, May to November 29, 1900, including action at Rhenoster River. Operations in the Transvaal May 1902. Despatches, *London Gazette*, September 10, 1901. Queen's medal with four clasps.

DRAGE, WILLIAM HENRY.—Ensign May 9, 1856; Adjutant March 2, 1858; Lieutenant March 4, 1861; Captain August 3, 1872; exchanged to 52nd Regiment.

DRURY, EDWARD.—Ensign December 3, 1802; Lieutenant 5th Foot August 28, 1804; Captain do. October 12, 1809; half-pay do. July 25 1816; exchanged to 6th Foot December 28, 1832; Brevet-Major July 22, 1830; died at Chatham March 10, 1837.

DUNDAS, JOHN HAMILTON.—Cornet 15th Hussars December 18, 1823; Lieutenant do. June 16, 1825; Captain unattached April 8, 1826; appointed to 1st Foot July 13, 1832; exchanged to half-pay unattached February 6, 1835; exchanged to 85th Regiment, October 18, 1839.

DUTHY, JOHN.—Ensign February 11, 1813; Lieutenant October 13, 1814; half-pay of the regiment March 25, 1817; exchanged to 4th Foot November 11, 1818; died at Tobago, March 8, 1820.

DUTTON, THE HON. CHARLES.—Ensign 25th Foot March 12, 1861; Lieutenant 85th Regiment March 3, 1865; Captain October 7, 1871; Major July 1, 1881; Lieut.-Colonel on half-pay May 15, 1889. *Afghan War*, 1879–80.—As Assistant Quartermaster-General 2nd Division Cabul Field Force; action of Shekabad. Despatches, *London Gazette*, July 30, 1880; medal; Brevet-Major.

DUTTON, ROBERT.—Ensign 85th Regiment November 21, 1811; Ensign 3rd Garrison Battalion January 25, 1813, made 1st Garrison Battalion 1814; half-pay of it September 25, 1817; appointed to 1st Veteran Battalion October 25, 1822; Lieutenant 98th Regiment October 14, 1824; New South Wales Veteran Corps September 24, 1825; exchanged to 27th Regiment March 16, 1826; retired March 30, 1826.

DUTTON, WILLIAM HOLMES.—Lieutenant 7th Fusiliers May 4, 1815; half-pay do. 1816; exchanged to 71st Regiment June 6, 1816; half-pay do. 1816; exchanged to 85th Regiment November 6, 1817; appointed to 4th Foot November 19, 1818; Captain do. August 15, 1822; Major do. March 20, 1827; exchanged to half-pay unattached July 5, 1827; Brevet Lieut.-Colonel November 23, 1841; Colonel June 20, 1854; Major-General October 26, 1858.

EDEN, SIR FREDERICK, BART.—Ensign July 22, 1813; killed on approach to New Orleans December 28, 1814.

EDWARDES, SIR HENRY HOPE, BART.—Ensign December 28, 1846; Lieutenant November 7, 1851; retired 1853; Cornet North Salop Yeomanry February 12, 1855; Lieutenant do. June 10, 1865.

EMERSON, EDWARD.—Captain Dublin Regiment July 31, 1795; regiment reduced but retained on full pay of it; placed on half-pay of it 1798; appointed to 85th Regiment May 20, 1800; retired June 26, 1803.

ENFIELD, G. S., VISCOUNT.—Ensign 29th Regiment May 2, 1822; Lieutenant 15th Foot April 28, 1825; appointed to 85th Regiment May 5, 1825; Captain unattached January 28, 1836; exchanged to Rifles May 11, 1826; exchanged to half-pay unattached June 18, 1830; exchanged to 47th Regiment November 26, 1830; exchanged to half-pay unattached April 5, 1833; exchanged to 60th Rifles July 2, 1847; retired May 29, 1856; appointed Colonel of the West Middlesex Militia.

ENGLISH, ERNEST ROBERT MALING.—Second-Lieutenant September 28, 1895; Lieutenant June 24, 1899; Captain August 19, 1905; Adjutant Territorial Force January 9, 1910. *South African War*, 1899–1902.—Operations in the Orange Free State February to May

1900, including operations at Paardeberg (February 17 to 26) (slightly wounded); actions at Poplar Grove, Dreifontein, Houtnek (Thoba Mountain), Vet River (May 5 and 6), and Zand River. Operations in the Transvaal in May and June 1900, including actions near Johannesburg and Pretoria. Operations in the Transvaal, east of Pretoria, July to November 29, 1900. Operations in the Transvaal, west of Pretoria, July to November 29, 1900, including actions at Elands River (August 4 to 16). Operations in Orange River Colony May to November 29, 1900, including action at Rhenoster River. Operations in Cape Colony, south of Orange River, 1899–1900. Operations in the Transvaal November 30, 1900, to June 1901, and March to May 31, 1902. Queen's medal with four clasps; King's medal with two clasps.

ERROLL, WILLIAM GEORGE, EARL OF, K.T., G.C.H.—Ensign 51st Regiment April 24, 1817; appointed to 85th Regiment January 8, 1818; Lieutenant 16th Foot May 3, 1821; exchanged to 12th Lancers June 28, 1821; exchanged to 38th Regiment January 16, 1823; exchanged to half-pay 45th Regiment January 23, 1823; died in London April 19, 1846.

ERSKINE, THE HON. H. F. ST. C.—Ensign August 10, 1820; Lieutenant Coldstream Guards February 20, 1823; Captain do. July 11, 1826; died in London May 24, 1829.

ETHERINGTON, GEORGE.—Lieutenant September 1, 1795; Captain June 25, 1803; died at Jamaica January 31, 1805.

EVANS, ROBERT.—Ensign February 4, 1800; Lieutenant May 28, 1801; half-pay of Regiment 1802; full pay December 3, 1802; half-pay of regiment 1806.

FAIRFAX, SIR HENRY, BART.—Ensign Canadian Fencibles April 21, 1808; Ensign 49th Regiment June 8, 1809; Lieutenant 41st Regiment November 23, 1809; appointed to Rifles June 21, 1810; exchanged to 85th Regiment February 2, 1813; Captain 85th Regiment, July 22, 1813; Major July 17, 1823; Lieut.-Colonel November 6, 1827, unattached; Colonel November 23, 1841; exchanged to 2nd Life Guards December 30, 1845; retired same date; died in Edinburgh February 3, 1860; served in the Peninsula.

FAWCETT, JOHN.—Ensign 25th Regiment March 20, 1799; Lieutenant 85th Regiment July 13, 1800.

FEILDING, THE HON. P. ROBERT B.—Ensign August 8, 1845; Lieutenant Coldstream Guards August 7, 1847; Captain do. August 21, 1851.

FERGUSSON, JAMES, C.B.—Ensign 18th Foot August 20, 1801; Lieutenant do. February 9, 1804; appointed to 43rd Regiment August 7, 1804; Captain do. December 11, 1806; Major 79th Regiment December 3,

1812; appointed to 85th Regiment January 25, 1813; Lieut.-Colonel 3rd Foot May 16, 1814; half-pay of Regiment February 25, 1816; appointed to 88th Regiment August 12, 1819; appointed to 52nd Regiment June 2, 1825; Colonel July 22, 1830; appointed Aide-de-Camp to the King same date; exchanged to half-pay of Coldstream Guards May 10, 1839; Major-General November 23, 1841; Colonel of 62nd Regiment March 9, 1850; Colonel of 43rd Regiment March 26, 1850; served on the Expedition to Walcheren; served in the Peninsula; actions at Pombal, Redinha, Miranda de Corvo, Foz d'Arouce, and Sabugal, action at San Munos, passage of the Bidassoa, and investment of Bayonne; was five times wounded, viz., at Vimiera (slightly), severely in the body and slightly in the foot at Ciudad Rodrigo, slightly at Badajos; gold medal for Badajos (assault and capture); silver medal for Vimiera, Corunna, Busaco, Fuentes d'Oñoro, Ciudad Rodrigo (assault and capture), Salamanca, Nivelle and Nive.

FILDER, WILLIAM ALEXANDER.—Ensign October 27, 1843; Lieutenant December 28, 1846; Captain January 30, 1852; exchanged to 37th Regiment June 11, 1852; died in Ceylon June 15, 1853.

FINN, JAMES GLOVER.—Ensign June 7, 1809; Lieutenant August 13, 1812; appointed to 56th Regiment January 25, 1813; half-pay of the Regiment August 25, 1817.

FISHER, CHARLES.—Ensign 60th Regiment March 20, 1811; Lieutenant do. June 18, 1812; appointed to 85th Regiment January 25, 1813; Captain do. June 22, 1815; exchanged to 39th Regiment September 28, 1815; half-pay of Regiment February 25, 1816; exchanged to York Chasseurs April 11, 1816; Regiment disbanded at Quebec August 24, 1819; retired July 1829; subsequently took Holy Orders; died in Yorkshire.

FITZGERALD, AUGUSTINE.—Ensign February 18, 1853; Lieutenant August 11, 1854; retired 1860.

FITZJAMES, MARK.—Ensign 52nd Regiment June 1800; Ensign 85th Regiment June 19, 1800.

FITZPATRICK, JOHN WILSON.—Ensign May 20, 1826; Lieutenant unattached June 26, 1827; retired August 1829.

FLOYD, ROBERT PEEL.—Ensign October 4, 1844; retired October 13, 1848; appointed Captain in the East Devon Militia June 7, 1854.

FORBES, THE HON. FREDERICK.—There was an officer of this name an Ensign in the 84th Foot April 27, 1820, who died April 23, 1826; connection with the 85th Regiment cannot be traced.

FORBES, JOHN GEORGE.—Lieutenant January 30, 1886; Adjutant December 21, 1891, to July 22, 1895; Captain February 11, 1894; Major

December 23, 1905. *South African War*, 1899–1902.—Operations in the Orange Free State February to May 1900, including operations at Paardeberg (February 17 to 26); actions at Poplar Grove, Dreifontein, and at Houtnek (Thoba Mountain). Operations in the Transvaal in May and June 1900. Operations in the Transvaal, east of Pretoria, July to November 29, 1900. Operations in the Transvaal, west of Pretoria, July to November 29, 1900, including actions at Elands River (August 4 to 16). Operations in Orange River Colony May to November 29, 1900. Operations in Cape Colony, south of Orange River, 1899–1900. Operations in the Transvaal November 30, 1900, to March 1902. Commandant Schoemans Kloof. Queen's medal with four clasps; King's medal with two clasps.

FORSTER, MATTHEW.—Ensign 46th Regiment December 19, 1811; Lieutenant 12th Foot December 24, 1812; appointed to 85th Regiment, January 25, 1813; Captain unattached December 26, 1822; exchanged back to 85th Regiment April 24, 1823; exchanged to half-pay of 49th Regiment June 8, 1830.

FORSTER, WILLIAM FREDERICK, K.H.—Ensign and Lieutenant, 3rd Foot Guards, May 26, 1813; Captain 85th Foot June 26, 1817; half-pay March 28, 1824; Major February 18, 1826; Lieut.-Colonel November 23, 1830.

FORT, RALPH.—Second-Lieutenant October 6, 1906; Lieutenant November 17, 1909.

FORTYE, FREDERICK JOHN CAMPBELL.—Ensign June 1, 1838; appointed to 44th Regiment November 8, 1838; Captain do. May 22, 1841; killed near Jugdullock, Afghanistan, January 11, 1842.

FOWLER, R. H.—Lieutenant from 69th Regiment March 2, 1878; Captain July 11, 1886; retired March 22, 1893. *Afghan War*, 1879–80.—With the Kuram Division, Yarmusht Expedition and assault of Zawa; medal.

FOX, CHARLES RICHARD.—Ensign June 29, 1815; Lieutenant West India Rangers November 5, 1818; placed on half-pay of the regiment December 25, 1818; exchanged back to 85th Regiment March 25, 1819; Captain Cape Corps Infantry August 9, 1820; exchanged to 15th Foot January 3, 1822; Major unattached November 6, 1824; exchanged to 85th April 14, 1825; Lieut.-Colonel unattached August 14, 1827; exchanged to 34th Regiment July 23, 1829; appointed to Grenadier Guards October 8, 1830; exchanged to half-pay unattached November 11, 1836; Colonel January 10, 1837; Aide-de-Camp to the King May 28, 1832; Major-General November 9, 1846; appointed Surveyor-General of the Ordnance.

FOX, JOHN.—Ensign 92nd Foot June 27, 1795; Lieutenant 85th Foot December 17, 1795; Lieutenant 60th Foot October 18, 1798; Captain (Army) October 3, 1799; Lieutenant and Captain Grenadier Guards May 3, 1800; Captain 68th Foot December 25, 1801; Half-pay do. 1804; Captain 36th Foot August 4, 1804; Brevet Major June 4, 1811; Major (substantive) October 1, 1812; retired on half-pay 1814; served with 36th Foot in Peninsula. He was the son of a blacksmith at Great Bolas, Shropshire and obtained his commission through the influence of Lord Essex; the father of Major Fox was much esteemed by the husband of the 'Cottage' or 'Peasant Countess.' There is an amusing story with regard to the circumstances under which this officer left the Guards (see *Notes and Queries*).

FRASER, THOMAS.—Captain from half-pay February 15, 1889; died November 1, 1891.

FRAZER (or FRASER) FREDERICK.—Ensign January 26, 1849; resigned August 21, 1849; appointed Lieutenant in Ross Militia, May 1855.

FRENCH, HENRY JOHN. — Ensign 90th Regiment August 27, 1812; Lieutenant 85th Regiment July 21, 1813; Captain September 25, 1823; Major May 23, 1836; Brevet Lieut.-Colonel and Deputy Quartermaster-General at Jamaica July 31, 1846; served in the Peninsula; silver medal for Nivelle and Nive.

GALBRAITH, WILLIAM.—Ensign June 1, 1855; Lieutenant November 30, 1855; Captain April 25, 1865; Major June 12, 1878; Lieut.-Colonel July 1, 1881. *Afghan War*, 1878–9–80.—Served in the Kuram Division as Assistant Adjutant-General from October 1878 to November 1879. Action of Peiwar Kotal, operations in Hariab and Khost Valleys, operations about Cabul, battle of Charasiah, and occupation of Cabul; with Kuram Division in Zaimusht Expedition. Despatches, *London Gazette*, February 4 and March 18, 1879, and January 16, 1880; Brevet Lieut.-Colonel; medal with two clasps.

GALLOWAY, FREDERICK WILLIAM.—Sub-Lieutenant September 10, 1875; retired May 23, 1877.

GAMMELL, WILLIAM.—Ensign 25th Regiment January 14, 1808; Lieutenant do. January 25, 1809; appointed to 95th Rifles January 25, 1813; Captain 104th Regiment April 18, 1846; regiment disbanded at Montreal May 24, 1817; half-pay; exchanged to 85th Regiment December 11, 1817; Major unattached August 29, 1826; exchanged to 87th Regiment October 11, 1833; retired same date; died at Plymouth February 20, 1853; slightly wounded at Badajos 1811.

GARNETT, FREDERICK WILLOCK.—Ensign November 29, 1859; Lieutenant April 9, 1864; retired August 14, 1867.

GARNHAM, W. H.—Ensign 40th Regiment October 30, 1799; placed on half-pay of 85th Regiment as Lieutenant July 15, 1800; omitted in Army List of 1825, no issue of pay having been made for seven years.

GARNIER, FRANCIS.—Ensign 64th Regiment August 16, 1827; Lieutenant do. November 19, 1830; exchanged to 77th Regiment October 1833; exchange cancelled, placed on half-pay of Staff Corps; appointed to 85th Regiment December 6, 1833; retired September 1, 1837.

GARRAWAY, WILLIAM.—Ensign February 18, 1800; Lieutenant June 25, 1801; placed on half-pay of Regiment 1802.

GASCOYNE, ERNEST FREDERICK.—Ensign 39th Regiment May 2, 1811; Lieutenant 85th Regiment May 13, 1813; Captain 3rd Garrison Battalion July 6, 1815; placed on half-pay of it September 20, 1816; exchanged to 54th Regiment December 24, 1818; Major unattached May 19, 1825; appointed to 32nd Regiment May 11, 1826; Lieut.-Colonel unattached June 3, 1828; appointed to a particular service January 1, 1838; placed on half-pay unattached February 25, 1839; exchanged to Grenadier Guards, August 7, 1840; Colonel November 23, 1841; exchanged to half-pay unattached November 15, 1850; Peninsula; silver medal for San Sebastian (assault and capture), Nivelle and Nive; slightly wounded at Bladensburg.

GELL, JAMES.—Ensign June 16, 1808; Lieutenant June 29, 1809; appointed to 6th Foot January 25, 1813; Captain 6th Foot September 15, 1827; died at Bombay, October 7, 1834.

GIBBES, JOHN GEORGE NATHANIEL.—Ensign 40th Regiment February 25, 1804; Lieutenant do. March 28, 1805; Captain 4th Garrison Battalion December 2, 1806; appointed to 85th Regiment April 14, 1808; appointed to 69th Regiment March 28, 1811; exchanged to Royal Regiment of Malta November 19, 1812; Brevet-Major August 12, 1819; omitted in Army List of 1828, no issue of pay having been made for seven years.

GLEIG, GEORGE ROBERT.—Ensign 3rd Garrison Battalion August 13, 1812; appointed to 85th Regiment January 25, 1813; Lieutenant do. July 20, 1813; retired on half-pay of 78th Regiment January 25, 1816; retired August 1825; returned to Balliol College, Oxford, B.A., M.A.; took Holy Orders; appointed Principal Chaplain to the Forces April 1, 1844; Chaplain-General July 2, 1846; author of 'The Subaltern' and 'Campaigns at Washington'; Peninsula; silver medal for San Sebastian (assault and capture), Nivelle and Nive; slightly wounded at Bladensburg; for other details of life see preceding pages in 'Military History.'

Glentworth, William Henry Edmond de Vere Sheaffe, Viscount.—Lieutenant November 12, 1884; transferred to the Rifle Brigade November 22, 1884.

Glew, Joseph Berry.—Ensign 48th Regiment April 5, 1798; Lieutenant 40th Regiment November 9, 1799; half-pay of it 1802; appointed to 53rd Regiment July 9, 1803; Captain 53rd Regiment May 18, 1809; exchanged to 4th Garrison Battalion October 12, 1809; exchanged to 69th Regiment October 11, 1810; exchanged to 85th Regiment March 28, 1811; appointed to 41st Regiment January 25, 1813; exchanged to half-pay, York Chasseurs, September 23, 1819; Brevet-Major June 10, 1815; died at Peel Cottage, Isle of Man, November 30, 1838; Peninsula; present at battles of Douro and Talavera; retreat to Badajos; American Campaign, including Niagara and the storming of Fort Erie, when he was severely wounded.

Glyn, Robert Carr.—Ensign December 13, 1851; Lieutenant February 18, 1853; Captain November 3, 1855; exchanged February 26, 1856.

Godfray, Herbert.—Sub-Lieutenant September 10, 1875; Lieutenant September 10, 1875; to Indian Staff Corps September 10, 1875.

Goodwike, H. H.—Lieutenant 85th Regiment February 21, 1855, to 90th Foot; did not join regiment.

Gordon, Alexander.—Ensign 15th Foot August 29, 1798; Lieutenant do. June 6, 1799; Captain 85th Regiment May 28, 1801; half-pay of regiment 1802; appointed to 69th Regiment May 25, 1803; appointed to 95th Rifles April 21, 1804; retired July 28, 1808.

Gordon, Hesse Maxwell.—Ensign May 3, 1821; died at Malta November 10, 1822.

Gordon, Sir James Willoughby, Bart., G.C.B., G.C.H.—Ensign 66th Regiment October 17, 1783; Lieutenant do. March 5, 1789; Captain do. September 2, 1795; Major do. November 9, 1797; Lieut.-Colonel 85th Regiment May 21, 1801; Deputy Barrackmaster-General, 1803; appointed to 92nd Regiment August 4, 1804; appointed Lieut.-Colonel Commandant of the Royal African Corps June 13, 1808; appointed Colonel do. July 25, 1810; Quartermaster-General 1811; Major-General June 4, 1813; Colonel of the 85th Regiment November 27, 1813; Colonel of the 23rd Fusiliers April 23, 1823; Lieut.-General May 27, 1825; died in London January 4, 1851; present as volunteer at the siege of Toulon 1793; at Bantry Bay in 1796; commanded the 85th when it took possession of Madeira July 24, 1801.

Gordon, Robert.—Ensign 60th Regiment June 27, 1803; appointed to 85th Regiment May 11, 1805; died 1810.

GORDON, WILLIAM ALEXANDER, C.B.—Ensign 112th Regiment October 2, 1794; Lieutenant do. December 29, 1794; appointed to 26th Regiment January 12, 1796; half-pay of the regiment 1798; appointed to 92nd Regiment April 12, 1799; Captain do. October 2, 1801; placed on half-pay of the regiment 1802; appointed to the 50th Regiment October 23, 1806; retired on half-pay 95th Regiment November 26, 1818; Brevet-Major June 4, 1813; Brevet Lieut.-Colonel December 26, 1813; Colonel July 22, 1830; Major-General November 23, 1841; appointed Colonel of the 54th Regiment August 15, 1850; Lieut.-General November 11, 1851; died August 10, 1856; served in the Peninsula; gold medal for Nive, silver medal for Fuentes d'Oñoro and Vittoria; served previously in Holland 1799; at Walcheren 1809; severely wounded in Peninsula at Arroyo del Molino; wounded and horse shot under him at St. Pierre; wounded in the foot at Hasparren February 14, 1814.

GORE, THE HON. CHARLES, C.B., K.H.—Cornet 16th Light Dragoons October 21, 1808; appointed to 6th Foot June 7, 1809; Lieutenant 43rd Regiment January 4, 1810; Captain York Chasseurs March 13, 1815; exchanged to half-pay of the 24th Foot April 20, 1815; exchanged to 85th Regiment June 15, 1815; Brevet-Major January 31, 1819; Major unattached August 20, 1825; Brevet Lieut.-Colonel September 19, 1832; Deputy Quartermaster-General in Canada April 20, 1826; Colonel June 10, 1837; Major-General November 9, 1846; served in Peninsula, silver medal for Ciudad Rodrigo (assault and capture), Badajos (assault and capture), Salamanca, Vittoria, Pyrenees, Nivelle, Nive, Orthes and Toulouse; was also present at the actions of San Milan, capture of Madrid, storming of the heights of Vera and the bridge of Zauzi; in the Waterloo campaign had one horse shot under him at Quatre Bras, and three at Waterloo.

GOULD, HENRY KIMBERLEY.—Ensign 96th Regiment July 18, 1862; Lieutenant 85th Regiment June 28, 1864; retired June 30, 1865.

GRANT, ALEXANDER G.—Cornet 8th Hussars October 4, 1833; Lieutenant do. December 18, 1835; exchanged to 85th Regiment February 23, 1838; Captain 85th Regiment August 5, 1842; Major do. November 10, 1848; Lieut.-Colonel 2nd West India Regiment August 31, 1855; retired April 29, 1856.

GRANT, DANIEL ALEXANDER.—Cornet 3rd Dragoon Guards January 24, 1865; Lieutenant do. November 7, 1868; Captain 85th Regiment November 13, 1872; Major July 1, 1881; Lieut.-Colonel August 19, 1885; retired Hon. Colonel October 27, 1886. *Afghan War*, 1879–80. —With the Kuram Division, Yarmusht Expedition and assault of Zawa. Mentioned in despatches; medal; Brevet-Major.

GRANT, G.—Lieutenant September 6, 1795 (Army); appointed to 85th Regiment December 16, 1795; no further information.

GRANT, JAMES.—Ensign September 22, 1808; Lieutenant October 24, 1809; superseded January 15, 1813; served in Peninsula, slightly wounded at Badajos 1811.

GRANT, JAMES MURRAY.—Ensign September 1, 1854; Lieutenant August 29, 1856.

GRANT, JOHN ALEXANDER.—Ensign 92nd Regiment November 12, 1794; Lieutenant do. August 5, 1795; appointed to 85th Regiment December 13, 1795; Captain do. 1804; died at Jamaica, August 8, 1806.

GRANT, WILLIAM.—Ensign August 30, 1805; apparently did not join the regiment.

GRAYDON, JOHN.—Ensign October 2, 1801; Lieutenant March 15, 1802; placed on half-pay of the regiment 1802; appointed to 16th Battalion of Reserve July 9, 1803; appointed to 88th Regiment August 5, 1804; killed in front of Talavera July 27, 1809.

GREENE, JOHN.—Ensign 90th Regiment March 12, 1812; Lieutenant 85th Regiment June 10, 1813; exchanged to half-pay of 22nd Light Dragoons June 24, 1821; died September 1840.

GREY, CHARLES.—Ensign 14th Foot January 5, 1809; Lieutenant 52nd Regiment May 24, 1810; Captain 85th Regiment January 28, 1813; killed near New Orleans December 23, 1814.

GREY, JOHN WILLIAM.—Ensign May 16, 1834; Lieutenant June 1, 1838; Captain March 31, 1843; Major December 17, 1852; Lieut.-Colonel April 28, 1857.

GREY, SAMUEL CLELAND.—Ensign December 24, 1805; Lieutenant 71st Regiment February 6, 1808; Captain do. February 23, 1815; half-pay of the regiment February 25, 1816; appointed to 29th Regiment August 13, 1829; retired August 20, 1829.

GRINSELL, SAMUEL DICKEN.—Ensign February 23, 1809; Lieutenant May 30, 1811; appointed to 50th Regiment January 25, 1813; half-pay of Regiment 1814; exchanged to 38th Regiment February 16, 1815; half-pay of regiment March 25, 1817; retired 1801.

GRISDALE, J. B.—Ensign 17th Foot; Lieutenant 85th Regiment July 14, 1800; died 1801.

GROVES, PERCY ROBERT CLIFFORD.—Second-Lieutenant (from Militia) October 18, 1899; Lieutenant March 9, 1901; West African Rifles March 7, 1903; Captain Shropshire Light Infantry March 29, 1909; Railway Staff Officer (Graded Staff-Lieutenant) South Africa December 6, 1900, to February 18, 1905. *South African War*, 1899–1902. On Staff. Operations in the Orange Free State February to May 1900, including operations at Paardeberg (February 17 to 26);

actions at Poplar Grove, Dreifontein, Houtnek (Thoba Mountain), Vet River (May 5 and 6), and Zand River. Operations in the Transvaal in May and June 1900, including actions near Johannesburg and Pretoria. Operations in the Transvaal, east of Pretoria, July to November 29, 1900. Operations in the Transvaal, west of Pretoria, July to November 29, 1900, including actions at Elands River (August 4 to 16). Operations in Orange River Colony May to November 29, 1900, including action at Rhenoster River. Operations in Cape Colony, south of Orange River, 1899–1900. Operations in the Transvaal November 30, 1900, to May 31, 1902. Queen's medal with four clasps; King's medal with two clasps.

GUBBINS, JAMES, C.B.—Ensign 48th Regiment December 19, 1845; appointed to 85th Regiment February 27, 1846; Lieutenant do. April 13, 1849; Captain do. February 18, 1853; Brevet-Major December 12, 1854; Major 23rd Foot August 29, 1856; Brevet Lieut.-Colonel April 26, 1859; Lieut.-Colonel June 28, 1862; Colonel April 20, 1869; Major-General July 11, 1879 (with honorary rank of Lieut.-General); retired July 1, 1881; served in the Eastern Campaign of 1854, as Aide-de-Camp to Sir De Lacy Evans, including the battles of Alma and Inkerman (severely wounded) and siege of Sebastopol; Brevet-Major; medal with clasps for Alma, Inkerman and Sebastopol; Knight of the Legion of Honour, 5th class of the Medjidie, Turkish war medal; 'C.B.'

GUBBINS, RICHARD, C.B.—Lieutenant 37th Foot February 3, 1803; exchanged to 24th Foot May 26, 1803; Captain do. December 22, 1804; appointed to 85th Regiment January 25, 1813; Brevet-Major June 4, 1814; Brevet Lieut.-Colonel September 29, 1814; Major June 15, 1815, 21st Fusiliers; half-pay of the regiment March 25, 1816; exchanged to 75th Regiment April 1, 1818; Lieut.-Colonel July 8, 1824 (67th Regiment); exchanged to half-pay 14th Foot May 25, 1826; died near Havant January 2, 1836; served in Peninsular War; commanded regiment at New Orleans; present at Bladensburg.

GUBBINS, RICHARD ROLLS, D.S.O.—Second-Lieutenant March 1, 1890; Lieutenant December 21, 1891; Captain June 24, 1899; Adjutant April 1, 1902, to March 31, 1905; Major June 17, 1908; Adjutant Militia September 28, 1905, to March 31, 1908. *South African War, 1899–1902.*—Operations in the Orange Free State February to May 1900, including operations at Paardeberg (February 17 to 26) (slightly wounded); actions at Houtnek (Thoba Mountain), Vet River (May 5 and 6), and Zand River. Operations in the Transvaal in May and June 1900, including actions near Johannesburg and Pretoria. Operations in the Transvaal, east of Pretoria, July to November 29, 1900. Operations in the Transvaal, west of Pretoria, July to

November 29, 1900, including actions at Elands River (August 4 to 16). Operations in Orange River Colony May to November 29, 1900, including action at Rhenoster River. Operations in Cape Colony, south of Orange River, 1899–1900. Served as Adjutant 2nd Battalion Shropshire Light Infantry May 11 to 31, 1902. Operations in the Transvaal November 30, 1900, to August 1901, and January to May 31, 1902. Despatches, *London Gazette*, September 10, 1901. Queen's medal with three clasps; King's medal with two clasps. 'D.S.O.'

GWYLLYM, THOMAS.—Ensign November 3, 1808; apparently did not join the regiment.

GWYNNE, NADOLIG XIMENES.—Lieutenant October 9, 1855; Substantive Lieut.-Colonel July 1, 1881; Lieut.-Colonel and Colonel to command Battalion June 21, 1885; placed on half-pay July 1, 1887; retired pay December 25, 1887.

HALLOWES, WILLIAM.—Ensign January 22, 1853; Lieutenant June 6, 1854; Captain March 4, 1861; Major October 7, 1871; Lieut.-Colonel October 10, 1877; retired June 12, 1878.

HAMILTON, CHRISTOPHER DOUGLAS.—Sub-Lieutenant from 80th Foot March 18, 1874; retired January 23, 1878.

HAMILTON, DOUGLAS JOHN.—Ensign 52nd Regiment July 28, 1808; Lieutenant do. February 1, 1810; Captain 85th Regiment January 28, 1813; killed at Bladensburg August 24, 1814.

HAMILTON, HENRY.—Ensign March 13, 1817; retired January 22, 1818.

HAMILTON, THOMAS M. M'NEILL.—Ensign April 24, 1828; Lieutenant February 1, 1833; Captain November 9, 1838; retired November 16, 1841; died August 30, 1862.

HANBURY, PHILIP LEWIS.—Second-Lieutenant October 18, 1899; Lieutenant January 22, 1901; Adjutant from April 1, 1905, to February 14, 1908; Captain March 1, 1909. *South African War*, 1899–1902.—Operations in the Orange Free State, February to May 1900, including operations at Paardeberg (February 17 to 26); actions at Poplar Grove, Dreifontein, Houtnek (Thoba Mountain), Vet River (May 5 and 6), and Zand River. Operations in the Transvaal in May and June 1900, including actions near Johannesburg and Pretoria. Operations in the Transvaal, east of Pretoria, July to November 29, 1900. Operations in the Transvaal, west of Pretoria, July to November 29, 1900, including actions at Elands River (August 4 to 16). Operations in Orange River Colony, May to November 29, 1900, including action at Rhenoster River. Operations in Cape Colony, south of Orange River, 1899–1900. Operations in the Transvaal November 30, 1900, to May 31, 1902. Queen's medal with four clasps; King's medal with two clasps.

HANCOCK, WILLIAM FREDERICK.—Ensign May 1, 1857; Lieutenant February 4, 1862; Captain June 12, 1867; retired October 14, 1871.

HANSON, EDWARD SYDENHAM GEORGE.—Ensign August 7, 1866; retired May 1, 1867.

HARFORD, CHARLES JOSEPH.—Ensign 82nd Regiment, January 16, 1846; Lieutenant do. October 27, 1848; Lieutenant 12th Foot July 29, 1853; Captain 12th Light Dragoons August 29, 1856; Captain 85th Regiment November 30, 1860; retired October 9, 1863.

HARRIS, ANTHONY.—Ensign February 4, 1800; Lieutenant May 14, 1801; half-pay of the regiment 1802; appointed to 69th Regiment July 9, 1803.

HARRIS, CHARLES.—Ensign 73rd Regiment January 11, 1809; Lieutenant do. August 16, 1810; exchanged to 3rd Light Dragoons August 29, 1811; Captain York Chasseurs November 10, 1813; exchanged to 85th Regiment June 23, 1814; killed in the night action on the left bank of the Mississippi, near New Orleans, December 23, 1814.

HARRIS, WILLIAM.—Ensign half-pay of 3rd Foot October 24, 1821; appointed to 85th Regiment April 7, 1825; Lieutenant do. April 8, 1826; retired July 6, 1829.

HARRIS-BURLAND, HARRY.—Ensign June 8, 1867; Adjutant August 3, 1872; Lieutenant; Captain June 10, 1874; died at Landour June 10, 1876.

HARWOOD, GEORGE.—Ensign February 4, 1800; Lieutenant May 27, 1801; half-pay of regiment 1802; reappointed to full pay October 25, 1802; died in Jamaica 1805.

HAVILAND, R. H.—From half-pay 76th Foot; Captain July 7, 1854; did not join regiment.

HAYDOCK, HENRY JAMES.—Ensign December 16, 1853; Lieutenant 90th Foot February 21, 1855.

HENDERSON, WILLIAM CHIPCHASE.—Ensign January 15, 1856; Lieutenant February 22, 1861; Captain March 6, 1867; retired October 31, 1877.

HENRY, CHARLES JOHN.—Ensign 95th Regiment August 26, 1824; Ensign 85th Regiment April 4, 1825; Lieutenant 56th Regiment July 11, 1826; Captain do. April 26, 1831; retired September 6, 1839.

HEWITT, ISAAC HENRY.—Lieutenant July 23, 1800; exchanged to 38th Regiment August 27, 1803; Captain 6th Foot June 4, 1807; Major do. June 2, 1814; served in Spain and Portugal attached to the Portuguese Army; Portuguese and Spanish Staff; half-pay Portuguese and Spanish Staff December 25, 1816; Brevet Lieut.-Colonel September 4, 1817; retired June 1825; died July 26, 1825.

HICKS, JAMES HORATIO.—Served in the ranks for seven and a half years; Lieutenant 1st Battalion Shropshire Light Infantry November 28, 1885; Captain 2nd Battalion March 22, 1893; retired June 11, 1901; Royal Garrison Regiment September 16, 1905; Major April 23, 1904; half-pay; South African War, 1879; Zulu Campaign; medal with clasp.

HICKS, JOHN DAVIES.—From the South Devon Militia; Captain 85th Regiment December 25, 1813; placed on half-pay of the regiment 1815; died in France January 8, 1827; wounded at Baltimore.

HICKSON, WILLIAM.—Ensign 46th Regiment September 24, 1812; exchanged to 85th, January 25, 1813; Lieutenant do. May 20, 1813; killed by a small rifle ball in the forehead during the night action on the left bank of the Mississippi, near New Orleans, December 23, 1814; wounded at Bladensburg August 24, 1814.

HIGGINSON, CECIL PICKFORD, D.S.O.—Lieutenant (from Militia) November 10, 1886; Captain November 27, 1895; Adjutant September 19, 1897; Major August 2, 1900; Brigade-Major South Africa; Deputy Assistant Adjutant-General South Africa June and July 1901; employed with the Rand Rifles 1902–3; Brigade-Major 2nd Brigade 1st Army Corps, Aldershot, February 10, 1903; Brigade-Major 2nd Brigade 1st Division, Aldershot, 1903–6; Deputy Assistant Adjutant and Quartermaster-General North China February 9, 1909. *South African War*, 1899-1902.—Served as Adjutant 2nd Battalion Shropshire Light Infantry from January 1, 1900, to January 21, 1901, and from July 10, 1901, to March 30, 1902. (Slightly wounded.) Operations in the Orange Free State, February to May 1900, including operations at Paardeberg (February 17 to 26, 1900); actions at Poplar Grove, Dreifontein, Houtnek (Thoba Mountain), Vet River (May 5 and 6) and Zand River. Operations in the Transvaal in May and June 1900, including actions near Johannesburg and Pretoria. Operations in the Transvaal, east of Pretoria, July to November 29, 1900. Operations in the Transvaal, west of Pretoria, July to November 29, 1900, including actions at Elands River (August 4 to 16). Operations in Orange River Colony, May to November 29, 1900, including action at Rhenoster River. Operations in Cape Colony, south of Orange River, 1899–1900. Operations in the Transvaal November 30, 1900, to May 31, 1902. Afterwards on Staff, and as Station Staff Officer. Despatches, *London Gazette*, February 8, 1901, and July 29, 1902. Queen's medal with four clasps; King's medal with two clasps. 'D.S.O.' Placed on the list of officers considered qualified for staff employment, in consequence of service on the staff in the field.

HILL, GEORGE.—Ensign 5th Foot April 14, 1788; Lieutenant do. October 26,

1786; Captain do. September 5, 1795; Major York Rangers July 2, 1803; appointed to 85th Regiment March 7, 1805; died 1809.

HOBSON, ROBERT (1).—53rd Regiment October 17, 1793; Lieutenant do. November 20, 1794; Captain do. October 21, 1795; Major 85th Regiment October 19, 1804; died in Jamaica January 25, 1808.

HOBSON, ROBERT (2).—Ensign July 11, 1805; Lieutenant December 25, 1805; appointed to Royal York Rangers March 9, 1809; appointed to 10th Veteran Battalion June 17, 1813; this was made 4th Battalion reduced September 24, 1816; placed on retired full pay of it; appointed to 8th Veteran Battalion November 1, 1819; placed on retired full pay of it 1810; died at William Henry, Quebec, December 13, 1832.

HODGSON, NATHANIEL HENRY.—Ensign March 25, 1862; Lieutenant July 25, 1865; retired July 20, 1867.

HOGG, GEORGE.—Ensign 39th Regiment May 29, 1806; appointed to 85th Regiment June 19, 1806; Lieutenant do. June 16, 1808; killed at Badajos June 10, 1811; severely wounded at Poza, near Fuentes d'Oñoro, May 5, 1811.

HOGG, JAMES.—Ensign October 8, 1799; Lieutenant February 22, 1800; Captain 7th West India Regiment May 4, 1808; appointed to 26th Regiment January 30, 1812; exchanged to half-pay of 27th Regiment April 25, 1816; exchanged to 86th Regiment February 20, 1825; retired.

HOGGE, CHARLES WELLS.—Ensign December 12, 1851; Lieutenant January 28, 1853; retired February 21, 1865.

HOGGINS, THOMAS.—Ensign 64th Regiment September 15, 1796; Lieutenant do. June 1, 1798; Captain do. September 28, 1799; half-pay 71st Regiment 1802; appointed to 85th Regiment April 4, 1805; killed in a duel at Brabourne Lees, Kent, January 5, 1810 (see Military History).

HOLMES, SAMUEL.—Ensign March 24, 1800; Lieutenant March 3, 1804; killed at Poza, near Fuentes d'Oñoro, May 5, 1811.

HOLMES-À COURT, RUPERT EDWARD.—Lieutenant July 5, 1905; Captain April 1, 1910; Adjutant 1911.

HOMPESCH, WILLIAM VINCENT.—Ensign 60th Regiment November 1795; Lieutenant do. January 1796; Captain 85th Regiment August 6, 1799; Major 25th Regiment February 28, 1805; Brevet Lieut.-Colonel January 1, 1812; exchanged to half-pay unattached November 3, 1825; Colonel July 22, 1830; Major-General June 28, 1838; died at Newcastle-on-Tyne November 28, 1849; served in the army of the States-General before he entered the British service; siege of Williamstadt 1793; Battle of Rouveroy; actions near Marchiene au Pont, Fleurus, and at the Dyle 1794; served in Holland

1799; present at the battle on landing August 25, and that of September 19, and those of October 2 and 6; present at Merxem and bombardment of Antwerp.

HOOD, WILLIAM EDWARD COMBER.—Second-Lieutenant from Norfolk Regiment July 4, 1894; exchanged to Bedfordshire Regiment September 26, 1894. *Operations in Chitral*, 1895.—With the Relief Force; storming of the Malakand Pass; action near Khar at descent into Swat Valley; medal with clasp.

HOOPER, JOHN CHARLES.—Second-Lieutenant October 18, 1899; Lieutenant January 30, 1901; Captain March 29, 1909; Adjutant January 11, 1911. *South African War*, 1899–1902.—Employed with Mounted Infantry. Operations in the Transvaal November 30, 1900, to May 31, 1902. Despatches, *London Gazette*, September 10, 1901. Queen's medal with six clasps; King's medal with two clasps.

HOPWOOD, EDWARD.—Ensign 72nd Regiment May 12, 1825; appointed to 29th Regiment June 9, 1825; Lieutenant 69th Regiment June 10, 1826; Captain unattached June 21, 1827; exchanged to 85th Regiment May 8, 1828; retired April 5, 1831.

HORROCKS, JOHN.—Ensign March 4, 1836; Lieutenant November 1, 1839; retired June 24, 1842.

HOWARD, THE HON. FREDERICK.—Ensign May 14, 1801; Lieutenant 10th Light Dragoons February 25, 1802; half-pay of regiment 1802; reappointed to full pay December 8, 1802; Captain 60th Regiment December 29, 1804; exchanged back to 10th Light Dragoons January 11, 1805; Major do. May 9, 1811; killed at Waterloo June 18, 1815.

HUGO, EDWARD.—Surgeon November 18, 1793.

HULTON, PRESTON.—Cornet 21st Light Dragoons, May 5, 1800; Lieutenant 85th Regiment September 24, 1802; reappointed to 21st Light Dragoons March 30, 1803; Captain do. October 10, 1805; retired April 4, 1811; appointed Captain in the South Hants Militia March 4, 1814; died at Barnfield near Southampton August 15, 1825.

HUMPHREYS, EDWARD.—Ensign July 5, 1829; Lieutenant April 26, 1834; retired May 27, 1836.

HUNT, WILLIAM.—Ensign 46th Regiment January 14, 1814; Ensign 85th Regiment September 8, 1814; Lieutenant do. June 29, 1815; half-pay of Regiment March 25, 1817; appointed to 2nd Foot January 25, 1825; retired on half-pay of Staff Corps February 5, 1829; died April 1850.

HUNT, WILLIAM THOMAS.—Ensign April 29, 1813; Lieutenant April 17, 1814; Captain April 9, 1825; Brevet-Major June 28, 1838; Major November 1, 1839; Lieut.-Colonel unattached August 5, 1842.

HUNTER, JAMES.—Ensign March 23, 1815; Lieutenant June 24, 1819; Captain August 20, 1825; Major March 4, 1836 (unattached); died in Edinburgh December 9, 1843.

HURFORD, A. V.—Ensign 38th Foot January 16, 1852; Lieutenant do. June 13, 1857; Captain do. December 2, 1862; Captain 85th Regiment March 3, 1873; died at Meerut; served in Burmah; medal and clasp for Pegu.

HUTH, PERCIVAL C., D.S.O.—Lieutenant November 1, 1906. *West Africa*, 1900.—Operations in Ashanti. Medal. *South African War*, 1901–2. —Severely wounded. Despatches, *London Gazette*, April 25, 1902. Queen's medal with five clasps. 'D.S.O.'

HYLTON, JOHN.—Ensign March 3, 1804; Lieutenant January 31, 1805; Captain June 27, 1811; cashiered, February 1813.

IBBETSON, CHARLES VILLIERS.—Ensign December 4, 1866; transferred to 54th Regiment January 11, 1867.

ILLINGWORTH, EDWARD ARTHUR.—Ensign 10th Foot, May 17, 1861; Lieutenant 85th Regiment May 31, 1864; Captain October 16, 1867; retired on half-pay March 10, 1874.

IMLACH, JAMES.—Ensign 81st Regiment February 2, 1809; Lieutenant 85th Regiment July 23, 1812; exchanged to 43rd Regiment January 25, 1813; half-pay 43rd Regiment September 4, 1817.

IRVINE, CHARLES.—Captain 68th Regiment March 1, 1800; appointed to 62nd Regiment December 9, 1800; half-pay of it 1800 and again 1802; appointed to 85th Regiment November 20, 1802; half-pay unattached 1807; Brevet Lieut.-Colonel March 9, 1803; Colonel January 1, 1812; Major-General June 4, 1814; died June 5, 1819.

IVES, EDWARD HARBORD.—Ensign May 1, 1867; Lieutenant January 25, 1871; Captain February 23, 1878; exchanged to Royal Fusiliers July 12, 1882.

JACKSON, FRANCIS.—From the North Mayo Militia; Captain 37th Regiment December 25, 1813; half-pay of it May 25, 1817; exchanged to 85th October 14, 1819; half-pay of it October 1821; reappointed to full pay April 8, 1825; Major November 6, 1827; retired November 1, 1839.

JACKSON, GEORGE.—Ensign 35th Regiment May 31, 1792; Lieutenant 35th Regiment April 30, 1793; Major 96th Regiment October 14, 1794; regiment reduced, but retained on full pay of it 1795; half-pay of it 1798; Brevet Lieut.-Colonel January 1, 1800; appointed to 85th Regiment May 17, 1800; exchanged to Scotch Brigade April 3, 1801.

JACKSON, OLIVER VAUGHAN.—Ensign March 2, 1832 ; Lieutenant May 27, 1836 ; exchanged to 33rd Regiment August 8, 1851 ; appointed Paymaster of the 6th Dragoon Guards, February 14, 1840 ; reverted to half-pay of the 85th Regiment March 20, 1846 ; retired August 8, 1851.

JAMES, JOHN.—Ensign July 7, 1825 ; Lieutenant November 6, 1827 ; retired August 9, 1833.

JAMES, THOMAS.—Ensign October 27, 1808 ; Lieutenant October 26, 1809 ; retired September 26, 1811.

JEBB, AVERY.—Ensign February 4, 1862 ; Lieutenant June 30, 1865 ; temporary half-pay February 9, 1870.

JENOUR, MATTHEW.—Captain half-pay 41st Foot December 24, 1787 ; Major 85th Foot May 4, 1795 ; Lieut.-Colonel 9th Garrison Battalion January 1, 1800 ; Lieut.-Colonel Garrison Reserve October 15, 1803.

JERVIS, GEORGE HENRY PARKER.—Ensign October 9, 1863 ; Lieutenant July 7, 1866 ; retired December 4, 1866.

JOHNSON, ARTHUR.—Ensign 81st Regiment July 11, 1811 ; Lieutenant 85th Regiment May 27, 1813 ; killed in the attack on the village of Urogne, near St. Jean de Luz, November 10, 1813.

From a Newspaper.

' Lieut. Arthur Johnson of the 85th, who fell in storming the French intrenchments on the 10th at St. Pé, was a youth of most admirable promise. He embarked in the military character at the early age of 15, and was severely wounded in the memorable battle of Talavera. On his recovery he was advanced to the infantry and fought at the battle of Busaco. He served also in Guernsey and Jersey ; and when promoted to the 85th, he embarked immediately for Spain, and was landed in the very scene of action near St. Sebastian, and shared in that memorable event. He was with the advance in all their movements into the South of France, and under the greatest deprivations and fatigues was never heard to murmur : his health and spirits were such as enabled him to encounter any hardships, though at the time of his death he was only just turned nineteen. By a letter from his worthy Colonel, enclosing some military journals, he adds :—" You have lost a son ; the regiment a member of its society, esteemed and beloved by every officer of the Corps and the Service ; an individual that I have no doubt, some day or other, had fate but spared him, would have been to the army a great and most useful ornament. He fell, poor fellow ! early in the engagement ; but his death was so easy and instantaneous that his countenance was retaining at the moment of being prepared for interment (which took place in the churchyard of Urogne) its former natural smile." '

JOHNSON, CHARLES CHRISTOPHER.—Cornet 17th Light Dragoons May 5 1804 ; Lieutenant do. September 12, 1805 ; Captain 60th Regiment

August 18, 1814; exchanged to 39th Regiment August 28, 1815; exchanged to 85th Regiment September 28, 1815; Major 93rd Regiment December 26, 1822; half-pay 10th Foot October 30, 1823; Brevet Lieut.-Colonel January 10, 1837; Colonel November 11, 1851; died September 30, 1854; present at storming of Monte Video and Expedition to Buenos Aires; Persian Order of the Lion and Sun, 2nd class.

JORDEN, JOHN.—Ensign 12th Foot March 5, 1796; Lieutenant 85th Regiment November 8, 1798; no further information.

JOUNCEY, A. W.—Cannot be identified; stated to have died November 22, 1884.

JUDGE, SPENCER F., D.S.O.—Captain from 1st Battalion November 8, 1897; retired. *Soudan,* 1888–9.—Action of Gamaizah; action of Arguin, action of Toski. Mentioned in despatches; medal with two clasps; bronze star; 'D.S.O.'; 4th class Osmanieh; 4th class Medjidie. *Expedition to Dongola,* 1896.—As Brigadier-Major 4th Brigade; despatches, *London Gazette,* November 3, 1896.

KAVANAGH, WALTER.—Ensign March 20, 1800; superseded November 1803,

KEATS, JOHN SMITH.—December 16, 1816, Royal Artillery; Lieutenant Royal Artillery March 2, 1825; appointed to 85th Regiment April 8, 1825; Captain unattached July 5, 1829; exchanged to 75th Regiment July 6, 1829; exchanged to half-pay of the waggon train May 31, 1833.

KELLY, PATRICK.—Ensign Newfoundland Fencibles March 5, 1807; Lieutenant do. April 26, 1810; exchanged to 85th February 28, 1811; appointed to 87th Regiment January 25, 1813; exchanged to 89th Regiment January 7, 1819; exchanged to half-pay of 60th Regiment June 24, 1819; retired May 1829; served in Peninsula; silver medal for Nivelle, Orthes and Toulouse.

KELSON, CHARLES.—Ensign 85th September 26, 1811; exchanged to 6th Foot January 25, 1812; Lieutenant 103rd Regiment March 4, 1813; regiment disbanded, placed on half-pay of it April 2, 1818; appointed to 97th Regiment March 25, 1824; Captain 97th Regiment June 10, 1832; appointed to Ceylon Rifles May 5, 1837; Brevet-Major November 9, 1846; Major unattached February 4, 1853; Brevet Lieut.-Colonel June 20, 1854; appointed to 3rd West India Regiment March 26, 1858; retired same day; died September 14, 1866, aged seventy-four; served in American Campaign; present at the action of Lundy's Lane, and the storming of Fort Erie.

KERR, CHARLES FORTESCUE (afterwards Viscount Dunluce June 30, 1834, when his mother succeeded to the peerage as Countess of Antrim); Ensign unattached July 1, 1828; appointed to Rifle Brigade July 2,

LIEUT. JOHN JORDEN
(*Militia Uniform.*)

1829; Lieutenant unattached March 9, 1832; exchanged to 99th Regiment March 23, 1832; appointed to 85th Regiment March 30, 1832; retired May 9, 1834; died at Holmwood near Henley July 28, 1834.

KERR, WILLIAM.—Ensign November 19, 1818; died at Holmwood near Henley February 19, 1819.

KETTLEWELL, HENRY WILDMAN.—Second-Lieutenant May 28, 1898; Lieutenant April 3, 1900; Captain August 8, 1906. *South African War,* 1899–1900.—Operations in the Orange Free State February to May 1900, including operations at Paardeberg (February 17 to 26) (slightly wounded). Operations in Cape Colony, south of Orange River, 1899–1900; Queen's medal with two clasps.

KETTLEWELL, WILLIAM WILDMAN.—Ensign from 61st Regiment June 8, 1867; Lieutenant July 13, 1870; transferred to 89th Foot.

KEYT, JOHN AUGUSTUS.—Ensign May 21, 1841; Lieutenant April 1, 1844; Captain March 14, 1851; retired July 15, 1853; died March 11, 1890.

KIDD, JOHN PARKER.—Ensign.—June 4, 1807; drowned on passage from Ireland December 1807.

KING, JOHN GRANT.—Ensign May 22, 1806; Lieutenant 48th Regiment February 8, 1808; died July 25, 1816; served in Peninsula; gold medal for Orthes.

KING, RICHARD N.—Lieutenant July 17, 1801; placed on half-pay of regiment 1802; appointed to 18th Regiment July 9, 1803.

KNIGHT, HENRY.—Ensign July 10, 1811; retired December 1812.

KNOX, SIR CHARLES EDMOND, K.C.B.—Ensign June 30, 1865; Lieutenant August 7, 1867; Captain June 11, 1876; Major July 1, 1883; Brevet Lieut.-Colonel 1885; Lieut.-Colonel December 6, 1885; Colonel December 9, 1889; Lieut.-Colonel to command February 11, 1890; Major-General December 4, 1899; Lieut.-General December 6, 1905; retired May 10, 1909; Distinguished Service Reward; commanded 4th Division 2nd Army Corps 1903. *Bechuanaland Expedition,* 1884–5.—In command of 4th Pioneers. Honourably mentioned. Brevet Lieut.-Colonel. *South African War,* 1899–1902.—On Staff (including command of 13th Brigade; afterwards of Mounted Mobile Columns). Relief of Kimberley. Operations in the Orange Free State February to May 1900, including operations at Paardeberg (February 17 to 18) (severely wounded, February 18). Actions at Poplar Grove and Dreifontein. Operations in Orange River Colony May to November 29, 1900, including actions at Bothaville and Caledon River (November 27 to 29). Operations in Orange River Colony November 30, 1900, to February 1901, March 1901 to May

1902. Operations in Cape Colony February to March 1901. Despatches, *London Gazette* February 3 and April 16, 1901, and July 29, 1902. Promoted Major-General for distinguished service. Queen's medal with four clasps; King's medal with two clasps. 'K.C.B.'

KNOX, CHARLES HENRY.—Ensign unattached April 8, 1826; exchanged to 95th Regiment December 14, 1826; Lieutenant 89th Regiment August 9, 1827; half-pay of regiment 1829; exchanged to 85th Regiment July 7, 1829; Captain 85th Regiment March 4, 1836; exchanged to 94th Regiment July 15, 1836; exchanged to half-pay unattached February 10, 1838; exchanged to 1st West India Regiment November 30, 1855; retired same day.

KNOX, FREDERICK CHARLES NORTHLAND.—Sub-Lieutenant and Lieutenant from the 106th Foot January 17, 1877; resigned November 11, 1878.

KNOX, THE HON. JOHN JAMES.—Ensign 52nd Regiment August 17, 1808; Lieutenant 19th Foot March 16, 1809; exchanged back to 52nd Regiment May 11, 1809; Captain 40th Regiment October 8, 1812; appointed to 85th Regiment January 25, 1813; Major do. May 22, 1817; Lieut.-Colonel June 24, 1819; half-pay of 4th West India Regiment; exchanged to 65th Regiment August 17, 1832; retired August 24, 1832; died July 9, 1856; served in the Peninsula; silver medal for Vimiera, Fuentes d'Oñoro, Ciudad Rodrigo (assault and capture), San Sebastian (assault and capture), Nivelle and Nive; also served in North America; wounded on approach to New Orleans; December 23, 1814.

KNOX, THOMAS EDWARD, C.B.—Ensign 98th Regiment January 26, 1838; appointed to 85th Regiment February 23, 1838; Lieutenant June 24, 1843; Captain July 31, 1846; Major West India Regiment August 17, 1852; Lieut.-Colonel 67th Regiment September 17, 1858; C.B. March 1, 1861; Colonel (Army) January 24, 1863; Lieut.-Colonel 9th Regiment October 18, 1864; Major-General June 13, 1876; Lieut.-General May 17, 1881; Honorary General July 1, 1881; Colonel Hampshire Regiment December 28, 1888.

KNOX, THE HON. WILLIAM STUART.—Ensign January 12, 1844; Lieutenant December 29, 1846; Captain 4th West India Regiment August 17, 1852; Major November 16, 1855.

LANGFORD, C. W. WILLIAM, LORD.—Ensign August 5, 1842; Lieutenant 7th Fusiliers February 16, 1844; retired February 27, 1846; died July 19, 1854.

LANGFORD, FRANCIS.—Ensign 1st West India Regiment September 5, 1862; Lieutenant do. August 4, 1863; Lieutenant 85th Regiment April 27, 1870; Captain do. February 2, 1878; retired as Honorary Major September 5, 1882.

[*By kind permission of Mrs. Warrenne Blake.*]

LIEUT.-COLONEL THE HONOURABLE JAMES KNOX.

LEACH, FRANCIS JAMES.—Second Lieutenant July 27, 1901; Lieutenant August 19, 1905; served previously in the ranks.—*South African War, 1899–1902.*—Relief of Kimberley. Operations in the Orange Free State February to May 1900, including operations at Paardeberg (February 17 to 26); actions at Poplar Grove and Zand River. Operations in the Transvaal in May and June 1900, including actions near Johannesburg, Pretoria, and Diamond Hill (June 11 and 12). Operations in Orange River Colony May to November 1900, including actions at Wittebergen (July 1 to 29). Operations in the Transvaal August 1901 to May 31, 1902. Operations in Orange River Colony, November 30, 1900, to August 1901; King's medal with two clasps.

LEE, RICHARD.—Ensign 17th Foot June 25, 1788; Lieutenant 12th Foot November 9, 1791; Major 124th Regiment August 11, 1794; half-pay of it 1798; Brevet Lieut.-Colonel January 1, 1800; appointed to 85th Regiment May 16, 1800; placed on half-pay of 135th Regiment 1801; exchanged to 63rd Regiment May 24, 1810; Colonel July 25, 1810; died June 10, 1811.

LEICESTER, FREDERICK.—Cornet 3rd Dragoons December 30, 1795; Lieutenant do. August 24, 1797; appointed to 85th Regiment June 26, 1801; half-pay of regiment 1802; Captain Staff Corps December 31, 1803; died 1812.

LE MARCHANT, SIR JOHN GASPARD.—Ensign 10th Foot October 26, 1820; Lieutenant half-pay 10th Foot October 24, 1821; exchanged to 57th Regiment February 7, 1822; Captain unattached June 30, 1825; appointed to 98th Regiment November 9, 1826; Major do. December 14, 1832; exchanged to half-pay unattached July 31, 1835; appointed Lieut.-Colonel 1st Regiment of British Legion in Spain July 11, 1835; also Adjutant-General; retired from Legion March 20, 1837; appointed to 20th Regiment November 10, 1837; exchanged to 99th Regiment January 4, 1839; Lieut.-Colonel 99th Regiment October 18, 1839; exchanged to Inspecting Field Officer of Cork Recruiting District September 27, 1842; appointed to 85th Regiment June 19, 1846; exchanged to half-pay unattached December 29, 1846; appointed Governor of Newfoundland 1847; medal from the British Legion in Spain.

LETHBRIDGE, THOMAS CHRISTOPHER M.—Ensign May 12, 1838; Lieutenant August 5, 1842; died at St. Kitt's March 31, 1844.

LEVINGE, SIR CHARLES, K.H.—Ensign 4th Foot November 18, 1813; Lieutenant 4th Foot October 5, 1815; half-pay of regiment 1816; exchanged to 85th Regiment March 21, 1816; half-pay of it December 25, 1818; exchanged to 10th Foot September 30, 1819; Captain half-pay 10th Foot October 24, 1821; appointed to 52nd Regiment August 14, 1823; Major July 14, 1825, unattached;

exchanged to 71st Regiment January 15, 1829; Brevet Lieut.-Colonel June 28, 1838; retired July 6, 1838; served at Waterloo with the 4th Foot.

LEWIN, FREDERICK CHALLONER.—Ensign July 16, 1861; Lieutenant April 25, 1865; died at Dalhousie May 14, 1870.

LINDSAY, OWEN.—Ensign April 4, 1805; superseded 1806.

LINTON, HENRY, M.V.O.—Second-Lieutenant November 3, 1906 (from the Militia); Lieutenant February 11, 1910.

LLUELLYN, RICHARD, C.B.—Entered the army in 1799 as Captain with temporary rank and served with the 52nd in Spain and the Mediterranean in 1800 and 1801; was reduced to half-pay on peace being concluded, but on the war breaking out again he relinquished his half-pay and purchased an ensigncy in the 85th July 24, 1803; Lieutenant half-pay 20th Foot April 7, 1804; exchanged to 88th Regiment May 24, 1804; Captain 28th Regiment February 28, 1805; half-pay of the regiment February 25, 1817; Brevet-Major April 23, 1812; Brevet Lieut.-Colonel June 18, 1815; Colonel January 10, 1837; Major unattached February 16, 1844; Major-General November 9, 1846; served in Peninsula; silver medal for Busaco and Albuera; served at Waterloo.

LONGFIELD, JOHN.—Ensign May 27, 1801; Lieutenant August 30, 1804; Captain 2nd Garrison Battalion May 16, 1805.

LONGFIELD, RICHARD.—Ensign March 9, 1800; probably never joined the regiment.

LUARD, EDWARD BOURRYAU.—Second Lieutenant April 3, 1891; Lieutenant February 11, 1894; Captain August 11, 1900; Major March 19, 1910. *South African War*, 1899–1902.—On Staff. Operations in the Transvaal December 1900 to May 31, 1902. Queen's medal with clasp; King's medal with two clasps.

LYLE, ACHESON FRANCIS ACHESON.—Sub-Lieutenant February 11, 1875; Lieutenant February 11, 1875; Captain May 11, 1884; Major November 30, 1892; died.

MACDONALD, JAMES.—Ensign 85th Regiment January 25, 1796; Lieutenant 19th Foot February 2, 1796; Captain do. September 10, 1803; Captain 5th Foot October 22, 1803; Major 78th Regiment April 17, 1804; Lieut.-Colonel do. Sept. 7, 1809; exchanged to 2nd Garrison Battalion February 21, 1811; exchanged to Coldstream Guards August 8, 1811; Colonel (army) August 12, 1819; Major Coldstream Guards July 25, 1821; Lieut.-Colonel Coldstream Guards May 27, 1825; Major-General July 22, 1830; Lieut.-General November 23, 1841; served in Expedition to Sicily and Calabria 1805–6; gold medal for Maida; served in Peninsula; silver medal for Salamanca, Vittoria, Nivelle and Nive.

M'DONNELL, CHARLES.—Ensign 85th Regiment June 8, 1815; appointed to 35th Regiment July 29, 1819; exchanged to half-pay of 32nd Regiment May 24, 1821; appointed to 50th Regiment February 14, 1840; retired same date.

M'DOUGALL, ARCHIBALD.—Lieutenant June 8, 1796; appointed to 85th Regiment October 5, 1796; no further information.

MACDOUGALL, DUNCAN.—Ensign 71st Regiment April 6, 1804; Lieutenant do. April 23, 1805; Captain Cape Regiment June 19, 1806; exchanged to 53rd Regiment February 6, 1812; appointed to 85th Regiment January 25, 1813; exchanged to half-pay of the regiment August 28, 1817; Brevet-Major October 20, 1814; Brevet-Lieut.-Colonel April 21, 1825; Major 79th Regiment July 16, 1830; Lieut.-Colonel do. September 6, 1833; retired March 13, 1835; appointed Colonel of the 9th Regiment of the British Legion of Spain; was Quartermaster-General; retired from the Legion in May 1836; fought in the Peninsula; silver medal for Salamanca, San Sebastian (assault and capture), Nivelle and Nive; severely wounded at Salamanca; received the medal from the British Legion of Spain.

MACDOUGALL, JAMES.—Ensign 23rd Fusiliers; Lieutenant do. September 8, 1813; half-pay of the regiment December 25, 1814; exchanged to 46th Regiment April 6, 1815; appointed to 85th Regiment May 25, 1815; half-pay of regiment December 25, 1818; exchanged to 42nd Regiment December 30, 1819; Captain do. September 10, 1825; Major do. October 23, 1835; Brevet Lieut.-Colonel November 1, 1842; Lieut.-Colonel April 14, 1846; retired February 15, 1850; served in Peninsula; silver medal for Nivelle, Nive, Orthes and Toulouse.

M'FADDEN, HUGH.—Ensign 63rd Regiment August 11, 1825; appointed to 85th Regiment June 7, 1827; Lieutenant do. December 30, 1831; retired October 22, 1833; appointed Governor of Longford Gaol.

M'GILLEWIE, JAMES.—Ensign February 18, 1813; Lieutenant October 14, 1814; died at Gibraltar April 12, 1827.

M'GREGOR, PETER.—Lieutenant 7th Garrison Battalion December 18, 1806; appointed to 85th Regiment October 19, 1809; Captain Royal York Rangers November 21, 1811; died March 23, 1815.

MCGRIGOR, JAMES RODERICK DUFF.—Second-Lieutenant September 5, 1877; transferred to Rifle Brigade May 25, 1878.

M'INTOSH, ÆNEAS.—Captain 2nd Battalion 71st Foot (half-pay) on its disbandment 1783; Captain 85th Regiment May 23, 1797; Major do April 7, 1808; Lieut.-Colonel (army) May 30, 1811; Major 79th Foot January 25, 1813.

From a Newspaper Cutting (January 1814).

'At Ardgowan, suddenly, on the 5th inst., Lieut.-Colonel Æneas M'Intosh, of the 79th Regiment, in the 54th year of his age. In him his brother Officers have lost a sincere and valuable friend, and his country an able and gallant defender. He entered the service at the early age of 17, and distinguished himself in various campaigns and expeditions in different quarters of the globe. In the year 1799, in the expedition to the Helder, he was twice wounded. At the attack on Flushing in 1809 he commanded the left wing of the 85th Regiment, and on that occasion his services were so highly appreciated, that he received in general orders the thanks of Lieut.-General Sir Eyre Coote. He also signalized himself in an eminent degree at Fuentes d'Onor, in 1811, under the immediate eye of Field-Marshal Wellington, and was promoted to the rank of Lieut.-Colonel. In the same year he volunteered and led the forlorn hope in the attack of Fort San Christoval.[1] These are but a few of the instances which might be given of the gallantry which shone so eminently conspicuous in this brave Officer. He was interred at Greenock on the 8th inst. with military honours and with every mark of public veneration to his memory, which a community deeply impressed with a conviction of his eminent worth could express. The Magistrates, the officers of the local Militia and the principal gentlemen of Greenock attended; and the solemn procession moved to the place of interment amidst an immense concourse of spectators, who unanimously participated in that mingled feeling of awe and regret which is more easily conceived than described.'

McKAY, WILLIAM ROWLAND.—Cornet 3rd Dragoon Guards May 8, 1867; Lieutenant 80th Foot April 3, 1869; Lieutenant 85th Regiment January 29, 1870; retired July 13, 1870.

MACKENZIE, FREDERICK HUGH.—Second Lieutenant November 16, 1887; Lieutenant January 25, 1896; Army Pay Department.

MACKENZIE, GEORGE DAVIS.—Ensign October 15, 1802; Lieutenant 5th Foot August 5, 1804; Captain do. August 30, 1809; died at Ilfracombe July 26, 1826.

M'KENZIE, JOHN (1).—Captain 103rd Regiment June 30, 1795; appointed to 85th Regiment November 11, 1795; Major do. August 28, 1801; half-pay of the regiment 1802; appointed to 5th Foot, August 1, 1804; Lieut.-Colonel 5th Foot May 8, 1806; killed at Corunna, January 1809.

M'KENZIE, JOHN (2).—Ensign Royal African Corps July 25, 1805; Lieutenant 85th Regiment April 22, 1806; retired 1810.

[1] This is incorrect. Both the forlorn hopes were led by Ensign Dyas.—ED.

MACLEOD, CHARLES MURRAY.—Ensign 78th Regiment March 11, 1813; Lieutenant do. February 17, 1814; appointed to 85th Regiment May 25, 1815; exchanged to half-pay 29th Regiment November 27, 1817.

M'LAUCHLAN, JOHN.—Ensign October 2, 1800; probably never joined regiment.

MCMAHON, KILLERMANN EYRE.—Lieutenant January 27, 1883; Captain February 11, 1890; Major March 8, 1901.

M'NEILL, HENRY.—Lieutenant July 18, 1800; retired Sept. 24, 1802.

MACPHERSON, DUNCAN ALEXANDER ALLAN.—Ensign August 18, 1869; Lieutenant October 20, 1871 to Indian Staff Corps October 28, 1876.

MAINWARING, GEORGE.—Ensign May 12, 1843; retired August 8, 1845.

MAITLAND, PATRICK.—Ensign February 20, 1823; Lieutenant July 7, 1825; Captain unattached April 8, 1826; exchanged to 51st Regiment June 23, 1843; Brevet-Major November 23, 1843; retired June 23, 1843.

MANNERS, HERBERT RUSSELL.—Ensign 37th Regiment August 28, 1838; Lieutenant do. January 7, 1842; Captain do. November 13, 1846; Captain 51st Foot December 19, 1851; Captain 85th Regiment August 10, 1855; Major unattached January 8, 1856; half-pay same date.

MANNERS, THOMAS.—Lieutenant 82nd Regiment June 11, 1794; Captain do. October 10, 1799; appointed to 46th Regiment May 4, 1800; exchanged to half-pay unattached June 15, 1801; appointed to 85th Regiment October 19, 1807; cashiered March 20, 1808.

MARDALL, C. E.—Second-Lieutenant January 14, 1880; Lieutenant July 1, 1881; to Indian Staff Corps.

MARESCAUX, OSCAR HYDE EAST.—Lieutenant August 29, 1885; Adjutant September 19, 1893; Captain October 4, 1893; Brevet-Major May 29, 1901; retired August 30, 1905. *Operations in Sierra Leone*, 1898–9. —Karene Expedition; also in Mendiland Expedition (including Song Town Kwalu Expedition) and Protectorate Expedition. Despatches, *London Gazette*, December 29, 1899; medal with clasp. *South African War*, 1899–1902.—On Staff. Relief of Kimberley. Operations in the Orange Free State February to May 1900, including operations at Paardeberg (February 17 to 26); actions at Poplar Grove and Dreifontein. Operations in Orange River Colony (May to November 29, 1900), including actions at Bothaville and Caledon River (November 27 to 29). Operations in Orange River Colony November 30, 1900, to February 1901, March 1901 to May 1902. Operations in Cape Colony February to March 1901. Despatches, *London Gazette*,

September 10, 1901. Brevet-Major. Queen's medal with four clasps; King's medal with two clasps.

MARSH, WILLIAM.—Ensign January 13, 1814; Lieutenant March 22, 1815; exchanged to half-pay 4th Foot March 21, 1816; died 1826.

MARSHAL, THE. REV. HENRY.—Chaplain November 18, 1793.

MARSHALL, CHARLES.—Lieutenant June 5, 1894; Captain June 6, 1901; retired.

MARSHALL, GEORGE JAMES.—Ensign 4th West India Regiment September 9, 1864; Lieutenant do. March 23, 1866; Lieutenant 85th Regiment February 9, 1870; exchanged to 1st West India Regiment April 27, 1870; served with British Honduras Expedition under Brigadier-General Harley in February and March 1867.

MARTIN, JAMES.—Ensign April 27, 1809; Lieutenant November 21, 1811; appointed to 38th Regiment January 25, 1813; half-pay of it March 25, 1817; died January 5, 1832.

MARTIN, RICHARD BARTH.—Ensign 20th Regiment April 12, 1812; appointed to 85th Regiment May 23, 1822; exchanged to 5th Dragoon Guards December 23, 1824; Lieutenant do. April 9, 1826; retired on half-pay unattached June 15, 1830.

MARTIN, ROBERT FANSHAWE.—Ensign 22nd Regiment June 26, 1823; appointed to 85th Regiment July 17, 1823; Lieutenant do, August 20, 1825; Captain unattached June 10, 1826; appointed to 76th Regiment November 7, 1826; Major do. September 17, 1839; Brevet Lieut.-Colonel September 16, 1845; half-pay of regiment November 11, 1845; died at Bombay July 13, 1846.

MASEFIELD, ROBERT.—Second-Lieutenant June 18, 1892; Lieutenant August 21, 1895; Captain June 9, 1901. *South African War*, 1902.—Operations in the Transvaal March to May 31, 1902. Queen's medal with two clasps.

MASSY, HUGH.—Ensign March 13, 1840; Lieutenant, January 12, 1844; Captain April 13, 1849; Major August 31, 1855.

MATHEW, WILLIAM HENRY W.—Ensign March 11, 1853; Lieutenant March 1, 1855; Captain September 23, 1862; exchanged August 11, 1864.

MAUNSELL, FREDERICK.—Ensign 18th Foot April 16, 1812; Lieutenant do. January 28, 1813; appointed to 85th Regiment March 18, 1813; Captain do. June 24, 1819; Major do. August 14, 1827; Lieut.-Colonel do. May 23, 1836; appointed Inspecting Field Officer of the Cork Recruiting District, June 19, 1846; served in the Peninsula; silver medal for San Sebastian (assault and capture), Nivelle and Nive; served also in North America (slightly wounded at Bladensburg and severely at New Orleans).

GENERAL FREDERICK MAUNSELL.
Colonel of the 85th King's Light Infantry from April 2nd, 1865, to 1876.

MAUNSELL, ROBERT.—Ensign June 24, 1842; Lieutenant March 28, 1845; Captain April 28, 1857.

MAUNSELL, ROBERT GEORGE STONE.—Ensign August 16, 1864; Lieutenant March 6, 1867; exchanged January 5, 1870.

MAUNSELL, WILLIAM WRAY.—Ensign September 1, 1837; Lieutenant November 16, 1841; Captain November 22, 1844; exchanged to 66th Regiment December 30, 1845; retired December 24, 1847; appointed Major of the East Kent Militia November 14, 1857; Lieut.-Colonel Commanding do. May 15, 1864; resigned February 27, 1865.

MAXWELL, JOSEPH.—Ensign March 7, 1811; exchanged to 46th Regiment January 25, 1813; Lieutenant 100th Regiment May 19, 1813; altered to 99th Regiment 1816; Regiment disbanded 1818; half-pay of it September 25, 1818.

MEIN, NICHOLAS ALEXANDER.—Ensign 52nd Regiment May 13, 1797; Lieutenant 12th Foot February 17, 1799; Captain do. June 4, 1801; half-pay of Regiment 1802; reappointed on full pay May 25, 1803; Major do. September 21, 1809; appointed to 43rd Regiment January 25, 1813; Brevet Lieut.-Colonel June 4, 1814; retired September 9, 1819.

MELLIFONT, DAVID.—Cornet 5th Dragoons June 29, 1780; Lieutenant 14th Light Dragoons August 26, 1786; Captain do. August 31, 1790; exchanged to 10th Foot, May 31, 1792; Brevet-Major Jan. 26, 1797; Major 10th Foot, December 27, 1799; Brevet Lieut.-Colonel April 29, 1802; Lieut-Colonel 85th Regiment July 16, 1803; died 1806.

MELLOR, GEORGE FREDERICK.—In ranks Imperial Yeomanry; commissioned Imperial Yeomanry; Second-Lieutenant 85th from Imperial Yeomanry; joined Indian Army as Lieutenant December 19, 1903; Captain Indian Army March 1, 1910; *South African War*, 1901–2. —Served with Imperial Yeomanry. Operations in the Transvaal November 1901 to May 31, 1902. Operations in Orange River Colony. King's medal with two clasps.

MERCER, EDWARD SMYTH.—Major 85th Regiment from 74th Regiment June 3, 1868; retired as Honorary Lieut.-Colonel April 9, 1870.

MEREDITH, CORTLAND, S.K.—Ensign October 19, 1799; Lieutenant May 16, 1800; Captain June 26, 1803; removed from service 1813.

METHOLD, FRANCIS.—Ensign November 10, 1808; died at Feversham, 1810.

MIDDLETON, RICHARD CARTHEW.—Second-Lieutenant 13th Provisional Battalion May 20, 1899; Lieutenant K.S.L.I. July 19, 1900; retired.

MILES, ROBERT PATRICK.—Second-Lieutenant August 12, 1899; Lieutenant January 6, 1901; Captain March 6, 1909; Superintendent Gymnasia, India, October 8, 1907. *South African War*, 1899–1902.—Operations in the Orange Free State February to May 1900, including operations at Paardeberg (February 17 to 26); actions at Poplar Grove, Dreifontein, Houtnek (Thoba Mountain), Vet River (May 5 and 6) and Zand River. Operations in Orange River Colony May to November 29, 1900. Operations in Cape Colony, south of Orange River, 1899–1900. Operations in the Transvaal March 1901 to May 31, 1902; Queen's medal with four clasps; King's medal with two clasps.

MINCHIN, GEORGE.—Ensign 18th Foot November 30, 1791; Lieutenant do. August 11, 1794; Captain 9th West India Regiment February 21, 1799; appointed to 85th Regiment December 25, 1802; died 1804.

MITCHELL, JAMES.—Ensign December 1, 1808; Lieutenant September 13, 1810; appointed to 24th Regiment January 25, 1813; appointed to 5th Garrison Battalion September 16, 1813; half-pay of it February 6, 1815; died at Inch, Queen's County, March 25, 1826.

MOLYNEUX, THE HON. HENRY R.—Ensign May 22, 1817; Lieutenant 39th Regiment October 24, 1821; half-pay of it same date; appointed to 6th Foot January 17, 1822; appointed to 10th Foot May 16, 1822; Captain 2nd Ceylon Regiment May 9, 1823; half-pay of it same year; exchanged to 32nd Regiment September 4, 1823; Major do. August 29, 1826; exchanged to 67th Regiment April 5, 1827; Lieut.-Colonel do. April 9, 1829; exchanged to 60th Regiment April 24, 1835; died in London, May 23, 1841.

MONCKTON, THE HON. CARLETON T.—Cornet 16th Light Dragoons September 17, 1812; Lieutenant do. July 8, 1813; half-pay of it 1814; re-appointed full pay March 30, 1815; exchanged to 1st Life Guards May 1, 1816; exchanged to 18th Hussars April 1, 1819; exchanged to half-pay 22nd Light Dragoons December 2, 1819; exchanged to 85th Regiment May 24, 1821; placed on half-pay of regiment October 1821; appointed to 38th Regiment May 26, 1822; Captain half-pay of 45th Regiment March 27, 1823; exchanged to Cape Corps Infantry April 10, 1823; appointed to 24th Regiment March 18, 1824; died at Quebec May 11, 1830.

MONEY, ERNLE WILLIAM KYRLE.—Lieutenant August 25, 1885; Captain September 13, 1893; Major May 11, 1904; Governor of Colchester Military Prison April 10, 1900; retired March 19, 1910. *Hazara Expedition*, 1888.—As Orderly Officer to Brigadier-General. Medal with clasp. *South African War*, 1899–1900.—Operations in Cape Colony, south of Orange River, 1899–1900. Queen's medal with clasp.

MOODY, GEORGE ROBERT BOYD.—Second-Lieutenant August 21, 1888; Lieutenant November 1, 1891; transferred to Army Service Corps October 1, 1893.

MOORE, STEPHEN GEORGE.—Lieutenant June 11, 1884; Captain January 1, 1892; Major August 19, 1901; Lieut.-Colonel August 19, 1909. *South African War*, 1902.—Operations in the Transvaal March to May 31, 1902. Queen's medal with two clasps.

MORETON, HARCOURT.—Ensign April 21, 1808; Lieutenant November 10 1808; appointed to 49th Regiment January 25, 1813; Captain 14th Foot January 12, 1814; half-pay of regiment March 25, 1816; appointed to 52nd Regiment October 26, 1841; Brevet-Major January 10, 1837; retired October 26, 1841; died June 4, 1854; Peninsula; slightly wounded at Badajos June 1811.

MORLAND, GEORGE.—Ensign 46th Foot February 23, 1855; Lieutenant 12th Foot March 9, 1858; Captain 85th Regiment March 22, 1864; retired July 25, 1865.

MOUNSEY, HEYSHAM ROBERT CECIL.—Lieutenant August 29, 1885; Captain November 30, 1892; retired June 25, 1898.

MUNDY, WILLOUGHBY F. M.—Ensign August 20, 1825; Lieutenant July 5, 1829; died at Nice April 25, 1834.

MUNRO, GEORGE ROSS.—Ensign 78th Regiment July 1, 1797; Lieutenant do. February 21, 1800; Captain 85th Regiment May 8, 1801.

MURRAY, JAMES.—Ensign July 16, 1803; Lieutenant November 16, 1804; exchanged to 2nd West India Regiment January 16, 1806.

MURRAY, PULTENEY HENRY.—From 1st Battalion, Major February 11, 1890; Lieut.-Colonel to command 2nd Battalion February 11, 1894; Colonel April 3, 1890; retired half-pay late Regimental District April 3, 1905. *Egyptian Expedition*, 1882.—Defence of Alexandria, occupation of Kafr Dowar, surrender of Damietta; medal; bronze star.

MUSHET, JOHN HUNTER.—From 37th Regiment; Captain 85th Regiment June 11, 1852; retired same date.

MYTTON, DEVEREUX HERBERT.—Ensign September 21, 1852; Lieutenant December 16, 1853; Captain April 30, 1858; retired October 16, 1867.

NAISH, JOHN.—Ensign August 8, 1799; Lieutenant April 23, 1800; Captain December 25, 1806; died in London February 12, 1812.

NAYLER, CHARLES.—Ensign June 24, 1813; Lieutenant June 2, 1814; retired on half-pay 63rd Regiment May 25, 1835; died April 1, 1856.

NIXON, BRINSLEY.—Ensign May 1801; Lieutenant October 2, 1801; half-pay of Regiment, 1802; reappointed to full pay November, 20, 1802; Captain May 18, 1809; appointed to 37th Regiment January 25, 1813; died November 18, 1813; Peninsula; slightly wounded at Fuentes d'Oñoro, May 5, 1811.

NORTON, WILLIAM.—Ensign August 23, 1799; Lieutenant April 25, 1800.

NOYES, KENNAWAY H. H.—Ensign June 8, 1855; Lieutenant January 15, 1856; Captain July 25, 1865; retired January 25, 1871.

NUGENT, SIR GEORGE, BART, G.C.B.—Ensign 39th Regiment July 5, 1773; Lieutenant 7th Fusiliers November 23, 1775; Captain 57th Regiment April 28, 1777; Major do. May 3, 1782; Lieut.-Colonel 97th Regiment September 8, 1783; half-pay of regiment; Lieut.-Colonel 13th Foot December 26, 1787; appointed to 4th Dragoon Guards June 16, 1789; exchanged to Coldstream Guards October 6, 1790; appointed Colonel of 85th Regiment March 1, 1794; Major-General May 3, 1796; Captain of St. Mawes November 2, 1796; Lieut.-Governor of Jamaica April 1, 1801; Lieut-General September 25, 1803; Colonel of 62nd Regiment December 27, 1805; Colonel of 6th Foot May 26, 1806; General June 4, 1813; Field-Marshal November 9, 1846; died at Westhorpe House, Marlow, March 11, 1849; served in North America in 1777; present at capture of Forts Montgomery and Clinton; campaign in Flanders 1793; siege of Valenciennes, Battle of St. Armand; action at Lincelles.

O'CONNELL, JAMES ROSS.—Lieutenant November 22, 1884; Captain November 3, 1892; Major March 1, 1907; retired March 6, 1909. *Expedition to Dongola*, 1896.—Operations of June 7 and September 19; despatches, *London Gazette*, November 3, 1896; 4th class Medjidie; Egyptian medal with two clasps. *Nile Expedition*, 1898.—Battle of Khartoum; relief of Gederef; despatches, *London Gazette*, September 30, 1898; 4th class of the Osmanieh; two clasps to Egyptian medal; medal. *Nile Expedition*, 1899.—Operations in first advance against Khalifa. Two clasps to Egyptian medal; 3rd class Medjidie.

O'CONNOR, GERALD F. G.—Ensign 73rd Regiment March 22, 1810; Lieutenant do. February 20, 1812; appointed to 85th Regiment April 15, 1813; Captain April 7, 1825; Brevet-Major June 28, 1838; dismissed the service January 17, 1840; served in North America; wounded at Bladensburg.

OGILVY, WALTER.—Ensign July 2, 1841; Lieutenant August 13, 1844; appointed to 49th Regiment January 31, 1845; retired July 7, 1846.

OLDFIELD, CHRISTOPHER CAMPBELL.—Ensign 10th Foot December 30, 1859; Lieutenant 85th Regiment September 11, 1863; Captain do. August 7, 1867; exchanged to 38th Foot March 3, 1873.

O'REILLY, DUVAL KNOX.—Ensign 69th Regiment July 18, 1834; Lieutenant do. October 7, 1836; Captain do. April 14, 1843; exchanged to 85th Regiment May 3, 1844; died at Barbadoes, August 12, 1844.

Orme, William Henry.—Cornet 3rd Light Dragoons January 24, 1845; Lieutenant do. April 3, 1846; Captain do. January 13, 1854; Captain 85th Regiment January 14, 1854; Major do. June 8, 1867; Major half-pay June 12, 1867.

Ormsby, Sir Thomas, Bart.—Ensign July 21, 1813; Captain March 9, 1815; Captain Cape Corps Infantry December 5, 1822; exchanged to 14th Light Dragoons, March 13, 1823; Major unattached December 30, 1824; died on board his yacht at Cowes August 9, 1833; served in North America; slightly wounded at New Orleans.

Orr, John.—Ensign 63rd Regiment July 14, 1808; Lieutenant 8th Garrison Battalion October 13, 1808; appointed to 85th Regiment April 27, 1809; appointed to 89th Regiment January 25, 1813; exchanged to half-pay of the regiment December 25, 1817; appointed to the 47th Regiment April 19, 1844; retired same date.

Osborn, Sir George Robert, Bart.—Ensign August 10, 1832; Lieutenant unattached November 7, 1834; appointed to 45th Regiment November 27, 1835; retired December 4, 1835.

Ottley, Robert William (1).—Ensign 53rd Regiment October 4, 1780; Lieutenant do. February 2, 1791; Captain 85th Regiment November 22, 1793; Captain-Lieutenant do. November 23, 1793; Major do. May 17, 1798; Brevet Lieutenant-Colonel January 1, 1800; severely wounded in Holland August 27, 1799.

Ottley, Robert William (2).—Ensign October 1, 1794; had been Adjutant from August 20, 1794; no further information.

Otway, George.—Ensign 70th Regiment March 31, 1791; Lieutenant 41st Regiment October 31, 1793; Captain do. June 3, 1795; Major 85th Regiment June 4, 1801; died at Stoneyhill Barracks, Jamaica, July 15, 1804.

Pack, Arthur John.—Ensign August 9, 1833; Lieutenant unattached May 5, 1837; exchanged to 7th Fusiliers July 26, 1838; Captain do. June 23, 1843; Brevet-Major June 20, 1854; Major 7th Fusiliers December 22, 1854; Lieutenant-Colonel 7th Fusiliers June 19, 1855.

Parkinson, Chaigneux Colvill.—Ensign May 19, 1857; Lieutenant March 25, 1862; exchanged May 31, 1864.

Parratt, Evelyn Latimer.—Ensign October 4, 1839; Lieutenant March 31, 1843; Captain December 28, 1846; retired 1852.

Patterson, Alexander.—Ensign 41st Regiment, September 23, 1836; appointed to 85th Regiment September 30, 1836; Lieutenant do. May 8, 1840; Captain do. January 13, 1846; retired December 28, 1846.

PAULET, LORD CHARLES (afterwards THE REV.).—Ensign June 24, 1819; retired February 1, 1821; afterwards took Holy Orders and was appointed a Prebendary of Salisbury 1833.

PAULET, LORD WILLIAM.—Ensign February 1, 1821; Lieutenant 7th Fusiliers May 23, 1822; Captain unattached February 12, 1825; exchanged to 85th Regiment April 21, 1825; retired on half-pay December 28, 1826; exchanged to 63rd Regiment March 29, 1827; appointed to 21st Fusiliers December 4, 1828; Major unattached September 10, 1830; exchanged to 68th Regiment January 18, 1833; Lieutenant-Colonel do. April 21, 1843; exchanged to half-pay unattached December 31, 1847; Colonel June 20, 1854; medal for Crimea.

PAYN, WILLIAM ARTHUR.—Second-Lieutenant Argyll and Sutherland Highlanders December 5, 1891; do. 85th Regiment December 16, 1891; Lieutenant do. October 25, 1894; Captain do. January 22, 1901; Major do. March 19, 1910; Staff Captain No. 4 District, Western Command, November 17, 1909. *South African War*, 1901–1902.—Operations in the Transvaal November 1901 to May 31, 1902. Queen's medal with four clasps.

PEARCE, CHARLES.—Lieutenant 35th Regiment October 21, 1795; appointed to 13th Foot January 5, 1799; Captain 85th Regiment May 22, 1800; appointed to 19th Foot April 24, 1801.

PEARSE, JAMES LANGFORD.—Second-Lieutenant 53rd Foot August 11, 1880; Lieutenant Shropshire Light Infantry July 1, 1881; Captain do. December 31, 1881; Major do. May 11, 1898; appointed Lieutenant-Colonel to command February 11, 1906; retired February 11, 1910.

PEEL, ARTHUR LENNOX.—Ensign September 2, 1842; appointed to 52nd Regiment May 12, 1843; Lieutenant do. May 1, 1846; Captain do. August 22, 1851.

PEEL, EDMUND.—Ensign March 15, 1815; Lieutenant March 13, 1817; half-pay of regiment December 25, 1818; exchanged to 49th Regiment February 18, 1819; exchanged to half-pay 25th Regiment May 2, 1822.

PEEL, EDMUND YATES.—Ensign 15th Foot January 31, 1845; appointed to 85th Regiment March 28, 1845; retired on half-pay of regiment May 27, 1856.

PENFORD, GEORGE.—Ensign August 6, 1799; Lieutenant February 21, 1800; no further information.

PERHAM, RICHARD.—Ensign 32nd Regiment November 10, 1807; Lieutenant do. July 27, 1809; exchanged to 85th Regiment September 16, 1809; appointed to 36th Regiment January 25, 1813; half-pay of it December, 1814; exchanged to 45th Regiment May 10, 1815; half-pay of it March, 1817 ; exchanged back to full pay of it May 15, 1817 ; Captain 45th Regiment January 23, 1825 ; died at Arnee, Madras, April 28, 1832.

PHIBBS, WILLIAM HARLOE.—Ensign 60th Regiment October 30, 1810 ; appointed to 27th Regiment March 14, 1811 ; Lieutenant do. June 24, 1813 ; half-pay of it April 25, 1817 ; appointed to 2nd Veteran Battalion January 15, 1824 ; appointed to 89th Regiment January 25, 1825 ; appointed to 25th Regiment May 5, 1825 ; Captain 85th Regiment June 3, 1836 ; exchanged to half-pay unattached May 11, 1838 ; Brevet-Major November 9, 1846 ; Staff Officer of Pensioners at Birr October 1845 ; Brevet Lieutenant-Colonel June 20, 1854 ; served in the American War ; present at the taking of Plattsburg ; lost his arm and was shot through both legs in Peninsula Campaign at Nivelle.

PIGOTT, JOHN.—Ensign 30th Regiment April 25, 1792 ; exchanged to 68th Regiment June 7, 1792 ; Lieutenant 85th Regiment March 5, 1794 ; Captain do. same date ; removed from the service 1801.

PIPON, JAMES KINNARD.—Ensign 94th Regiment August 3, 1826 ; Lieutenant do. December 9, 1828 ; Captain do. March 6, 1835 ; exchanged to 85th Regiment July 15, 1836 ; exchanged to half-pay of 68th Regiment March 31, 1843 ; Brevet-Major November 9, 1846.

PLENDERLEATH, GEORGE.—Ensign 87th Regiment July 26, 1802 ; placed on half-pay of it 1802 ; appointed to 85th Regiment November 24, 1803 ; resigned June, 1804.

POWELL, EDWARD.—Ensign June 24, 1804 ; Lieutenant August 30, 1805 ; appointed to 57th Regiment January 25th 1813 ; half-pay of it March 25, 1817 ; exchanged back to full pay July 24, 1817 ; Captain 9th Foot April 8, 1825 ; exchanged to half-pay unattached January, 1828 ; retired April, 1831.

POWER, MANLEY.—Ensign 68th Regiment June 24, 1819 ; exchanged to 32nd Regiment December 30, 1830 ; Lieutenant 38th Regiment April 17, 1823 ; exchanged to 85th Regiment September 4, 1823 ; Captain unattached June 30, 1825 ; appointed to 32nd Regiment June 8, 1826 ; appointed to 85th regiment December 28, 1826 ; Brevet-Major June 28, 1838 ; Major 85th Regiment August 5, 1842 ; Lieutenant-Colonel May 19, 1846, unattached ; exchanged back to 85th regiment April 13, 1852 ; commanded regiment ; Colonel June 20, 1854 ; seized with paralysis at the Cape, returned to England and died at Bath April 27, 1857.

POYNTZ, ROBERT HUGH.—Second-Lieutenant October 22, 1902 ; Lieutenant May 18, 1907.

PREVOST, GEORGE PHIPPS.—Ensign August 26, 1853 ; Lieutenant 18th Foot January 15, 1855.

PRINCE, WILLIAM HENRY.—Major 85th Regiment July 1, 1859.

PRINCE, PEREGRINE.—Second-Lieutenant January 18, 1902; Lieutenant September 15, 1906. *South African War*, 1902.—Operations in the Transvaal April to May 31, 1902; Queen's medal with four clasps.

PURDON, ROWLAND.—Lieutenant from 81st Foot August 26, 1867; Captain 85th Regiment October 31, 1877; retired 2 February, 1878.

RADCLIFFE, WILLIAM SCOTT WARLEY.—Lieutenant August 25, 1886; Captain June 5, 1894; Brevet-Major November 29, 1900; retired pay September 15, 1906; Hon. Colonel Cheshire Imperial Yeomanry. *South African War*, 1900–1902.—On Staff; also performed duties of Acting Chief Staff Officer, Belfast. Afterwards Commandant Helvetia, and at Dalmanutha. Operations in the Orange Free State February to May 1900, including actions at Vet River (May 5 and 6) and Zand River. Operations in the Transvaal in May and June 1900, including actions near Johannesburg and Pretoria. Operations in the Transvaal, east of Pretoria, July to November 29, 1900. Operations in the Transvaal, west of Pretoria, July to November 29, 1900, including actions at Elands River (August 4 to 16). Operations in Orange River Colony, May to November 29, 1900, including action at Rhenoster River. Operations in the Transvaal November 30, 1900, to May 31, 1902. Despatches, *London Gazette*, September 10, 1901. Queen's medal with three clasps; King's medal with two clasps; Brevet-Major.

RAMSBOTTOM, GEORGE ROBERT SOMERVILLE.—Ensign April 1, 1859; retired February 26, 1864.

RAMSAY, SIR ALEXANDER, BART.—Ensign June 11, 1830; Lieutenant May 30, 1834; retired November 24, 1835.

RAVENHILL, EDWARD HARRY GORING.—Ensign April 25, 1865; Lieutenant July 20, 1867; Captain February 10, 1876; Major September 30, 1882; Lieutenant-Colonel to command 1st Battalion August 19, 1889. *Afghan War*, 1879-1880.—With the Kuram Division in the Yarmusht expedition; medal.

READE, RAYMOND NORTHLAND REVELL, C.B.—Second-Lieutenant January 14, 1880; Lieutenant July 1, 1881; Captain October 19, 1887; Major April 6, 1898; Brevet Lieutenant-Colonel November 29, 1900; Brevet-Colonel February 10, 1904; Colonel February 23, 1907; Deputy Assistant Adjutant-General, Egypt, 1889–1893; A.D.C. to Major-General, Aldershot, 1893–1896; Special Service, Ashanti, 1895–1896; West African Frontier Force 1897–1899; Deputy Assistant Adjutant-General for Instruction, Dublin District, 1899; Special Service, South Africa, 1899; Deputy Assistant Adjutant-General (Intelligence), South Africa, 1899–1901; Commandant Royal Military College, Kingston, Canada, 1901–1905; Assistant Adjutant and Quartermaster-General, Malta, 1907–1910; Brigadier-General in charge of Administration Northern Command 1910. *Afghan War*, 1879–1880.—With the Kuram Division; medal. *Ashanti Expedition*, 1895–1896.—Star. *West Africa*, 1898.—Expedition to Anam; medal with clasp. *South African War*, 1889–1900.—Special Service Officer. Afterwards on Staff. Advance on Kimberley, including actions at Belmont, Enslin, Modder River, and Magersfontein. Operations in the Orange Free State, February to May, 1900. Operations in the Transvaal in May and June, 1900. Operations in the Transvaal, east of Pretoria, July to November 29, 1900, including actions at Lydenberg (September 5 to 8). Operations in the Transvaal, west of Pretoria, July to November 29, 1900, including action at Zilikats Nek. Operations in Cape Colony, north of Orange River, including action at Ruidam. Despatches, *London Gazette*, April 16, 1901. Brevet of Lieutenant-Colonel; Queen's medal with four clasps.

REEVES, BOLES.—Lieutenant October 9, 1855; retired February 4, 1862.

REILLY, JOHN L.—Lieutenant August 10, 1796; Captain December 11, 1799; wounded near Bergen, October 2, 1799.

RICHARDSON, RICHARD.—Ensign August, 1799; Lieutenant December 5, 1799; exchanged to 31st Regiment February 7, 1800; placed on half-pay of the Corsican Regiment; died 1817.

RICKETTS, WILLIAM.—Ensign 37th Regiment November 18, 1795; Lieutenant do. November 2, 1796; exchanged to 85th Regiment February 20, 1800; exchanged to half-pay of Irish Brigade, May 8, 1801.

RIDGWAY, SAMUEL.—From 2nd Lancashire Militia, Captain 85th Regiment December 25, 1813; placed on half-pay of the regiment 1814.

RIVETT-CARNAC, ERNEST HENRY.—Sub-Lieutenant and Lieutenant February 12, 1876; to Bengal Staff Corps July 19, 1880.

ROBERTS, HODDER WILLIAM.—Ensign October 26, 1807; Lieutenant October 25, 1808; exchanged to 32nd Regiment September 16, 1809; half-pay of 8th Garrison Battalion March 22, 1810; retired June 1830.

ROBERTS, WILLIAM.—Ensign 21st Fusiliers October 25, 1792; Lieutenant Independent Company November 11, 1793; Captain do. February 16, 1794; Major late 135th Regiment December 26, 1794; Brevet Lieutenant-Colonel January 1, 1800; appointed to 85th Regiment April 4, 1801.

ROBERTSON, DONALD.—Ensign June 8, 1809; Lieutenant January 7, 1813; appointed to 82nd Regiment January 25, 1813; placed on half-pay of it March 25, 1817; died November 1, 1848.

ROBESON, ARTHUR HEMMING.—Lieutenant August 23, 1884; Captain April 13, 1892; Captain Wiltshire Regiment July 2, 1892; Adjutant 4th Volunteer Battalion Cheshire Regiment May 1, 1893.

ROBINSON, EDWARD F.—Second-Lieutenant November 29, 1905; retired.

ROBINSON, FRANCIS WINGFIELD.—Ensign July 21, 1865; Lieutenant August 17, 1867; Captain June 22, 1876; Major May 15, 1884; Lieutenant-Colonel to command 1st Battalion August 19, 1893. *Afghan War*, 1879–1880.—With the Kuram Division, Yarmusht Expedition and assault of Zawa. Medal.

ROGERS, HENRY.—Ensign March 1800; Lieutenant August 28, 1800; half-pay of the regiment 1802; reappointed to full pay of it October 15, 1802; Captain 2nd West India Regiment May 12, 1805; appointed to 6th Foot March 26, 1807; Brevet-Major August 12, 1819; Major unattached July 20, 1826; exchanged back to 6th Foot June 11, 1829; died in Liverpool December 23, 1831; served in the Peninsula; wounded at Echalar and at Orthes.

ROLLESTON, EDMUND.—Ensign 24th Regiment 1794; Lieutenant 29th Regiment May 6, 1795; appointed to 85th Regiment November 1, 1797; no further information.

ROOPER, HENRY GODOLPHIN.—Ensign April 13, 1849; Lieutenant August 17, 1852; Adjutant October 29, 1852; Captain September 1, 1854; retired October 2, 1855.

ROSS, SIR CHARLES, BART.—Cornet 7th Light Dragoons January 5, 1780; Lieutenant do. April 14, 1783; Captain 3rd Dragoon Guards May 31, 1784; Major 37th Regiment June 13, 1787; Lieutenant-Colonel do. March 16, 1791; Colonel August 21, 1795; appointed to 116th Regiment September 1, 1795; regiment reduced but retained on full pay of it; Major-General June 18, 1798; Colonel-Commandant of 85th Regiment May 9, 1800; half-pay of the regiment 1802; Lieutenant-General October 30, 1805; Colonel of the 85th Regiment December 27, 1805; appointed Colonel of the 86th Regiment October 30, 1806; appointed Colonel of the 37th Regiment June 25, 1816; died at Balnagown Castle, Ross-shire, February 8, 1814.

ROSS, JOHN.—Ensign 25th Regiment November 13, 1779; Lieutenant do. September 1, 1781; Captain do. April 26, 1793; Major 85th Regiment March 5, 1794; Lieutenant-Colonel do. June 15, 1796; wounded at Bergen October 2, 1799.

ROSS, PATRICK W—— SYDENHAM.—Ensign November 9, 1838; Lieutenant September 2, 1842; Captain January 16, 1849; exchanged to 53rd Regiment December 6, 1850.

ROTHWELL, WILLIAM.—Ensign July 10, 1811; appointed to 90th Regiment January 25, 1819; Lieutenant York Chasseurs November 25, 1813; regiment disbanded, placed on half-pay of it December 11, 1819;

appointed to 62nd regiment April 7, 1825; exchanged to half-pay unattached October 12, 1826.

ROWE, FRANCIS RICHARD.—Ensign June 25, 1804; Lieutenant 89th Regiment March 21, 1805; Captain do. December 17, 1812; died March 28, 1813.

ROWLES, HENRY.—Ensign September 9, 1828; Lieutenant August 9, 1833; exchanged to 8th Hussars February 23, 1838; retired June 30, 1843.

RUDKIN, HENRY WILLIAM.—Ensign January 11, 1867; Lieutenant October 16, 1867; Abyssinian Campaign 1868; transport train, medal.

RUNDLE, ARTHUR THOMAS CURGENVEN.—Second-Lieutenant (from Militia) May 29, 1895; Lieutenant March 15, 1899; C.R.I. January 30, 1901 to February 9, 1904; Captain March 5, 1905; Adjutant March 5, 1905, to August 30, 1907; Adjutant Militia August 31, 1907, to July 14, 1908; Adjutant Territorial Force July 15, 1908 to January 4, 1910.

RYCROFT, SIR NELSON, BART.—Ensign October 13, 1848; Lieutenant January 30, 1852; retired 1853; appointed Captain on the Staff June 1, 1854.

ST.QUINTIN, FRANCIS JOHN.—Ensign 22nd Regiment April 8, 1825; Lieutenant unattached January 28, 1826; exchanged to 36th Regiment April 22, 1826; Captain do. November 21, 1828; half-pay 9th Regiment July 9, 1830; exchanged to 85th Regiment February 24, 1832; Brevet-Major November 23, 1841; retired November 22, 1844; died in Canada February 7, 1847.

SADLER, RALPH HENRY HAYES.—Sub-Lieutenant 37th Regiment November 12, 1873; do. 85th Regiment July 11, 1874; resigned February 26, 1877.

SALMON, THOMAS.—Ensign 9th Foot August 26, 1799; Lieutenant 85th Regiment May 21, 1801; half-pay of the regiment 1802; re-appointed to full pay November 25, 1802; Captain 7th Fusiliers March 21, 1805; died January 31, 1811.

SANKEY, B——.—Captain 92nd Regiment September 9, 1795; appointed to 85th Regiment December 19, 1795; no further information.

SCHALCH, VERNON ANSDELL.—Ensign May 29, 1869; Lieutenant October 28, 1871; to Bengal Staff Corps October 28, 1871.

SCHAW, CHARLES.—Ensign 55th Regiment September 10, 1807; Lieutenant do. May 3, 1809; Captain 60th Regiment January 10, 1811; appointed to 85th Regiment January 25, 1813; exchanged to half-pay 37th Regiment October 14, 1819; exchanged to 33rd Regiment December 30, 1824; appointed to 21st Fusiliers July 27, 1832; Brevet-Major July 22, 1830; retired April 24, 1835; died March 5, 1874, aged 80; was for many years Police Magistrate at Richmond, Tasmania.

SCOTT, MICHAEL THOMAS.—Lieutenant 7th Fusiliers July 29, 1795; exchanged to half-pay 60th Regiment February 9, 1797; appointed to 85th Regiment May 31, 1800; appointed to 23rd Fusiliers 1803.

SCOTT, JAMES.—Ensign January 23, 1853; Lieutenant June 6, 1854; retired September 14, 1855.

SETON, CHARLES.—Ensign August 7, 1867; retired August 18, 1869.

SETON, MILES CHARLES.—Ensign 25th Regiment March 30, 1826; appointed to 85th Regiment May 21, 1826; Lieutenant 85th Regiment April 5, 1831; exchanged to half-pay unattached August 17, 1832; appointed to 99th Regiment February 17, 1837; retired March 3, 1837; died September 17, 1877.

SEYMOUR, CONWAY F. CHARLES.—Ensign February 21, 1840; Lieutenant October 27, 1843; Captain November 12, 1848.

SHEEHAN, WILLIAM.—Ensign 69th Regiment January 27, 1808; appointed to 85th Regiment October 20, 1808; Lieutenant October 25, 1806; exchanged to Newfoundland Fencibles February 28, 1811; cashiered 1812.

SHEPHERD, J——.—Lieutenant July 25, 1800; no further information.

SHIPLEY, CHARLES ORBY.—Second-Lieutenant February 11, 1888; resigned April 26, 1890; joined the 4th Battalion Herefordshire Militia and commanded when this Battalion was disbanded in 1908.

SIMPSON, TOM THORPE.—Second-Lieutenant May 20, 1899; died of dysentery at Schoeman's Kloof, South Africa, April 30, 1901.

SITWELL, CLAUDE GEORGE HENRY, D.S.O.—From Militia, Second-Lieutenant September 14, 1878; Lieutenant July 1, 1881; Captain September 13, 1886; Captain Manchester Regiment February 13, 1889; employed with Egyptian Army September 28, 1892; Major Royal Dublin Fusiliers October 19, 1898; employed in Uganda Protectorate May 11, 1895; Lieutenant-Colonel (Brevet) October 4, 1899; died 1901. *Afghan War*, 1879-1880.—With the Kuram Division, Yarmusht Expedition; medal. *Egyptian Expedition*, 1882.—Defence of Alexandria; occupation of Kafr Dowar; surrender of Damietta; medal; bronze star. *East Africa*, 1895-6-7-8.—Expeditions against Kitosh, Kabras and Kikelwa Tribes, 1895; in command; Naudi Expedition 1895-6; mentioned in Despatches. Uganda 1897-1898; commanded operations against Mwanga; action near Katonga River and other engagements; mentioned in despatches. 'D.S.O.'

SITWELL, FREDERICK (1).—Lieutenant (from 11th Dragoons) March 1, 1794.

SITWELL, FREDERICK (2).—Ensign 12th Foot September 8 1846; appointed to 85th Regiment September 9, 1846; Lieutenant do. August 1, 1851; Captain July 15, 1853; exchanged to 3rd Light Dragoons.

SKOULDING, ALDERMAN.—Ensign June 26, 1801; Lieutenant 9th Foot December 18, 1801; placed on half-pay of it 1802; appointed to 36th Regiment November 19, 1807; retired June 29, 1809.

SMITH, GEORGE WASHINGTON.—From 98th Foot; Captain 85th Regiment August 21, 1869; Major do. July 1, 1881; Brigade-Major Meean Meer; served with the 98th Regiment in the Peshawar Expeditionary Force on Euzofzie frontier 1858; heights of Sittana; medal and clasp.

SMITH, HORACE MACKENZIE, D.S.O.—Second-Lieutenant (from Militia) April 9, 1892; Lieutenant September 22, 1894; Captain March 9, 1901; employed with West African Frontier Force July 3, 1898, to October 13, 1899. *South African War*, 1899–1902.—(Employed with Mounted Infantry). Operations in the Orange Free State, February to May, 1900, including action at Poplar Grove (wounded). Operations in the Transvaal, November 30, 1900, to May 31, 1902. Despatches, *London Gazette*, September 10, 1901. Queen's Medal with three clasps; King's medal with two clasps. 'D.S.O.'

SMITH, SIR LIONEL, BART., G.C.B., G.C.H.—Ensign 24th Foot March 1795; Lieutenant do. October 28, 1795; Captain 85th Regiment May 22, 1801; Major do. April 22, 1802; appointed to the 16th Foot June 24, 1802; Lieutenant-Colonel Nova Scotia Fencibles June 6, 1805; appointed to 18th Foot August 1, 1805; appointed to 65th Regiment November 25, 1806; Colonel June 4, 1813; Major-General August 12, 1819; appointed Colonel of the 96th Regiment April 9, 1832; do. 78th Regiment October 10, 1834; Lieutenant-General January 10, 1837; appointed Colonel of 40th Regiment February 9, 1837; appointed Governor of the Mauritius; died there January 2, 1842; sent in charge of the Jamaica Maroons to the Coast of Africa; expedition against Surinam under Sir Charles Green; commanded an expedition to the Persian Gulf against the pirates; wounded by a sabre cut at the cavalry action at Ashta, February 21, 1818.

SMITH, ROBERT ASTLEY.—Lieutenant (from Militia) May 6, 1885; Adjutant December 24, 1890, to September 18, 1893; Captain November 30, 1892; Brevet-Major November 29, 1900; Major August 19, 1905; Lieutenant-Colonel February 11, 1910; Adjutant Militia September 19, 1893 to September 18, 1898. *South African War*, 1899–1902.—Operations in the Orange Free State, February to May, 1900, including operations at Paardeberg (February 17 to 26) (slightly wounded); actions at Poplar Grove, Dreifontein, Houtnek (Thoba Mountain), Vet River (May 5 and 6) and Zand River. Operations in the Transvaal in May and June, 1900, including actions near Johannesburg and Pretoria. Operations in the Transvaal, east of Pretoria, July to November 29, 1900. Operations in the Transvaal, west of Pretoria, July to November 29, 1900, including actions at Elands River (August 4 to 16). Operations in Orange River Colony May to November 29, 1900, including action at Rhenoster River. Operations in Cape Colony, south of Orange River, 1899–1900. Operations in the Transvaal, November 30, 1900, to July, 1901, and April to May 31, 1902. Despatches, *London Gazette*, September 10, 1901; Queen's medal with four clasps; King's medal with two clasps. Brevet of Majority.

SMYTHE, INGOLDSBY WILLIAM THOMAS SOMERSET.—Ensign (from 16th Foot) June 12, 1867; Lieutenant October 28, 1871; Captain September 5, 1882; Major March 21, 1886; Lieutenant-Colonel August 4, 1903; transferred to Army Pay Department April 1, 1904.

SODEN, JAMES NICHOLSON.—Ensign 32nd Regiment November 25, 1795; Lieutenant 85th Regiment July 11, 1800; Captain do. March 20, 1808; exchanged to 24th Regiment January 25, 1813; placed on half-pay of it 1814; died January 27, 1832.

SOWRAY, GERALD RUSSELL.—Second Lieutenant (from Militia) January 17, 1891; Lieutenant January 1, 1894; Captain Manchester Regiment May 26, 1900; Adjutant Hereford Volunteer Rifles April 2, 1901; died at Lichfield. *South African War*, 1899–1900.—Operations in the Orange Free State February to May, 1900, including operations at Paardeberg (February 17 to 26), actions at Poplar Grove and Dreifontein. Operations in Cape Colony, south of Orange River, 1899–1900. Queen's medal with three clasps.

SPENCER, LORD CHARLES.—Possibly intended for Charles Henry Spencer Churchill of the Rifle Brigade and 60th Rifles. The connection of this officer with the 85th, however, is not apparent. To his name in MS. note is affixed 'See "Rifle Brigade."'

SPENCER-SMITH, GILBERT JOSHUA.—Ensign March 24, 1863; Lieutenant July 21, 1865; Captain October 14, 1871; retired August 9, 1867.

SPENS, JAMES, C.B.—Sub-Lieutenant May 29, 1872; Lieutenant May 29, 1872; Captain July 1, 1883; Adjutant February 21, 1884; Major April 7, 1886; Lieut.-Colonel February 11, 1898; Lieut.-Colonel in command (Brevet-Colonel) November 29, 1900; Colonel August 12, 1903; Major-General December 1, 1906; Instructor Royal Military College 1886–1893; Commanding Infantry Brigade (graded Colonel on the Staff), South Africa, January 1901 to May 1901; Brigadier-General Commanding Mobile Column, South Africa, May 1901 to September 1902; Second-Class District, Indian Command (Brigade-Commander), August 2, 1903, to August 11, 1908; General Officer Commanding Lowland Division, Scottish Command, March 21, 1910. *Afghan War*, 1879–1880.—With the Kuram Division, Yarmusht Expedition and assault of Zawa; medal. *South African War*,

1899–1902.—In command 2nd Battalion Shropshire Light Infantry from January 1 to April 30, and from June 5 to November 29, 1900; afterwards on Staff. Operations in the Orange Free State, February to May, 1900, including operations at Paardeberg (February 17 to 26), actions at Poplar Grove, Dreifontein, Houtnek (Thoba Mountain), Vet River (May 5 and 6) and Zand River. Operations in the Transvaal in May and June 1900, including actions near Johannesburg and Pretoria. Operations in Transvaal, east and west of Pretoria, July to November 29, 1900, including actions at Elands River (August 4 to 16). Operations in Orange River Colony, May to November 29, 1900, including action at Rhenoster River. Operations in Cape Colony, south of Orange River, 1899–1900. Operations in the Transvaal and Orange River Colony, November 30, 1900, to May 31, 1902. Operations on the Zululand Frontier of Natal in September and October 1901. Despatches, *London Gazette*, February 8, April 16, and May 7, 1901, and July 29, 1902. A.D.C. to the King, with Brevet of Colonelcy. Queen's medal with four clasps; King's medal with two clasps. 'C.B.'

SPOONER, JOHN.—Lieutenant October 27, 1808; appointed to 21st Fusiliers January 25, 1813; half-pay of regiment January 25, 1816; died February 3, 1834.

SPROT, EDWARD MARK.—Second-Lieutenant November 19, 1892; Lieutenant December 3, 1895; resigned July 19, 1899.

STACE, GEORGE HENRY.—Ensign May 9, 1856; Lieutenant March 4, 1861; exchanged September 11, 1863.

STANNARD, RICHARD.—Ensign 5th West India Regiment December 1800; Lieutenant 4th West India Regiment January 1, 1801; exchanged to 17th Foot April 1801; half-pay of it 1802; exchanged to 85th Regiment November 5, 1803; Captain do. September 26, 1809; appointed to 84th Regiment January 25, 1813; exchanged to half-pay 2nd Garrison Battalion August 3, 1815; died February 5, 1829.

STEPHENSON, ARTHUR THOMAS.—Ensign 1st Foot February 12, 1806; Lieutenant do. March 23, 1807; appointed to 85th Regiment January 25, 1809; placed on half-pay of it June 25, 1817; appointed to 76th Regiment April 8, 1825; Captain unattached April 3, 1827; appointed to 2nd West India Regiment February 14, 1828; died 1828.

STEVENSON, WILLIAM PRINCE.—Ensign July 20, 1867; retired May 29, 1869.

STILL, THOMAS WALKER.—Ensign August 31, 1804; Lieutenant May 13, 1805; resigned May 1, 1808.

STIRLING, CHARLES JAMES ROBERT.—Sub-Lieutenant and Lieutenant September 11, 1876; resigned November 1, 1879.

STORER, ——.—Ensign August 30, 1804; Lieutenant May 12, 1805; resigned April 22, 1806.

STUART, THE HON. JAMES.—Ensign January 11, 1821; Lieutenant April 7, 1825; Captain November 6, 1827; exchanged to half-pay unattached October 18, 1839; died in London December 12, 1840.

STUDDERT, JOHN FITZGERALD.—Ensign May 19, 1846; retired November 5, 1847.

SURTEES, ROBERT LAMBTON.—Embodied Militia 137 days; Lieutenant June 9, 1901. *South African War*, 1899–1902.—Employed with Mounted Infantry; operations in Transvaal November 1900 to May 1902; Queen's medal with clasps.

SUTHERLAND, JOHN.—Ensign February 12, 1801; Lieutenant March 6, 1804; died at Jamaica November 20, 1806.

SWINEY, SHAPLAND WILLIAM.—Ensign May 16, 1811; appointed to 6th Foot January 25, 1813; Lieutenant do. April 29, 1813; exchanged to half-pay 39th Regiment July 11, 1816.

TALBOT, THOMAS.—Lieutenant (from 24th Foot) March 1, 1794.

TAYLOR, BROOK.—Ensign 81st Regiment May 15, 1827; Lieutenant do. June 15, 1830; Captain do. November 28, 1834; appointed to 85th Regiment January 17, 1840; Major do. July 31, 1846; Lieutenant-Colonel do. December 29, 1846.

TAYLOR, HERBERT EDWARD.—Ensign January 22, 1824; Lieutenant September 10, 1825; Captain June 11, 1830; retired October 27, 1843.

TAYLOR, ROBERT KIRKPATRICK.—Ensign July 15, 1853; Lieutenant August 31, 1855; Captain February 21, 1865; retired March 10, 1875.

TAYLOUR, LORD JOHN HENRY.—Ensign October 14, 1849; Lieutenant December 3, 1852; Captain August 31, 1855; Major August 7, 1867; exchanged to 94th Foot June 3, 1868.

TENNANT, GEORGE.—Ensign July 6, 1829; Lieutenant May 9, 1834; Captain October 4, 1839; Major May 19, 1846; retired November 10, 1848; appointed Major 1st Stafford Militia; resigned April 1, 1853.

THISTLETHWAYTE, AUGUSTUS FREDERICK.—Ensign December 29, 1846; exchanged to 26th Regiment February 9, 1849; retired March 15, 1850.

THOMAS, FRANCIS ARTHUR.—Ensign November 29, 1855; Lieutenant April 30, 1858; placed on half-pay February 18, 1859.

THOMPSON, GEORGE.—Ensign February 23, 1844; Lieutenant March 31, 1847; Captain October 29, 1852; Major July 21, 1865; retired Honorary Lieutenant-Colonel June 10, 1874.

THOMPSON, GEORGE ASH.—Ensign March 16, 1815; Lieutenant May 22, 1817; half-pay of regiment December 25, 1818; appointed Paymaster of the Regiment January 27, 1820; retired on half-pay June 6, 1851, with the rank of Major.

THOMPSON, JAMES.—Lieutenant July 19, 1800; appointed to 67th Regiment December 8, 1803.

THOMPSON, M. J.—From 24th Regiment, Second-Lieutenant May 25, 1878; Lieutenant July 1, 1881; to Bengal Staff Corps.

THORNTON, SIR WILLIAM, K.C.B.—Ensign 89th Regiment March 31, 1796; Lieutenant 46th Regiment March 1, 1797; Captain do, June 25, 1803; appointed to 53rd Regiment December 17, 1803; Major Royal York Rangers November 13, 1806; appointed Lieutenant-Colonel and Inspecting Field Officer of Militia in Canada January 28, 1808; appointed to 34th Regiment August 1, 1811; appointed to Duke of York's Greek Light Infantry, January 23, 1812; appointed to 85th Regiment January 25, 1813; Colonel June 4, 1814; Deputy Adjutant-General in Ireland August 12, 1819; exchanged to half-pay of 85th Regiment September 30, 1819; Major-General May 27, 1825; Lieutenant-Governor of Jersey August 8, 1830; Colonel 96th Regiment October 10, 1834; Lieutenant-General June 28, 1838; Colonel of 85th Regiment April 9, 1839; died at Greenford, near Hanwell, Middlesex, March 30, 1840; served in Peninsula; present at San Sebastian and the investment of Bayonne; gold medal for Nive; served in North America; severely wounded at Bladensburg, where being left was taken prisoner, but was released in exchange for Commodore Barney; present at New Orleans, where he was also severely wounded.

THURLOW, THE HON. JOHN EDMUND HOVEL.—Ensign 60th Regiment September 4, 1835; Lieutenant do. March 6, 1841; Captain do. August 2, 1844; exchanged to 85th Regiment May 2, 1845; retired December 3, 1852; appointed Captain in East Kent Militia.

TODD, WILLIAM.—Ensign August 30, 1827; Lieutenant March 2, 1832; Captain August 13, 1844; died at Waterford January 15, 1849.

TOMLIN, EDWARD.—Quartermaster November 18, 1793; Lieutenant August 12, 1795; Captain August 5, 1799.

TOMLIN, ROGER.—Ensign June 24, 1801; half-pay of the Regiment 1802; appointed to the Queen's German Regiment May, 1803; Lieutenant 52nd Regiment November 3, 1803; Captain Maltese Regiment February 13, 1805; appointed to 35th Regiment June 6, 1805; died at Messina July 28, 1806; present at Maida.

TOVIN, LIONEL HENRY.—Second-Lieutenant September 12, 1906; Lieutenant March 6, 1909.

TOWNSHEND, CHARLES VERE FERRER, C.B., D.S.O.—Lieutenant Royal Marines February 1, 1881; Indian Staff Corps January 15, 1886; Captain Indian Staff Corps February 1, 1892; Brevet-Major July 10, 1895; Royal Fusiliers July 28, 1900; Shropshire Light Infantry March 7, 1906; Brevet Lieutenant-Colonel November 18, 1896; Brevet-Colonel February 10, 1904; Substantive Colonel August 2, 1907; employed with Egyptian Army 1896–1898; Special Service South Africa February and March, 1900; Staff of Military Governor, Bloemfontein (graded Assistant Adjutant-General) March to September 1900; Assistant Adjutant-General India August 2, 1907, to April 8, 1909; Officer Commanding District, South Africa, April 9, 1909; temporary Brigadier-General. *Soudan Expedition*, 1884–1885.—Suakin and Nile. Operations round Suakin in 1884. Mentioned in despatches. *Nile Expedition*.—With Royal Marines attached to Guards' Camel Regiment. Actions at Abu Klea and El Gubat, and reconnaissance on Metammeh. Medal with two clasps; bronze star. *Hunza-Nagar Expedition*, 1891–1892.—Taking of Hilt. Despatches, *London Gazette*, June 21, 1892. Medal with clasp. *Operations in Chitral*, 1895.—Commanded the Garrison of Chitral during the siege of the Fort. Thanked by Government of India. Despatches, *London Gazette*, July 16, 1895. Brevet of Majority. Medal with clasp. 'C.B.' *Expedition to Dongola*, 1896.—Operations of June 7 and September 19. Despatches, *London Gazette*, November 3, 1896; Brevet of Lieutenant-Colonelcy. Egyptian medal with two clasps. *Nile Expedition*, 1897.—Clasp to Egyptian medal. *Nile Expedition*, 1898.—Battles of the Atbara and Khartoum. Despatches, *London Gazette*, May 24 and September 30, 1898. Two clasps to Egyptian medal. Medal. 'D.S.O.' *South African War*, 1899–1900.—Special Service Officer; afterwards on Staff. Operations in the Orange Free State February to May 1900. Operations in Orange River Colony May to September 1900. Operations in Cape Colony, 1900. Queen's medal with two clasps.

TRAVERS, ARTHUR.—Ensign May 17, 1794; Lieutenant December 30, 1795; removed from the service 1801.

TRAVERS, JOSEPH OATES.—Ensign September 19, 1795; appointed to the Rifle Corps March 3, 1803; retired 1804; appointed Barrack-Master at Portsmouth, with rank of Major, May, 1818; severely wounded in the action near Helder August 27, 1799; present at the defence of the British Lines September 10; present at the battles of September 19, October 2, 6, and 10; wounded at Bergen October 2.

TRAVERS, ROBERT.—Ensign February 14, 1800; Lieutenant June 24, 1801; placed on half-pay of Regiment 1802; appointed to 55th Regiment November 20, 1802.

TYNTE, CHARLES KEMEYS K.—Ensign November 10, 1838; appointed to 11th Hassars, March 13, 1840; Lieutenant do. April 16, 1842; Captain do. September 26, 1845; exchanged to Grenadier Guards June 9, 1846; exchanged to half-pay 5th Foot November 8, 1850; appointed Colonel of the 1st Somerset Militia October 20, 1857.

URQUHART, BEAU COLCLOUGH.—Ensign 6th Foot April 17, 1811; Lieutenant 85th Regiment May 6, 1813; exchanged to half-pay 71st Regiment November 6, 1817; retired December 1829; died October 5, 1861; severely wounded at New Orleans.

URQUHART, JAMES.—Ensign 49th Regiment; Lieutenant do. February 2, 1796; Captain 85th Regiment May 23, 1801; died at Port Royal, Jamaica, August, 1802. Served in Holland, where he was twice wounded, and also in the Expedition to Copenhagen.

URQUHART, WILLIAM HENRY.—Ensign December 7, 1855; Lieutenant January 24, 1860.

VANCE, JAMES YOUNG.—Ensign 40th Regiment April 21, 1837; Lieutenant do. June 5, 1840; exchanged to 85th Regiment October 29, 1841; retired October 20, 1843; served in Scinde with the 40th Regiment.

VANDELEUR, ROBERT.—Ensign March 22, 1815; Lieutenant 18th Hussars October 19, 1815.

VANDELEUR, THOMAS PAKENHAM.—Ensign April 27, 1815; Lieutenant December 5, 1822; half-pay of the regiment; exchanged to 27th Regiment May 22, 1823; Captain do. December 16, 1824; appointed to 21st Fusiliers April 7, 1825; exchanged to half-pay unattached January 15, 1829; exchanged to 3rd West India Regiment June 25, 1847; Brevet-Major June 28, 1838; retired June 25, 1847; died February 25, 1877.

VASSAR-SMITH, CHARLES MARTIN.—Second-Lieutenant May 5, 1901; Lieutenant May 6, 1905; Captain March 19, 1910. *South African War*, 1901–1902.—Operations in the Transvaal August 1901 to May 31, 1902; Queen's medal with four clasps.

VAUGHAN, THOMAS WRIGHT.—No information.

VEITCH, WILLIAM.—Ensign January 30, 1812; appointed to 48th Regiment January 25, 1813; exchanged to half-pay of 58th Regiment May 30, 1816; died March 25, 1867; served in Peninsula; silver medal for Nive and Toulouse.

VICE, CHARLES FREDERICK.—Ensign December 14, 1870; Lieutenant November 1, 1871; exchanged to Bengal Staff Corps January 5, 1876.

VIVIAN, JOHN HAINES.—Ensign August 9, 1867; Lieutenant October 14, 1871; Captain November 28, 1880; transferred to 90th Light Infantry November 28, 1880.

WALKER, WILLIAM.—Ensign 49th Regiment July 6, 1809; Lieutenant do. June 13, 1811; exchanged to 85th Regiment January 25, 1813; Captain 85th Regiment March 22, 1815; exchanged to half-pay of 24th Regiment June 15, 1815; died September 8, 1867.

WALLER, *see* ARMSTRONG, A. S.

WALLSCOURT, JOSEPH, LORD.—Ensign January 28, 1813; Lieutenant February 24, 1814; exchanged to half-pay West India Rangers March 25, 1819; exchanged to 18th Regiment December 19, 1822; appointed to 98th Regiment March 25, 1824; exchanged to half-pay 52nd Regiment May 13, 1824; exchanged to 62nd Regiment July 10, 1828; Captain unattached June 17, 1829; appointed to 33rd Regiment July 27, 1832; retired August 10, 1832; died May 28, 1849.

WALTERS, ROBERT HOLE.—Ensign October 20, 1843; Lieutenant May 19, 1846; appointed to Cape Mounted Rifles April 1, 1847; appointed to 73rd Regiment December 22, 1848; Captain 73rd Regiment November 14, 1851; exchanged to 31st Regiment July 15, 1853; died 1854.

WARBURTON, AUGUSTUS.—Ensign 4th Foot August 28, 1799; Lieutenant do. February 27, 1800; Captain 60th Regiment December 12, 1801; appointed to 57th Regiment May 25, 1803; Major 91st Regiment December 18, 1806; Lieutenant-Colonel and Inspecting Field Officer of Militia in Canada August 1, 1811; appointed to 41st Regiment September 3, 1813; appointed to 85th Regiment May 11, 1815; placed on half-pay of the regiment December 25, 1818; exchanged back to full pay September 30, 1819; Colonel May 27, 1825; died at Cheltenham May 22, 1836.

WARBURTON, CHARLES.—Ensign March 31, 1843; Lieutenant January 31, 1846; Captain December 12, 1851; retired January 30, 1852; died August 15, 1859.

WARD, RICHARD OCTAVIUS.—Ensign June 10, 1826; Lieutenant February 12, 1828, unattached; exchanged to 12th Foot February 14, 1828; exchanged to 10th Hussars December 10, 1829; Captain 10th Hussars August 30, 1833; retired May 16, 1834; died at Nice March 30, 1844.

WARD, THOMAS HOBBINS.—Cornet 11th Hussars September 10, 1841; Lieutenant do. July 8, 1842; Captain do. July 10, 1846; appointed to 85th Regiment October 27, 1848; retired December 16, 1853; died July 11, 1874, aged 53.

WARDE, GEORGE.—Ensign November 22, 1844; Lieutenant January 13, 1846; Captain December 17, 1852; exchanged to 51st Regiment.

WARDE, LAMBERT HOULTON.—Ensign March 9, 1855; Lieutenant September 14, 1855; appointed to 2nd Life Guards February, 1861.

WATSON, BENJAMIN AIREY.—Ensign 38th Regiment July 20, 1797; Lieutenant do. March 18, 1798; exchanged to 85th Regiment August 27, 1803; Captain do. September 21, 1809; appointed to 53rd Regiment January 25, 1813; exchanged to 4th West India Regiment July 8, 1813; regiment disbanded at Sierra Leone April 24, 1819; placed on half-pay of it; retired April, 1826.

WATSON, STEPHEN.—Sub-Lieutenant 50th Foot August 3, 1872; do. 85th Regiment November 23, 1872; Lieutenant do. November 23, 1873; transferred to 59th Foot November 14, 1874.

WATTON, JOHN BUDDER.—Cornet 2nd Dragoon Guards June 1, 1796; Lieutenant do. September 6, 1798; exchanged to 85th Regiment November 14, 1798; no further information.

WATTS, JOHN.—Ensign 21st Fusiliers February 1, 1810; Lieutenant do. November 12, 1812; appointed to 85th Regiment January 25, 1813; Captain do. November 6, 1823; exchanged to half-pay unattached May 8, 1828; exchanged to 72nd Regiment June 14, 1831; retired June 21, 1831; appointed a Military Knight of Windsor; died October 2, 1873.

WELBY, ALFRED CHOLMELEY EARLE.—Ensign October 16, 1867; transferred to 56th Regiment January 8, 1868.

WELLINGS, GEORGE.—Ensign 72nd Regiment February 14, 1811; Lieutenant 57th Regiment same date; appointed to 85th Regiment January 25, 1813; Captain do. March 23, 1815; exchanged to half-pay unattached June 25, 1825; died May 18, 1861; served in Peninsula; silver medal for San Sebastian (assault and capture) and Nive; served in North America, wounded at Baltimore and on the approach to New Orleans.

WELMAN, WELLESLEY.—Ensign from 100th Foot August 7, 1867; Lieutenant October 7, 1871; Captain June 12, 1878; retired July 27, 1881.

WHARTON, WILLIAM.—Ensign 5th Foot December 4, 1806; Lieutenant 7th Garrison Battalion December 11, 1806; appointed to 35th Regiment July 28, 1807; appointed to 85th Regiment October 6, 1808; Captain 73rd Regiment August 13, 1812; placed on half-pay of Regiment June 25, 1817; appointed Sub-Inspector of Militia in Ionian Islands December 25, 1817; exchanged to half-pay of 43rd Regiment June 1, 1820; died February 5, 1855; served in Walcheren Expedition of 1809; present at siege of Flushing; served in Peninsula; present at siege of Badajos; silver medal for Fuentes d'Oñoro; served in the Expedition to Stralsund and Hanover, 1813, under Major-General Gibbs; Waterloo Campaign; severely wounded at Waterloo.

WHISH, CHARLES FRANCIS DIXON.—Ensign 50th Foot September 20, 1864; Lieutenant do. December 2, 1868; Captain 85th Regiment January 5, 1870; retired February 10, 1876.

WHITAKER, RICHARD.—Ensign 34th Regiment May 4, 1805; Lieutenant 4th Garrison Battalion December 24, 1806; exchanged to 85th Regiment September 3, 1808; superseded February, 1809.

WHITE, CHARLES.—Ensign June 14, 1794; no further information.

WHITE, FINCH.—Ensign April 30, 1855; Lieutenant November 16, 1855; Captain July 21, 1865; Major December 12, 1877; Lieutenant-Colonel September 3, 1882; retired Honorary Colonel May 20, 1885.

WHITE, JOHN JOSEPH.—Second-Lieutenant February 5, 1887; Lieutenant February 11, 1890; Captain June 8, 1898; retired. *South African War*, 1899–1900.—Employed with Mounted Infantry. Relief of Kimberley. Operations in the Orange Free State, February to May, 1900, including operations at Paardeberg (February 17 to 26), actions at Poplar Grove, Vet River (May 5 and 6) and Zand River. Operations in the Transvaal in May and June, 1900, including actions near Johannesburg, Pretoria and Diamond Hill (June 11 and 12). Operations in the Transvaal, east of Pretoria, July to October, 1900, including action at Belfast (August 26 and 27). Despatches, *London Gazette*, September 10, 1901. Queen's medal with seven clasps.

WHITE, RICHARD.—Ensign April 6, 1809; Lieutenant July 10, 1811; died at Hythe January 5, 1813.

WHYTE, GEORGE WARBURTON.—Ensign December 10, 1795; Lieutenant 31st Regiment October 8, 1799; exchanged back to 85th Regiment February 7, 1800; Captain do. May 11, 1805; appointed to 40th Regiment January 25, 1813; Major 4th Garrison Battalion September 2, 1814; placed on half-pay of it December 1814; died 1820.

WILBRAHAM, H—— V——.—From 69th Regiment, Second-Lieutenant March 2, 1878; Lieutenant February 26, 1881; Captain September 1, 1886; Major February 11, 1898; retired May 11, 1898; Lieutenant-Colonel, Reserve of Officers. *Afghan War*, 1879–1880.—With the Kuram Division; medal.

WILKIN, SIR GEORGE, C.B., K.H.—Ensign 82nd Regiment September 14, 1794; Lieutenant January 7, 1795; appointed to 31st Regiment September 10, 1795; Captain M'Donnell's Regiment April 30, 1795; corps reduced, but retained on full pay of it; placed on half-pay of it 1798; appointed Captain to 85th Regiment May 18, 1800; Brevet-Major April 25, 1808; Major May 10, 1808, 95th Rifles; Brevet Lieutenant-Colonel June 4, 1814; retired December 23, 1819. Served in the Peninsula; gold medal for Salamanca; silver medal for Vittoria and Pyrenees.

WILKINSON, CLEMENT ARTHUR.—Second-Lieutenant December 23, 1893; Lieutenant January 8, 1898; Captain January 22, 1902; Brevet-Major August 22, 1902; Adjutant Special Reserve March 29, 1909. *West Africa*, 1897–1898.—Lagos. Employed in Hinterland. Medal with clasp. *West Africa (Northern Nigeria)*, 1900.—Munshi Expedition; clasp. *West Africa*, 1900.—Operations in Ashanti. Despatches, *London Gazette*, December 4, 1900; medal with clasp. *South African War*, 1901–1902.—Employed with Mounted Infantry. Despatches, *London Gazette*, April 25, 1902; Queen's medal with five clasps; Brevet of Majority.

WILKINSON, THOMAS.—Ensign 4th Garrison Battalion September 1, 1808; Lieutenant 43rd Regiment January 18, 1810; appointed to 85th Regiment January 25, 1813; Captain do. October 13, 1814; killed at New Orleans January 1815.

WILLIAMS, SIR WILLIAM FREKE, K.H.—Ensign African Corps August 30, 1810; Lieutenant do. June 10, 1811; appointed to 85th Regiment January 25, 1813; Captain do. October 13, 1814; Major do. April 6, 1825; exchanged to half-pay unattached April 14, 1825; appointed to a particular service January 1, 1838; Brevet Lieutenant-Colonel June 28, 1838; died December 12, 1860, aged 68. Served in Peninsula; silver medal for San Sebastian (assault and capture), Nivelle and Nive; served in North America; severely wounded at Bladensburg.

WILLIAMSON, WILLIAM.—Ensign November 16, 1841; Lieutenant November 22, 1844; Captain August 1, 1851; Lieutenant-Colonel October 10, 1865; died February 25, 1873.

WILMER, HENRY CHUDLEIGH.—Ensign March 9, 1860; Lieutenant August 16, 1864; retired June 8, 1867.

WILMOT, EARDLEY.—Ensign 35th Regiment March 11, 1819; Lieutenant 7th Fusiliers July 11, 1822; Captain unattached May 12, 1826; exchanged to 85th Regiment June 25, 1826; Major unattached June 21, 1831; Brevet Lieutenant-Colonel November 9, 1846.

WILSON, STEPHEN K. H.—Ensign August 3, 1855; Lieutenant April 28, 1857; exchanged June 28, 1864.

WINTERSCALE, CYRIL FRANCIS BARONMEAN.—Second-Lieutenant (from Militia) January 29, 1902; Lieutenant September 22, 1906. *South African War*, 1901–1902.—Operations in the Transvaal, November, 1901, to May 31, 1902; Queen's medal with five clasps.

WOOD, RICHARD.—Ensign 31st Regiment December 2, 1813; placed on half-pay of it 1814; appointed to 85th Regiment May 8, 1834; retired May 16, 1834.

WOOD, SIR WILLIAM, K.C.B., K.H.—Ensign 14th Foot January 27, 1797; Lieutenant do. December 27, 1797; Captain do. December 3, 1802; Major do. May 14, 1807; Lieutenant-Colonel 85th Regiment April 8, 1813; retired on half-pay of 41st Regiment May 11, 1815; Colonel July 22, 1830; Major-General November 23, 1814; appointed

Colonel of 3rd West India Regiment February 8, 1849. Served in North America; severely wounded at Bladensburg; had served in Peninsula with 14th Regiment; silver medal for Corunna.

WOODS, GEORGE JOHN.—Ensign August 14, 1867; exchanged to 81st Foot August 26, 1867.

WORSLEY, HENRY, C.B.—Ensign 5th Foot August 13, 1799; Lieutenant 96th Regiment April 24, 1800; Captain do. September 7, 1804; appointed to 89th Regiment December 8, 1804; appointed to 85th Regiment August 9, 1806; Major 4th Garrison Battalion June 13, 1811; appointed to 34th Regiment January 23, 1812; Brevet Lieutenant-Colonel June 21, 1813; appointed Captain of Yarmouth Castle, Isle of Wight, 1818; died at Newport, Isle of Wight, May 13, 1820; carried one of the Colours of the 5th Foot at Bergen September 19, 1799; served in Walcheren Expedition 1809; present at the siege of Flushing; served in Peninsula; present at Fuentes d'Oñoro, Badajos and Vittoria; had his horse shot under him in the affair at the Pass of Donna Maria July 31, 1813; gold medal for Pyrenees, Nivelle, Nive, and Orthes.

WRENCH, FRANK EVELYN NEVILL.—Second-Lieutenant (from Militia) April 21, 1900; Lieutenant October 4, 1901; half-pay. *South African War*, 1899–1902.—Operations in the Transvaal, east of Pretoria, July to November 29, 1900. Operations in the Transvaal November 30, 1900, to May 31, 1902. Queen's medal with three clasps; King's medal with two clasps.

WRIGHT, CHARLES.—Gazetted Ensign 1805, but apparently did not join the regiment.

WRIGHT, WALTER CECIL.—Second-Lieutenant May 18, 1892; Lieutenant April 24, 1895; transferred to Northumberland Fusiliers. *South African War*, 1899–1902.—Operations in the Orange Free State, February to May, 1900, including operations at Paardeberg (February 17 to 26); actions at Houtnek (Thoba Mountain), Vet River (May 5 and 6) and Zand River. Operations in the Transvaal in May and June, 1900, including actions near Johannesburg and Pretoria. Operations in the Transvaal, west of Pretoria, May to November, 1900. Operations in Orange River Colony, May to November 29, 1900, including action at Rhenoster River. Operations in Cape Colony, south of Orange River, 1899–1900. Operations in the Transvaal November 30, 1900, to August, 1901. Operations in Orange River Colony August, 1901, to May 31, 1902. Despatches, *London Gazette*, July 29, 1902. Queen's medal with three clasps; King's medal with two clasps.

WYNNE, JOHN.—Lieutenant 85th Regiment May 25, 1795; appointed to 53rd Regiment September 16, 1795; exchanged back to 85th

CAPTAIN ROBERT HENRY WYNYARD, C.B.
From a Painting in the possession of the Officers' Mess.

Regiment October 23, 1800; placed on half-pay of Regiment 1802; appointed to 44th Regiment July 9, 1803; Captain 63rd Regiment August 3, 1804; Brevet-Major June 4, 1814; retired November 18, 1819.

WYNYARD, HENRY B. JENNER.—Ensign April 9, 1825; Lieutenant May 21, 1826; Captain 89th Regiment May 21, 1841; exchanged to half-pay unattached January 23, 1846; appointed Adjutant of the North Mayo Militia June 23, 1846.

WYNYARD, ROBERT HENRY, C.B.—Ensign February 25, 1819; Lieutenant July 17, 1823; Captain unattached May 20, 1826; appointed to 85th Regiment June 8, 1826; Major do. July 25, 1841; Lieutenant-Colonel do. December 30, 1842.

WYSE, JAMES FRANCIS.—Lieutenant Irish Brigade October 1, 1794; placed on half-pay of it 1798; appointed to 85th Regiment May 10, 1799; Captain do. March 7, 1800.

YORKE, GRANTHAM MUNTON.—Ensign 52nd Regiment August 10, 1826; Lieutenant 94th Regiment January 15, 1829; exchanged to 85th Regiment January 29, 1829; placed on half-pay unattached March 30, 1832; retired June 1833; took Holy Orders subsequently.

YOUNG, B.S.—No information.

YOUNG, JAMES BROWN.—Ensign 26th Regiment October 15, 1847; exchanged to 85th Regiment February 9, 1849; exchanged to 51st Regiment.

APPENDIX IV

COLONELS

NUGENT, SIR GEORGE, BART., G.C.B.—Ensign July 5, 1773, 39th Regiment; Lieutenant November 23, 1775, 7th Fusiliers; Captain April 28, 57th Regiment; Major May 3, 1782, 57th Regiment; Lieutenant-Colonel September 8, 1783, 97th Regiment; placed on half-pay of the Regiment (wanting); appointed to 13th Foot, December 26, 1787; appointed to 4th Dragoon Guards June 16, 1789; exchanged to Coldstream Guards October 6, 1790; appointed Colonel of the 85th March 1, 1794; Major-General May 3, 1796; appointed Captain of St. Mawes, November 2; appointed Lieutenant-Governor of Jamaica April 1, 1801; Lieutenant-General September 25, 1803; appointed Colonel of the 62nd Regiment December 27, 1805; appointed Colonel of the 6th Foot May 26, 1806; appointed General June 4, 1813; Field-Marshal November 9, 1846; died at Westhorpe House, Marlow, March 11, 1849. Served in North America in 1777; present at the capture of Forts Montgomery and Clinton; campaign in Flanders, 1793; present at the siege of Valenciennes, battle of St. Armand, and action at Lincelles.

ROSS, SIR CHARLES, BART.—Cornet January 5, 1780, 7th Light Dragoons; Lieutenant April 14, 1783, 7th Light Dragoons; Captain May 31, 1784, 3rd Dragoon Guards; Major June 13, 1787, 37th Regiment; Lieutenant-Colonel March 16, 1791, 37th Regiment; Colonel August 21, 1795; appointed to 116th Regiment September 1, 1795; regiment reduced, but retained on full pay of it; Major-General June 18, 1798; appointed Colonel Commanding of 85th May 9, 1800; placed on half-pay of the regiment 1802; Lieutenant-General October 30, 1805; appointed Colonel of the 85th December 27, 1805; appointed Colonel of the 68th Regiment October 30, 1806; appointed Colonel of the 37th Regiment June 25, 1810; died at Balnagown Castle, Ross-shire, February 8, 1814.

ASGILL, SIR CHARLES, BART., G.C.H.—Ensign February 27, 1778, 1st Foot Guards; Lieutenant and Captain February 3, 1781, 1st Foot Guards; Captain and Lieutenant-Colonel March 3, 1790, 1st Foot Guards;

Colonel February 26, 1795; Major-General January 1, 1798; appointed Colonel Commanding of the 46th Regiment May 9, 1800; Lieutenant-General January 1, 1805; appointed Colonel of the 5th West India Regiment February 8, 1806; appointed Colonel of the 85th October 30, 1806; appointed Colonel of the 11th Foot February 25, 1807; General June 4, 1814; died in London, July, 1832. Taken prisoner at the siege of York Town, Virginia, in October, 1781; one of the thirteen Captains who were ordered by General Washington to draw lots who should suffer death, when it fell on him, and for six months he daily expected his execution, but was released by an act of Congress, passed at the intercession of France; returned to England on parole, and shortly afterwards went to France to thank the Queen for having saved his life; served in the campaign in Flanders and was present during the retreat through Holland in 1794. See *United Service Magazine* for November, 1834, p. 317.

STANWIX, THOMAS SLAUGHTER.—Ensign September 26, 1766, Coldstream Guards; Lieutenant and Captain June 6, 1773, Coldstream Guards; Captain and Lieutenant-Colonel February 27, 1779, Coldstream Guards; Colonel and Aide-de-Camp to the King April 9, 1789; Major-General December 20, 1793; appointed Major Coldstream Guards April 1, 1795; appointed Lieutenant-Colonel Coldstream Guards December 2, 1795; Lieutenant-General June 26, 1799; appointed Colonel of the 5th Battalion of the 60th Regiment May 9, 1806; appointed Colonel of the 97th Regiment September 8, 1806; appointed Colonel of the 85th Regiment February 25, 1807; General April 25, 1808; died at Sunbury, Middlesex, November 25, 1815.

GORDON, SIR JAMES WILLOUGHBY, BART., G.C.B., G.C.H.—Ensign October 17, 1783, 66th Regiment; Lieutenant March 5, 1789, 66th Regiment; Captain September 2, 1795, 66th Regiment; Major November 9, 1797, 66th Regiment; Lieutenant-Colonel May 21, 1801, 85th Regiment; appointed Deputy Barrackmaster-General, 1803; appointed to 92nd Regiment August 4, 1804; appointed Lieutenant-Colonel Commanding of the Royal African Corps, June 13, 1808; appointed Colonel of the same July 25, 1810; appointed Quartermaster-General 1811; Major-General June 4, 1813; appointed Colonel of the 85th Regiment November 27, 1813; appointed Colonel of the 23rd Fusiliers April 23, 1823; Lieutenant-General May 27, 1825; died in London January 4, 1851. Present as a volunteer at the siege of Toulon, 1793; served in the Peninsula; present at the capture of Madrid; at Burgos.

TAYLOR, SIR HERBERT, G.C.B., G.C.H.—Cornet March 25, 1794, 2nd Dragoon Guards; Lieutenant July 17, 1794, 2nd Dragoon Guards; Captain May 6, 1795, 2nd Dragoon Guards; Major January 22, 1801, 2nd Dragoon Guards; Lieutenant-Colonel December 26, 1801, 9th West

India Regiment; placed on half-pay of the regiment March 18, 1803; appointed to Coldstream Guards May 26, 1803; Colonel July 25, 1810; Major-General June 4, 1813; appointed Colonel of the 85th April 23, 1823; Lieutenant-General May 27, 1825; died at Rome March 20, 1839. Served as a volunteer in 1793 at the sieges of Valenciennes and Dunkirk; after obtaining his commission in 1794 was present at the battles of the 17th, 22nd, and 26th of April, near Cateau; and on May 10, 22, and 27, near Tournay; present at the actions in Holland on September 19 and October 2 and 6, 1799.

THORNTON, SIR WILLIAM, K.C.B.—Ensign March 31, 1796, 89th Regiment; Lieutenant March 1, 1797, 46th Regiment; Captain June 25, 1803, 46th Regiment; appointed to 53rd Regiment December 17, 1803; Major November 13, 1806, Royal York Rangers; appointed Lieutenant-Colonel and Inspecting Field Officer of Militia in Canada January 28, 1808; appointed to 34th Regiment August 1, 1811; appointed to the Duke of York's Greek Light Infantry January 23, 1812; appointed to 85th Regiment January 25, 1813; Colonel June 4, 1814; appointed Deputy Adjutant-General in Ireland August 12, 1819; exchanged to half-pay of the regiment September 30, 1819; Major-General May 27, 1825; appointed Lieutenant-Governor of Jersey August 8, 1830; appointed Colonel of the 96th Regiment October 10, 1834; Lieutenant-General June 28, 1838; appointed Colonel of the 85th Regiment April 9, 1839; died at Greenford, near Hanwell, Middlesex, March 30, 1840. Gold medal for Nive; severely wounded at Bladensburg and at New Orleans.

FITZGERALD, SIR JOHN FORSTER, K.C.B.—Ensign October 29, 1793, Independent Company; Lieutenant January 31, 1794, Independent Company; Captain May 9, 1794, 79th Regiment; placed on half-pay of it 1794; appointed to 46th Regiment October 31, 1800; placed on half-pay of the regiment 1802; appointed to New Brunswick Fencibles July 9, 1803; Brevet-Major September 25, 1803; Major November 9, 1809, 60th Regiment; Brevet Lieutenant-Colonel July 25, 1810; Lieutenant-Colonel August 12, 1819; Colonel August 9, 1823, 60th Regiment; exchanged to 20th Regiment February 5, 1824; Major-General July 22, 1830; Lieutenant-General November 23, 1841; appointed Colonel of the 85th Regiment April 4, 1840; appointed Colonel of the 62nd Regiment November 21, 1843; appointed Colonel of the 18th Regiment March 9, 1850. Gold medal for Badajos (assault and capture), Salamanca, Vittoria and Pyrenees.

PEARSON, SIR THOMAS, C.B., K.C.H.—Ensign October 2, 1796, 23rd Fusiliers; Lieutenant April 25, 1719, 23rd Fusiliers; Captain August 7, 1800, 23rd Fusiliers; Major December 8, 1804, 23rd Fusiliers; Brevet Lieutenant-Colonel June 4, 1811; appointed Lieutenant-Colonel

GENERAL SIR H. P. DE BATHE, BART.
Colonel of the 85th King's Light Infantry from April 25, 1880, to January, 1907.

and Inspecting Field Officer of Militia in Canada February 28, 1812; appointed to 43rd Regiment November 16, 1815; placed on half-pay of the regiment February, 1817; exchanged to 23rd Fusiliers July 24, 1817; Colonel July 19, 1821; Major-General July 22, 1830; Lieutenant-General November 23, 1841; Colonel of the 85th Regiment November 21, 1843; died in Bath May 20, 1847. Gold medal for Albuhera and Chrystler's Farm, America. Expedition to Ostend, 1798; the Helder, 1799; Expedition to Ferrol, 1800; Egyptian Campaign of 1801, severely wounded at Aboukir; siege and capture of Copenhagen, 1807; Expedition to Martinique, 1809, wounded; severely wounded at the siege of Olivenca; American Expedition, horse shot under him at Chrystler's Farm; severely wounded at the siege of Fort Erie.

GUISE, SIR JOHN WRIGHT, BART., K.C.B.—Ensign November 4, 1794, 70th Regiment; appointed to 3rd Foot Guards March 4, 1795; Lieutenant and Captain October 25, 1795, 3rd Foot Guards; Captain and Lieutenant-Colonel July 25, 1805, 3rd Foot Guards; Colonel June 4, 1813; appointed 1st Major of the 3rd Foot Guards July 25, 1814; Major-General August 12, 1819; Lieutenant-General January 10, 1837; appointed Colonel of the 85th June 1, 1847. Medal for Egypt; gold medal for Fuentes d'Oñoro, Salamanca, Vittoria, and Nive; silver medal for Busaco; served at Ferrol, Vigo, and Cadiz, 1800; Expedition to Hanover; investment of Bayonne, and repulse of the sortie; died at Elmore Court, near Gloucester, April 1, 1865.

MAUNSELL, FREDERICK.—Ensign 18th Foot April 16, 1812; Lieutenant 18th Foot January 28, 1813; Lieutenant 85th Foot March 18, 1813; Captain do. June 24, 1819; Major do. August 14, 1827; Lieutenant-Colonel do. May 23, 1836; Colonel (Army) November 9, 1846; Major-General (Army) June 20, 1854; Lieutenant-General June 1, 1862; Colonel 85th Light Infantry April 2, 1865. Inspecting Field Officer Cork Recruiting District June 19, 1846. Served with the 85th Light Infantry in the Peninsula and the South of France from August 1813 to the end of that war in 1814, including the siege of San Sebastian, passage of the Bidassoa; battles of Nivelle November 10, and Nive December 9, 10, and 11, and investment of Bayonne; war medal with three clasps—San Sebastian (assault and capture), Nivelle and Nive; served also in the American War, and was slightly wounded at Bladensburg August 24, and severely at New Orleans December 23, 1814, and was at the taking of Fort Bourjee; he subsequently commanded the 85th during the Canadian Rebellion; died 1876.

ERRINGTON, ARNOLD CHARLES.—Ensign 51st Foot February 4, 1826; Lieutenant do. September 13, 1831; Adjutant do. April 29, 1836 to July 13, 1837; Captain do. July 14, 1837; Major do. July 25, 1845;

Brevet Lieutenant-Colonel December 9, 1853; Lieutenant-Colonel 51st Foot February 13, 1855; Brevet-Colonel November 28, 1854; Major-General March 6, 1868; Lieutenant-General February 10, 1876; General August 10, 1878; Colonel 85th Regiment December 23, 1876; Colonel 51st Foot September 27, 1879; Brigadier-General Bengal December 13, 1861, to March 23, 1866. Served with the 51st Light Infantry in Burmah from April to December, 1852; was on board the Honourable East India Company's steam sloop *Sesostris* during the naval action and destruction of the enemy's stockades on the Rangoon River; served during the succeeding three days' operations in the vicinity, and commanded the regiment at the assault and capture of Rangoon; he also commanded the troops engaged in the assault and capture of Bassein, May 19 (wounded), and received the especial approbation of the Governor-General in Council for his services upon this occasion; medal and clasp for Pegu, and Brevet of Lieutenant-Colonel.

HILL, PERCY, C.B.—Ensign April 24, 1835; Ensign 68th Regiment June 26, 1835; Lieutenant do. October 5, 1838; Captain do. January 23, 1846; Captain Royal Canadian Rifle Regiment September 11, 1846; Major do. October 22, 1850; Lieutenant-Colonel do. December 29, 1854; Lieutenant-Colonel 2nd Battalion Rifle Brigade June 22, 1855; Colonel December 29, 1857; Major-General March 6, 1868; Lieutenant-General October 1, 1877; Colonel 85th Light Infantry September 27, 1879. General Hill commanded the 2nd Battalion of the Rifle Brigade in the Crimea, subsequent to the fall of Sebastopol, from October 14, 1855, till the end of the war in 1856; also commanded the Battalion throughout the whole of its services in the suppression of the Indian Mutiny, including the capture of Lucknow and numerous affairs during the Oude and Trans Gogra Campaigns, including the capture of Baree and Sirsee, and capture of the Fort of Majidia; commanded the force at the affair of Hyderghur, and held a detached command which captured 15 guns at Sitka Ghats; was frequently mentioned in despatches, and especially in Sir Hope Grant's of the action of Nawabgunge for repulsing the enemy's rear attack; also in Lord Clyde's at the operations on the Kaptee on December 31, 1858. 'C.B.' Medal and clasp.

DE BATHE, SIR HENRY PERCIVAL, BART.—Ensign and Lieutenant Scots Fusilier Guards November 11, 1839; Lieutenant and Captain do. February 17, 1854; Brevet-Colonel November 28, 1854; Major-General March 6, 1868; Major-General Northern District July 2, 1874, to March 31, 1878; Lieutenant-General October 8, 1876; Colonel 89th Foot December 4, 1877; General January 1, 1879; Colonel 85th Regiment April 25, 1880. Served with the Scots Fusilier Guards in the Crimean Campaign from November 17, 1854,

LIEUT.-GENERAL SIR C. E. KNOX, K.C.B.
(Colonel of the Regiment, 6th January, 1907.)

and was present at the siege and fall of Sebastopol. Medal with clasp, 5th class of the Medjidie and Turkish medal.

KNOX, SIR CHARLES EDMOND, K.C.B.—Ensign 85th Regiment June 30, 1865; Lieutenant do. August 7, 1867; Captain do. June 11, 1876; Major do. July 1, 1883; Brevet Lieutenant-Colonel December 9, 1885; Colonel (Army) December 9, 1889; Lieutenant-Colonel 85th Regiment February 11, 1890; half-pay February 11, 1894; Regimental District July 29, 1895; Staff employment November 29, 1899; Major-General December 4, 1899; Lieutenant-General December 6, 1905; Colonel Shropshire Light Infantry January 6, 1907; Major-General Infantry Brigade, Aldershot, November 29, 1899, to December 3, 1899; Major-General South Africa December 4, 1899, to September 18, 1902; Major-General 4th Division 2nd Army Corps and Major-General 4th Division, Southern Command, October 30, 1902, to May 9, 1906. Retired May 10, 1909. Bechuanaland Expedition 1884–1885; in command of the Pioneers; honourably mentioned; Brevet of Lieutenant-Colonel. *South African War*, 1899–1902. On Staff (including command of 13th Brigade; afterwards of Mounted Mobile Columns). Relief of Kimberley. Operations in the Orange Free State February to May, 1900, including operations at Paardeberg February 17 and 18 (severely wounded February 18). Actions at Poplar Grove and Driefontein. Operations in Orange River Colony May to November 29, 1900, including actions at Bothaville and Caledon River (November 27 to 29); operations in Orange River Colony November 30, 1900, to February, 1901, and March, 1901, to May, 1902; operations in Cape Colony February to March, 1901; despatches, *London Gazette*, February 8 and April 16, 1901, and July 29, 1902; promoted Major-General for Distinguished Service. Queen's medal with four clasps; King's medal with two clasps. 'K.C.B.'

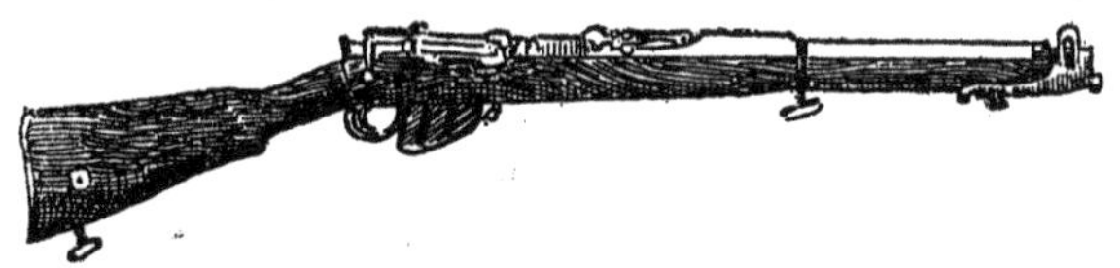

APPENDIX V

QUARTERMASTERS

BARRY, WILLIAM.—September 1809; died 1809.

COLLINS, WILLIAM.—October 26, 1826; retired June 30, 1834.

DAVIDSON, ——.—October 26, 1809; appointed Lieutenant in 4th Veteran Battalion March 18, 1813; battalion reduced 1814; placed on retired full pay of it; died 1816.

DUXBURY, THOMAS.—From the 3rd Lancashire Militia March 18, 1813; exchanged to half-pay of the 14th Foot May 9, 1816; died in Dublin May 11, 1843.

EDWARDS, GEORGE.—Quartermaster 3rd Foot October 25, 1827; retired on half-pay of 2nd Garrison Battalion December 14, 1832; appointed to 85th Regiment June 13, 1834; retired on half-pay of the regiment April 25, 1841; Honorary Captain; died March 12, 1866, aged 82.

FORREST, JAMES.—Honorary Lieutenant September 6, 1899.

HAWKINS, T.—Quartermaster December 11, 1872; retired 1881.

HILL, WILLIAM.—Quartermaster July 22, 1859.

HUMPHREYS, JOHN.—From 94th Foot.

IRWIN, HENRY.—Quartermaster May 20, 1795.

KING, ——.—Quartermaster August 7, 1800.

LEWIS, EDWIN.—Honorary Lieutenant March 20, 1901.

MARSHALL, J K.—Sergeant 53rd Regiment; Quartermaster 7th Garrison Battalion January 1, 1807; Quartermaster 85th Regiment March 25, 1808; deserted 1809.

NEIL, WILLIAM.—Quartermaster April 25, 1841; retired on half-pay December 12, 1851; Honorary Captain July 1, 1859.

PIERCY, JOSEPH.—Quartermaster June 25, 1801; appointed to 60th Regiment November 25, 1802; reappointed to 85th Regiment December 1802.

ROSS, ALEXANDER.—Quartermaster 14th Foot January 20, 1814; placed on half-pay of regiment April 1816; exchanged to 85th Regiment May 9, 1816; died at Malta August 22, 1826. Present at Waterloo.

ROUSE, JOHN R.—Quartermaster 85th Regiment December 12, 1851; retired June 3, 1859; Honorary Major March 30, 1867.

WILDIG, THOMAS.—Honorary Lieutenant April 23, 1902.

WILMOTT, A. E.—Quartermaster 85th Regiment September 13, 1881; Captain September 11, 1891; retired September 6, 1899.

SURGEONS

ABERCROMBIE, M.—Assistant-Surgeon 1799; Surgeon June 19, 1800.

ADOLPHUS, SIR JACOB, M.D.—Assistant-Surgeon 60th Regiment October 10, 1802; Surgeon do. October 5, 1804; exchanged to 85th Regiment November 3, 1808; Staff Surgeon June 29, 1809; Deputy-Inspector, by Brevet, July 17, 1817; Deputy-Inspector November 25, 1819; Inspector by Brevet May 27, 1825; half-pay 1827; Inspector-General November 15, 1827; died at Cheltenham January 2, 1845; served in Ireland during the Rebellion in 1798 and with the Walcheren Expedition in 1809.

ARMSTRONG, H.—From the 75th Regiment; Surgeon May 15, 1855; died 1856.

BAIN, DAVID STUART ERSKINE.—Assistant-Surgeon 80th Regiment January 23, 1846; do., appointed to 85th Regiment August 16, 1850; appointed to Staff June 27, 1851.

BARTLEY, ALEXANDER FISHER.—Assistant-Surgeon January 14, 1853; Surgeon December 7, 1865.

BOLSTER, T. G.—Assistant-Surgeon February 25, 1872.

BOURNES, D.C.G.—Assistant-Surgeon August 12, 1870.

BROWN, DAVID.—Assistant-Surgeon July 8, 1824; died at Malta June 6, 1825.

CAMPBELL, GEORGE McIVOR.—No information.

CARSON, WILLIAM, M.D.—Assistant-Surgeon October 7, 1836; Surgeon 1st Foot August 28, 1846; exchanged to Staff September 11, 1846.

CARTER, JOHN.—Assistant-Surgeon July 31, 1806; Surgeon 60th Regiment May 31, 1809; reappointed to 85th Regiment July 13, 1809; exchanged to 50th Regiment February 25, 1813; half-pay of Regiment 1814; died March 25, 1829.

CLERIHEW, GEORGE, M.D.—Assistant-Surgeon (Staff) November 2, 1832; appointed to 1st Foot January 11, 1833; Surgeon 85th Regiment December 5, 1843.

D'ARCEY, THOMAS, M.D.—Assistant-Surgeon (Staff) December 29, 1837; do. 85th Regiment October 5, 1841; do. 56th Regiment April 3, 1846; Surgeon 30th Regiment October 27, 1846; died April 11, 1855.

DOUGHTY, EDWARD.—Assistant-Surgeon 60th Regiment October 30, 1800; Surgeon 85th Regiment November 25, 1802; exchanged back to 60th Regiment November 3, 1808.

DOYLE, CHARLES SIMON, M.D.—Assistant-Surgeon January 8, 1807; Surgeon 35th Regiment March 31, 1808; Staff Surgeon September 7, 1815; placed on half-pay; died at Demerara, October 1836. Present at Waterloo.

FIDDES, THOMAS.—Assistant-Surgeon 9th Foot January 4, 1810; Surgeon 38th Regiment May 26, 1814; half-pay 1814; Surgeon 85th Regiment August 17, 1815; appointed to 8th Hussars October 8, 1830; appointment cancelled November 2, 1830; died on board the Transport *Stakesby* August 7, 1836, on the passage from Cork to Halifax, Nova Scotia.

GRIFFIN, GEORGE.—Assistant-Surgeon 8th Veteran Battalion April 28, 1814 (made 2nd Veteran Battalion in 1815); placed on half-pay of it May 24, 1816; appointed to 32nd Regiment May 22, 1817; Surgeon 34th Regiment September 26, 1834; appointed to 85th Regiment December 29, 1837; retired on half-pay of the regiment December 5, 1843.

HERBET, HENRY C.—From the 45th Regiment, Assistant-Surgeon December 24, 1858.

HOME, GEORGE.—Assistant-Surgeon August 18, 1825; Surgeon September 23, 1836; half-pay of the Regiment December 29, 1837; Surgeon 67th Regiment November 19, 1841; exchanged to Staff March 15, 1844; Staff-Surgeon July 13, 1849; died September 28, 1857.

HUMPHREY, WILLIAM CHARLES.—Assistant-Surgeon 35th Regiment January 1827; appointed to 95th Regiment November 19, 1830; appointed to 85th Regiment April 29, 1836; Staff Surgeon (Second Class) July 2, 1841.

JOHNSTON, T. W.—Surgeon July 22, 1856.

KERANS, WILLIAM ROBERT.—Assistant-Surgeon October 12, 1867.

LEET, JOHN KNOX.—Assistant-Surgeon June 27, 1851.

LOINSWORTH, FREDERICK ALBERT.—Assistant-Surgeon July 27, 1804; do. 14th Foot May 15, 1806; Surgeon 96th Regiment June 8, 1809; Staff Surgeon July 7, 1814; Deputy Inspector-General July 22, 1830; died at Calcutta August 21, 1842; present at Lugo and Corunna, also at Guadaloupe in 1815.

LUKIS, THOMAS.—Assistant-Surgeon April 9, 1812; do. 9th Light Dragoons February 25, 1813; do. 55th Regiment December 16, 1813; exchanged to half-pay of 50th Regiment January 14, 1819; died October 24, 1848.

M'DIARMID, DUNCAN.—Assistant-Surgeon 6th Veteran Battalion July 2, 1807; do. 85th Regiment June 1, 1809; Surgeon 60th Regiment January 24, 1811; do. 73rd Regiment September 5, 1811; Staff Surgeon September 7, 1815; half-pay 1816; died October 30, 1830; present at Waterloo.

MACFARLANE, JAMES.—Assistant-Surgeon June 19, 1800; Surgeon 85th Regiment January 29, 1802; half-pay of regiment 1802; Surgeon 93rd Regiment March 24, 1803; died at the Cape of Good Hope June 10, 1810.

M'NEIL, ——.—Surgeon September 16, 1795.

MARTIN, JOHN WOODHOUSE.—From the Cornwall Militia, Assistant-Surgeon 85th Regiment July 27, 1809; do. 22nd Regiment October 18, 1810; appointed Supernumerary Assistant Surgeon in the East Indies December 16, 1819; died at Ryepore February 11, 1825.

NORRIS, N.—Assistant-Surgeon December 31, 1858.

REID, ——.—Assistant-Surgeon June 19, 1800; Surgeon 2nd West India Regiment December 8, 1802.

RICHARDS, CHARLES FREDERICK, M.B.—Assistant-Surgeon February 17, 1867.

SAMPSON, JAMES.—Assistant-Surgeon June 27, 1811; do. Newfoundland Fencibles March 26, 1812; do. 104th Regiment August 31, 1815; regiment disbanded, placed on half-pay of it July 25, 1817; Staff July 25, 1826.

SKEEN, WILLIAM, M.D.—Surgeon April 3, 1867.

SMYTH, ROBERT DUNKIN.—Assistant-Surgeon October 6, 1825; half-pay of regiment 1831; appointed to Royal Military College March 1, 1833; Surgeon 76th Regiment November 6, 1840; exchanged to 17th Foot December 15, 1840; appointed to 14th Light Dragoons July 21, 1843; do. to 87th Regiment March 2, 1846; do. to 2nd Dragoons October 1, 1847; retired on half-pay July 21, 1854; died September 6, 1866.

STOREY, ——.—Assistant-Surgeon October 12, 1802.

TEDLIE, BERESFORD, M.D.—Assistant-Surgeon July 8, 1813; half-pay of regiment December 25, 1818; Surgeon 2nd West India Regiment September 18, 1823; killed in action with the Ashantees January 21, 1824.

THOMPSON, J. ALEXANDER W., M.D.—Assistant-Surgeon 3rd West India Regiment November 19, 1844; do. 85th Regiment August 28, 1846; do. 24th Regiment August 16, 1850.

WEIR, GORDON.—Assistant-Surgeon November 15, 1810; died on the passage from Lisbon November 10, 1811.

WILLIAMS, REES.—Assistant-Surgeon 20th Regiment March 25, 1807; Surgeon 50th Regiment March 2, 1809; do. 85th Regiment February 25, 1813; died on board H.M.S. *Diadem* on the Potomac River, 1814.

WITNEY, LUKE.—Assistant-Surgeon 46th Regiment October 25, 1810; do. 85th Regiment May 27, 1813; Surgeon 90th Regiment June 17, 1824; died in the Ionian Islands, 1825.

PAYMASTERS

AUSTIN, JOHN.—Lieutenant July 22, 1800; appointed Paymaster August 20, 1803; Captain 58th Regiment November 28, 1805; Brevet-Major September 4, 1807; Brevet Lieutenant-Colonel February 25, 1813; Major October 25, 1814, Spanish Staff; placed on half-pay of it December 25, 1816; appointed to 97th Regiment March 25, 1824; Lieutenant-Colonel June 8, 1826, unattached; exchanged to 17th Foot June 30, 1829; retired August 13, 1829.

BIGGAR, ALEXANDER.—Paymaster 1813; cashiered July 1819.

HOGG, GEORGE.—Paymaster August 8, 1798; died 1803.

HUGHES, WILLIAM.—Paymaster (from 24th Regiment) November 20, 1867; appointed to 19th Foot September 27, 1876.

LONGLEY, A.—Paymaster March 20, 1879.

MANBY, JOSEPH LANE.—Paymaster April 14, 1808; dismissed the service February 1813; died November 22, 1846.

MINCHIN, JAMES WHITE.—From 62nd Regiment; Paymaster March 30, 1867; transferred back to 62nd Regiment October 30, 1867.

PECHELL, WILLIAM MORTIMER.—Ensign 33rd Regiment August 5, 1842; Lieutenant do. December 8, 1846; appointed Paymaster of the 85th Regiment June 6, 1851; died February 18, 1859.

Extract from Regimental Orders, February 19, 1859.

'GRAHAM'S TOWN.

'The Commanding Officer takes this opportunity of expressing his deep regret (which he feels sure is shared by all ranks in the Regt.) at the Death on the 18th inst. of the late Paymaster Pechell, an officer who has been 16 years in the Service, 7 of which he served in the 85th, and whose honourable character and kind disposition caused him to be esteemed by all who knew him. The Commanding Officer requests that the officers will wear crape on their left arm on all occasions when in uniform for 6 weeks from this date.'

THOMPSON, GEORGE ASH.—Ensign March 16, 1815; Lieutenant May 22, 1817; placed on half-pay of the Regiment December 25, 1818; appointed Paymaster of the Regiment January 27, 1820; retired on half-pay of it June 6, 1851; Honorary Major September 9, 1859; died at Toronto, Canada, June 26, 1861.

THE STAFF OF THE BUGLE-MAJOR OF THE 85TH (DUKE OF YORK'S OWN) LIGHT INFANTRY

THIS Staff is mounted in silver and engraved as the sketch shows. It was presented to the Royal United Service Institution on April 16, 1913, and is here reproduced by their kind permission. Length, 5 feet; hall mark, Roman capital 'D' (1799); London make; maker's initials, 'G. D'

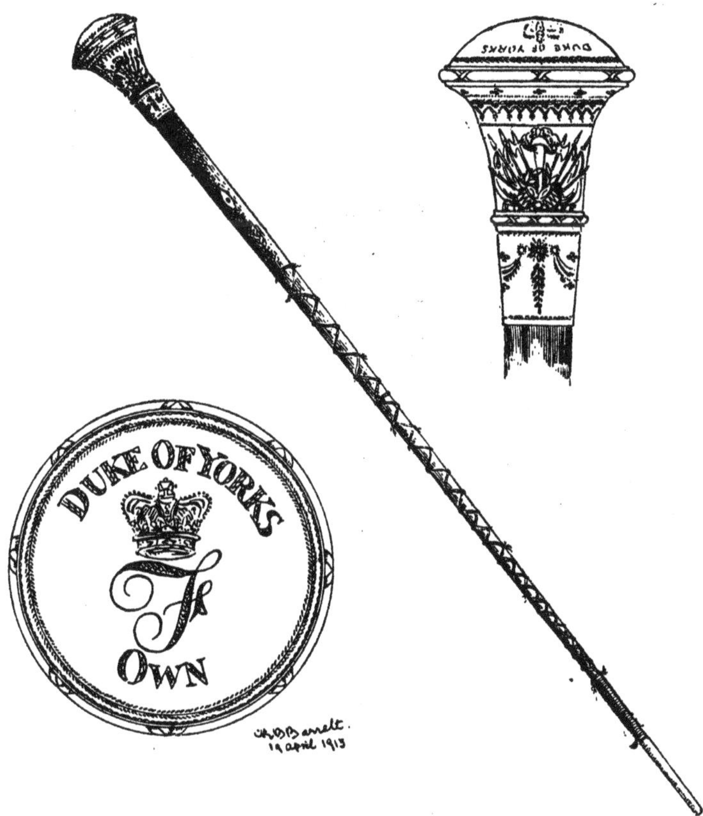

Probably was in stock when the regiment received its new title of 'Duke of York's Own' in 1815—a title only retained until 1821.

Some of the original silver braid still remains, but the tassels are lacking.

The 'F' stands for Frederick (Augustus) Duke of York. At present the Bugle-Major does not carry a staff, but a walking-stick. How the regiment came to possess this Staff, or when it fell into disuse, is not known, but its authenticity is indubitable.

www.ingramcontent.com/pod-product-compliance
Ingram Content Group UK Ltd.
Pitfield, Milton Keynes, MK11 3LW, UK
UKHW050146280726
14058UKWH00007B/866